Studies in Systems, Decision and Control

Volume 34

About this Series

The series "Studies in Systems, Decision and Control" (SSDC) covers both new developments and advances, as well as the state of the art, in the various areas of broadly perceived systems, decision making and control- quickly, up to date and with a high quality. The intent is to cover the theory, applications, and perspectives on the state of the art and future developments relevant to systems, decision making, control, complex processes and related areas, as embedded in the fields of engineering, computer science, physics, economics, social and life sciences, as well as the paradigms and methodologies behind them. The series contains monographs, textbooks, lecture notes and edited volumes in systems, decision making and control spanning the areas of Cyber-Physical Systems, Autonomous Systems, Sensor Networks, Control Systems, Energy Systems, Automotive Systems, Biological Systems, Vehicular Networking and Connected Vehicles, Aerospace Systems, Automation, Manufacturing, Smart Grids, Nonlinear Systems, Power Systems, Robotics, Social Systems, Economic Systems and other. Of particular value to both the contributors and the readership are the short publication timeframe and the world-wide distribution and exposure which enable both a wide and rapid dissemination of research output.

More information about this series at http://www.springer.com/series/13304

Zhanshan Wang · Zhenwei Liu
Chengde Zheng

Qualitative Analysis and Control of Complex Neural Networks with Delays

Science Press
Beijing

Springer

Zhanshan Wang
College of Information Science and Engineering
Northeastern University
Shenyang
China

Zhenwei Liu
College of Information Science and Engineering
Northeastern University
Shenyang
China

Chengde Zheng
Department of Mathematics
Dalian Jiaotong University
Dalian
China

ISSN 2198-4182 ISSN 2198-4190 (electronic)
Studies in Systems, Decision and Control
ISBN 978-3-662-47483-9 ISBN 978-3-662-47484-6 (eBook)
DOI 10.1007/978-3-662-47484-6

Jointly published with Science Press, Beijing
ISBN: 978-7-03-045218-4 Science Press, Beijing

Library of Congress Control Number: 2015942213

Springer Heidelberg New York Dordrecht London

Printed on acid-free paper

Springer-Verlag GmbH Berlin Heidelberg is part of Springer Science+Business Media (www.springer.com)

Preface

Background of This Book With the development of neural networks theory, many neural network models and stability concepts have been extended and upgraded. For example, it is well known that recurrent neural networks (RNNs) can be used to realize associate memory and information storage. The fundamental explanation of the statement is based on the fact that artificial neural networks (ANNs) are from the biological neural networks (BNNs). In fact, this explanation seems too far-fetched. Although ANNs models are the reduction of the real biological neural network models in function, this does not mean that ANNs have the same characteristics as BNNs in nature. In other words, ANNs are an implementation of partial functions of BNNs in engineering applications. Therefore, based on the existing research results of dynamical systems, we can regard RNNs as special dynamical systems that have the fading memory function. In this way, there will be an excellent explanation that RNNs are dynamical systems with memory and storage function. Moreover, with the assembly of many RNNs, the traditional neural network models are coupled together and a new name called complex neural networks (CNNs) emerges, which can be regarded as an upgraded version of RNNs. CNNs have more complex dynamics than RNNs due to the different coupling strengths and topology structures. For such kind of CNNs, under some restrictions on the coupling matrices and topology structures, synchronization problem now being hot research, is an upgraded version of the classical stability conception.

As one of the most important qualitative characteristics of a dynamical system, stability problem is a long-term continuous research topic in many fields such as mechanics, mathematics, control theory and neural networks. Stability characteristic or stability concept may have different meanings due to different application problems, such as stability of fixed point, structure stability, stability in the sense of Lyapunov, input to output stability, input to state stability and invariant sets. Every stability definition conforms to the requirements of the engineering practice or reality. Meanwhile, different requirements of the engineering practice or reality may produce or promote the development of stability theory, in which some different branches of stability theory may come into the world. Every theory research has its own practical background, and the real-life world provides the opportunities of emergence and establishment of new theory. From this point of view, almost all the hot academic topics in the present are in accordance with the current requirement of the economic development and science and technology innovation. That is, the hot academic topic is to solve the present problems encountered in reality.

Along this line, this book aims to investigate some stability problems for recurrent neural networks with delays, in which the main purpose of the research is to reduce the conservativeness of the stability criteria. Specifically, the book is mainly focused on the qualitative stability analysis and synthesis of continuous-time real-valued recurrent neural networks with delays (a special case of additive neural network models). The discussed stability concept is in the sense of Lyapunov, and naturally the proof method is based on Lyapunov stability theory. For a fixed/specified activation function and external constant input, the concerned stability problems fall into the fields of Lyapunov stability. If the activation functions belong to a special class of functions, the concerned stability problems, strictly speaking, fall into the scope of absolute stability. In this sense, most of the existing stability results for RNNs are absolutely stable for any activation function belonging to a specified function class. Therefore, different stability definitions can lead to different understandings on the dynamics of RNNs. Meanwhile, some other qualitative characteristics of RNNs such as passivity, dissipativity, invariant set and synchronization are also discussed. Based on the qualitative analysis, two kinds of control schemes are designed to realize the stabilization and controlled synchronization for the concerned complex dynamical networks.

Why to Write This Book? The first author Zhanshan Wang began formally to contact stability theory of RNNs at the end of 2002. It was an opportunity for him to join the teams to translate the excellent monograph "A.N. Michel and D. Liu, *Qualitative Analysis and Synthesis of Recurrent Neural Networks*, New York: Marcel Dekker, 2002" into a Chinese book. The Chinese version was published by Science Press of China in 2004. It was in early 2005 that Zhanshan Wang began to dedicate his life to the stability research of RNNs. Zhenwei Liu in 2007 and Chengde Zheng in 2009 began to engage in the stability research of RNNs under the direction of Prof. Huaguang Zhang, who is a professor at Northeastern University of China, under the assistance of Zhanshan Wang. Through 10-years' research on the stability theory of RNNs, we have cooperated on many academic treatises and achieved some valuable results, and especially improved a lot in the aspects of understanding the meaning of stability theory. Therefore, it is necessary to write a book to introduce our results.

Although a number of monographs on stability and neural networks have appeared, this book has its untouchable features which distinguishes it from others.

First, the historical and logical development of stability theory of RNNs involved in this monograph is rather complete. From the point of view of system, readers can find not only the origin of stability theory of RNNs and some well-known models of RNNs, additive neural networks, Hopfield neural networks and Cohen-Grossberg neural networks, but also some new insights into the relations among these models of RNNs and the different evolutions of stability analysis. From the point of view of practicability, one can find many relations among different stability definitions, stability theory and stability analysis methods.

Second, since the monograph is a summary of the study results of the authors, the methods proposed here for stability analysis, stabilization and synchronization to a great degree benefit from the theory of nonlinear control systems and dynamical systems, and are more advanced than those appearing in other introductory books. We only mention a few of them as examples: To present a detailed review of the stability research of Cohen-Grossberg-type RNNs (i.e., a kind of additive RNNs), at least 17 aspects of RNNs have been introduced, which will be helpful for readers to further investigate. How to study the effects of delay on the stability of RNNs, weighting-delay method and secondary delay partitioning are proposed, which forms the novel delay partitioning method. To demonstrate an evolution of stability method, stability results have been studied for RNNs from fixed point to invariant set, from global stability to local stability, from absolute stability to relative stability, and from self-stability to controlled-stability. Some insightful comments are presented for different kinds of stability definitions. Because there are many excellent papers scattered in books, conferences and Internet, we have collected a lot of classical and excellent books and papers for further reading on the stability research of RNNs in a systematic manner in this monograph, which may show their great roles in the development of stability theory of RNNs.

Last but not the least, some rather unique contributions are included in this monograph. For example, the relations and meanings between Hopfield neural networks and Cohen-Grossberg neural networks are discussed; the reasons why Lyapunov stability theory is significantly popular in the scientific community are presented; some comparisons among absolute stability, complete stability and global stability are provided. These statements are first discussed by the authors and their merits both in neural network system theory and stability theory will be interesting.

The Audiences of This Book The book is suitable for a formal graduate course in stability theory of neural networks or dynamical systems, or for self-study by researchers and practitioners with an interest in system theory in the following areas: all engineering disciplines, stability theory, control engineering, dynamical system, computer science and applied mathematics. It is assumed that the reader of this book has some background in neural networks, ordinary differential equations, matrix theory and automatic control theory.

The Content of This Book This book is divided into 12 chapters. Chapter 1 provides the background knowledge on the origin of artificial neural networks, especially the relations among the associate memory networks, Hopfield neural networks and Cohen-Grossberg neural networks. Furthermore, as some dynamical systems have information processing capability, it is reasonable to understand that, as a special case of dynamical systems, RNNs have some computation and storage ability. For dynamical neural networks, one of the fundamental qualitative properties is stability, which is not only related to the external structure of the networks, but also related to the signal transmission delay. In this case, some summaries about the delay effects on the stability of neural networks and the linear matrix inequality (LMI)-based analysis method are provided, which are the strategic insights of the authors' research in the past 10 years.

Chapter 2 reviews the history of dynamical systems and stability theory. Different kinds of definitions of dynamical systems are compared. The well-known Lyapunov stability theory is revisited, and the general stability theory is also introduced. Meanwhile, the applications of dynamical system theory are simply summarized. Finally, some comments on the evolution of different stability definition are provided, which will help readers to understand the stability definition in a different sense. This chapter mainly presents the preliminaries of dynamical systems and the corresponding stability theory. Looking through these preliminaries, one can find the evolutionary trajectory of the research on the dynamical system and the stability theory, from which one can find some meaningful inspiration and excite someone to further extend the cognitive ability on the stability concept.

The literature on the stability research of recurrent neural networks is presently scattered throughout journals and conference proceedings. Consequently, to become reasonably proficient in the stability analysis of recurrent neural networks may require considerable investment of time. This book aims to fill this void. To accomplish this, Chap. 3 presents a detailed review of the development of stability of Cohen-Grossberg type neural networks (a special kind of RNNs). The contents include the research directions of stability of RNNs, stability analysis for Cohen-Grossberg type RNNs, and the sufficient and necessary stability conditions of RNNs. In each section, there are many insightful comments on the concerned problems by the authors.

Chapters 4 and 5 present two kinds of delay-dependent stability results for RNNs with time-varying delay on the basis of delay partitioning method. The main method in Chap. 4 is, in the case of fixed interval terminal of time-delay, to insert many virtual points in this interval, and by optimizing these dynamical subintervals partitioned by the virtual points, some novel delay-dependent stability criteria are established. In contrast, the main method in Chap. 5 is, in the case of flexible interval terminal of time delay, to adjust the variable terminal parameter to change the length of the subinterval, and by constructing some novel Lyapunov functions with variable upper and lower integral term, some delay-dependent stability criteria are established. These two methods of partitioning the delay interval are different, which are based on different insight into the flexible change in delay interval.

In Chap. 6, delay-dependent exponential stability criteria for delayed static neural networks (a special kind of RNNs) are established on the basis of LMI method. Static neural networks are often used in the optimization problem. Therefore, the established stability result plays an important role in judging the convergence of designed optimization neural networks.

In Chap. 7, local stability or multiple stability criteria are established for a class of RNNs with discontinuous activation functions and time-varying delay, in which multiple equilibrium points exist. Such RNNs are usually used in the associative memory and pattern recognition. The present local stability result has large basin of attraction of the fixed point, and this feature can keep the stable memory of RNNs longer.

Some stability analysis methods of RNNs have been extended to the more general qualitative cases of RNNs, such as passivity, dissipativity and invariant sets, and synchronization, which have formed Chaps. 8, 9 and 10, respectively. These results will provide profound insight into the dynamics of RNNs with delays.

Based on the above qualitative analysis results, controller design problems are considered for the stabilization and controlled synchronization of RNNs, which form Chaps. 11 and 12, respectively. From these two chapters, one can see that stability analysis is the fundamental of the controller synthesis, and makes the controller design more convenient.

The fundamental knowledge of artificial neural networks in Chap. 1 is mainly cited from the classical textbook (Rojas, Springer-Verlag, 1996). The background materials in Chap. 2 are mainly from Wikipedia–the free encyclopedia on the Internet. However, some remarkable comments on the evolution of Hopfield model and Cohen-Grossberg model, and stability conception in Chaps. 1 and 2 are presented by the authors, which make the contents of this book more systematic and complete. Without the background materials on the history of artificial neural networks and dynamical systems, many insightful comments cannot exhibit their powerful effectiveness. The materials from Chaps. 3 to 12 are from the combined research results of Zhanshan Wang, Zhenwei Liu and Chengde Zheng, respectively.

Acknowledgments The authors would like to acknowledge all of the help and encouragement received in the development of this book. Many our graduate students and colleagues have contributed to the materials of the book. They are Y. Huang, Q. Shan, S. Ding, H. Tian, Z. Huang, B. Huang, J. Wang, Z. Shen, J. Wang, H. Liang, C. Cai and L. Liu. Some of them are the coauthors of some cited papers, which form several chapters of this book. Some of them helped to collect the materials. Moreover, there are many books and academic papers cited and referred, therefore, the authors of the book would like to thank so many excellent scholars and experts, R. Rojas, J. Hale, X. Liao, T. Chen, J. Park, J. Jian, D. Xu, A. Michel, M. Forti, J. Wang, D. Liu, H. Zhang, S. Grossberg, G. Chen, L. Pecora, S. Strogatz, L. Chua and so on for their excellent books and papers on the theory of dynamical systems, neural networks and stability. The entire contents of the book were mainly compiled by Zhanshan Wang. Finally, Zhanshan Wang, Zhenwei Liu and Chengde Zheng have checked the whole book carefully.

We are grateful to the National Natural Science Foundation of China (Grant Nos. 61473070, 61433004, 61273022, 61074073, 61203046), the Fundamental Research Funds for the Central Universities of China (Grant Nos. N130504002, N110504001, N140406001), and SAPI Fundamental Research Funds of China (Grant No. 2013ZCX01), Program for New Century Excellent Talents in University of China under Grant NCET-10-0306, which gave the best supports for the writing of this book.

The authors are indebted to Haina Zhang, China Science Press's Editor, for the consideration, support, and professionalism that she rendered during the preparation and production of this book. The authors would also like to thank their families for their understanding during the writing of this book.

Shenyang
April 2015

Zhanshan Wang
Zhenwei Liu
Chengde Zheng

Contents

1 **Introduction to Neural Networks** 1
1.1 Natural and Artificial Neural Networks 2
1.2 Models of Computation 3
1.3 Networks of Neurons 7
1.4 Associative Memory Networks 12
1.5 Hopfield Neural Networks 14
1.6 Cohen–Grossberg Neural Networks 22
1.7 Property of Neural Network 26
1.8 Information Processing Capacity of Dynamical Systems 27
1.9 Stability of Dynamical Neural Networks 28
1.10 Delay Effects on Dynamical Neural Networks 30
1.11 Features of LMI-Based Stability Results 31
1.12 Summary 33
References 33

2 **Preliminaries on Dynamical Systems and Stability Theory** 37
2.1 Overview of Dynamical Systems 37
2.2 Definition of Dynamical System and Its Qualitative Analysis 41
2.3 Lyapunov Stability of Dynamical Systems 44
2.4 Stability Theory 47
2.5 Applications of Dynamical Systems Theory 61
2.6 Notations and Discussions on Some Stability Problems 63
2.6.1 Notations and Preliminaries 64
2.6.2 Discussions on Some Stability Definitions 72
2.7 Summary 85
References 86

3 Survey of Dynamics of Cohen–Grossberg-Type RNNs 91
3.1 Introduction 91
3.2 Main Research Directions of Stability of RNNs 95
3.2.1 Development of Neuronal Activation Functions 95
3.2.2 Evolution of Uncertainties in Interconnection Matrix 98
3.2.3 Evolution of Time Delays 100
3.2.4 Relations Between Equilibrium and Activation Functions 101
3.2.5 Different Construction Methods of Lyapunov Functions 102
3.2.6 Expression Forms of Stability Criteria 108
3.2.7 Domain of Attraction 109
3.2.8 Different Kinds of Neural Network Models 110
3.3 Stability Analysis for Cohen–Grossberg-Type RNNs 112
3.3.1 Stability on Hopfield-Type RNNs 112
3.3.2 Stability on Cohen–Grossberg-Type RNNs 113
3.3.3 The Case with Nonnegative Equilibria 121
3.3.4 Stability via M-Matrix or Algebraic Inequality Methods 131
3.3.5 Stability via Matrix Inequalities or Mixed Methods 143
3.3.6 Topics on Robust Stability of RNNs 148
3.3.7 Other Topics on Stability Results of RNNs 150
3.3.8 Qualitative Evaluation on the Stability Results of RNNs 151
3.4 Necessary and Sufficient Conditions for RNNs 153
3.5 Summary 159
References 159

4 Delay-Partitioning-Method Based Stability Results for RNNs 173
4.1 Introduction 173
4.2 Problem Formulation 175
4.3 GAS Criteria with Single Weighting-Delay 178
4.3.1 Weighting-Delay-Independent Stability Criterion 178
4.3.2 Weighting-Delay-Dependent Stability Criterion 184
4.4 GAS Criteria with Multiple Weighting-Delays 189
4.5 Implementation of Optimal Weighting-Delay Parameters 195
4.5.1 The Single Weighting-Delay Case 195
4.5.2 The Multiple Weighting-Delays Case 196
4.6 Illustrative Examples 196
4.7 Summary 202
References 202

5 Stability Criteria for RNNs Based on Secondary Delay Partitioning 205
5.1 Introduction 205
5.2 Problem Formulation and Preliminaries 207
5.3 Global Asymptotical Stability Result 210
5.4 Illustrative Example 220
5.5 Summary 222
References 222

6 LMI-Based Stability Criteria for Static Neural Networks 225
6.1 Introduction 225
6.2 Problem Formulation 226
6.3 Main Results 227
6.4 Illustrative Example 235
6.5 Summary 236
References 236

7 Multiple Stability for Discontinuous RNNs 239
7.1 Introduction 239
7.2 Problem Formulations and Preliminaries 241
7.3 Main Results 243
7.4 Illustrative Examples 252
7.5 Summary 255
References 256

8 LMI-based Passivity Criteria for RNNs with Delays 259
8.1 Introduction 259
8.2 Problem Formulation 261
8.3 Passivity for RNNs Without Uncertainty 262
8.4 Passivity for RNNs with Uncertainty 269
8.5 Illustrative Examples 272
8.6 Summary 274
References 275

9 Dissipativity and Invariant Sets for Neural Networks with Delay 277
9.1 Delay-Dependent Dissipativity Conditions for Delayed RNNs 277
9.1.1 Introduction 277
9.1.2 Problem Formulation 279
9.1.3 θ-dissipativity Result 281

9.2 Positive Invariant Sets and Attractive Sets of DNN . . . 288
9.2.1 Introduction . . . 288
9.2.2 Problem Formulation and Preliminaries . . . 289
9.2.3 Invariant Set Results . . . 291
9.3 Attracting and Invariant Sets of CGNN with Delays . . . 300
9.3.1 Introduction . . . 300
9.3.2 Problem Formulation and Preliminaries . . . 301
9.3.3 Invariant Set Result. . . . 304
9.4 Summary . . . 307
References. . . . 307

10 Synchronization Stability in Complex Neural Networks . . . 311
10.1 Introduction . . . 311
10.2 Problem Formulation and Preliminaries . . . 313
10.3 Synchronization Results. . . . 317
10.4 Illustrative Example . . . 324
10.5 Summary . . . 329
References. . . . 329

11 Stabilization of Stochastic RNNs with Stochastic Delays . . . 333
11.1 Introduction . . . 333
11.2 Problem Formulation and Preliminaries . . . 335
11.3 Stabilization Result . . . 339
11.4 Illustrative Examples . . . 354
11.5 Summary . . . 358
References. . . . 358

12 Adaptive Synchronization of Complex Neural Networks . . . 361
12.1 Introduction . . . 361
12.2 Problem Formulation and Preliminaries . . . 363
12.3 Adaptive Synchronization Scheme . . . 367
12.4 Illustrative Example . . . 373
12.5 Summary . . . 379
References. . . . 380

Index . . . 385

Chapter 1
Introduction to Neural Networks

Analog circuits have played a very important role in the development of modern electronic technology. Even in our digital computer era, analog circuits still dominate such fields as communications, power, automatic control, audio, and video electronics because of their real-time signal processing capabilities. Conventional digital computation methods have run into a serious speed bottleneck due to their serial nature. To overcome this problem, a new computation model as an alternative, called "neural networks" has been proposed, which is based on some aspects of neurobiology and adapted to integrated circuits. The key features of neural networks are asynchronous parallel processing, continuous-time dynamics, and global interaction of network elements. Some encouraging if not impressive applications of neural networks have been proposed for various fields such as optimization, linear and nonlinear programming, associative memory, pattern recognition, and computer vision. Since 1943, when Warren McCulloch and Walter Pitts [1] presented the first model of artificial neurons, new and more sophisticated proposals have been made from decade to decade. Mathematical analysis has solved some of the mysteries posed by the new models but has left many questions open for future investigations. Needless to say, the study of neurons, their interconnections, and their role as the brain's elementary building blocks is one of the most dynamic and important research fields in modern biology. It is not an exaggeration to say that researchers have learned more about the nervous system in the past 70 years than ever before. In this chapter we will deal with artificial neural networks, and therefore the first question to be clarified is their relation to the biological paradigm. What do we abstract from real neurons for our models? What is the link between neurons and artificial computing units? In the following we will present a preliminary answer to these important questions, in which some contents are mainly cited from the book [2].

Z. Wang et al., *Qualitative Analysis and Control of Complex Neural Networks with Delays*, Studies in Systems, Decision and Control 34,
DOI 10.1007/978-3-662-47484-6_1

1.1 Natural and Artificial Neural Networks

Artificial neural networks are an attempt at modeling the information processing capabilities of nervous systems. Thus, one needs to consider the essential properties of *biological neural networks* from the viewpoint of information processing. This will allow us to design abstract models of artificial neural networks, which can then be simulated and analyzed. Although the models were proposed to explain the structure of the brain and the nervous systems of some animals are different in many respects, there is a general consensus that the essence of the operation of neural ensembles is "control through communication" [3]. Animal nervous systems are composed of thousands or millions of interconnected cells. Each of them is a very complex arrangement that deals with incoming signals in many different ways. However, neurons are rather slower when compared to electronic logic gates. These can achieve switching times of a few nanoseconds, whereas neurons need several milliseconds to react to a stimulus. Nevertheless, the brain is capable of solving problems that no digital computer can yet efficiently deal with.

Massive and hierarchical networking of the brain seems to be the fundamental precondition for the emergence of consciousness and complex behavior [4]. So far, however, biologists and neurologists have concentrated their research on uncovering the properties of individual neurons. Today, mechanisms for the production and transport of signals from one neuron to the other are well-understood physiological phenomena, but how these individual systems cooperate to form complex and massively parallel systems capable of incredible information processing feats has not yet been completely elucidated. Mathematics, physics, and computer science can provide invaluable help in the study of these complex systems. It is not surprising that the study of the brain has become one of the most interdisciplinary areas of scientific research in recent years. However, we should be careful with the metaphors and paradigms commonly introduced when dealing with the nervous system. It seems to be a constant in the history of science that the brain has always been compared to the most complicated contemporary artifact produced by human industry [5]. In ancient times the brain was compared to a pneumatic machine, in the Renaissance to a clockwork, and to the telephone network. There are some today, who consider computers the paradigm par excellence of a nervous system. It is rather paradoxical that when John von Neumann wrote his classical description of future universal computers, he tried to choose terms that would describe computers in terms of brains, not brains in terms of computers. The nervous system of an animal is an information processing totality. The sensory inputs, i.e., signals from the environment, are coded and processed to evoke the appropriate response. Biological neural networks are just one of many possible solutions to the problem of processing information. The main difference between neural networks and conventional computer systems is the massive parallelism and redundancy they exploit in order to deal with the unreliability of the individual computing units. Moreover, biological neural networks are self-organizing systems and each individual neuron is also a delicate self-organizing structure capable of processing information in many different ways.

One of the interesting and important topics is to study the information processing capabilities of complex hierarchical networks of simple computing units. We deal with systems whose structure is only partially predetermined. Some parameters modify the capabilities of the network and it is our task to find the best combination for the solution of a given problem. The adjustment of the parameters will be done through a learning algorithm, i.e., not through explicit programming but through an automatic adaptive method. Artificial neural networks have aroused much interest in recent years, not only because they exhibit interesting properties, but also because they try to mirror the kind of information processing capabilities of nervous systems. Biological neural networks have given us a clue regarding the properties that would be interesting to include in artificial networks.

Recurrent neural networks (RNNs) are a kind of networks whose neurons send feedback signals to each other. This concept includes a huge number of possibilities. Typically, in the community of science and technology, engineers always consider RNNs that are artificial neural networks useful in technological applications. To complement these contributions, it is necessary to present a brief summary focusing on biological recurrent neural networks (bRNNs) that are found in the brain. Since feedback is ubiquitous in the brain, this task could include most of the brain's dynamics. The current review divides bRNNs into those in which feedback signals occur in neurons within a single processing layer, which occurs in networks for such diverse functional roles as storing spatial patterns in short-term memory, winner-take-all decision making, contrast enhancement and normalization, hill climbing, oscillations of multiple types (synchronous, traveling waves, chaotic), storing temporal sequences of events in working memory, and serial learning of lists. These feedback signals occur between multiple processing layers, such as when bottom-up adaptive filters activate learned recognition categories and top-down learned expectations focus attention on expected patterns of critical features and thereby modulate both types of learning.

In the following sections, we present a review for bRNNs in the aspects of models of computation, networks of neurons, associative memory dynamical networks, Hopfield's networks, and Cohen-Grossberg networks. Meanwhile, some properties of neural networks, information processing capacity of dynamical, stability of RNNs, delay effects on networks, and features of LMI-based stability are presented, respectively.

1.2 Models of Computation

Artificial neural networks can be considered as just another approach to the problem of computation. The first formal definitions of computability were proposed in the 1930s and 1940s and at least five different alternatives were studied at the time. The computer era was started, not with one single approach, but with a contest of alternative *computing models*. It is well known that the von Neumann computer emerged as the undisputed winner in this confrontation, but its triumph did not lead to the dismissal of the other computing models.

(1) The mathematical model Mathematicians avoided dealing with the problem of a function's computability until the beginning of the twentieth century. This happened not just because existence theorems were considered sufficient to deal with functions, but mainly because nobody had come up with a satisfactory definition of computability, certainly a relative concept that depends on the specific tools that can be used. If we want to talk about computability we must specify which tools are available. We can start with the idea that some primitive functions and composition rules are "obviously" computable. All other functions that can be expressed in terms of these primitives and composition rules are then also computable. David Hilbert, the famous German mathematician, was the first to state the conjecture that a certain class of functions contains all intuitively computable functions. Hilbert was referring to the primitive recursive functions, the class of functions which can be constructed from the zero and successor function using composition, projection, and a deterministic number of iterations (primitive recursion). However, in 1928, Wilhelm Ackermann was able to find a computable function which was not primitive recursive. This led to the definition of the general recursive functions. In this formalism, a new composition rule has to be introduced, the so-called μ operator, which is equivalent to an indeterminate recursion or a lookup in an infinite table. At the same time Alonzo Church and his collaborators developed the lambda calculus, another alternative to the mathematical definition of the *computability concept* [6]. In 1936, Church and Kleene were able to show that the general recursive functions can be expressed in the formalism of the lambda calculus. This led to the Church thesis that computable functions are the general recursive functions. David Deutsch has added that this thesis should be considered to be a statement about the physical world and be given the same status as a physical principle. He thus speaks of a "Church principle" [7].

(2) The logic-operational model or Turing machines In his classical paper "On Computable Numbers with an Application to the Entscheidungsproblem," Alan Turing introduced another kind of computing model. The advantage of his approach is that it consists of an operational, mechanical model of computability. A Turing machine is composed of an infinite tape, in which symbols can be stored and read again. A read-write head can move to the left or to the right according to its internal state, which is updated at each step. The Turing thesis states that *computable functions* are those that can be computed with this kind of device. It was formulated concurrently with the Church thesis and Turing was able to show almost immediately that they are equivalent [8]. The Turing approach made clear for the first time what "programming" means, curiously enough at a time when no computer had yet been built.

Note that, according to the introduction of Turing machine we can know that before Turing machine, the computation problem is focused on the function or problem itself. All the efforts are devoted to the internal research of computational problem or the computational model, which is based on the accurate analytical mathematical method instead of approximation method. After Turing machine, the computational problem can be done by external auxiliary tool. It is the emergence of Turing machine that changes the way of solving the computational problem from the analytical function itself to the external auxiliary tool. This is a great creativity in the history of solving computational problems. Since then, many computational models, such as digital computer and analog computer based on hardware implementation, are frequently proposed and developed.

(3) The computer model The first electronic computing devices were developed in the 1930s and 1940s. Since then, "computation-with-the-computer" has been regarded as computability itself. However, the first engineers developing computers were for the most part unaware of Turing's or Church's research. Konrad Zuse, for example, developed in Berlin between 1938 and 1944 the computing machines Z_1 and Z_3, which were programmable but not universal, because they could not reach the whole space of the computable functions. Zuse's machines were able to process a sequence of instructions but could not iterate. Other computers of the time, like the Mark *I* built at Harvard, could iterate a constant number of times but were incapable of executing open-ended iterations (WHILE loops). Therefore, the Mark *I* could compute the primitive but not the general recursive functions. Also the ENIAC, which is usually hailed as the world's first electronic computer, was incapable of dealing with open-ended loops, since iterations were determined by specific connections between modules of the machine. It seems that the first universal computer was the Mark *I* built in Manchester [9, 10]. This machine was able to cover all *computable functions* by making use of conditional branching and self-modifying programs, which is one possible way of implementing indexed addressing [2].

(4) Cellular Automata The history of the development of the first mechanical and electronic *computing devices* shows how difficult it was to reach a consensus on the architecture of universal computers. Aspects such as the economy or the dependability of the building blocks played a role in the discussion, but the main problem was the definition of the minimal architecture needed for universality. In machines like the Mark *I* and the ENIAC there was no clear separation between memory and processor, and both functional elements were intertwined. Some machines still worked with base 10 and not 2, some were sequential, and others parallel. John von Neumann, who played a major role in defining the architecture of sequential machines, analyzed at that time a new computational model which he called cellular automata. Such automata operate in a "computing space" in which all data can be processed simultaneously. The main problem for cellular automata is communication and coordination between all the computing cells. This can be guaranteed through certain algorithms and conventions. It is not difficult to show that all computable functions, in the sense of Turing, can also be computed with cellular automata, even of the one-dimensional type, possessing only a few states. Turing himself considered this kind of *computing model* at one point in his career [11]. Cellular automata as computing model resemble massively parallel multiprocessor systems of the kind that has attracted considerable interest recently.

(5) The biological model or neural networks The explanation of important aspects of the physiology of neurons set the stage for the formulation of *artificial neural network* models which do not operate sequentially, as Turing machines do. Neural networks have a hierarchical multilayered structure which sets them apart from cellular automata, so that information is transmitted not only to the immediate neighbors but also to more distant units. In artificial neural networks one can connect each unit to any other. In contrast to conventional computers, no program is handed over to the hardware—such a program has to be created, that is, the free parameters of the network have to be found adaptively. Although neural networks and cellular automata are potentially more efficient than conventional computers in certain application areas, at the time of their conception they were not yet ready to take center stage. The necessary theory for harnessing the dynamics of complex parallel systems is still being developed right before our eyes. In the meantime, conventional computer technology has made great strides. There is no better illustration for the simultaneous and related emergence of these various computability models than the life and work of John von Neumann himself. He participated in the definition and development of at least three of these models: in the architecture of sequential computers [12], the theory of cellular automata and the first neural network models. He also collaborated with Church and Turing in Princeton [11]. Artificial neural networks have, as initial motivation, the structure of biological systems, and constitute an alternative computability paradigm. For this reason we will review some aspects of the way in which biological systems perform information processing. The fascination that still pervades this research field has much to do with the points of contact with the surprisingly elegant methods used by neurons in order to process information at the cellular level. Several million years of evolution have led to very sophisticated solutions to the problem of dealing with an uncertain environment. In the following section we discuss some elements of these strategies in order to determine what features we want to adopt in our abstract models of neural networks.

What are the elementary components of any conceivable *computing model*? In the theory of general recursive functions, for example, it is possible to reduce any computable function to some composition rules and a small set of primitive functions. For a universal computer, we ask about the existence of a minimal and sufficient instruction set. For an arbitrary computing model the following metaphoric expression has been proposed:

$$computation = storage + transmission + processing.$$

The mechanical computation of a function presupposes that these three elements are present, that is, data can be stored, communicated to the functional units of the model, and transformed. It is implicitly assumed that a certain coding of the data has been agreed upon. Coding plays an important role in information processing because, as Claude Shannon showed in 1948, when noise is present information can still be transmitted without loss, if the right code with the right amount of redundancy is chosen. Modern computers transform storage of information into a form of information transmission. Static memory chips store a bit as a circulating current until the bit is read. Turing machines store information in an infinite tape, whereas transmission is performed by the read-write head. Cellular automata store information in each cell, which at the same time is a small processor.

1.3 Networks of Neurons

In biological neural networks information is stored at the contact points between different neurons, the so-called synapses. These elements play for the storage, transmission, and processing of information. Other forms of storage are also known, because neurons are themselves complex systems of self-organizing signaling. More details on how neurons compute can be found in [2].

Nervous systems possess global architectures of variable complexity, but all are composed of similar building blocks, the neural cells or neurons. They can perform different functions, which in turn leads to a very variable morphology. Dendrites, synapses, cell body, and axon are the minimal structures we adopt from the biological model. Artificial neurons for computing have input channels, a cell body, and an output channel. Synapses will be simulated by contact points between the cell body and input or output connections, i.e., a weight will be associated with these points.

(1) Transmission of information The fundamental problem of any information processing system is the transmission of information, as data storage can be transformed into a recurrent transmission of information between two points. Biologists have known for more than 100 years that neurons transmit information using electrical signals. Because we are dealing with biological structures, this cannot be done by simple electronic transport as in metallic cables. Evolution arrived at another solution involving ions and semipermeable membranes. The British scientists Alan Hodgkin and Andrew Huxley were able to show that it is possible to build an electric model of the cell membrane based on very simple assumptions. The membrane behaves as a capacitor made of two isolated layers of lipids. It can be charged with positive or negative ions. The different concentrations of several classes of ions in the interior and exterior of the cell provide an energy source capable of negatively polarizing the interior of the cell.

Hodgkin–Huxley differential equation describes the instantaneous variation of the cell's potential V as a function of the conductances of sodium, potassium, and leakages (g_{Na}, g_K, g_L) and of the equilibrium potentials for all three groups of ions called V_{Na}, V_K and V_L with respect to the current potential:

$$\frac{\mathrm{d}V}{\mathrm{d}t} = \frac{1}{C_m}(I - g_{Na}(V - V_{Na}) - g_K(V - V_K) - g_L(V - V_L)), \tag{1.1}$$

where C_m is the capacitance of the cell membrane, the terms $V - V_{Na}$, $V - V_K$, $V - V_L$ are the electromotive forces acting on the ions. The conductances g_{Na}, g_K, and g_L reflect the permeability of the membrane to sodium, potassium, and leakages, i.e., the number of open channels of each class. Any variation of the conductances translates into a corresponding variation of the cell's potential V. The variations of g_{Na} and g_K are given by differential equations which describe their oscillations. The conductance of the leakages, g_L, can be taken as a constant. A signal can be produced by modifying the polarity of the cell through changes in the conductances g_{Na} and g_K. The conductance and resistance of a cell membrane in relation to the different classes of ions depend on its permeability. This can be controlled by opening or closing excitable ionic channels. In addition to the static ionic channels already mentioned, there is another class that can be electrically controlled. These channels react to a depolarization of the cell membrane. When the potential of the interior of the cell in relation to the exterior reaches a threshold, the sodium-selective channels open automatically and positive sodium ions flow into the cell making its interior positive. This in turn leads to the opening of the potassium-selective channels and positive potassium ions flow to the exterior of the cell, restoring the original negative polarization.

A neuron codes its level of activity by adjusting the frequency of the generated impulses. This frequency is greater for a greater stimulus. In some cells the mapping from stimulus to frequency is linear in a certain interval. This means that information is transmitted from cell to cell using what engineers call frequency modulation. This form of transmission helps to increase the accuracy of the signal and to minimize the energy consumption of the cells.

(2) Information processing at the neurons and synapses Neurons transmit information using action potentials. The processing of this information involves a combination of electrical and chemical processes, regulated for the most part at the interface between neurons, the synapses. Neurons transmit information not only by electrical perturbations. Although electrical synapses are also known, most synapses make use of chemical signals. When an electric impulse arrives at a synapse, the synaptic vesicles fuse with the cell membrane. The transmitters flow into the synaptic gap and some attach themselves to the ionic channels. If the transmitter is of the right kind, the ionic channels are opened and more ions can now flow from the exterior to the interior of the cell. The cell's potential is altered in this way. If the potential in the interior of the cell is increased, this helps prepare an action potential and the synapse causes an excitation of the cell. If negative ions are transported into the cell, the probability of starting an action potential is decreased for some time and we are dealing with an inhibitory synapse. Synapses determine a direction for the transmission of information. Signals flow from one cell to the other in a well-defined manner. This will be expressed in artificial neural networks models by embedding the computing elements in a directed graph. A well-defined direction of information flow is a basic element in every computing model, and is implemented in digital systems by using diodes and directional amplifiers.

The interplay between electrical transmission of information in the cell and chemical transmission between cells is the basis for neural information processing. Cells process information by integrating incoming signals and by reacting to inhibition. The flow of transmitters from an excitatory synapse leads to a depolarization of the attached cell. The depolarization must exceed a threshold, that is, enough ionic channels have to be opened in order to produce an action potential. This can be achieved by several pulses arriving simultaneously or within a short time interval at the cell. If the quantity of transmitters reaches a certain level and enough ionic channels are triggered, the cell reaches its activation threshold. At this moment an action potential is generated at the axon of this cell. In most neurons, action potentials are produced at the so-called axon hillock, the part of the axon nearest to the cell body. In this region of the cell, the number of ionic channels is larger and the cell's threshold is lower. The dendrites collect the electrical signals which are then transmitted electronically through the cytoplasm [12]. The transmission of information at the dendrites makes use of additional electrical effects. Streams of ions are collected at the dendrites and brought to the axon hillock. There is spatial summation of information when signals coming from different dendrites are collected, and temporal summation when signals arriving consecutively are combined to produce a single reaction. In some neurons not only the axon hillock but also the dendrites can produce action potentials. In this case information processing at the cell is more complex than in the standard case.

It can be shown that digital signals combined in an excitatory or inhibitory way can be used to implement any desired logical function. The number of computing units required can be reduced if the information is not only transmitted but also weighted. This can be achieved by multiplying the signal by a constant. Such is the kind of processing we find at the synapses. Each signal is an all-or-none event but the number of ionic channels triggered by the signal is different from synapse to synapse. It can happen that a single synapse can push a cell to fire an action potential, but other synapses can achieve this only by simultaneously exciting the cell. With each synapse $i(1 \leq i \leq n)$ we can therefore associate a numerical weight w_i. If all synapses are activated at the same time, the information that will be transmitted is $w_1 + w_2 + \cdots + w_n$. If this value is greater than the cell's threshold, the cell will fire a pulse. It follows from this description that neurons process information at the membrane. The membrane regulates both transmission and processing of information. Summation of signals and comparison with a threshold is a combined effect of the membrane and the cytoplasm. If a pulse is generated, it is transmitted and the synapses set some transmitter molecules free. From this description an abstract neuron [3] can be modeled which contains dendrites, a cell body and an axon. The same three elements will be present in the artificial computing units.

(3) Storage of information—learning In neural networks information is stored at the synapses. Some other forms of information storage may be present, but they are either still unknown or not very well understood. A synapse's efficiency in eliciting the depolarization of the contacted cell can be increased if more ionic channels are opened. In the past several years, N-methyl-D-aspartic acid (NMDA) receptors have been studied because they exhibit some properties that could help explain some forms of learning in neurons [2, 3]. NMDA receptors are just one of the mechanisms used by neurons to increase their plasticity, i.e., their adaptability to changing circumstances. Through the modification of the membrane's permeability a cell can be trained to fire more often by setting a lower firing threshold. NMDA receptors also offer an explanation for the observed phenomenon that cells that are not stimulated to fire tend to set a higher firing threshold. The stored information must be refreshed periodically in order to maintain the optimal permeability of the cell membrane. This kind of information storage is also used in artificial neural networks. Synaptic efficiency can be modeled as a property of the edges of the network. The networks of neurons are thus connected through edges with different transmission efficiencies. Information flowing through the edges is multiplied by a constant which reflects their efficiency. One of the most popular learning algorithms for artificial neural networks is Hebbian learning. The efficiency of synapses is increased any time when the two cells that are connected through this synapse fire simultaneously. The efficiency of synapses is decreased when the firing states of the two cells are uncorrelated. The NMDA receptors act as coincidence detectors of presynaptic and postsynaptic activity, which in turn leads to greater synaptic efficiency.

(4) The neuron—a self-organizing system The short review of the properties of biological neurons in the above is necessarily incomplete and can offer only a rough description of the mechanisms and processes by which neurons deal with information. Nerve cells are very complex self-organizing systems which have evolved in the course of millions of years. The information processing capabilities of neurons depend essentially on the characteristics of the cell membrane. Ionic channels appeared very early in evolution to allow unicellular organisms to get some kind of feedback from the environment. Consider the case of a paramecium, a protozoan with cilia, which are hairlike processes which provide it with locomotion. A paramecium has a membrane cell with ionic channels and its normal state is one in which the interior of the cell is negative with respect to the exterior. In this state the cilia around the membrane beat rhythmically and propel the paramecium forward. If an obstacle is encountered, some ionic channels sensitive to contact open, let ions into the cell, and depolarize it. The depolarization of the cell leads in turn to a reversing of the beating direction of the cilia and the paramecium swims backward for a short time. After the cytoplasm returns to its normal state, the paramecium swims forward, changing its direction of movement. If the paramecium is touched from behind, the opening of ionic channels leads to a forward acceleration of the protozoan. In each case, the paramecium escapes its enemies.

From these humble origins, ionic channels in *neurons* have been perfected over millions of years of evolution. In the protoplasm of the cell, ionic channels are produced and replaced continually. They attach themselves to those regions of the neurons where they are needed and can move laterally in the membrane. The regions of increased neural sensitivity to the production of action potentials are thus changing continuously according to experience. The electrical properties of the cell membrane are not totally predetermined. They are also a result of the process by which action potentials are generated. Now let us also consider the interior of the neurons. The number of biochemical reaction chains and the complexity of the mechanical processes occurring in the neuron at any given time have led some researchers to look for its control system. Stuart Hameroff, for example, has proposed that the cytoskeleton of neurons does not just perform a static mechanical function, but in some way provides the cell with feedback control. It is well known that the proteins that form the microtubules in axons coordinate to move synaptic vesicles and other materials from the cell body to the synapses. This is accomplished through a coordinated movement of the proteins, configured like a cellular automaton [14, 15].

Consequently, transmission, storage, and processing of information are performed by neurons exploiting many effects and mechanisms that we still do not understand fully. Each individual neuron is as complex or more complex than any of our computers. For this reason, we will call the elementary components of artificial neural networks simply "computing units" and not neurons. In the mid-1980s, the PDP (parallel distributed processing) group already agreed to this convention at the insistence of Francis Crick [2, 16].

The aforementioned discussion is only an illustration of how important it is to define the primitive functions (or activation functions) and composition rules of the computational model. If we are computing with a conventional von Neumann processor, a minimal set of machine instructions is needed in order to implement all computable functions. In the case of artificial neural networks, the primitive functions are located in the nodes of the network and the composition rules are contained implicitly in the interconnection pattern of the nodes, in the synchrony or asynchrony of the transmission of information, and in the presence or absence of cycles.

In the structure of an abstract neuron with n inputs, each input channel i can transmit a real value x_i, and the primitive function or activation function $f(\cdot)$ computed in the body of the abstract neuron can be selected arbitrarily. Usually the input channels have an associated weight, which means that the incoming information x_i is multiplied by the corresponding weight w_i, which is formulated by the well-known fundamental neuron's model. The transmitted information is integrated at the neuron (usually just by adding the different signals) and the primitive function is then evaluated. If we conceive of each node in an artificial neural network as a primitive function capable of transforming its input in a precisely defined output, then *artificial neural networks* are nothing but networks of primitive functions. Different models of artificial neural networks differ mainly in the assumptions about the primitive functions used, the interconnection pattern, and the timing of the transmission of information. Therefore, typical artificial neural networks have the following three parts: the structure of the nodes, the topology of the network, and the learning algorithm used to find the weights of the network. Different selections of the weights produce different network outputs.

In brief, neural networks, or artificial neural networks to be more accurate, represent a technology that is rooted in many disciplines: neurosciences, mathematics, statics, physics, compute sciences, and control engineering. Neural networks find many applications in such diverse fields as modeling, time series analysis, optimal computation, and optimal control by virtue of an important property: the ability to learn from input data with or without supervision, the dynamical features to run in the circuit implementation without programming complex numerical algorithms.

Work on artificial neural networks, commonly referred to as "neural networks," has been motivated right from its inception by the recognition that the human brain computer works in an entirely different way from the conventional digital computer. The brain is a highly complex and parallel computer (information-processing system). It has the ability to organize its structural constituents, known as neurons, so as to perform certain computations (e.g., pattern recognition, perception, motor control, and optimal computation) many times faster than the fastest digital computer in existence today.

A developing neuron is synonymous with a plastic brain: plasticity permits the developing nervous system to adapt to its surrounding environment. Just as plasticity appears to be essential to the functioning of neurons as information-processing units in the human brain, so is it with neural networks made up of artificial neurons. In its most general form, a neural network is a machine that is designed to model the way in which the brain performs a particular task or function of interest; the network is usually implemented by using electronic components or is simulated in software on a digital computer. An interest of this book is confined largely to an important class of neural networks that perform useful computations through a process of dynamical evolution or learning. To achieve good performance, neural networks employ a massive interconnection of simple computing cells referred to as neurons or processing units. Thus, we can offer another definition of a neural network viewed as an adaptive machine: neural network is a massively parallel distributed processor made up of simple processing units, which has a natural propensity for storing experimental knowledge and making it available for use. It resembles the brain in two respects: (1) Knowledge is acquired by the network from its environment through a learning precess; (2) Interneurons connection strengths, known as synaptic weights, are used to store the acquired knowledge.

Finally we must keep in mind: in the theory of artificial neural networks we do not consider the whole complexity of real biological neurons. We only abstract some general principles and content ourselves with different levels of detail when simulating neural ensembles. The general approach is to conceive each neuron as a primitive function producing numerical results at some points in time. These will be the kinds of model of polynomial equations, which can be used to curve fitting and spline interpolation. However, we can also think of artificial neurons as computing units which produce pulse trains in the way that biological neurons do. We can then simulate this behavior and look at the output of simple networks. This kind of approach, although more closely related to the biological paradigm, is still a very rough approximation of the biological processes.

1.4 Associative Memory Networks

The perceptron learning algorithm is an example of supervised learning. This kind of approach does not seem very plausible from the biologist's point of view, since a teacher is needed to accept or reject the output and adjust the network weights if necessary. Some researchers have proposed alternative learning methods in which the network parameters are determined as a result of a self-organizing process. In unsupervised learning, corrections to the network weights are not performed by an external agent, because in many cases we do not even know what solution we should expect from the network. The network itself decides what output is best for a given input and reorganizes accordingly. We will make a distinction between two classes of unsupervised learning: reinforcement and competitive learning. In the first method each input produces a reinforcement of the network weights in such a way as to enhance the reproduction of the desired output. Hebbian learning is an example of a reinforcement rule that can be applied in this case. In competitive learning, the elements of the network compete with each other for the "right" to provide the output associated with an input vector. Only one element is allowed to answer the query and this element simultaneously inhibits all other competitors.

In the case of unsupervised learning, the n-dimensional input is processed by exactly the same number of computing units as there are clusters to be individually identified. The inputs are processed by the neural units. Each unit computes its weighted input, but only the unit with the largest excitation is allowed to fire a 1 or high level. The other units are inhibited by this active element through the lateral connections. Therefore, deciding whether or not to activate a unit requires global information about the state of each unit. The firing unit signals that the current input is an element of the cluster of vectors it represents. We could also think of this computation as being performed by perceptrons with variable thresholds. The thresholds are adjusted in each computation in such a way that just one unit is able to fire. Neural networks without feedback can be capable of mapping an input space into an output space using only feedforward computations. In the case of backpropagation networks we demanded continuity from the activation functions at the nodes. The neighborhood of a vector x in input space is therefore mapped to a neighborhood of the image y of x in output space. It is this property that gives its name to the continuous mapping networks.

Another class of neural systems is known generally as associative memories. The goal of learning is to associate known input vectors with given output vectors. Contrary to continuous mappings, the neighborhood of a known input vector x should also be mapped to the image y of x, that is, if $B(x)$ denotes all vectors whose distance from x (using a suitable metric) is less than some positive constant ϵ, then we expect the network to map $B(x)$ to y. Noisy input vectors can then be associated with the correct output. Associative memories can be implemented using networks with or without feedback, but the latter produce better results. However, as we will see not all networks converge to a stable state after having been set in motion. Some restrictions on the network architecture are needed. The function of an associative memory is to recognize previously learned input vectors, even in the case where some noise has been added. The advantage of associative memories is that only the local information stream must be considered. The response of each unit is determined exclusively by the information flowing through its own weights. If we take the biological analogy seriously or if we want to implement these systems in very large-scale integration circuits (VLSI), locality is always an important goal. And as we will see a learning algorithm derived from biological neurons can be used to train associative networks: it is called Hebbian learning. The associative networks should not be confused with conventional associative memory of the kind used in digital computers, which consists of content addressable memory chips. Associative networks can be regarded as dynamical systems, whose attractors are exactly those vectors one would like to store. In the case of linear eigenvector automaton, unfortunately just one vector absorbs almost the whole of input space. The secret of associative network design is locating as many attractors as possible in input space, each one of them with a well-defined and bounded influence region. To do this one must introduce a nonlinearity in the activation of the network units so that the dynamical system becomes nonlinear.

Associative networks have been studied for a long time. Donald Hebb considered in the 1950s how neural assemblies could self-organize into feedback circuits capable of recognizing patterns [17]. Hebbian learning has been interpreted in different ways and several modifications of the basic algorithm have been proposed, but they all have three aspects in common: Hebbian learning is a local, interactive, and time-dependent mechanism. A synaptic phenomenon in the hippocampus, known as long-term potentiation, is thought to be produced by Hebbian modification of the synaptic strength. There were some other experiments with associative memories in the 1960s, for example, the hardware implementations by Karl Steinbuch of his "learning matrix." A precise mathematical description of associative networks was given by Kohonen in the 1970s [18]. His experiments with many different classes of associative memories contributed enormously to the renewed surge of interest in neural models. Some researchers tried to find biologically plausible models of associative memories following his lead [19]. Some discrete variants of correlation matrices were analyzed in the 1980s, as done for example by Kanerva who considered the case of weight matrices with only two classes of elements, 0 or 1 [20]. In the 1990s much work had been done on understanding the dynamical properties of recurrent associative networks [21]. It is important to find methods to describe the changes in the basins of attraction of the stored patterns. Haken has proposed his model of a synergetic computer, a kind of associative network with a continuous dynamics in which synergetic effects play the crucial role [22]. The fundamental difference from conventional models is the continuous, instead of discrete, dynamics of the network, regulated by some differential equations.

1.5 Hopfield Neural Networks

One of the milestones for the great renaissance in the field of neural networks was the associative model proposed by Hopfield (born on July 15, 1933, his research fields include physics, molecular biology and neuroscience) at the beginning of the 1980s. Hopfield's approach illustrates the way theoretical physicists like to think about ensembles of computing units. No synchronization is required, each unit behaving as a kind of elementary system in complex interaction with the rest of the ensemble. An energy function must be introduced to harness the theoretical complexities posed by such an approach.

(1) Synchronous and asynchronous networks A relevant issue for the correct design of recurrent neural networks is the adequate synchronization of the computing elements. In the case of McCulloch-Pitts networks we solved this difficulty by assuming that the activation of each computing element consumes a unit of time. The network is built taking this delay into account and by arranging the elements and their connections in the necessary pattern. When the arrangement becomes too contrived, additional units can be included which serve as delay elements. What happens when the synchronization of the computing elements is eliminated? The synchronization of the output was achieved by requiring that all computing elements evaluate their inputs and compute their output simultaneously. Under this assumption the operation of the associative memory can be described with simple linear algebraic methods. The excitation of the output units is computed using vector-matrix multiplication and evaluating the sign function at each node. The methods we have used before to avoid dealing explicitly with the synchronization problem have the disadvantage, from the point of view of both biology and physics, that global information is needed, namely a global time. Whereas in conventional computers synchronization of the digital building blocks is achieved using a clock signal, there is no such global clock in biological systems. In a more biologically oriented simulation, global synchronization should thus be avoided. Networks in which the computing units are activated at different times and which provide a computation after a variable amount of time are stochastic automata. Networks built from this kind of units behave like stochastic dynamical systems.

Above, we have already discussed recurrent associative networks in which the output of the network is fed back to the input units using additional feedback connections. In this way we designed recurrent dynamical systems and tried to determine their fixed points. However, there is another way to define a recurrent associative memory made up of two layers which send information recursively between them. The input layer contains units that receive the input to the network and send the result of their computation to the output layer. The output of the first layer is transported by bidirectional edges to the second layer of units, which then return the result of their computation back to the first layer using the same edges. As in the case of associative memory models, we can ask whether the network achieves a stable state in which the information being sent back and forth does not change after a few iterations [23]. Such a network is known as a resonance network or bidirectional associative memory (BAM). The activation function of the units is the sign function and information is coded using bipolar values. The BAM is thus a generalization of a unidirectional associative memory. An input vector, the "key," can be presented to the network from the left or from the right and, after some iterations, the BAM finds the corresponding complementary vector. As can be seen, no external feedback connections are necessary. The same edges are used for the transmission of information back and forth.

So far only conventional or bidirectional associative memories working with synchronized units have been considered. Dropping the assumption of simultaneous firing of the computing elements leads to the appearance of novel network properties. In 1982 the American physicist John Hopfield proposed an asynchronous neural network model (i.e., a kind of additive neural network model. For convenience to cite the work by Hopfield, this kind of additive neural model is also called Hopfield neural model) which made an immediate impact in the AI community. It is a special case of a bidirectional associative memory, but chronologically it was proposed before the BAM.

In the *Hopfield model* it is assumed that the individual units preserve their individual states until they are selected for a new update. The selection is made randomly. A Hopfield network consists of n totally coupled units, that is, each unit is connected to all other units except itself. The network is symmetric because the weight w_{ij} for the connection between unit i and unit j is equal to the weight w_{ji} of the connection from unit j to unit i. This can be interpreted as meaning that there is a single bidirectional connection between both units. The absence of a connection from each unit to itself avoids a permanent feedback of its own state value [24].

The symmetry of the weight matrix and a zero diagonal are thus necessary conditions for the convergence of an asynchronous totally connected network to a stable state. These conditions are also sufficient. The units of a Hopfield network can be assigned a threshold θ different from zero. In this case each unit selected for a state update adopts the state 1 if its total excitation is greater than θ, otherwise the state -1. This is the activation rule for perceptrons, so that we can think of Hopfield networks as asynchronous recurrent networks of perceptrons. The energy function of a Hopfield network composed of units with thresholds different from zero can be defined in a similar way as for the BAM. Note that from the derivation relation of the energy function of *Hopfiled neural networks* and BAM neural networks, Hopfield neural network is a special kind of bidirectional associative memory neural network [2]. Therefore, Hopfield neural networks have a close relation with associative memory networks.

The energy function of a Hopfield network is a quadratic form. A Hopfield network always finds a local minimum of the *energy function* (i.e., local stable equilibrium point in the sense of Hopfield energy function) (Note that, this kind of stability is not the stability in the sense of Lyapunov). Hopfield networks can also be used to compute logical functions. Conjunction, for example, can be implemented with a network of three units. The states of two units are set and remain fixed during the computation (clamping their states). Only the third unit can change its state. If the network weights and the unit thresholds have the appropriate values, the unconstrained unit will assume a state that corresponds to the conjunction of the two clamped states. Because Hopfield used an additive neural network to solve a kind of practical problem and then made a theoretical analysis on the additive networks by involving energy function, especially the test result was satisfied, this work provided a new way to handle the difficulties encountered at that time. Different assumptions on the connection weights and activation function led to different energy function shapes, which is directly related to the stability of the solutions. This problem is directly relevant to the stability property of neural networks, which has been the hot topic in the community of control theory and neural networks since the pioneering work of Hopfield.

Observe the *Lyapunov function* of a Hopfield network, it is only a function of the input and output. Although it does not possess the complete information contained in the state variables, we can nevertheless derive the steady-state properties of the state variables from the properties of this energy function. The so-called Lyapunov function defined by Hopfield can be interpreted as the "generalized energy" of the neural network, although its exact physical meaning is not very clear (along the same routine to Hopfield's analysis method, some stability analysis results are proposed for cellular neural networks) [25]. However, for the class of neural networks with Sigmoid *activation function* or with piecewise linear function, the defined energy function or Lyapunov function does always converge to a local minimum, where the concerned neural network produces the desired output.

(2) Isomorphism between Hopfield model and Ising models This is a significant discovery of J.J. Hopfield, who is both a biologist and a physicist in the world! Therefore, physicists have analyzed the *Hopfield model* in such exquisite detail because it is isomorphic to the Ising model of magnetism (at temperature zero) [26]. Ising proposed the model that now bears his name nearly 90 years ago in order to describe some properties of ensembles of elementary magnets [27]. In general, the *Ising model* can be used to describe those systems made of particles capable of adopting one of two states. In the case of ferromagnetic materials, their atoms can be modeled as particles of spin 1/2 (up) or spin –1/2 (down). The spin points in the direction of the magnetic field. All tiny magnets interact with each other. This causes some of the atoms to flip their spin until equilibrium is reached and the total magnetization of the material reaches a constant level, which is the sum of the individual spins. With these few assumptions we can show that the energy function deduced from the Ising model has the same form as the energy function of Hopfield networks [2].

The total magnetic field h_i sensed by the atom i in an ensemble of particles is the sum of the fields induced by each atom and the external field h^* (if present), that is,

$$h_i = \sum_{j=1}^{n} w_{ij} x_j + h^*, \tag{1.2}$$

where w_{ij} represents the magnitude of the magnetic coupling between the atoms labeled i and j. The magnetic coupling changes according to the distance between atoms and the magnetic permeability of the environment. The potential energy E of a certain state $(x_1, x_2, \ldots, x_n)$ of an Ising material can be derived from (1.2) and has the form

$$E = -\frac{1}{2} \sum_{i=1}^{n} \sum_{j=1}^{n} w_{ij} x_i x_j + \sum_{i=1}^{n} (-h^* x_i), \tag{1.3}$$

In paramagnetic materials the coupling constants are zero. In ferromagnetic materials the constants w_{ij} are all positive, which leads in turn to a significant coupling of the spin states. Equation (1.3) is isomorphic to the *energy function* of Hopfield networks (i.e., $E(x) = -\frac{1}{2} \sum_{i=1}^{n} \sum_{j=1}^{n} w_{ij} x_i x_j + \sum_{i=1}^{n} (\theta_i x_i)$ or $E(x) = -\frac{1}{2} X^T W X + \theta^T X$, $W = (w_{ij})_{n \times n}$). This is why the term energy function is used in the first place. Both systems are dynamically equivalent, but only in the case of zero temperature, since the system behaves deterministically at each state update. On the contrary, when we consider Boltzmann machines, we will accept a time-varying temperature and stochastic state updates as in the full Ising model. Nearly all the neural network models are inspired by the biological networks and the physical world!

(3) Convergence to stable states It is easy to show that Hopfield models always converge to stable states. The proof of this fact relies on analysis of the new value of the *energy function* after each state updates. Now we will state a stability of Hopfield neural networks as follows.

Proposition 1 *A Hopfield network with n units and asynchronous dynamics, which starts from any given network state, eventually reaches a stable state at a local minimum of the energy function.*

The wording of the Proposition 1 has been carefully chosen. That the network "eventually" settles in a stable state means that the probability of not reaching such a state approaches zero as the number of iterations increases. It would be possible to select always one and the same unit for computation of the excitation, and in this case the network would stay in deadlock. Since the units are selected randomly, the probability of such pathological behavior falls to zero as time progresses. Note that, in the proof of the Proposition 1 only the symmetry and the zero diagonal of the weight matrix were used. The proof of convergence is very similar to the proof of convergence for the bidirectional associative memory networks (BAM) [2]. However, in the case of a BAM the decisive property was the independence of a unit's state from its own excitation. This is also the case for Hopfield networks, since no unit feeds its own state back into itself, i.e., the diagonal of the weight matrix is zero.

There is a simpler proof of the Proposition 1, which has the advantage of offering a nice visualization of the dynamics of a Hopfield network [31]. Assume that we classify the units of a network according to their states: the first set contains the units with state 1, the second set contains the units with state -1. There are edges linking every unit with all the others, so that some edges go from one set to the other. We now randomly select one of the units and compute its "attraction" by the units in its own set and the *attraction* by the units in the other set. The "attraction" is the sum of the weights of all edges between a unit and the units in its set or in the other one. If the attraction from the outside is greater than the attraction from its own set, the unit changes sides by altering its state. If the external attraction is lower than the internal, the unit keeps its current state. This procedure is repeated several times, each time selecting one of the units randomly. It corresponds to the updating strategy of a Hopfield network. The selected unit must change sides. It is clear that the network must eventually reach a stable state, because the sum of the weights of all edges connecting one set to the other can only become lower in the course of time. Since the number of possible network states is finite, a global state must be reached in which the attraction of one set by the other cannot be further reduced. This is the task known in combinatorics as the minimal cut problem, in which we want to find a cut of minimal flow in a graph. The procedure described always finds a locally minimal cut.

(4) The limits of Hopfield networks The first article of Hopfield and Tank on parallel solutions to combinatorial problems received a lot of attention [29, 30]. The theoretical question was whether this could be a method to solve NP-hard (i.e., nondeterministic polynomial-time hard) problems or at least to get an approximate solution in polynomial time. In the following years many other researchers tried to extend the range of combinatorial problems that could be solved by Hopfield's technique, trying to improve the quality of the results at the same time. It emerged that well-behaved average problems could be solved efficiently. However, these average results should be compared to the expected running time for the worst case. Bruck and Goodman [31] showed that a polynomially bounded network (on the size of the problem) is unable to find the global minimum of the energy function of NP-complete problems (encoded as Hopfield networks) with a 100 % guarantee. Stated in another way: if we try to transform all local minima of the Hopfield network into an optimal solution of the combinatorial problem, the size of the network explodes exponentially.

The author in [2] proceeds to prove the result of Bruck and Goodman, but he introduces an additional complexity class: the complement of the class NP. The class NP of nondeterministic problems solvable in polynomial time is different from the class P of problems solvable in polynomial time. If a problem is a member of the class P, the same is true for the complementary problem. The complement of the decision problem "For the problem instance I, is X true for I?" is just "For the problem instance I, is X false for I?" A deterministic polynomial time algorithm terminates on each of the two questions. It is only necessary to substitute "true" for "false" to transform a polynomial time algorithm for a problem in P in an algorithm for its complement. But this is not necessarily so for problems in NP. A solution for the Traveling Salesman Decision Problem (TSDP), that is, the computation of the tour's length and the comparison with the decision's threshold, can be verified in polynomial time. However, the complementary problem has the wording "Is there no tour with a total length smaller than R?" If the answer is "yes," no polynomial time algorithm is known that could verify this assertion. It would be necessary to propose a data structure on which to perform some computations that could convince us of the truth of the assertion. Theoreticians assume that the complement of the TSDP probably does not belong to the class NP. The class that contains the complement of all NP problems is called co-NP. It is generally assumed that $NP \neq$ co-NP. This inequality is somewhat strong, since it implies that $P \neq NP$. Otherwise, one would have co-$P =$ co-$NP = P = NP$, i.e., the equality $NP =$ co-NP would be valid. Yet theoreticians expect that eventually it will be proved that $NP \neq$ co-NP.

The following lemma determines under what conditions equality of the classes NP and co-NP would be possible. One can assume that this condition cannot be fulfilled [2].

Lemma 1 *If there is an NP-complete problem X whose complement X_c belongs to NP, then $NP =$ co-NP.*

Lemma 1 is true because any problem Y in NP can be reduced in polynomial time into X. The complement of Y can therefore be transformed in polynomial time into X_c. Since a solution of X_c can be verified in polynomial time, the same is true for any solution of Y_c. This and some additional technical details would imply that $NP =$ co-NP. Neural networks are just a subset of the algorithmic world. Since it is suspected that there is no polynomial time algorithm for the problems in the class NP, it should be possible to prove that Hopfield networks of bounded size are subjected to the same limitations. The following proposition settles this question [31].

Proposition 2 *Let L be an NP-complete decision problem and H a Hopfield network with a number of weights bounded by a polynomial on the size of the problem. If H can solve L (100 % success rate), then NP = co-NP.*

Proposition 2 can be proved as follows. The problem L has a certain size defined by an appropriate coding. Since we must compute the energy function and from it derive the necessary weights for H, a polynomial bound on the total number of weights is necessary. A Hopfield network always finds a local minimum of its energy function. In this case, a 100 % hit rate means that all local minima of the energy function should make possible a decision on the truth or falsity of the decision problem L. The Hopfield network can be considered a data structure that makes possible the verification of the found solution. It is only necessary to verify whether the solution found by the network is indeed a local minimum of the energy function. The polynomial size of the net makes this verification possible in polynomial time. The decision problem and the complement are, in this case, completely symmetric. The TSDP can be answered with "yes" if the tour found by the network is shorter than the decision threshold. But the complement of the TSDP can be decided also just by comparing the length of the optimal tour found with the decision threshold. Therefore, the complement of L is a member of the class NP and it follows from Lemma 1 that NP = co-NP. Since it is generally assumed that this cannot be so, there should be a contradiction in the premises. The network H does not exist unless NP = co-NP.

Even if we content ourselves with a polynomially bounded network that can provide approximate solutions (for example, traveling salesman problem (TSP) roundtrips not larger by a given ϵ than the optimal tour), no such network can be built. It is because of this inherent limitation that some researchers have sought to introduce stochastic factors into the networks, as it may be falling into the scope of Boltzmann machines. *Hopfield networks* as massively parallel systems are only interesting if they can be implemented in hardware and not just simulated in a sequential computer. Some proposals have been made for special chips capable of simulating Hopfield networks but the most promising approach are optical computers, capable of solving the connectivity problem of neural networks. For other topics on the equivalence of Hopfield and perceptron learning, complexity of learning in Hopfield models, and parallel combinatorics, implementation of Hopfield neural networks, readers can further refer to [2] for more details.

Generally speaking, with the introduction in 1982 of the model named after him, John Hopfield established the connection between neural networks and physical systems of the type considered in statistical mechanics. This in turn gave computer scientists a whole new arsenal of mathematical tools for the analysis of neural networks. Other researchers had already considered more general associative memory models in the 1970s, but by restricting the architecture of the network to a symmetric connection matrix with a zero diagonal, it was possible to design recurrent networks with stable states. With the introduction of the concept of the energy function, the convergence properties of the networks could be more easily analyzed. The Hopfield network also has the advantage, in comparison with other models, of a simple technical implementation using electronic or optical devices [32]. The properties of Hopfield networks have been investigated since 1982 using the theoretical tools of statistical mechanics [33]. Gardner [34] published a classical treatise on the capacity of the perceptron and its relation to the *Hopfield model*. The total field sensed by particles with a spin can be computed using the methods of mean field theory. This simplifies a computation which is hopeless without the help of some statistical assumptions [35]. Using these methods Amit et al. [36] showed that the number of stable states in a Hopfield network of n units is bounded by $0.14n$. A recall error is tolerated only 5 % of the time. This upper bound is one of the most cited numbers in the theory of Hopfield networks. In 1988 Kosko proposed the BAM model, which is a kind of "missing link" between conventional associative memories and Hopfield networks. Many other variations have been proposed since, some of them with asynchronous, others with synchronous dynamics [21].

Hopfield networks have also been studied from the point of view of dynamical systems. In this respect spin glass models play a relevant role. These are materials composed of particles with spin and mutual interactions [37, 38]. Combinatorial problems have a long tradition, but a really systematic theory capable of unifying the myriad of heuristic methods developed in the past was first developed in the 1960s and 1970s [39]. The important point was the increasingly important role played by computers and the emergence of a new attitude which tried to reach whole classes of problems and not just individual cases. An important research theme that remains is how to split a combinatorial problem into subtasks which can be assigned to different processors [40]. The efforts of Hopfield and Tank with the TSP led to many other similar experiments in related fields. Wilson and Pawley [41] repeated their experiments but they could not confirm the optimistic results of the former authors. The main difficulty is that complex combinatorial problems produce an exponential number of local minima of the energy function. In sequential computers, Hopfield models cannot compete with conventional methods [42]. Many heuristics have been proposed for the TSP, starting with the classical work of Kernighan and Lin [43]. The only way to make Hopfield models competitive is through the use of special hardware. Sheu et al. [44] obtained interesting results and significant speedup in comparison with sequential computers by a technique they called hardware annealing. One of the first to deal with the intrinsic limits of the Hopfield model for the solution of the TSP was Abu-Mostafa [45, 46], who nevertheless considered only the case of networks of constant size. Bruck and Goodman [31] considered networks of variable but polynomially bounded size and obtained the same negative results. Although this almost meant the "death of the traveling salesman" [33], the Hopfield model and its stochastic variants have been applied in many other fields, such as psychology, simulation of ensembles of biological neural networks, and chaotic behavior of neural circuits. The optical implementation of Hopfield networks is a promising field of research. Other than masks, holograms can also be used to store the network weights [47]. The main technical problem is still the size reduction of the optical components, which could make them a viable alternative to conventional electronic systems.

1.6 Cohen–Grossberg Neural Networks

In the 1950s, Grossberg (Stephen Grossberg, born on December 31, 1939, is a cognitive scientist, neuroscientist, mathematician, biomedical engineer, and neuromorphic technologist) introduced the paradigm of using nonlinear systems of differential equations to show how brain mechanisms can give rise to behavioral functions. The laws were derived from an analysis of how psychological data about human and animal learning can arise in an individual learner adapting autonomously in real time. This paradigm is helping to solve the classical mind/body problem, and is the basic mathematical formalism that is used in biological neural network research today.

In the 1960s and 1970s, Grossberg generalized the additive and shunting models to a class of dynamical systems that included these models as well as nonneural biological models, and proved content addressable memory theorems for this more general class of models. As part of this analysis, he introduced a *Lyapunov function* method to help classify the limiting and oscillatory dynamics of competitive systems by keeping track of which population is winning through time. This Lyapunov method led him and Michael Cohen to discover in 1981 and publish in 1982 and 1983 a Lyapunov function, which they used to prove that global limits exist in a class of dynamical systems with symmetric interaction coefficients, which includes the additive and shunting models. John Hopfield published this Lyapunov function for the additive model in 1984. Some scientists started to call Hopfield's contribution "the Hopfield model." In an attempt to correct this historical error, other scientists called the more general model and Lyapunov function "the Cohen–Grossberg model." Still other scientists call it "the Cohen–Grossberg–Hopfield model" (this assertion can be found in Wikipedia encyclopedia). In 1987, Bart Kosko (born on February 7, 1960) adapted the Cohen–Grossberg model and Lyapunov function, which proved global convergence of short-term memory (STM), to define an adaptive bidirectional associative memory that combines STM and long-term memory (LTM) and which also globally converges to a limit [48].

Early applications of the additive model included computational analysis of vision, learning, recognition, reinforcement learning, and learning of temporal order in speech, language, and sensory-motor control. The additive model has continued to be a cornerstone of neural network research to the present time in decision making. Physicists and engineers unfamiliar with the classical status of the additive model in neural networks called it the *Hopfield model* after the first application of this equation in Hopfield [29]. Grossberg [49] summarizes historical factors that contributed to their unfamiliarity with the neural network literature. The additive model can be generalized in many ways, including the effects of delays and other factors.

A much more general class of systems has this property:

$$\dot{x}_i = a_i(x)[b_i(x_i) - c(x)], \tag{1.4}$$

where $x = (x_1, x_2, \ldots, x_n)^T$, $a_i(x)$ is a state-dependent amplification function, $b_i(x_i)$ is a self-signal function, and $c(x_i)$ is the state-dependent adaptation level against which each $b_i(x_i)$ is compared. With different parameters and signal functions for each cell in (1.4), system (1.4) may represent for arbitrarily many cells, for example, recurrent competitive field. Grossberg [50] proved that all trajectories in such systems are "stored in STM"; that is, converge to equilibrium values as $t \to \infty$, even in systems that possess infinitely many equilibrium points. The proof shows how each $x_i(t)$ gets trapped within a sequence of decision boundaries that get laid down through time at the abscissa values of the peaks in the graphs of the signal functions $b_i(x_i)$, starting with the highest peaks and working down. Multiple peaks correspond to multiple cooperating subpopulations. These graphs may thus be very complex if each population contains multiple cooperating subpopulations. After all the decision boundaries get laid down, each $x_i(t)$ is trapped within a single valley of its b_i graph. After this occurs for all the x_i variables, the function $B(x(t)) = \max[b_i(x(t)) : i = 1, 2, \ldots, n]$ is a *Lyapunov function*, whose Lyapunov property is then used to complete the proof of the theorem. A special case of the theorem concerns a competitive market with an arbitrary number of competing firms. Each firm can choose one of infinitely many production and savings strategies that are unknown to the other firms. The firms know each other's behaviors only through their effect on a competitive market price, and they produce more goods at any time only if application of their own firm's production and savings strategy will lead to a net profit with respect to that market price. The theorem proves that the price in such a market is stable and that each firm balances its books. The theorem does not, however, determine which firms become rich and which go broke.

Due to the importance of symmetry in proving global approach to equilibria, as in the adaptation level systems (1.4), Cohen and Grossberg attempted to prove that all trajectories of systems of the Cohen–Grossberg form:

$$\dot{x}_i = a_i(x_i)\left[b_i(x_i) - \sum_{j=1}^{n} c_{ij} d_j(x_j)\right], \tag{1.5}$$

with symmetric interaction coefficients $c_{ij} = c_{ji}$ and weak assumptions on their defining functions, approach equilibria as $t \to \infty$. Systems (1.5) include both Additive Model and Shunting Model networks with distance-dependent, and thus symmetric, interaction coefficients, the Brain-State-in-a-Box model [**?**], the continuous-time version of the McCulloch and Pitts [1] model, the Boltzmann Machine equation [53**?**] model, the Volterra-Lotka model [54], the Gilpin and Ayala model [55], the Eigen and Schuster model [56], the Cohen and Grossberg [57, 58] Masking Field model, and so on. More details can be found by searching the keyword "recurrent network networks" in the Wikipedia on the internet.

Cohen and Grossberg first attempted to prove global equilibrium by showing that all Cohen–Grossberg systems generate jump trees, and thus no jump cycles, which would immediately prove the desired result. This hypothesis still stands as an unproved conjecture.While doing this, inspired by the use of Lyapunov methods for more general competitive systems, Cohen and Grossberg [59, 60] discovered the Cohen–Grossberg Lyapunov function that they used to prove that global equilibria exist:

$$V = -\sum_{i=1}^{n} \int_{0}^{x_i} b_i(\xi_i) d_i^{'}(\xi_i) d\xi_i + \frac{1}{2} \sum_{j=1}^{n} \sum_{k=1}^{n} c_{jk} d_j(x_j) d_k(x_k). \tag{1.6}$$

Equation (1.6) defines a Lyapunov function because integrating V along trajectories implies that:

$$\dot{V}(x) = -\sum_{i=1}^{n} a_i(x_i) d_i^{'}(x_i) \left[b_i(x_i) - \sum_{k=1}^{n} c_{ik} d_k(x_k) \right]^2. \tag{1.7}$$

If $a_i(x_i)d_i^{'}(x_i) \geq 0$, then (1.7) implies that $\dot{V}(x) \leq 0$ along trajectories, where $d_i^{'}(t) = \frac{\mathrm{d}d_i(t)}{\mathrm{d}t}$. Once this basic property of a Lyapunov function is in place, it is a technical matter to rigorously prove that every trajectory approaches one of a possibly large, or infinite, number of equilibrium points. As noted above, the Lyapunov function (1.6) proposed in Cohen and Grossberg [59] includes both the Additive Model and Shunting Model, among others. A year later, Hopfield [29] published the special case of the Additive Model and Lyapunov function and asserted, without proof, that trajectories approach equilibria. Based on this 1984 article, the Additive Model has been erroneously called the Hopfield network by a number of investigators, despite the fact that it was published in multiple articles since the 1960s and its Lyapunov function was also published in 1982–1983. A historically more correct name, if indeed names must be given, is the Cohen–Grossberg–Hopfield model, which is the name already used in the literature. More details can be found by searching Stephen Grossberg in Wikipedia, the free encyclopedia on the Internet.

Symmetry does not imply convergence: Synchronized oscillations. Cohen [61] showed that symmetric coefficients are not sufficient to ensure global convergence by constructing distance-dependent (hence symmetric) on-center off-surround networks that support persistent oscillations. These networks can send excitatory feedback signals to other populations than themselves. It has long been known that shunting networks with slow inhibitory interneurons can persistently oscillate, e.g., Ellias and Grossberg [62]. This observation led to the prediction that neural networks can undergo synchronized oscillations, first called order-preserving limit cycles by Grossberg [63, 64], during attentive resonant states. The early articles concerning synchronized oscillations during attentive brain dynamics have been followed by hundreds more. Persistent oscillations can also occur in recurrent competitive field (RCF) defined by Grossberg [65].

Note that the contents in Sects. 1.5 and 1.6 are mainly from the Wikipedia on the Internet. By comparing the evolution of *Hopfield model* and *Cohen–Grossberg model*, we can find many interesting findings.

(1) In fact, the neural model used by Hopfield is an additive neural model (the additive neural-network model includes, as special cases, the Hopfield neural network, the cellular neural networks (CNNs), and several other classes of extensively employed neural networks), which has been studied for a long time in the community of neural science. Grossberg had also studied this kind of additive neural model in the aspects of biology. According to the introduction in this section, Grossberg has proposed many different kinds of neural models, in which the Cohen–Grossberg model is one of the most important model. However, since the publications of Hopfield's work in 1982 and 1984, respectively, a kind of additive neural model was renamed by Hopfield model, which seemed to be a new kind of neural model.

(2) Comparing the work of Hopfield with that of Grossberg in neural networks, Grossberg may have more contribution in the field of biological neural networks. However, Hopfield may be more famous than Grossberg for his considered neural model in the engineering field. This may raise a delusion that the Cohen–Grossberg model is an extension of the Hopfield model. That is, Hopfield model is earlier than the Cohen–Grossberg model, and Cohen–Grossberg's work is on the basis of Hopfield's work. There is no literature to explain the original relations between the Hopfield model and the Cohen–Grossberg model. Here, to the best of the authors' knowledge, it is the first time that a relation is presented between these two models in a monograph. In the aspect of mathematical model of neural networks, Hopfield model is not new, and Cohen–Grossberg model has no inheritance relationship to the so-called Hopfield model.

(3) Hopfield and Grossberg (born in 1939) are all outstanding scholars in the field of biological neural networks. Meanwhile Hopfield was also a physicist, who can use the physical thinking to consider the relation and phenomena in the biological neural networks. In contrast, Grossberg is an excellent expert in the biological neural networks, and he was used to considering the neuronal phenomena and biological relations from the opinion of a biologist. Therefore, different viewpoints may lead to different ways to further study the same phenomena. In this aspect, Hopfield is superior to Grossberg, which leads to the famous findings by using the simple neural network model! In another way, Hopfieds' work also discovers that a simple additive neural network has many fascinating features in engineering applications, let alone the complex neural models, e.g., Cohen–Grossberg neural model.

(4) The reason that Hopfield is more famous than Grossberg, which is given by the authors, is based on the fact that Hopfield has solved the challenging application problems using the theoretical neural model, while Grossberg did not. When we recall the time in the 1980s, the industrial development was one of the urgent problems in the world. Technology applications are the hot and important projects in any country. Just under such a background, fundamental theory research may be a bit weaker than the technical applications. Hopfield' work using the neural model to solve the optimization problems just met the demands of the times. Therefore, more scholars from the world began to study Hopfield's work. For brief citing of the model in Hopfield's work, a natural way is to name the model by the author's name, which is similar to the origin of the TS fuzzy model. This term is not decided by Hopfield,

but by the academic market. Thinking in this way, it seems doomed that Hopfield should appear in that era.

(5) Hopfield and Grossberg are both excellent scientists. However, different research directions may lead to different results. Grossberg seems absorbed in the fundamental theory research of one field. Therefore, Grossberg's work may be highly recognized in the local fields of his speciality. In contrast, Hopfield used the existing theory to solve the engineering problems. Hopfield on his own may not be famous at that time, while the solved engineering problems may be a significant project and hot topic at that time. In this way, Hopfield is hot! It is the time that chooses Hopfield.

1.7 Property of Neural Network

The process used to perform the learning process is called a learning algorithm, the function of which is to modify the synaptic weight of the network in an orderly fashion to obtain a desired design objective. The modification of the synaptic weights provides the traditional method for the design of neural network. Such an approach is the closest to linear adaptive filter theory, which is already well established and successfully applied in many diverse fields. However, it is also possible for a neural network to modify its own topology, which is motivated by the facts that neurons in the brain can die and that new synaptic connections can grow.

For the neural networks in a general sense, it is apparent that a neural network derives its *computing ability* through, first, its massively parallel distributed structure and, second, its ability to learn and therefore generalize. Generalization means the neural network producing reasonable outputs for inputs not encountered during training or learning. These two information-processing capabilities make it possible for neural networks to solve complex or large-scale problems that are currently intractable. In practice, neural network cannot provide the solution by working individually. Rather, they need to be integrated into a consistent system. Specifically, a complex problem of interest is decomposed into a number of relatively simple tasks, and neural networks are assigned a subset of the tasks that match their inherent capabilities. It is important to recognize that we have a long way to go before we can build a computer architecture that mimics a human brain.

Neural networks offer the following useful properties and capabilities for engineering applications.

(1) Nonlinearity. An artificial neuron can be linear or nonlinear. A neural network, made up of interconnections of nonlinear neurons, is itself nonlinear. Moreover, the nonlinearity is of weak approximation in the sense that it is distributed throughout the networks. Nonlinearity is a highly important property, particularly if the underlying physical mechanism responsible for generalization of the input signal is inherently nonlinear.

(2) Input–output mapping. The network is presented with an example picked at random from the set, and the synaptic weights of the network are modified to minimize the difference between the desired response and the actual response of the

network produced by the input signal in accordance with an appropriate statistical criterion. The training of the network is repeated for many examples in the set until the network reaches a steady state where there are no further significant changes in the synaptic weights.

(3) Adaptivity. Neural networks have a built-in capability of adapting their synaptic weights to changes in the surroundings. The natural structure of a neural network for pattern classification, signal processing, and control applications, coupled with the adaptive capability of the network, makes it a useful tool in adaptive pattern classification, adaptive signal processing, and adaptive control applications. As a general rule, it may be said that the more adaptive we make a system, all the time ensuring that the system remains stable, the more robust its performance will likely be when the system is required to operate in a nonstationary environment. It should be emphasized that adaptivity does not always lead to robustness; indeed it may do the very opposite. This problem is referred to as the stability-plasticity dilemma.

1.8 Information Processing Capacity of Dynamical Systems

Many dynamical systems, both natural and artificial, are driven by external input signals and can be seen as performing nontrivial, real-time computations of these signals. It is an interesting topic to study the online information processing that takes place within these dynamical systems. The importance of this topic is underlined by the multitude of examples from nature and engineering which fall into this category. These include the activation patterns of biological neural circuits to sensory input streams, the dynamics of many classes of artificial neural network models used in artificial intelligence (including the whole field of "reservoir computing" as well as its recent implementation using time delay systems), and systems of interacting chemicals, such as those found within a cell, that display intracellular control mechanisms in response to external stimuli. Recent insight into robotics has demonstrated that the control of animal and robot locomotion can greatly be simplified by exploiting the physical body's response to its environment [66–71]. Additional systems are population dynamics responding to, e.g., changes in food supply, ecosystems responding to external signals such as global climate changes, and the spread of information in social network graphs. All these exemplary systems display completely different physical implementations.

How to present a general framework that allows to compare the computational properties of a broad class of dynamical systems is the main work in [66]. One is able to characterize the information processing that takes place within these systems in a way that is independent of their physical realization, using a quantitative measure. It is normalized appropriately such that completely different systems can be compared and it allows us to characterize different computational regimes (linear vs. nonlinear; long vs. short memory).

Some initial steps in this direction are provided by the linear memory capacity (later theoretically extended to linear systems in discrete time and continuous time systems, and using Fischer information). Although it has been argued that short-term memory of a neural system is crucial for the brain to perform useful computation on sensory input streams, it is a belief that the general paradigm underlying the brain's computation cannot solely rely on maximizing linear memory. The neural circuits have to compute complex nonlinear and spatiotemporal functions of the inputs. Prior models focusing on linear memory capacity in essence assumed the dynamic system to only implement memory, while the complex nonlinear mapping is off-loaded to an unspecified readout mechanism. The authors in [66] proposed to endow the dynamic system with all the required computational capacity, and used only simple linear read-out functions. The capacity measures introduced in [66] are therefore of great interest since they quantify all the information processing that takes place within the dynamical system, and don't introduce an artificial separation between linear and nonlinear information processing.

One of the startling results of the work in [66] with potentially far reaching consequences is that all dynamical systems, provided they obey the condition of having a fading memory and have linearly independent internal variables, have in principle the same total normalized capacity to process information. The ability to carry out useful information processing is therefore an almost universal characteristic of dynamical systems. This result provides theoretical justification for the widely used paradigm of reservoir computing with a linear readout layer. Indeed, it confirms that such systems can be universal for computation of time invariant functions with fading memory. The main contribution in [66] is to uncover the (potentially universal) trade-off between memory depth and nonlinear computations performed by the dynamic system itself, and also discuss how the influence of noise decreases the *computational capacity* of a dynamical system.

Inspired by the work in [66], recurrent neural networks as a special *dynamical system*, its computational capability, storage capability, and learning ability fall into the notation framework of the fading memory, Hilbert space, and fading memory dynamical system defined in [66]. Therefore, many features of different dynamical systems and their applications are subject to the understanding of the essentials of the internal memory capacity of the concerned systems. No matter what the features are, stability as one of the fundamental properties of dynamical systems is still indispensable.

1.9 Stability of Dynamical Neural Networks

According to different measures, neural networks can be divided into different kinds of models. For example, neural networks can be classified into feedforward neural networks and recurrent neural networks according to different connection forms. It is well known that feedforward neural networks have been studied deeply more than 30 years. The advantages of feedforward neural network are approximation capability and learning mechanism. However, recurrent neural networks have also shown that they own the computational capability and memory capability besides approximation capability and learning mechanism. Therefore, recurrent neural networks have been re-examined with the development of feedforward neural networks. The pioneer work of recurrent neural networks is contributed to the great biophysician Hopfield J. J., who first used the simple recurrent neural networks, now called Hopfield neural networks, solved the optimal Problem of Traveling Salesman and implemented the 4 digital A/D convertor via recurrent neural networks in the 1980s. Hopfield's work pioneered the analog computation of continuous time recurrent neural networks instead of digital computation using complex numerical algorithms on a digital computer. In the Hopfield's research of recurrent neural network, one of the most important problems is stability of the designed recurrent neural networks, which can guarantee stable operation of the hardware circuits. At the same time, biologist Grossberg and his student Cohen proposed a class of competitive-cooperative neural networks, which is widely called Cohen–Grossberg neural networks (CGNN) (note that, plus "+" in interconnection coefficients means competitive and minus "–" in interconnection coefficients means cooperation), and it has been shown that they can represent a large kind of biological neural networks and can realize the pattern formation. In some cases of network parameters, Hopfield neural networks are a special case of Cohen–Grossberg neural networks. For the case of strictly positive amplification, stability analysis of Cohen–Grossberg neural networks is along the same routine as that of Hopfield neural networks. In this case, the equilibrium point of recurrent neural networks can be positive, negative, zero, or their mixtures. However, for the case of nonnegative amplification, stability analysis of Cohen–Grossberg neural networks is a different routine from that of Hopfield neural networks, and the equilibrium point must be nonnegative, which represents the survival and extinction of species. Therefore, in the aspect of engineering applications, CGNN and Hopfield neural networks are different except that they have the similar mathematical model.

Because of the simplicity of circuit implementation of the recurrent neural networks, the following work is on how to reduce the conservatism of the stability criterion because of the intervention of different kinds of delays in the circuit implementation. Time delays usually destroy the stability property of neural networks and produce oscillation, bifurcation, and chaos. According to the environment of implementation, such delays as concentrated or discrete delay(s), multiple different delays, neutral type delay, distributed delay, and their time-varying correspondences can be induced. In the stability analysis of recurrent neural networks, delays are usually regarded as one of the disadvantageous factors even though some delays may have active factors in the design and implementation of neural circuits. In general, the constant delay is rather easier to be tackled than time-varying delay. For the time-varying delay, the changed rate of time-varying delay is usually restricted to be less than 1 in the earlier stability study, not due to the practical consideration but due to the technical difficulty in analytical derivation of stability result. At present, the restriction of changed rate of time-varying delay to be less than 1 is canceled because some useful inequalities and decomposition methods (e.g., free weight matrix method, delay partitioning method) are involved. Theory research serves practical application. How to reduce the gap between theory and application is a long-term and difficult task.

Since then, a lot of stability results have been reported in journals and conferences by scholars and researchers all over the world, for example, America, Italy, China, and Canada. Among these stability results, almost all the earlier results are based on the norm theory, M-matrix theory, measure theory, and algebraic inequality methods. The common feature is that all the stability results take absolute operation on the interconnection weights, which ignored the inhibitory effects (e.g., negative interconnection weight) of neurons on the whole neural networks. Therefore, the conservatism of the stability results are much great. With the emergence of LMI, another kind of stability criteria are proposed in the recent 20 years. This kind of LMI-based stability results not only consider the inhibitory effects of neurons in the analysis, but also consider the connection strengths between neurons. Therefore, LMI method reduces the conservativeness of stability criteria significantly and becomes an important method in the qualitative analysis and synthesis of nonlinear control systems.

1.10 Delay Effects on Dynamical Neural Networks

For practical neuronal networks, there are many factors to be considered in investigating the features of dynamical behavior, which are related to many biological knowledge. Inspired by the biological neurons, some functions of neurons can be imitated in the application of industrial technology. On the other hand, in order to study the underlying mechanism of neural networks in theory, some analysis methods must be utilized, such as mathematical analytical model, symbol model, relation model, and so on. Among these mathematical models, the commonly used model is the mathematical models based on the differential equations (e.g., ordinary differential equations and partial differential equations), in which the model based on ordinary differential equations (ODE) are widely researched.

In order to use the ODE model to describe the real applications or plants, some modifications of ODE models are necessary. For example, some external perturbations, which are unavoidable, are often modeled by the additive or multiplicative disturbance. For the uncertainties of the connection relations or the variations of model structure of the concerned systems, the connection coefficients or the connection matrix may be modeled as the uncertainty coefficients or matrices. These uncertainties are usually divided into two parts: the deterministic and nondeterministic parts. In general, how to describe the nondeterministic parts of the structure uncertainties forms the different research directions in the related topics. In fact, the structure uncertainties may arise from internal and external factors. However, in the uncertain mathematical model based on ODE, these structure uncertainties are completely combined into the connection matrices.

Besides the structure uncertainties, signal transmission delays cannot be neglected. Signal transmission delay may arise from the concerned systems themselves or from the external sources communicating through long distance. Therefore, the delays may produce many different effects on the concerned systems. For example, a delay involved in a nonlinear system may produce complex behaviors such as chaos, periodic solution, stability equilibrium, and saddle points.

Now we will state the fundamental reasons why the delayed systems or neural networks with time delays are widely studied in recent years. From the above sections' introduction, one can see that the development of neural networks theory is executed in the exploration of biological neural networks. All the key points are in the internal of neuronal network itself. In the early time of theoretical research, much emphasis is placed on the interval interconnection relations of neurons, which is about the external strength information. In the current stage, as external information has been fully developed, the research focus is aimed at the internal information. One of these internal information is the signal transmission and the action delay of the synapse. Therefore, different kinds of delay information are considered in the neural network model, and many surprising phenomena are found. How to obtain or use some of the surprising phenomena is the main motivation for studying the delayed neural networks. For more details on the evolution of time delays, readers can refer to Sect. 3.2.3.

1.11 Features of LMI-Based Stability Results

Before 2002, the concept of Lyapunov diagonal stability (LDS) was applied to establish the stability criteria in neural networks community. The advantage of LDS lies in that an unknown variable is involved in the stability criteria, in which the plus and minus signs of interconnection weights are considered. Therefore, the LDS-based stability criterion considers the excitatory (i.e., plus sign of interconnection weight) and inhibitory (i.e., minus sign of interconnection weight) effects of neurons on the whole neural networks. Since 2002, *linear matrix inequality* (LMI) method was introduced into the neural network community, which improve the LDS-based stability results. The advantages of LMI consist in the fact that there are many degrees of freedom in the stability criteria, and the excitatory and inhibitory effects of neurons on the whole neural networks are also considered. LDS-based stability result is a special case of stability results based on LMI. Recently, LMI method is greatly developed in many scientific fields, such as automatic control, signal processing, and stability analysis of nonlinear system. The main reasons why LMI method is greatly appreciated, the authors think, lie in the following aspects.

(1) The LMI Toolbox in MATLAB is freely available, which can be mastered in a very short time. Therefore, LMI-based results can be easily checked using MATLAB software.

(2) In essence, the LMI tool bridges the absolute value methods (e.g., norm theory, M-matrix theory, measure theory, and algebraic inequality methods) and LDS method. That is to say, LMI method is easier to check than algebraic inequality methods, and has more freedom than those methods such as norm theory, M-matrix theory, measure theory, and LDS methods. Therefore, in the total performance evaluation of the desired results, LMI-based results are more perfect than the others.

(3) Intuitionally, LMI-based method is most suitable for the model or system described in state-space equation. Therefore, LMI-based method is mostly welcome in the fields of modern control system theory. The advantages of modern control system theory are that they can reflect the inner variables or factors of the investigated systems and the physical meanings of some state variables are clear. For the classical control theory, for example, transfer function methods, it can be converted into H-infinity problem using the H-infinity norm and realization theory. Therefore, in some cases and for some specific problems, LMI method can also be applied to classical control theory. That is, both modern control system theory and classical control theory can make a space for LMI method.

(4) Technically, many matrix theory methods can be incorporated into the LMI-based methods. Therefore, like algebraic inequality methods (which mainly deal with the scalar space or dot measure, and almost all scalar inequalities can be used in the algebraic inequality methods), many matrix inequalities can be used in the LMI-based method and many kinds of LMI-based stability results can be presented. Especially, LMI-based method deals directly with the two-dimensional vector space, which extends the application space of algebraic inequality methods. Therefore, more information on the system can be contained in LMI-based results than those of algebraic inequality methods.

(5) Besides the LMI method, there are many other methods to study the stability of the neural networks with time delays, for example, bilinear matrix inequality, algebraic inequality, measure theory, norm theory, M-matrix, etc. Different methods correspond to different mathematical models. For some kinds of mathematical models, some methods may be invalid. Therefore, there are many factors to effect the choice of the adopted method, such as the types of the concerned mathematical models and the hobbies of the researchers.

(6) LMI method is more suitable for the systems described by matrix-vector form. Meanwhile, during the analysis and synthesis of the concerned problems, some mathematical difficulties can also arise. Therefore, similar to the algebraic inequality methods, there are some constraints for the applications of LMI method. For example, with more free weighted matrices being involved in the stability criteria, the representation complexity of the stability criteria hindered the development of LMI method, which have deprived of the straight advantages of LMI method. Meanwhile, the feasibility and computational burden of LMI-based stability criteria will be increased. Thus, any mathematical method has its own features, both in advantages and disadvantages.

In this book, the reason why we mainly adopt the LMI method is based on the following consideration. First, both artificial neural networks and recurrent dynamical neural networks originate from the biological neuronal networks, in which the inhibitory action and excitability action are the two main response mechanisms. These bipolar actions can be well described by the matrix description, while other nonmatrix forms always neglect bipolar actions. Second, the concerned RNNs are expressed in the form of state-space equations, which is also suitable for the LMI method. Third, Lyapunov stability theory is more suitable for the concerned RNNs, and many matrix inequality methods can be easily incorporated into the qualitative analysis procedure. Finally, parallel to the algebraic inequality methods, it is necessary to build some stability criteria based on LMI methods in order to enrich the neural networks stability theory.

1.12 Summary

In this chapter, the evolution of artificial neural networks is simply introduced. The main function of neurons is the computation and memory, which may concern the iterated processing of information. For understanding the mechanism of neurons, many kinds of neural models are exploited, among these models are some famous models, for example, Hopfield network model, Cohen–Grossberg neural network model, and bidirectional associative network model. For details on bidirectional associative network model and Hopfield network model, readers can refer to the classical book [2]. For details on Cohen–Grossberg network model, readers can refer to "recurrent neural networks" in Wikipedia on the Internet. Some fundamental knowledge of this chapter is based on the reference [2], and others are from the Wikipedia—the free encyclopedia on the Internet. This chapter mainly presents a background on the recurrent neural networks, for example, evolution history of neural networks, stability of dynamical neural networks, relations between neural networks and dynamical systems. These contents are always scattered in many books, published papers, or Wikipedia encyclopedia. It is also another purpose of this book to collect and run through these good materials (including the interested topic and the references herein) for the readers to study the stability of neural networks conveniently.

References

1. W. McCulloch, W. Pitts, A logical calculus of the ideas immanent in nervous activity. Bull. Math. Biophys. **5**, 115–133 (1943)
2. R. Rojas, Neural networks: A systematic introduction, Springer (1996)
3. A. Brown, *Nerve Cells and Nervous Systems* (Springer, Berlin, 1991)
4. J. Horgan, Can science explain consciousness? Sci. Am. **271**(1), 88–94 (1994)
5. P. McCorduck, Machines Who Think: A Personal Inquiry into the History and Prospects of Artificial Intelligence, CRC Press (2004)
6. J. Rosser, Highlights of the history of the Lambda-calculus. Ann. Hist. Comput. **6**(4), 337–349 (1984)
7. D. Deutsch, Quantum theory, the Church-Turing principle and the universal quantum computer. Proc. R. Soc. Lond. Ser. A Math. Phys. Sci. **400**(1818), 97–117 (1985)
8. A. Turing, On computable numbers, with an application to the Entscheidungs problem. Proc. Lond. Math. Soc. **42**, 230–265 (1937)
9. R. Rojas, Who invented the computer?-The debate from the viewpoint of computer architecture, in W. Gautschi (ed.), Mathematics of Computation 1943–1993, pp. 361–366 (1994). (AMS, Proceedings of Symposia on Applied Mathematics, 1994)
10. M. Croarken, *Early Scientific Computing in Britain* (Clarendon Press, Oxford, 1990)
11. A. Hodges, *Alan Turing: The Enigma of Intelligence* (Counterpoint, London, 1983)

12. N. Stern, John von Neumann's influence on electronic digital computing, 1944–1946. Ann. Hist. Comput. **2**(4), 349–361 (1980)
13. O. Steward, *Principles of Cellular, Molecular, and Developmental Neuroscience* (Springer, New York, 1989)
14. S. Hameroff, *Ultimate Computing-Biomolecular Consciousness and Nanotechnology* (North-Holland, Amsterdam, 1987)
15. S. Hameroff, J. Dayhoff, R. Lahoz-Beltra, A. Samsonovich, S. Rasmussen, Conformational automata in the cytoskeleton. Computer **25**(11), 30–39 (1992)
16. F. Crick, *Astonishing Hypothesis: The Scientific Search for the Soul* (Charles Scribner's Sons, New York, 1994)
17. P. Milner, The Mind and Donald O. Hebb, Sci. Am. **268**(1), 124–129 (1993)
18. T. Kohonen, Correlation matrix memories. IEEE Trans. Comput. **21**, 353–359 (1972)
19. L. Cooper, A possible organization of animal memory and learning, in *Proceedings of the Nobel Symposium on Collective Properties of Physical Systems*, ed. by B. Lundquist, S. Lundquist (Academic Press, New York, 1973), pp. 252–264
20. P. Kanerva, *Sparse Distributed Memory* (MIT Press, Cambridge, 1988)
21. Y. Kamp, M. Hasler, *Recursive Neural Networks for Associative Memory* (Wiley, New York, 1990)
22. H. Haken, *Information and Self-Organization* (Springer, Berlin, 1988)
23. B. Kosko, Bidirectional associative memories. IEEE Trans. Syst. Man Cybern. **18**, 49–60 (1988)
24. J. Hopfield, Neural networks and physical systems with emergent collective computational abilities. Proc. Natl. Acad. Sci. **79**, 2554–2558 (1982)
25. L. Chua, L. Yang, Cellular neural networks: theory. IEEE Trans. Circuits Syst. **35**(10), 1257–1272 (1988)
26. D. Amit, *Modeling Brain Function: The World of Attractor Neural Networks* (Cambridge University Press, Cambridge, UK, 1989)
27. E. Ising, Beitrag zur theorie des ferromagnetismus. Zeitschrift fur Physik **31**(253), 253–258 (1925)
28. J. Bruck, On the convergence properties of the Hopfield model. Proc. IEEE **78**(10), 1579–1585 (1990)
29. J. Hopfield, Neurons with graded response have collective computational properties like those of two-state neurons. Proc. Natl. Acad. Sci. **81**, 3088–3092 (1984)
30. J. Hopfield, D. Tank, Neural computations of decisions in optimization problems. Biol. Cybern. **52**, 141–152 (1985)
31. J. Bruck, J. Goodman, On the power of neural networks for solving hard problems. J. Complex. **6**, 129–135 (1990)
32. N. Farhat, D. Psaltis, A. Prata, E. Paek, Optical implementation of the Hopfield model. Appl. Opt. **24**, 1469–1475 (1985)
33. B. Muller, J. Reinhardt, T.M. Strickland, *Neural Networks: An Introduction*, 2nd edn. (Springer-Verlag, Berlin, 1995)
34. E. Gardner, Maximum storage capacity in neural networks. Europhys. Lett. **4**, 481–485 (1987)
35. J. Hertz, A. Krogh, R. Palmer, *Introduction to the Theory of Neural Computation* (Addison-Wesley, Redwood City, CA, 1991)
36. D. Amit, H. Gutfreund, H. Sompolinsky, Storing infinite numbers of patterns in a spin-glass model of neural networks. Phys. Rev. Lett. **55**(14), 1530–1533 (1985)
37. D. Stein, *Lectures in the Sciences of Complexity* (Addison-Wesley, Redwood City, CA, 1989)
38. D. Stein, Spin glasses. Sci. Am. **261**(1), 36–43 (1989)
39. E. Reingold, J. Nievergelt, N. Deo, *Combinatorial Algorithms: Theory and Practice* (Prentice-Hall, Englewood Cliffs, 1977)
40. A. Gibbons, W. Rytter, *Efficient Parallel Algorithms* (Cambridge University Press, Cambridge, 1988)
41. G. Wilson, G. Pawley, On the stability of the traveling salesman problem algorithm of Hopfield and tank. Biol. Cybern. **58**, 63–70 (1988)

42. D. Johnson, More approaches to the traveling salesman guide. Nature **330**(6148), 525–525 (1987)
43. S. Lin, B. Kernighan, An effective heuristic algorithm for the traveling salesman problem. Operations Res. **21**, 498–516 (1973)
44. B. Sheu, B. Lee, C.F. Chang, Hardware annealing for fast retrieval of optimal solutions in Hopfield neural networks. International Joint Conference on Neural Nctworks, Seattle, IEEE Press **II**, 327–332 (1991)
45. Y. Abu-Mostafa, Neural networks for computing?, in Neural Networks for Computing, J.S. Denker, (ed.), America Institute of Physics, New York, vol. 151, pp. 1–6 (1986)
46. Y. Abu-Mostafa, J. St Jacques, Information capacity of the Hopfield model. IEEE Trans. Inform. Theory **31**(4), 461–464 (1985)
47. D. Psaltis, K. Wagner, D. Brady, Learning in optical neural computers. SCIVAL **III**, 549–555 (1987)
48. B. Kosko, Adaptive bidirectional associative memories. Appl. Opt. **26**(23), 4847–4860 (1987)
49. S. Grossberg, Nonlinear neural networks: principles, mechanisms, and architectures. Neural Netw. **1**, 17–61 (1988)
50. S. Grossberg, Behavioral contrast in short term memory: serial binary memory models or paraller continuous memory models? J. Math. Psychol. **3**, 199–219 (1978)
51. J.A. Anderson, J.W. Silverstein, S.R. Ritz, R.S. Jones, Distinctive features, categorical perception, and probability learning: some applications of a neural model. Psychol. Rev. **84**, 413–451 (1977)
52. D.H. Ackley, G.E. Hinton, T.J. Sejnowski, A learning algorithm for Boltzmann machines. Cogn. Sci. **9**, 147–169 (1985)
53. F. Ratliff, H.K. Hartline, W.H. Miller, Spatial and temporal aspects of retinal inhibitory interactions. J. Opt. Soc. Am. **53**, 110–120 (1963)
54. A.J. Lotka, *Elements of mathematical biology* (Dover, New York, 1956)
55. M.E. Gilpin, F.J. Ayala, Global models of growth and competition, in *Proceedings of the National Academy of Sciences*, vol. 70, pp. 3590–3593 (1973)
56. M. Eigen, P. Schuster, The hypercycle: A principle of natural self-organization, B: The abstract hypercycle. Naturwissenshaften **65**, 7–41 (1978)
57. M.A. Cohen, S. Grossberg, Neural dynamics of speech and language coding: developmental programs, perceptual grouping, and competition for short-term memory. Hum. Neurobiol. **5**, 1–22 (1986)
58. M.A. Cohen, S. Grossberg, Masking fields: A massively parallel neural architecture for learning, recognizing, and predicting multiple groupings of patterned data. Appl. Opt. **26**, 1866–1891 (1987)
59. M. Cohen, S. Grossberg, Absolute stability of global pattern formation and paralled memory storage by competitive neural networks. IEEE Trans. Syst. Man Cybern. **13**, 815–826 (1983)
60. S. Grossberg, Associative and competitive principles of learning and development: The temporal unfolding and stability of STM and LTM patterns, in *Competition and Cooperation in Neural Networks*, ed. by S.I. Amari, M. Arbib (Springer, New York, 1982), p. 1982
61. M. Cohen, Sustained oscillations in a symmetric cooperative-competitive neural network: Disproof of a conjecture about content addressable memory. Neural Netw. **1**, 217–221 (1988)
62. S. Ellias, S. Grossberg, Pattern formation, contrast control, and oscillations in the short-term memory of shunting on-center off-surround networks. Biol. Cybern. **20**, 69–98 (1975)
63. S. Grossberg, Adaptive pattern classification and universal recoding, I: Parallel development and coding of neural feature detectors. Biol. Cybern. **23**, 121–134 (1976)
64. S. Grossberg, Adaptive pattern classification and universal recoding, II: Feedback, expectation, olfaction, and illusions. Biol. Cybern. **23**, 187–202 (1976)
65. S. Grossberg, Contour enhancement, short-term memory, and constancies in reverberating neural networks. Stud. Appl. Math. **52**, 213–257 (1973)
66. J. Dambre, D. Verstraeten, B. Schrauwen, S. Massar, Information processing capacity of dynamical systems. Sci. Rep., **2**, 00514-13 (2012)

67. O. White, D. Lee, H. Sompolinsky, Short-term memory in orthogonal neural networks. Phys. Rev. Lett. **92**(14), 148102 (2002)
68. M. Hermans, B. Schrauwen, Memory in linear recurrent neural networks in continuous time. Neural Netw. **23**(3), 341–355 (2010)
69. S. Gangulia, D. Huhc, H. Sompolinsky, Memory traces in dynamical systems. Proc. Natl. Acad. Sci. USA **105**, 18970–18975 (2008)
70. S. Boyd, L. Chua, Fading memory and the problem of approximating nonlinear operators with Volterra series. IEEE Trans. Circuits Syst. **32**(11), 1150–1161 (1985)
71. L. Appeltant, M. Soriano, Q. Van der Sande, J. Danckaert et al., Information processing using a single dynamical node as complex system. Nat. Commun. **2**, 468–472 (2011)

Chapter 2
Preliminaries on Dynamical Systems and Stability Theory

In this chapter, we will review some fundamental concepts on dynamical systems and stability theory, and give some evaluations and opinions on some problems, which will be helpful for understanding the main contents of this book.

2.1 Overview of Dynamical Systems

The concept of a dynamical system has its origins in Newtonian mechanics. There, as in other natural sciences and engineering disciplines, the evolution rule of dynamical systems is an implicit relation that gives the state of the system for only a short time into the future. (The relation is either a differential equation, difference equation, or other timescale.) To determine the state for future time requires iterating the relation many times, each advancing time a small step. The iteration procedure is referred to as solving the system or integrating the system. If the system can be solved, given an initial point it is possible to determine all its future positions, a collection of points known as a trajectory or orbit.

Before the advent of computers, finding an orbit required sophisticated mathematical techniques and could be accomplished only for a small class of dynamical systems. Numerical methods implemented on electronic computing machines have simplified the task of determining the orbits of a dynamical system.

For simple dynamical systems, knowing the trajectory is often sufficient, but most dynamical systems are too complicated to be understood in terms of individual trajectories. The difficulties are listed as follows:

(1) The systems studied may only be known approximately—the parameters of the system may not be known precisely or terms may be missing from the equations. The approximations used bring into question the validity or relevance of numerical solutions. To address these questions several notions of stability have been introduced in the study of dynamical systems, such as Lyapunov stability, Lagrange stability,

Z. Wang et al., *Qualitative Analysis and Control of Complex Neural Networks with Delays*, Studies in Systems, Decision and Control 34,
DOI 10.1007/978-3-662-47484-6_2

Hurwitz stability, and structural stability or connective stability. The stability of dynamical systems implies that there is a class of models or initial conditions for which the trajectories would be equivalent. The operation for comparing orbits to establish their equivalence changes with the different notions of stability.

(2) The type of trajectory may be more important than one particular trajectory. Some trajectories may be periodic, whereas others may wander through many different states of the system. Applications often require enumerating these classes or maintaining the system within one class. Classifying all possible trajectories has led to the qualitative study of dynamical systems, that is, properties that do not change under coordinate transformations. Linear dynamical systems and systems that have two numbers describing a state are examples of dynamical systems where the possible classes of orbits are understood.

(3) The behavior of trajectories as a function of a parameter may be what is needed for an application. As a parameter varies, the dynamical systems may have bifurcation points where the qualitative behavior of the dynamical system changes. For example, it may go from having only periodic motions to having apparently erratic behavior, as in the transition to turbulence of a fluid.

(4) The trajectories of the system may appear erratic randomly. In these cases it may be necessary to compute averages using one very long trajectory or many different trajectories. The averages are well defined for ergodic systems and a more detailed understanding has been worked out for hyperbolic systems. Understanding the probabilistic aspects of dynamical systems has helped to establish the foundations of statistical mechanics and of chaos.

Many people regard Henri Poincare as the founder of dynamical systems. Poincare published two classical monographs, "New Methods of Celestial Mechanics" (1892–1899) and "Lectures on Celestial Mechanics" (1905–1910). In the monographs, he successfully applied the results of his research to the problem of the motion of three bodies and studied in detail the behavior of solutions (frequency, stability, asymptotic property, and so on). These papers included the Poincare recurrence theorem, which states that certain systems will, after a sufficiently long but finite time, return to a state very close to the initial state.

Aleksandr Lyapunov developed many important approximation methods. His methods, which was developed in 1899, make it possible to define the stability of sets of ordinary differential equations. He created the modern theory of the stability of a dynamic system. In 1913, George David Birkhoff [1] proved Poincare's "Last Geometric Theorem," a special case of the three-body problem, a result that made him world famous. In 1927, he published his "Dynamical Systems." Birkhoff's most durable result was his 1931 discovery of what is now called the ergodic theorem. Combining insights from physics on the ergodic hypothesis with measure theory, this theorem solved, at least in principle, a fundamental problem of statistical mechanics. The ergodic theorem has also had repercussions for dynamics. Stephen Smale made significant advances as well. His first contribution is the Smale horseshoe that jump started significant research in dynamical systems. He also outlined a research program carried out by many others. Oleksandr Mykolaiovych Sharkovsky developed Sharkovsky's theorem on the periods of discrete dynamical systems in 1964. One of the implications of the theorem is that if a discrete dynamical system on the real line has a periodic point of period 3, then it must have periodic points of every other period.

Some theories related to dynamical systems are simply described as follows:

(1) Arithmetic dynamics

Arithmetic dynamics is a field emerged in the 1990s that amalgamates two areas of mathematics, dynamical systems and number theory. Classically, discrete dynamics refers to the study of the iteration of self-maps of the complex plane or real line. Arithmetic dynamics is the study of the number-theoretic properties of integer, rational, p-adic, and/or algebraic points under repeated application of a polynomial or rational function.

(2) Chaos theory

Chaos theory describes the behavior of certain dynamical systems, that is, systems whose state evolves with time, that may exhibit dynamics that are highly sensitive to initial conditions (popularly referred to as the butterfly effect). As a result of this sensitivity, which manifests itself as an exponential growth of perturbations in the initial conditions, the behavior of chaotic systems appears random. This happens even though these systems are deterministic, meaning that their future dynamics are fully defined by their initial conditions, with no random elements involved. This behavior is known as deterministic chaos or simply chaos.

(3) Complex systems

Complex systems are a scientific field, which study the common properties of systems considered complex in nature, society and science. It is also called complex systems theory, complexity science, study of complex systems and/or sciences of complexity. The key problems of such systems are difficulties with their formal modeling and simulation. From such perspective, in different research contexts complex systems are defined on the basis of their different attributes. The study of complex systems is bringing new vitality to many areas of science where a more typical reductionist strategy has fallen short. Complex systems are, therefore, often used as a broad term encompassing a research approach to problems in many diverse disciplines including neurosciences, social sciences, meteorology, chemistry, physics, computer science, psychology, artificial life, evolutionary computation, economics, earthquake prediction, molecular biology, and inquiries into the nature of living cells themselves.

(4) Control theory

Control theory is an interdisciplinary branch of engineering and mathematics, which deals with how to influence the behavior of dynamical systems subjectively.

(5) Ergodic theory

Ergodic theory is a branch of mathematics that studies dynamical systems with an invariant measure and related problems. Its initial proposal was motivated by problems of statistical physics.

(6) Functional analysis

Functional analysis is the branch of mathematics, and specifically of analysis concerned with the study of vector spaces and operators acting upon them. It has its historical roots in the study of functional spaces, in particular transformations of functions, such as the Fourier transform, as well as in the study of differential and integral equations. This usage of the word "functional" goes back to the calculus of variations, implying a function whose argument is a function. Its use in general has been attributed to mathematician and physicist Vito Volterra and its founding is largely attributed to mathematician Stefan Banach.

(7) Graph dynamical systems

The concept of graph dynamical systems (GDS) can be used to capture a wide range of processes taking place on graphs or networks. A major theme in the mathematical and computational analysis of GDS is to relate their structural properties (e.g., the network connectivity) and the global dynamics that result.

(8) Projected dynamical systems

Projected dynamical systems are a mathematical theory investigating the behaviour of dynamical systems where solutions are restricted to a constraint set. The discipline shares connections to and applications with both the static world of optimization and equilibrium problems and the dynamical world of ordinary differential equations. A projected dynamical system is given by the flow to the projected differential equation.

(9) Symbolic dynamics

Symbolic dynamics are the practice of modeling a topological or smooth dynamical system by a discrete space consisting of infinite sequences of abstract symbols, each of which corresponds to a state of the system, with the dynamics (evolution) given by the shift operator.

(10) System dynamics

System dynamics are an approach to understanding the behavior of complex systems over time. It deals with internal feedback loops and time delays that affect the behavior of the entire system. What makes system dynamics different from other approaches to studying complex systems is the use of feedback loops and stocks and flows. These elements help describe how even seemingly simple systems display baffling nonlinearity.

(11) Topological dynamics

Topological dynamics is a branch of the theory of dynamical systems in which qualitative, asymptotic properties of dynamical systems are studied from the viewpoint of general topology.

2.2 Definition of Dynamical System and Its Qualitative Analysis

A dynamical system is a concept in mathematics where a fixed rule describes how a point in a geometrical space depends on time. Examples include the mathematical models that describe the swinging of a clock pendulum, the flow of water in a pipe, and the number of fish each spring time in a lake. At any given time a dynamical system has a state given by a set of real numbers (a vector) that can be represented by a point in an appropriate state space (a geometrical manifold). Small changes in the state of the system create small changes in the numbers. The evolution rule of the dynamical system is a fixed rule that describes what future states follow from the current state. The rule is deterministic; in other words, for a given time interval only one future state follows from the current state.

In the following, several types of definitions of dynamical system are provided.

(1) A *dynamical system* is a four-tuple $\{T, X, A, S\}$ where T denotes time set, X is the state space (a metric space with metric d), A is the set of initial states, and S denotes a family of motions. When $T = \mathbb{R}^+ = [0, \infty)$, we speak of a continuous-time dynamical system; and when $T = N = \{0, 1, 2, \ldots, \}$, we speak of a discrete-time dynamical system. For any motion $x(\cdot; x_0, t_0) \in S$, we have $x(t_0; x_0, t_0) = x_0 \in A \subset X$ and $x(t, x_0, t_0) \in X$ for all $t \in [t_0, t_1] \cap T$, $t_1 > t_0$, where t_1 may be finite or infinite. The set of motions S is obtained by varying (t_0, x_0) over $T \times A$. A dynamical system is said to be *autonomous*, if every $x(\cdot, x_0, t_0) \in S$ is defined on $T \cap [t_0, \infty)$ and if for each $x(\cdot, x_0, t_0) \in S$ and for each τ such that $t_0 + \tau \in T$, there exists a motion $x(\cdot, x_0, t_0 + \tau) \in S$ such that $x(t + \tau; x_0, t_0 + \tau) = x(t; x_0, t_0)$ for all t and τ satisfying $t + \tau \in T$.

A set $M \subset A$ is said to be *invariant* with respect to the set of motions S if $x_0 \in M$ implies that $x(t, x_0, t_0) \in M$ for all $t \geq t_0$, for all $t_0 \in T$, and for all $x(\cdot; x_0, t_0) \in S$. A point $p \in X$ is called an *equilibrium* for the dynamical system $\{T, X, A, S\}$ if the singleton $\{p\}$ is an invariant set with respect to the motion S. The term stability (more specially, Lyapunov stability) usually refers to the qualitative behavior of motions relative to an invariant set (resp. an equilibrium), whereas the term boundedness (more specially, Lagrange stability) refers to the (global) boundedness properties of the motions of a dynamical system. Of the many different types of *Lyapunov stability* that have been considered in the literature, perhaps the most important ones include stability, uniform stability, asymptotic stability, uniform asymptotic stability, exponential stability, asymptotic stability in the large, uniform asymptotic stability in the large, exponential stability in the large, instability, and complete instability. The most important *Lagrange stability* types include boundedness, uniform boundedness, and uniform ultimate boundedness of motions.

(2) A *dynamical system* (geometrical definition) is the tuple $\langle M, f, T\rangle$, with M a manifold (locally a Banach space or Euclidean space), T the domain for time (nonnegative reals, the integers) and f an evolution rule $t \to f(t)$ (with $t \in T$) such that $f(t)$ is a diffeomorphism of the manifold to itself. So, f is a mapping of the time

domain into the space of diffeomorphisms of the manifold to itself. In other terms, $f(t)$ is a diffeomorphism, for every time t in the domain T.

(3) A *dynamical system* (measure theoretical definition) may be defined formally, as a measure-preserving transformation of a sigma-algebra, the quadruplet (X, Σ, μ, τ). Here, X is a set, and Σ is a sigma-algebra on X, so that the pair (X, Σ) is a measurable space. μ is a finite measure on the sigma-algebra, so that the triplet (X, Σ, μ) is a probability space. A map $\tau : X \to X$ is said to be Σ-measurable if and only if, for every $\sigma \in \Sigma$, one has $\tau^{-1}\sigma \in \Sigma$. A map τ is said to preserve the measure if and only if, for every $\sigma \in \Sigma$, one has $\mu(\tau^{-1}\sigma) = \mu(\sigma)$. Combining the above, a map τ is said to be a measure-preserving transformation of X, if it is a map from X to itself, it is Σ-measurable, and is measure-preserving. The quadruple (X, Σ, μ, τ), for such a τ, is then defined to be a dynamical system.

The map τ embodies the time evolution of the dynamical system. Thus, for discrete dynamical systems the iterations $\tau^n = \tau \circ \tau \circ \cdots \circ \tau$ for integer n are studied. For continuous dynamical systems, the map τ is understood to be a finite time evolution map and the construction is more complicated.

(4) A *dynamical system* is a manifold M called the phase (or state) space endowed with a family of smooth evolution functions $\Phi(t)$ that for any element of $t \in T$, the time, map a point of the phase space back into the phase space. The notion of smoothness changes with applications and the type of manifold. There are several choices for the set T. When T is taken to be the reals, the dynamical system is called a flow, and if T is restricted to the nonnegative reals, then the dynamical system is a semi-flow. When T is taken to be the integers, it is a cascade or a map, and the restriction to the nonnegative integers is a semi-cascade.

For example, the evolution function $\Phi(t)$ is often the solution of a differential equation of motion $\dot{x}(t) = v(x(t))$ or $\dot{x} = v(x)$ for brevity. The equation gives the time derivative, represented by the dot, of a trajectory $x(t)$ on the phase space starting at some point x_0. The vector field $v(x)$ is a smooth function that at every point of the phase space M provides the velocity vector of the dynamical system at that point. (These vectors are not vectors in the phase space M, but in the tangent space $T_x M$ of the point x.) Given a smooth function $\Phi(t)$, an autonomous vector field can be derived from it. There is no need for higher order derivatives in the equation, nor for time dependence in $v(x)$ because these can be eliminated by considering systems of higher dimensions. Other types of differential equations can be used to define the evolution rule: $G(x, \dot{x}) = 0$ is an example of an equation that arises from the modeling of mechanical systems with complicated constraints.

The differential equations determining the evolution function $\Phi(t)$ are often ordinary differential equations: in this case the phase space M is a finite dimensional manifold. Many of the concepts in dynamical systems can be extended to infinite-dimensional manifolds—those that are locally Banach spaces—in which case the differential equations are partial differential equations. In the late twentieth century the dynamical system perspective to partial differential equations started gaining popularity.

The *qualitative properties* of dynamical systems do not change under a smooth change of coordinates (this is sometimes taken as a definition of qualitative): a singular point of the vector field (a point where $v(x) = 0$) will remain a singular point under smooth transformations; a periodic orbit is a loop in phase space and smooth deformations of the phase space cannot alter it to be a loop. It is in the neighborhood of singular points and periodic orbits that the structure of a phase space of a dynamical system can be well understood. In the qualitative study of dynamical systems, the approach is to show that there is a change of coordinates (usually unspecified, but computable) that makes the dynamical system as simple as possible.

A similar concept to the qualitative properties of dynamical systems is the *rectification*. A flow in most small patches of the phase space can be made very simple. If y is a point where the vector field $v(y) \neq 0$, then there is a change of coordinates for a region around y where the vector field becomes a series of parallel vectors of the same magnitude. This is known as the rectification theorem. The rectification theorem says that away from singular points the dynamics of a point in a small patch is a straight line. The patch can sometimes be enlarged by stitching several patches together, and when this works out in the whole phase space M the dynamical system is integrable. In most cases the patch cannot be extended to the entire phase space. There may be singular points in the vector field (where $v(x) = 0$) or the patches may become smaller and smaller as some point is approached. The more subtle reason is a global constraint, where the trajectory starts out in a patch, and after visiting a series of other patches comes back to the original one. If the next time the orbit loops around phase space in a different way, then it is impossible to rectify the vector field in the whole series of patches.

Dynamical systems theory is an area of mathematics used to describe the behavior of complex dynamical systems, usually by employing differential equations or difference equations. When differential equations are employed, the theory is called continuous dynamical systems. When difference equations are employed, the theory is called discrete dynamical systems. When the time variable runs over a set that is discrete over some intervals and continuous over other intervals or is any arbitrary time—set such as a cantor set—one gets dynamic equations on timescales. Some situations may also be modeled by mixed operators, such as differential–difference equations. This theory deals with the long-term qualitative behavior of dynamical systems, and studies the solutions of the equations of motion of systems that are primarily mechanical in nature, although this includes both planetary orbits as well as the behavior of electronic circuits and the solutions to partial differential equations that arise in biology. Much of modern research is focused on the study of chaotic systems. This field of study is also called just dynamical systems, mathematical dynamical systems theory, and mathematical theory of dynamical systems.

Dynamical systems theory and chaos theory deal with the long-term qualitative behavior of dynamical systems. Here, the focus is not on finding precise solutions to the equations defining the dynamical system (which is often hopeless), but rather to answer questions like, "Will the system settle down to a steady state in the long term, and if so, what are the possible steady states?" or "Does the long-term behavior of the system depend on its initial condition?" An important target is to describe the fixed points, or steady states of a given dynamical system; these are values of the variable that do not change over time. Some of these fixed points are attractive, meaning that if the system starts out in a nearby state, it converges toward the fixed point.

Similarly, one is interested in periodic points, states of the system that repeat after several time steps. Periodic points can also be attractive. Sharkovskii's theorem is an interesting statement about the number of periodic points of a one-dimensional discrete dynamical system. Even simple nonlinear dynamical systems often exhibit seemingly random behavior that has been called chaos. The branch of dynamical systems that handles the clear definition and investigation of chaos is called chaos theory.

2.3 Lyapunov Stability of Dynamical Systems

Various types of stability may be discussed for the solutions of differential equations describing dynamical systems. The most important type is that concerning the stability of solutions near to a point of equilibrium. This may be discussed by the theory of Lyapunov. In simple terms, if all solutions of the dynamical system that start out near an equilibrium point x_e stay near x_e forever, then x_e is *Lyapunov stable*. More strongly, if x_e is Lyapunov stable and all solutions that start out near x_e converge to x_e, then x_e is asymptotically stable. The notion of exponential stability guarantees a minimal rate of decay, i.e., an estimate of how quickly the solutions converge. The idea of Lyapunov stability can be extended to infinite-dimensional manifolds, where it is known as structural stability, which concerns the behavior of different but "nearby" solutions to differential equations. Input-to-state stability (ISS) applies Lyapunov notions to systems with inputs.

Lyapunov stability is named after Aleksandr Lyapunov, a Russian mathematician who published his book "The General Problem of Stability of Motion" in 1892 [2]. Lyapunov was the first to consider the modifications necessary in nonlinear systems to the linear theory of stability based on linearizing near a point of equilibrium. His work, initially published in Russian and then translated to French, received little attention for many years. Interest in it started suddenly during the Cold War (1953–1962) period when the so-called "Second Method of Lyapunov" was found to be applicable to the stability of aerospace guidance systems which typically contain strong nonlinearities not treatable by other methods. A large number of publications appeared then and since in the control and systems literature [3–7]. More recently, the concept of the Lyapunov exponent (related to Lyapunov's First Method of discussing stability) has received wide interest in connection with chaos theory. Lyapunov stability methods have also been applied to finding equilibrium solutions in traffic assignment problems [8].

(1) Definition of Lyapunov stability for continuous-time systems

Consider an autonomous nonlinear dynamical system

$$\dot{x}(t) = f(x(t)) \text{ with } x(0) = x_0, \tag{2.1}$$

where $x(t) \in D \subseteq \mathbb{R}^n$ denotes the system state vector, D an open set containing the origin, and $f : D \to \mathbb{R}^n$ continuous on D. Suppose f has an equilibrium at x_e so that $f(x_e) = 0$, then
(1) This equilibrium is said to be *Lyapunov stable*, if, for every $\epsilon > 0$, there exists a $\delta = \delta(\epsilon)$ such that, if $\|x(0) - x_e\| < \delta$, then for every $t \geq 0$ we have $\|x(t) - x_e\| < \epsilon$.
(2) The equilibrium of the above system is said to be *asymptotically stable* if it is Lyapunov stable and there exists $\delta > 0$ such that if $\|x(0) - x_e\| < \delta$, then $\lim_{t\to\infty} \|x(t) - x_e\| = 0$.
(3) The equilibrium of the above system is said to be *exponentially stable* if it is asymptotically stable and there exist $\alpha > 0$, $\beta > 0$, $\delta > 0$ such that if $\|x(0) - x_e\| < \delta$, then $\|x(t) - x_e\| \leq \alpha\|x(0) - x_e\|e^{-\beta t}$, for $t \geq 0$.

Conceptually, the meanings of aforementioned terms are the following:
(1) Lyapunov stability of an equilibrium means that solutions starting "close enough" to the equilibrium (within a distance δ from it) remain "close enough" forever (within a distance ϵ from it). Note that this must be true for any ϵ that one may want to choose.
(2) Asymptotic stability means that solutions that start close enough not only remain close enough but also eventually converge to the equilibrium.
(3) Exponential stability means that solutions not only converge, but in fact converge faster than or at least as fast as a particular known rate $\alpha\|x(0) - x_e\|e^{-\beta t}$.

The trajectory x is (locally) *attractive* if $\|y(t) - x(t)\| \to 0$ (where $y(t)$ denotes the system output) for $t \to \infty$ for all trajectories that start close enough, and globally attractive if this property holds for all trajectories. That is, if x belongs to the interior of its stable manifold, it is asymptotically stable if it is both attractive and stable. (There are counterexamples showing that attractivity does not imply asymptotic stability.)

(2) *Lyapunov's second method* for stability

Lyapunov, in his original 1892 work, proposed two methods for demonstrating stability. The first method developed the solution in a series which was then proved convergent within limits. The second method, which is almost universally used nowadays, makes use of a Lyapunov function $V(x)$ which has an analogy to the potential function of classical dynamics. It is introduced as follows for a system having a point of equilibrium at $x = 0$.

Consider a function $V(x) : \mathbb{R}^n \to \mathbb{R}$ such that
(1) $V(x) \geq 0$ with equality if and only if $x = 0$ (positive definite).
(2) $\dot{V}(x(t)) = \frac{\mathrm{d}V(x(t))}{\mathrm{d}t} \leq 0$ with equality not constrained to only $x = 0$ (negative semidefinite. Note: for asymptotic stability, $\dot{V}(x(t))$ is required to be negative definite!).
Then $V(x)$ is called a Lyapunov function candidate and the system is stable in the sense of Lyapunov. Furthermore, the system is asymptotically stable, in the sense of Lyapunov, if $\dot{V}(x(t)) \leq 0$ with equality if and only if $x = 0$. Global asymptotic stability (GAS) follows similarly.

Note that, (1) $V(0) = 0$ is required; otherwise for example $V(x) = \frac{1}{1+|x|}$ would "prove" that $\dot{x}(t) = x(t)$ is locally stable. (2) An additional condition called "properness" or "radial unboundedness" is required in order to conclude global stability.

It is easier to visualize this method of analysis by thinking of a physical system (e.g., vibrating spring and mass) and considering the energy of such a system. If the system loses energy over time and the energy is never restored, then eventually the system must grind to a stop and reach some final resting state. This final state is called the attractor. However, finding a function that gives the precise energy of a physical system can be difficult, and for abstract mathematical systems, economic systems or biological systems, the concept of energy may not be applicable. *Lyapunov's realization* is that stability can be proven without requiring knowledge of the true physical energy, provided that a Lyapunov function can be found to satisfy the above constraints.

Lyapunov stability method is mainly focused on the system (2.1), i.e., a system with zero input. In fact, many systems have external control inputs. If the control law is in the form of state feedback, then the closed-loop systems is equivalent to the system (2.1). In this case, Lyapunov stability theory can be directly applied. If the external input exists and is different from the system states, some variants of Lyapunov stability theory should be investigated. In the following, we will introduce some other stability analysis methods.

(3) Stability for systems with inputs

A system with inputs (or controls) has the form

$$\dot{x}(t) = f(x(t), u(t)), \tag{2.2}$$

where the (generally time-dependent) input $u(t)$ may be viewed as a control, external input, stimulus, disturbance, or forcing function. The study of such systems is the subject of control theory and applied in control engineering. For systems with inputs, one must quantify the effect of inputs on the stability of the system. The main two approaches to this analysis are BIBO stability (for linear systems) and input-to-state (ISS) stability (for nonlinear systems)

(4) *Barbalat's lemma* and stability of time-varying systems

Assume that $f(t)$ is function of time only.

(1) Having $\dot{f}(t) \to 0$ does not imply that $f(t)$ has a limit at $t \to \infty$. For example, $f(t) = \sin(ln(t)), t > 0$.

(2) Having $f(t)$ approaching a limit as $t \to \infty$ does not imply that $\dot{f}(t) \to 0$. For example, $f(t) = \frac{\sin(t^2)}{t}$.

(3) Having $f(t)$ lower bounded and decreasing ($\dot{f}(t) \leq 0$) implies it converges to a limit. But it does not say whether or not $\dot{f}(t) \to 0$ as $t \to \infty$.

Barbalat's Lemma says: If $f(t)$ has a finite limit as $t \to \infty$ and if $\dot{f}(t)$ is uniformly continuous (or $\ddot{f}(t)$ is bounded), then $\dot{f}(t) \to 0$ as $t \to \infty$.

Usually, it is difficult to analyze the asymptotic stability of time-varying systems because it is very difficult to find Lyapunov functions with a negative definite derivative. We know that in case of autonomous (time-invariant) systems, if $\dot{V}(x(t))$ is negative semidefinite, then also, it is possible to know the asymptotic behavior by invoking invariant set theorems. However, this flexibility is not available for time-varying systems. This is where "Barbalat's lemma" comes into picture. It says:

If $V(x(t), t)$ satisfies following conditions: (1) $V(x(t), t)$ is lower bounded. (2) $\dot{V}(x(t), t)$ is negative semidefinite. (3) $\dot{V}(x(t), t)$ is uniformly continuous in time (satisfied if $\ddot{V}(x(t), t)$ is finite), then $\dot{V}(x(t), t) \to 0$ as $t \to \infty$.

(5) Differential inequality methods

This kind of stability analysis methods generally do not need the so-called Lyapunov functions, and can also determine the stability properties of the concerned system. Such kinds of methods include, but not limited to, Halanay inequality, contraction principle, Gronwall's inequality, comparison theorems, and so on [9–13]. Even by constructing Lyapunov functions, except the Lyapunov stability theory, there are still many methods to derive the stability criteria for the concerned systems, e.g., Barbalat's lemma. However, a key point should be kept in mind, that is, the definition of stability property must be declared beforehand. In this way, there are many different senses of stability definition in the literature. In the research of neural network stability theory, more emphasis is placed on the stability of Lyapunov sense. Fortunately, other stability definitions in the Hopfield sense and Lagrange sense are also paid attention. Except Lyapunov function method, input–output method and differential inequality methods are also powerful to analyze the qualitative characteristics of dynamical systems.

2.4 Stability Theory

In mathematics, *stability theory* addresses the stability of solutions of differential equations and of trajectories of dynamical systems under small perturbations of initial conditions. The heat equation, for example, is a stable partial differential equation because small perturbations of initial data lead to small variations in temperature at a later time as a result of the maximum principle. One must specify the metric used to measure the perturbations when claiming a system is stable. In partial differential equations one may measure the distances between functions using L_p norms or the sup norm, while in differential geometry one may measure the distance between spaces using the Gromov–Hausdorff distance.

In dynamical systems, an *orbit* is called Lyapunov stable if the forward orbit of any point is in a small enough neighborhood or it stays in a small (but perhaps, larger) neighborhood. Various criteria have been developed to prove stability or instability of an orbit. Under favorable circumstances, the question may be reduced to a well-studied problem involving eigenvalues of matrices. A more general method involves Lyapunov functions. In practice, any one of different stability criteria is applied.

Many parts of the qualitative theory of differential equations and dynamical systems deal with asymptotic properties of solutions and the trajectories—what happens with the system after a long period of time. The simplest kind of behavior is exhibited by equilibrium points or fixed points, and by periodic orbits. If a particular orbit is well understood, it is natural to ask next whether a small change in the initial condition will lead to similar behavior. *Stability theory* addresses the following questions: (1) will a nearby orbit indefinitely stay close to a given orbit? (2) will it converge to the given orbit? (this is a stronger property). In the former case, the orbit is called stable and in the latter case, asymptotically stable, or attracting.

Stability means that the trajectories do not change too much under small perturbations. The opposite situation, where a nearby orbit is getting repelled from the given orbit, is also of interest. In general, perturbing the initial state in some directions results in the trajectory asymptotically approaching the given one and in other directions to the trajectory getting away from it. There may also be directions for which the behavior of the perturbed orbit is more complicated (neither converging nor escaping completely), and then stability theory does not give sufficient information about the dynamics.

One of the key ideas in stability theory is that the *qualitative behavior* of an orbit under perturbations can be analyzed using the linearization of the system near the orbit. In particular, at each equilibrium of a smooth dynamical system with an n-dimensional phase space, there is a certain $n \times n$ matrix A whose eigenvalues characterize the behavior of the nearby points (Hartman–Grobman theorem). More precisely, if all eigenvalues are negative real numbers or complex numbers with negative real parts, then the point is a stable attracting fixed point, and the nearby points converge to it at an exponential rate. If none of the eigenvalues is purely imaginary (or zero) then the attracting and repelling directions are related to the eigenspaces of the matrix A with eigenvalues whose real part is negative and, respectively, positive. Analogous statements are known for perturbations of more complicated orbits.

(1) *Stability of fixed points*

The simplest kind of an orbit is a fixed point or an equilibrium. If a mechanical system is in a stable equilibrium state, then a small push will result in a localized motion, for example, small oscillations as in the case of a pendulum. In a system with damping, a stable equilibrium state is moreover asymptotically stable. On the other hand, for an unstable equilibrium, such as a ball resting on a top of a hill, certain small pushes will result in a motion with a large amplitude that may or may not converge to the original state. There are useful tests of stability for the case of a linear system. Stability of a nonlinear system can often be inferred from the stability of its linearization.

Let $f : \mathbb{R} \to \mathbb{R}$ be a continuously differentiable function with a *fixed point* a, $f(a) = a$. Consider the dynamical system obtained by iterating the function f:

$$x_{n+1} = f(x_n), n = 0, 1, 2, 3, \ldots, \tag{2.3}$$

The fixed point a is stable if the absolute value of the derivative of f at a is strictly less than 1, and unstable if it is strictly greater than 1. This is because near the point a, the function f has a *linear approximation* with slope $\dot{f}(a)$:

$$f(x) \approx f(a) + \dot{f}(a)(x - a). \tag{2.4}$$

Thus

$$\frac{x_{n+1} - a}{x_n - a} = \frac{f(x_n) - a}{x_n - a} = \dot{f}(a), \tag{2.5}$$

which means that the derivative measures the rate at which the successive iterates approach the fixed point a or diverge from it. If the derivative at a is exactly 1 or -1, then more information is needed in order to decide stability.

There is an analogous criterion for a continuously differentiable map $f : \mathbb{R}^n \to \mathbb{R}^n$ with a fixed point a, expressed in terms of its Jacobian matrix at a, $J = J_a(f)$. If all eigenvalues of J are real or complex numbers with absolute value strictly less than 1 then a is a stable fixed point; if at least one of them has absolute value strictly greater than 1 then a is unstable. Just as for $n = 1$, the case of the largest absolute value being 1 needs to be investigated further—the Jacobian matrix test is inconclusive. The same criterion holds more generally for diffeomorphisms of a smooth manifold.

For the linear autonomous systems, the stability of fixed points of a system of constant coefficient linear differential equations of first order can be analyzed using the eigenvalues of the corresponding matrix. An autonomous system $\dot{x} = Ax$, where $x(t) \in \mathbb{R}^n$ and A is an $n \times n$ matrix with real entries, has a constant solution $x(t) = 0$. (In a different language, the origin $0 \in \mathbb{R}^n$ is an equilibrium point of the corresponding dynamical system.) This solution is *asymptotically stable* as $t \to \infty$ ("in the future") if and only if for all eigenvalues λ of A, $Re(\lambda) < 0$ (which means the real part of eigenvalue λ is negative). Similarly, it is asymptotically stable as $t \to -\infty$ ("in the past") if and only if for all eigenvalues λ of A, $Re(\lambda) > 0$. If there exists an eigenvalue λ of A with $Re(\lambda) > 0$ then the solution is unstable for $t \to \infty$. Application of this result in practice, in order to decide the stability of the origin for a linear system, is facilitated by the Routh–Hurwitz stability criterion. The eigenvalues of a matrix are the roots of its characteristic polynomial. A polynomial in one variable with real coefficients is called a Hurwitz polynomial if the real parts of all roots are strictly negative. The Routh–Hurwitz theorem implies a characterization of Hurwitz polynomials by means of an algorithm that avoids computing the roots.

For a nonlinear autonomous systems, asymptotic stability of fixed points can often be established using the *Hartman–Grobman theorem*. Suppose that v is a C^1-vector field in $\mathbb{R}^n$ which vanishes at a point p, $v(p) = 0$. Then the corresponding autonomous system $\dot{x} = v(x)$ has a constant solution $x(t) = p$. Let $J = J_p(v)$ be the $n \times n$ Jacobian matrix of the vector field v at the point p. If all eigenvalues of J have strictly negative real part, then the solution is asymptotically stable. This condition can be tested using the Routh–Hurwitz criterion.

For the general dynamical systems, a general way to establish Lyapunov stability or asymptotic stability of a dynamical system is by means of Lyapunov functions, which has been introduced above.

(2) *Structural stability*

Let G be an open domain in $\mathbb{R}^n$ with compact closure and smooth $(n - 1)$-dimensional boundary. Consider the space $X^1(G)$ consisting of restrictions to G of C^1 vector fields on $\mathbb{R}^n$ that are transversal to the boundary of G and are inward oriented. This space is endowed with the C^1 metric in the usual fashion. A vector field $F \in X^1(G)$ is weakly structurally stable if for any sufficiently small perturbation F_1, the corresponding flows are topologically equivalent on G: there exists a homeomorphism $h : G \to G$ which transforms the oriented trajectories of F into the oriented trajectories of F_1. If, moreover, for any $\epsilon > 0$ the homeomorphism h may be chosen to be C^0_ϵ-close to the identity map when F^1 belongs to a suitable neighborhood of F depending on ϵ, then F is called (strongly)structurally stable.

These definitions are extended in a straightforward way to the case of n-dimensional compact smooth manifolds with boundary. Andronov and Pontryagin originally considered the strong property. Analogous definitions can be given for diffeomorphisms in place of vector fields and flows: in this setting, the homeomorphism h must be a topological conjugacy.

In mathematics, structural stability is a fundamental property of a dynamical system which means that the qualitative behavior of the trajectories is unaffected by small perturbations (to be exact C^1-small perturbations). Examples of such qualitative properties are numbers of fixed points and periodic orbits (but not their periods). Unlike Lyapunov stability, which considers perturbations of initial conditions for a fixed system, structural stability deals with perturbations of the system itself. Variants of this notion apply to systems of ordinary differential equations, vector fields on smooth manifolds and flows generated by them, and diffeomorphisms.

Structurally stable systems were introduced by Andronov and Pontryagin [14] under the name "systemes grossiers" or rough systems [14, 15]. They announced a characterization of rough systems in the plane, the Andronov–Pontryagin criterion. In this case, structurally stable systems are typical, they form an open dense set in the space of all systems endowed with appropriate topology. In higher dimensions, this is no longer true, indicating that typical dynamics can be very complex (e.g., strange attractor). An important class of structurally stable systems in arbitrary dimensions is given by Anosov diffeomorphisms and flows.

Structural stability of the system provides a justification for applying the qualitative theory of dynamical systems to analysis of concrete physical systems. The idea of such qualitative analysis goes back to the work of Henri Poincare on the three-body problem in celestial mechanics. Around the same time, Aleksandr Lyapunov rigorously investigated stability of small perturbations of an individual system. In practice, the evolution law of the system (i.e., the differential equations) is never known exactly, due to the presence of various small interactions. It is, therefore, crucial to know that basic features of the dynamics are the same for any small perturbation of the "model" system, whose evolution is governed by a certain known physical law. Qualitative analysis was further developed by George Birkhoff in the 1920s [16], but was first formalized with introduction of the concept of rough system by Andronov and Pontryagin in 1937 [14]. This was immediately applied to analysis of physical systems with oscillations by Andronov, Witt, and Khaikin. The term "structural stability" is due to Solomon Lefschetz, who oversaw translation of their monograph into English. Ideas of structural stability were taken up by Stephen Smale and his school in the 1960s in the context of hyperbolic dynamics [17]. Earlier, Marston Morse and Hassler Whitney initiated and Rene Thom developed a parallel theory of stability for differentiable maps, which forms a key part of singularity theory.

Thom envisaged applications of this theory to biological systems. Both Smale and Thom worked in direct contact with Mauricio Peixoto, who developed Peixoto's theorem in the late 1950s.

When Smale started to develop the theory of hyperbolic dynamical systems, he hoped that structurally stable systems would be "typical." This would have been consistent with the situation in low dimensions: dimension two for flows and dimension one for diffeomorphisms. However, he soon found examples of vector fields on higher dimensional manifolds that cannot be made structurally stable by an arbitrarily small perturbation. This means that in higher dimensions, structurally stable systems are not dense. In addition, a structurally stable system may have transversal homoclinic trajectories of hyperbolic saddle closed orbits and infinitely many periodic orbits, even though the phase space is compact. The closest higher dimensional analog of structurally stable systems considered by Andronov and Pontryagin is given by the Morse–Smale systems.

(3) *Rough system* (structurally stable dynamical system)

A smooth dynamical system is called rough systems if the following properties hold. For any $\epsilon > 0$, there is a $\delta > 0$ such that for any perturbation of the system by not more than δ in the C^1-metric, there exists a homeomorphism of the phase space which displaces the points by not more than ϵ and converts the trajectories of the unperturbed system into trajectories of the perturbed system.

Formally, this definition assumes that a certain Riemannian metric is given on the phase manifold. In fact, one speaks of a structurally stable system when the phase manifold is closed, or else if the trajectories form part of some compact domain G with a smooth boundary not tangent to the trajectories; here the perturbation and the homeomorphism are considered on G only. In view of the compactness, the selection of the metric is immaterial.

Thus, a small (in the sense of C^1) perturbation of a structurally stable system yields a system equivalent to the initial one as regards all its topological properties (however, this definition comprises one additional requirement, i.e., this equivalence must be realized by a homeomorphism close to the identity). The terms "roughness" and "(structural) stability" are used in a broader sense, e.g., to mean merely the preservation of some property of the system under a small perturbation (in such a case it is preferable to speak of the structural stability of the property in question).

As said above, *structurally stable systems* were introduced by A.A. Andronov and L.S. Pontryagin. If the dimension of the phase manifold is small (one for discrete time and one or two for continuous time), structurally stable systems can be simply characterized in terms of the qualitative properties of behavior of trajectories (then they are the so-called Morse–Smale systems, cf. Morse–Smale system); in that case they form an open everywhere-dense set in the space of all dynamical systems, provided with the C^1-topology. Thus, systems whose trajectories display a behavior which is more complex and more sensitive to small perturbations are considered here as exceptional. If the dimensions are larger, none of these facts hold, as was established by S. Smale. He advanced the hypothesis according to which, irrespective of all these complications, it is possible in the general case to formulate the following necessary and sufficient conditions for structural stability in terms of a qualitative picture of the behavior of the trajectories: (1) the non-wandering points (cf. Non-wandering point) should form a hyperbolic set Ω, in which the periodic trajectories are everywhere dense (the so-called Smale's Axiom A); and (2) the stable and unstable manifolds of any two trajectories from Ω should intersect transversally (the strong transversality condition). That these conditions are sufficient have now been proved in almost all cases; as regards their necessity, proof is only available if the definition of structural stability is somewhat changed.

(4) *Absolute stability*

Absolute stability practically means that a system is convergent for any choice of parameters and nonlinear functions, within specified and well-characterized sets.

In 1944, when studying the stability of an autopilot, Lur'e and Postnikov introduced the concept of absolute stability and the Lur'e problem [18]. Since then, the problem of absolute stability for Lur'e-type systems has received considerable attention and many fruitful results, such as Popov's criterion, circle criterion, and Kalman–Yakubovih–Popov (KYP) lemma have been proposed [19–21]. From the view of modern robustness theory, absolute stability theory can be considered as the first approach to robust stability of nonlinear uncertain systems [22].

Absolute stability theory guarantees stability of feedback systems whose forward path contains a dynamic linear time-invariant system and whose feedback path contains a memoryless (possibly time-varying) nonlinearity. These stability criteria are generally stated in terms of the linear system and apply to every element of a specified class of nonlinearities. Hence, absolute stability theory provides sufficient conditions for robust stability with a given class of uncertain elements. The literature on absolute stability is extensive. A convenient way to distinguish these results is to focus on the allowable class of feedback nonlinearities. Specifically, the small-gain, positivity, and circle theorems guarantee stability for arbitrarily time-varying nonlinearities, whereas the Popov criterion does not. This is not surprising since the Lyapunov function upon which the small-gain, positivity, and circle theorems are based is a fixed quadratic Lyapunov function which permits arbitrary time variation of the nonlinearity. Alternatively, the Popov criterion is based on a Lur'e-Postnikov Lyapunov function which explicitly depends on the nonlinearity, thereby restricting its allowable time variation.

(5) *Complete stability*

A system is said to be completely stable if each trajectory of the system converges toward an equilibrium point, as $t \to \infty$. Complete stability of neural networks is one of the most important dynamical properties in view of practical applications to solve a large number of signal processing tasks, including image processing, pattern recognition, and optimization problems.

The standard Lyapunov–Krasovskii functional methods or Lyapunov–Razumikhin function methods are usually used for global asymptotic stability analysis of delayed neural networks with a unique equilibrium point. For the analysis of complete stability, the Lyapunov method and the classic LaSalle approach are no longer effective because of the multiplicity of attractors [23]. For example, for PWL neuron activations a nonstrict energy is present both in the generic case where the neural network equilibrium points are isolated, as well as in the degenerate case where there are infinite nonisolated equilibrium points [24]. It is well known that to prove complete stability using LaSalle approach for nonstrict energy functions, it is required to characterize the invariant sets contained in the set where the energy is constant on orbits. As a matter of fact, such a characterization seems hard to accomplish for the NNs with PWL function or cellular neural network (CNN), since the sets involved have a complex structure even in the simplest case where there are finitely many equilibrium points. The situation is further complicated in degenerate cases where there are infinitely many nonisolated equilibrium points. In [25], an example shows that the existence of a stable equilibrium point does not imply complete stability of a CNN.

In [24], a new method is proposed to study the complete stability of cellular neural network (CNN). It allows one to completely sidestep the analysis of the neural network invariant sets. The method is based on a fundamental limit theorem for the length of the neural network output trajectories. Namely, it has been shown that the symmetry of the CNN interconnection matrix implies that the total length of the CNN output trajectories is necessarily finite. This in turn ensures convergence of the outputs, and also the state variables, toward an equilibrium point. Furthermore, this result is true regardless of the nature of the set of the CNN equilibrium points, so that complete stability is naturally proved not only for isolated equilibrium points, but also when the equilibrium points are not isolated.

(6) Input-to-state stability (ISS)

The *input to state stability* property provides a natural framework in which to formulate notions of stability with respect to input perturbations. It is generally known that ISS and set-ISS are powerful tools in the analysis of the stability and robustness of control systems [26]. Seminal works on ISS and set-ISS include [27] which provides a converse Lyapunov theory for set stability, and [28–31] which introduce ISS, extend the notion to noncompact sets, and generalize to arbitrary closed invariant sets, respectively. Input-to-state stability was introduced in [32], and has proved to be a very useful paradigm in the study of nonlinear stability, see for instance [33–45], as well as its variants such as integral ISS and input/output stability [46–56]. The notion of ISS takes into account the effect of initial states in a manner fully compatible with Lyapunov stability, and incorporates naturally the idea of "nonlinear gain" functions. Roughly speaking, a system is ISS provided that, no matter what is the initial state, if the inputs are small, then the state must eventually be small. Dualizing this definition one arrives at the notion of detectability which is the main subject of input/output-to-state stability (IOSS). A system $\dot{x} = f(x, u)$ with measurement "output" map $y = h(x)$ is IOSS if there are some functions $\beta \in \mathcal{KL}$ and $\gamma_1, \gamma_2 \in \mathcal{K}_\infty$ such that the estimate:

$$|x(t)| \leq \max\{\beta(x(0), t), \gamma_1(\|u_{[0,t]}\|), \gamma_2(\|y_{[0,t]}\|)\},$$

holds for any initial state $x(0)$ and any input $u(\cdot)$, where $x(\cdot)$ is the ensuing trajectory and $y(t) = h(x(t))$ the respective output function. (States $x(t)$, input values $u(t)$, and output values $y(t)$ lie in appropriate Euclidean spaces. We use $|\cdot|$ to denote Euclidean norm and $\|\cdot\|$ for supremum norm.) The terminology IOSS is self-explanatory: formally there is "stability from the I/O data to the state." The term was introduced in the paper [57], but the same notion had appeared before: it represents a natural combination of the notions of "strong" observability [32] and ISS, and was called simply "detectability" in [58, 59] and was called "strong unboundedness observability" in [35].

Roughly, a system is *output stable* if, for any initial state, the output converges to zero as $t \to \infty$. Inputs may influence this stability in different ways, for instance, one may ask that output approaches to zero only for those inputs for which input approaches to zero, or just that output remains bounded whenever input is bounded. The notion of output stability is also related to that of stability with respect to two measures [60]. As in the corresponding ISS paper [61], there are close relationships between output stability with respect to inputs, and robustness of stability under output feedback. This suggests the study of yet another property, which is obtained by a "*small gain*" argument from IOS: there must exist some $\mathcal{X} \in \mathcal{KL}$ so that $|y(t)| \leq \beta(x(0), t)$ if $|u(t)| \leq \mathcal{X}(|y(t)|), \forall t$. Combining the traditional stability definition and input/output stability conception, there may form may new stability conceptions such as output Lagrange stability, output-Lagrange input-to-output stable, state-independent IOS, robustly output stable, and so on. Such behavior is of central interest in control theory [60]. For the stability research of neural networks, these concepts are also very important.

The relation between *stability and ISS* can be briefly stated as follows. There are two very conceptually different ways of formulating the notion of stability of control systems. One of them, which we may call the input/output approach, relies on operator-theoretic techniques. In this approach, a "system" is a causal operator F between spaces of signals, and "stability" is taken to mean that F maps bounded inputs into bounded outputs, or finite-energy inputs into finite-energy outputs. More stringent typical requirements are that the gain of F be finite (in more classical mathematical terms that the operator be bounded), or that it have finite incremental gain (mathematically, that it be globally Lispchitz). The input/output approach has been extremely successful in the robustness analysis of linear systems subject to nonlinear feedback and mild nonlinear uncertainties, and in general in the area that revolves around the various versions of the small-gain theorem. Moreover, geometric characterizations of robustness (gap metric and the like) are elegantly carried out in this framework. Finally, I/O stability provides a natural setting in which to study the classification and parameterization of dynamic controllers. On the other hand, there is the model-based, or state-space approach to systems and stability, where the basic object is a forced dynamical system, typically described by differential or difference equations. In this approach, there is a standard notion of stability, namely Lyapunov asymptotic stability of the unforced system. Associated to such a system, there is an operator F mapping inputs (forcing functions) into state trajectories (or into outputs, if partial measurements on states are of interest). It becomes of interest then to ask to what extent Lyapunov-like stability notions for a state-space system are related to the stability of the associated operator F. It is well known that [62], in contrast to the case of linear systems, where there is an equivalence between state-space and I/O stability, for nonlinear systems the two types of properties are not so closely related. Even for the very special and comparatively simple case of feedback linearizable systems, this relation is far more subtle than it might appear at first sight: if one first linearizes a system and then stabilizes the equivalent linearization, in terms of the original system one does not in general obtain a closed-loop system that is input/output stable in any reasonable sense. However, it is always possible to make a choice of a feedback law that achieves such stability, in the linearizable case as well as for all other stabilizable systems [63].

A system that is ISS exhibits low overshoot and low total energy response when excited by uniformly bounded or energy- bounded signals, respectively. These are highly desirable qualitative characteritics. However, it is sometimes the case that feedback design does not render ISS behavior, or that only a weaker property than ISS is verified in a step in recursive design. Input-to-state stability concept gives a link between the two alternative paradigms of stability, I/O and state space. This notion differs fundamentally from the operator-theoretic ones that have been classically used in control theory, first of all because it takes account of initial states in a manner fully compatible with Lyapunov stability. Second, *boundedness* (finite gain) is far too strong a requirement for general nonlinear operators, and it must be replaced by nonlinear gain estimates, in which the norms of output signals are bounded by a nonlinear function of the norms of inputs, the definition of ISS incorporates such gains in a natural way. The iss notion was originally introduced in [32] and has since been employed by several authors in deriving results on control of nonlinear systems. It can be stated in several equivalent manners, which indicates that it is at least a mathematically natural concept: dissipation, robustness margins, and classical Lyapunov-like definitions. The dissipation characterizations are closely related to the pioneering work of Willems in 1976, who introduced an abstract concept of energy *dissipation* in order to unify I/O and state space stability, and in particular with the purpose of understanding conceptually the meaning of Kalman–Yakubovich positive-realness (passivity), and frequency domain stability theorems in a general nonlinear context. Four natural definitions of input-to-state stability are proposed, that is, from GAS to ISS, from lyapunov to dissipation, gain margins and estimates, all these case are equivalent for some kind of nonlinear system. More details can refer to famous reference [63].

(7) Practical stability

For a practical system, engineers concern not only stability in the sense of Lyapunov but also boundedness properties of the system responses. This is because a system might be stable or asymptotically stable in theory, while it is actually unstable in practice because the stable domain of the desired attractor is not large enough. On the other hand, sometimes the desired states of a system may be mathematically unstable, and yet the system may oscillate sufficiently near its state such that its performance is acceptable. That it, it is stable in practice. Taking into this fact, researchers have introduced the notion of *practical stability* [64–68]. Some important results in practical stability analysis over a finite time interval has been obtained in [68], which is related to the stability definition over a finite time. From this point of view, *finite-time stability* is also investigated for dynamical systems [69–72].

(8) Estimation of *domain of attraction*

The study of the determination of stable region for nonlinear systems is one of the most interesting aspects [36, 73, 74]. For this reason in the last 20 years several efforts have been made on the subject, generally arising from Lyapunov stability theory [75–78]. Among these studies, to construct a Lyapunov function to estimate domain of attraction is usually adopted. Genesio, etc., used time reversing method (or trajectory reversing method) and demonstrated the results by a two-dimensional system [76]. Vidyasagar and Vannelli modified the Zubov's theorem to compute the stable region [78]. They proposed a method to construct a rational function for Lyapunov function with the help of Taylor series expansion approximation. Reference [79] discussed how to maximize the estimation of the domain of attraction by choosing linear state feedback control law for nonlinear control systems.

In general, the origin of a given nonlinear system is not globally asymptotically stable, instead, it is locally asymptotically stable. Thus, it is important to know the stable region (or "domain of attraction") of the operation point of the system. The domain of attraction of the origin is defined as $S = \{x_0 : x(t, x_0) \to 0 \text{ as } t \to \infty\}$, where $x(\cdot, x_0)$ denotes the solution of the system corresponding to the initial condition $x(0) = x_0$. In addition, the domain of attraction S is also called "region of attraction", "basin" or "stable region (margin)." All trajectories starting within this neighborhood converge to the origin. For technical systems the knowledge of the size of such a region is very important because it contains all the initial states which lead to an asymptotically stable system behavior. Unfortunately, in general, an algebraic description of this region is not available [80, 81].

As is well known, Hopfield-type neural networks are mainly applied as either associative memories (pattern recognition) or optimization solvers. When applied as associative memories, the equilibrium points of the neural networks represent the stored patterns. The attraction domain of each equilibrium point coincides with the region from which the corresponding stored pattern can be retrieved even in the existence of noise, that is, the attraction domain of a stable equilibrium point characterizes the error-correction capability of the corresponding stored pattern. When applied as an optimization solver, the equilibrium points of the neural networks characterize all possible optimal solutions of the optimization problem. The attraction domain of each equilibrium point then coincides with the region that the network, starting from any initial guess in it, will evolve to the optimal solution. Therefore, identifying the attraction domain is important in the application of neural networks.

The approaches extensively used in the existing investigation into this field of neural networks are mainly based on Lyapunov direct method and so depend on the construction of Lyapunov function. However, there is no general rule guiding us to construct an optimal Lyapunov function for a given system, that is, constructing a Lyapunov function requires skill. Meanwhile, in the existing results, the Lyapunov functions used to characterize the attraction domain are mostly constructed by the method of characteristics, which strongly depends on the solutions of system [82].

Therefore, how to propose simple algorithms to estimate the domain of attraction for nonlinear system with and without control is still a challenging topic.

Note that, some contents in above two sections are from the Wikipedia on the internet. Following the trajectories of development of Lyapunov stability theory and other stability methods, some brief comments can be provided by the authors.

(1) Although the *Lyapunov stability theory* was proposed in his Ph.D. dissertation in 1892, its popularization just began in the 1950s, which is a delay over half a century. Before the 1950s, Lyapunov was a famous scientist only in Russia, while after 1950s, Lyapunov was a world famous scientist! This phenomenon is mainly due to the successful application of Lyapunov stability theory to the stability research of aerospace guidance systems which typically contain strong nonlinearities not treatable by other methods in the Cold War (1953–1962) period. Therefore, it is the famous application that promotes the popularization of one potential theoretical achievement, along with the discovery's name.

(2) Almost all theoretical findings and achievements have potential application value in practice. One theoretical achievement can be significantly recognized after a long time, for example, some decades or centuries. This may be attributed to the assonance or synchronization of theory and application. After all, all the theoretical achievements are used to deal with the matters encountered in the real life world. Significant and large-scale application projects may produce or promote the rapid development of some kinds of technical theories. Small scale applications may keep the development of some technical theories gradually. Almost all the theories in the application fields belong to the technical theory, or more accurately the techniques. Therefore, it is not reasonable to expect one theoretical achievement to have significant scientific value in real world in a short time. How to evaluate the theoretical achievement should be systematically considered, for example, by the peer review of authorities and public evaluations.

(3) No matter what the theoretical researchers do, their projects or interests must have some relations to the reality, for example, social science and natural science. Therefore, theoretical research achievement may not be a direct application in industrial fields, but it can be very useful in social science. Keeping in mind, research or problem is doomed to come from the reality, the mixed world of livings and nature. There are many phenomena to be discovered and explained. How to build or find a relation or bridge to the real word is a key way to demonstrate the value of the theoretical research.

(4) If one researcher has no a specific research direction in his speciality, he can trace the plan and demand of the state's project to find a topic, which is better to relate to the field of his research interest. Therefore, with the development of the state's project, theoretical problems occurred in the real world are indispensable and need to be solved. Under such kind of background, some theories associated with the projects or the problems may be emphasized gradually. Thus, it is not difficult to understand that much theoretical research has its historical roles in the developments of human society.

(5) Along with the same line as above arguments, we can consider the importance of the stability research of recurrent neural networks. In the 1980s, the digital and

electronic computers encountered many difficulties to be dealt with. In this case, analog computer began to appear, and neural computer as one of typical representatives began to be studied intensively. One of the fundamental problem of neural computer is the calculation ability. In this circumstance, Hopfield studied a kind of additive neural networks model and found their calculating capability. Then in 1984, Hopfield designed a circuit implementation of neural optimizer, which lay a solid foundation for the analog computer. Because the neural optimizer is a dynamical system, and its dynamics effect the stability of the equilibrium point, which is directly related to the optimal solution of the corresponding optimization problems, the stability analysis of recurrent neural networks began to be developed since 1982 or 1983. Up to now, it has gone thirty years for the research of stability theory of recurrent neural networks. There are many different achievements to be obtained.

(6) For a given equilibrium point of a dynamical system (i.e., the considered equilibrium point must be known in advance), how the initial values or the boundary values of the dynamical systems affect the stability property of the given equilibrium point is the main topic of Lyapunov's stability theory. Interestingly, the size of the stability degree ϵ or the initial size of neighbor δ in the definition of stability is not deterministic. Therefore, there are more space to define different kind of stability concept. In contrast, Hopfield studied the stability problem of the so-called Hopfield neural networks in the sense of *Hopfield's stability* instead of Lyapunov's stability (for example, the energy function introduced by Hopfield is not a standard Lyapunov function, which has been pointed by X. Liao in [83]). In fact, no matter what kind of definitions of stability, the common purpose of the definition is to solve the practical problems both in engineering and theory. Therefore, there are many different kind of stability concepts being proposed such as structure stability, practical stability, connective stability, synchronization stability, periodic stability, and so on. Keeping in mind, there is no fixed form of stability definition (correspondingly, the stability theory), all the theoretical researches must adapt to the different demands of practical applications.

(7) *Lyapunov second method* in essence is the gradient descent method of solving the numerical computation. This method is fundamental in the numerical analysis, which can be equivalent to the Euler method or the tangents method. All these methods are the local methods, which means the boundedness of the initial space or universe. According to the construction of Lyapunov's energy function, many optimization problems can be solved based on the gradient descent algorithms. Hopfield neural network itself is just a kind of gradient dynamical system, which can be directly applied to the optimization problems. How to build the relationship between dynamical neural network model and its energy function is the key problems. It is Hopfield who creatively proposed the *energy function* in the sense of Hopfield's definition, and gave the stability analysis of the concerned neural network in the sense of Hopfield's stability. Note that, as pointed out in Chap. 1, Professor Grossberg used the concept of energy function in neural networks field more earlier than hopfield did. However, since Professor Hopfield used a kind of additive neural networks to solve the optimization problems and, meanwhile, used the energy function concept to analyze the stability property of the concerned neural networks successfully in

1982–1984, researches on neural networks began to recover. It is the Hopfield's pioneering work that initiated the new era of optimization computation in neural network community. As far as this point is concerned, Hopfield is more excellent than Grossberg in control and optimization engineering fields. For convenience to mention the work by Hopfield, the additive neural network model studied by Hopfield is called Hopfiled neural networks, while the energy function used by Hopfield is called Hopfield energy function, and the corresponding stability concept is also called Hopfield stability by the later researchers. Due to the introduction of Hopfield's energy function into recurrent neural networks, optimization problems based on RNNs in many engineering fields such as mechanics, dynamic engineering, architecture, operation research, computational science, and so on have been solved or promoted significantly.

(8) In *Lyapunov stability theory*, the concerned system is in the absence of inputs, i.e., $\dot{x}(t) = f(x(t))$, and the equilibrium point is usually assumed to be zero, that is, $f(0) = 0$. This requirement in fact is the most important assumption in the application of Lyapunov method. How to guarantee the zero solution being the equilibrium point was not discussed in the Lyapunov stability theory. After all, in 1892, the considered system was only limited to the *isolated systems* (respect to the concept of complex systems or complex networks at present), and the stability of the concerned system on its own is existent or objective. There is no need to discuss the existence and uniqueness of the equilibrium point of the concerned systems. Therefore, from the viewpoint of energy, a system must be stable when the energy approached to the minimum, i.e., zero. Therefore, zero, as an objective existence, is believed to be a natural way to understand the world. However, with the emergence of complex systems or complex phenomena, how to determine the stable states is not easy, how to find the minimal point is not easy either. Equivalence is not necessarily equivalent. Hence, Lyapunov's stability theory falls into the scope of Newton mechanism, which is based on the reference frame or *reference coordinate system*. Different selection of reference frame may have different influence on the stability analysis of the concerned system. On the Earth, it is natural to choose the Earth as the reference frame. However, when studying the relative motion among different objects, how to select the frame is important for the considered problems. For example, *synchronization stability* as an extension of classical stability, is now intensively emphasized by the researchers due to the emergence of world wide web, interconnection networks, internet of things, and so on. In fact, in the research of synchronization problem, Lyapunov stability theory is still valid. Recalling the Lyapunov stability theory again, we can find that the nonlinear systems studied by Lyapunov is just an *error system*. This is the underlying reason why Lyapunov stability theory can be used in the observer design, filter design and synchronization! Therefore, Lyapunov stability is about the stability of error system. Since the intrinsic equilibrium point of the nonlinear system $\dot{x}(t) = f(x(t))$ is zero, then the error system between the state and the intrinsic equilibrium point is just the nonlinear system itself $\dot{x}(t) = f(x(t))$. The zero in $f(0) = 0$ is just the relative distance or the error between one state and an reference point. The reference point can be the intrinsic equilibrium point of the nonlinear system itself or the desired target out side the nonlinear system. If the *relative motion*

is static or the zero solution of error system is stable, it is the fundamental meaning of Lyapunov stability theory (i.e., *relative stability* or stability is just the relatively static movement). This also implies that idea of relativity has been used in Lyapunov stability theory, which is the fundamental reason of the universality of Lyapunov stability theory. When the input is considered in the system, i.e., $\dot{x}(t) = f(x(t), u(t))$ and $y(t) = h(x(t))$, stability with respect to different variables may form different stability concepts [84], for example input-state stability, integral input-to-state stability, and so on. Therefore, in the new environments, Lyapunov stability theory should be kept with the times, which can make the idea or thoughts of Lyapunov stability theory be carried forward and further developed. For example, Lyapunov synchronization stability theory (LSST) should be regarded as an upgrade of classical Lyapunov stability theory (LST) in the networks era. The most outstanding features of LSST include: 1) The equilibrium point of each node system is not required. The synchronous state or synchronous target can be the dynamics of any node system or their combinations, or the external specified target. 2) Synchronization stability is a relative stability. Many existing stability definitions such as Lyapunov stability, ISS, IOSS, stability of fixed points, stability of orbits, stability of sets can be unified in the framework of synchronization stability. 3) Many synthesis problems such as regulation/tracking problems, observer, filter, master-slave synchronization, drive-response synchronization, state/parameter estimation, system/parameter identification can be unified in the frameworks of synchronization stability theory. While classical Lyapunov stability theory has already promoted the development of automatic control theory for isolated systems, Lyapunov synchronization stability theory would be sure to promote the development of automatic control theory for complex interconnected networks.

(9) Referring to the stability definition in the sense of Lyapunov, there may have many different stability definitions existing in practice according to different applications. One of remarkable stability concepts is the stability of living systems. This kind of system has many stably periodic trajectories, limited cycles or other complex dynamics, except the fixed equilibrium points. For example, the heart rate and biological cycle in a living creature are all sinusoidal wave. This kind of sinusoidal signal may include two parts: frequency and amplitude. Therefore, stability definition may be a mixed concept of time domain, frequency domain, and space domain. Inspired by the *Lyapunov stability theory*, many different kinds of stability definition can be proposed. So does the stability theory of complex neural networks.

(10) In the qualitative research of dynamical systems, most efforts are placed on the external evolutionary dynamics of dynamical systems, for example, the infinite state behavior of dynamical systems as time approaches to infinity. This leads to the different definitions of stability and their corresponding stability results. It is well known that most fixed point or equilibrium point is locally stable. Starting from different *domain of attraction* of equilibrium may lead to different dynamical trajectories. This feature of initial domain is especially important in the learning of neural networks when choosing the initial weight values, as well as in the associative memory and pattern formation. Therefore, there are two directions in the qualitative analysis of dynamical systems. One is concerned with the ultimate state behavior

as time approached to infinity, such as many kinds of qualitative characteristic researches as global asymptotical stability, input-to-state stability, passivity, domain of attraction of equilibrium and dissipativity. The other is concerned with the initial condition set as time just begins, which is a inverse mapping of the infinite dynamical behavior of a dynamical system. The origin or the beginning of the initial condition is the definition domain of the concerned problem, from which the ultimate dynamic behavior of the concerned system begins to evolve as time approaches to infinity. In general, a good initial condition may trigger a better solution of the concerned problem and a less cost of the design procedure.

(11) Nowadays, more emphasis is placed on the stability of dynamical systems. This kind of system can be modeled by differential equation, difference equation, or other kind of recurrent forms. However, in the real world, too many systems can not be defined or modeled by the mathematical analytical equations. For this kind of nonanalytical systems, the results for the dynamical systems can not be used. For example, for a complex *data-driven system*, it is impossible to model it by mathematical model accurately. Any approximated description method including differential equation and difference equation can be suitable for the case of small scale systems, and not suitable for the large-scale complex systems. This evolution is similar to the principle of using distributed delay to replace the discrete delay for the large-scale circuits. Therefore, how to establish data-driven stability theory for the large-scale complex system is an urgent project in the contemporary era. This will be similar to the emergence of Lyapunov stability theory in the 1900s.

In brief, doing research with great concentration is a basic principle. One cannot expect his findings to be popularized in a short time. A good attitude for a researcher is a prerequisite for his great achievement, no matter when his achievements have been recognized and popularized.

2.5 Applications of Dynamical Systems Theory

In the following, some application fields are listed for the significant contribution of dynamical systems theory.

(1) In biomechanics

In sports biomechanics, dynamical systems theory has emerged in the movement sciences as a viable framework for modeling athletic performance. From a dynamical systems perspective, the human movement system is a highly intricate network of codependent subsystems (e.g., respiratory, circulatory, nervous, skeletomuscular, perceptual) that are composed of a large number of interacting components (e.g., blood cells, oxygen molecules, muscle tissue, metabolic enzymes, connective tissue and bone). In dynamical systems theory, movement patterns emerge through generic processes of self-organization found in physical and biological systems.

(2) In cognitive science

Dynamical system theory has been applied in the field of neuroscience and cognitive development, especially in the neo-Piagetian theories of cognitive development. It is the belief that cognitive development is best represented by physical theories rather than theories based on syntax and artificial intelligence (AI). It also believed that differential equations are the most appropriate tool for modeling human behavior. These equations are interpreted to represent an agent's cognitive trajectory through state space. In other words, dynamicists argue that psychology should be the description (via differential equations) of the cognitions and behaviors of an agent under certain environmental and internal pressures. The language of chaos theory is also frequently adopted. In it, the learner's mind reaches a state of disequilibrium where old patterns have broken down. This is the phase transition of cognitive development. Self-organization (the spontaneous creation of coherent forms) sets in as activity levels link to each other. Newly formed macroscopic and microscopic structures support each other, speeding up the process. These links form the structure of a new state of order in the mind through a process called scalloping (the repeated building up and collapsing of complex performance). This new state is progressive, discrete, idiosyncratic, and unpredictable. Dynamical systems theory has recently been used to explain a long-unanswered problem in child development referred to as the A-not-B error.

(3) In human development

Dynamical systems theory is a psychological theory of human development. Unlike dynamical systems theory, which is a mathematical construct, dynamical systems theory is primarily nonmathematical and driven by qualitative theoretical propositions. This psychological theory does, however, apply metaphors derived from the mathematical concepts of dynamical systems theory to attempt to explain the existence of apparently complex phenomena in human psychological and motor development.

As it applies to developmental psychology, this psychological theory was developed by Esther Thelen, Ph.D. at Indiana University Bloomington [85]. Thelen became interested in developmental psychology through her interest and training in behavioral biology. She wondered if "fixed action patterns," or highly repeatable movements seen in birds and other animals, were also relevant to the control and development of human infants.

According to Miller [86], dynamical systems theory is the broadest and most encompassing of all the developmental theories. Theory attempts to encompass all the possible factors that may be in operation at any given developmental moment, i.e., it considers development from many levels (from molecular to cultural) and timescales (from milliseconds to years). Development is viewed as constant, fluid, emergent or nonlinear, and multidetermined. Dynamical systems theory's greatest impact lies in early sensorimotor development. However, researchers working in fields closely related to (developmental) psychology such as linguistics have built upon Thelen's work in order to, for example, model the development of language in an individual using dynamic systems theory by linking language development to overall cognitive development.

Esther Thelen believed that development involved a deeply embedded and continuously coupled dynamic system. It is unclear, however, if her utilization of the concept of "dynamic" refers to the conventional dynamics of classical mechanics or to the metaphorical representation of "something that is dynamic" as applied in the colloquial sense in common speech, or both. The typical view presented by R.D. Beer showed that information from the world goes to the nervous system, which directs the body, which in turn interacts with the world. Esther Thelen instead offers a developmental system that has continual and bidirectional interaction between the world, nervous system and body. The exact mechanisms for such interaction, however, remain unspecified.

The dynamical systems view of development has three critical features that separate it from the traditional input–output model. Firstly, the system must be multiply causal and self-organizing. This means that behavior is a pattern formed from multiple components in cooperation with none being more privileged than another. The relationship between the multiple parts is what helps provide order and pattern to the system. Why this relation would provide such order and pattern, however, is unclear. Secondly, a dynamic system depends on time making the current state a function of the previous state and the future state a function of the current state. The third feature is the *relative stability* of a dynamic system. For a system to change, a loose stability is needed to allow for the components to reorganize into a different expressed behavior. What constitutes a stability as being loose or not loose, however, is not specified. Parameters that dictate what constitutes one state of organization versus another state are also not specified, as a generality, in dynamical systems theory. The theory contends that development is a sequence of times where stability is low allowing for new development and where stability is stable with less pattern change. The theory contends that to make these movements, you must scale up on a control parameter to reach a threshold (past a point of stability). Once that threshold is reached, the muscles begin to form the different movements. This threshold must be reached before each muscle can contract and relax to make the movement. The theory can be seen to present a variant explanation for muscle length-tension regulation but the extrapolation of a vaguely outlined argument for muscle action to a grand theory of human development remains unconvincing and unvalidated.

2.6 Notations and Discussions on Some Stability Problems

This section is divided into two parts. One is for the symbol notations, basic lemmas, basic definition of stability and equilibrium point. The other is for the discussions on some stability concepts, which will show the diversity of stability definitions.

2.6.1 Notations and Preliminaries

Throughout this book, the following notations are used if no confusion occurs.

Let $\mathbb{R}^n$ denote the n-dimensional Euclidean space, and $\mathbb{R}$ denote the real space. Let $A = [a_{ij}]_{n\times n}$ denote an $n \times n$ matrix. Let W^T, W^{-1} denote the transpose and the inverse of a square matrix W, respectively. Let $W > 0 (< 0)$ denote a positive (negative) definite symmetric matrix. I denote an identity matrix with compatible dimension. $A = \text{diag}(a_i)$ denotes the diagonal matrix. Matrices, if not explicitly stated, are assumed to have compatible dimensions. The symbol $*$ is used to denote a matrix which can be inferred by symmetry. If A is a matrix, $||A||$ denotes its operator norm or Euclidean norm. $\lambda_{\max}(A)$, $\lambda_{\min}(A)$ or $\lambda_M(A)$ and $\lambda_m(A)$ mean the maximum/largest and minimum/smallest eigenvalue of A respectively. $Re(\lambda) < 0$ means the real part of eigenvalue λ is negative, where λ is the eigenvalue of a square matrix A. For $h > 0$, $\mathbb{C}([-h, 0]; \mathbb{R}^n)$ denotes the family of continuous functions ϕ from $[-h, 0]$ to $\mathbb{R}^n$ with the norm $||\phi|| = \sup_{-h\le s\le 0} |\phi(s)|$, where $|\cdot|$ is the Euclidean norm in $\mathbb{R}^n$.

Definition 2.1 (*Positive (semi)definite function*) A function $V : \mathcal{D} \to \mathbb{R}$, where $D \subseteq \mathbb{R}^n$ is said to be positive semidefinite if $V(0) = 0$ and for every $x \in \mathcal{D}$ it holds that $V(x) \ge 0$. It is called positive definite if additionally for every $x \in \mathcal{D} - \{0\}$ it is true that $V(x) > 0$. A function V is called negative (semi)definite if $-V$ is positive (semi)definite.

Definition 2.2 (*Positive (semi)definite matrix*) A matrix $Q \in M_n(\mathbb{R})$ is called *positive (semi)definite* if the corresponding quadratic function $V(x) = x^T Qx$ is positive (semi)definite.

An algebraic criterion exists to test whether a given symmetric matrix is positive definite or positive semidefinite.

Definition 2.3 (*Criterion for positive semidefiniteness*) Let Q be a symmetric matrix with appropriate dimension. Then, it is *positive (semi)definite* if and only if all its eigenvalues are (nonnegative)positive.

Definition 2.4 (*Derivative along the trajectories of a system, see* [87]) Given a dynamical system $\Sigma : \dot{x}(t) = f(x(t))$, where $f : \mathbb{R}^n \to \mathbb{R}^n$, and a function $V : \mathbb{R}^n \to \mathbb{R}$, we define the derivative of V along the trajectories of the system Σ as: $\frac{\mathrm{d}V}{\mathrm{d}t}|_\Sigma \equiv \frac{\mathrm{d}V}{\mathrm{d}t}|_{x(t)} = \frac{\partial V}{\partial z} f(z)|_{z=x(t)}$. $(L_f V)(x)$ or $\dot{V}(x)$ is widely used to denote the derivative of V.

Lemma 2.5 (Lyapunov stability theorem, see [87]) *Let $x = 0$ be an equilibrium point for the system $\dot{x} = f(x)$ with $x \in \mathbb{R}^n$ and $\mathcal{D} \subseteq \mathbb{R}^n$ with $\mathcal{D} \ni 0$. The vector field f is locally Lipschitz (so that the differential equation admits unique solutions). Let $V : \mathcal{D} \to [0, \infty)$ be a continuously differentiable function such that: 1) V is positive definite in $\mathcal{D}$. 2) $L_f V$ is negative semidefinite in $\mathcal{D}$. Then $x = 0$ is stable. If additionally $L_f V$ is negative definite in $\mathcal{D}$, then the origin is locally asymptotically stable.*

Lemma 2.6 (Krasovskii–LaSalle principle) *The Krasovskii–LaSalle principle (also known as the invariance principle) is a criterion for the asymptotic stability of an autonomous (possibly nonlinear) dynamical system.*

The global Krasovskii–LaSalle principle: given a representation of the system $\dot{x} = f(x)$, where x is the vector of variables, with $f(0) = 0$. If a C^1 function $V(x)$ can be found such that (1) $V(x) > 0$ for all $x \neq 0$ (positive definite); (2) $\dot{V}(x) \leq 0$ for all x(negative seminegative); (3) $V(x) \to \infty$ if $x \to \infty$ and $V(0) = \dot{V}(0) = 0$ (Such functions can be thought of as "energy-like").

Let $\mathcal{I}$ be the union of complete trajectories contained entirely in the set $\{x : \dot{V}(x) = 0\}$. Then the set of accumulation points of any trajectory is contained in $\mathcal{I}$. In particular, if $\mathcal{I}$ contains no trajectory of the system except the trivial trajectory $x(t) = 0$ for $t \geq 0$, then the origin is globally asymptotically stable.

Local version of the Krasovskii–LaSalle principle: if $V(x) > 0$ when $x \neq 0$, $\dot{V}(x) \leq 0$ holds only for x in some neighborhood D of the origin, and the set $\{\dot{V}(x) = 0\} \cap D$ does not contain any trajectories of the system besides the trajectory $x(t) = 0, t \geq 0$, then the local version of the Krasovskii–LaSalle principle states that the origin is locally asymptotically stable.

Relation to Lyapunov theory. If $\dot{V}(x)$ is negative definite, the global asymptotic stability of the origin is a consequence of Lyapunov's second theorem. The Krasovskii–LaSalle principle gives a criterion for asymptotic stability in the case when $\dot{V}(x)$ is only negative semidefinite.

Definition 2.7 (*Lyapunov–Krasovskii functional, see* [88–91]) Generalizations of the Lyapunov method to delay differential equations have been found, notably by Krasovskii [90]. As an example in [91] (see [Sect. 5.3, Corollary 3.1]), for a general system

$$\dot{x} = f(x_t),\ f : \mathcal{C}([-\tau, 0], \mathbb{R}^n) \to \mathbb{R}^n, \tag{2.6}$$

under usual regularity assumptions, the existence of a so-called Lyapunov–Krasovskii functional $V : \mathcal{C}([-\tau, 0], \mathbb{R}^n) \to \mathbb{R}$ and of $\alpha_1, \alpha_2 : \mathbb{R}^+ \to \mathbb{R}^+$, α_1 is unbounded, and α_2 is positive definite, such that

$$\alpha_1(|x(t)|) \leq V(x_t),\ \frac{\mathrm{d}V(x_t)}{\mathrm{d}t} \leq \alpha_2(|x(t)|), \tag{2.7}$$

along the trajectories, ensures asymptotic stability of the origin. Here, one defines as usual $x_t(s) = x(t+s), \tau \leq s \leq 0$.

In particular, simple quadratic Lyapunov–Krasovskii functionals of the type

$$V(x_t) = x^T(t)Px(t) + \int_{t-\tau}^{t} x^T(\eta)Qx(\eta)\mathrm{d}\eta, \tag{2.8}$$

for positive definite matrices $P, Q \in \mathbb{R}^{n\times n}$ have been used early [91, 92].

Definition 2.8 (*infinitesimal generator, see* [93]) Consider the stochastic system

$$\mathrm{d}x(t) = f(x(t))\mathrm{d}t + g(x(t))\mathrm{d}\omega, \tag{2.9}$$

where $x(t) \in \mathbb{R}^n$ is the system state, ω is a r-dimensional standard Wiener process, and $f(\cdot), g(\cdot) : \mathbb{R}^n \to \mathbb{R}^n$ are locally Lipschitz functions and satisfy $f(0) = g(0) = 0$.

For any given $V(x) \in \mathcal{C}^2$, associated with the stochastic system (2.9), the *infinitesimal generator* $\mathcal{L}$ is defined as follows:

$$\mathcal{L}V(x) = \frac{\partial V}{\partial x} f(x) + \frac{1}{2}\mathrm{Tr}\left\{g^T(x)\frac{\partial^2 V}{\partial x^2} g(x)\right\}, \tag{2.10}$$

where $\mathrm{Tr}(A)$ is the trace of a matrix $A = (a_{ij})_{n\times n}$, i.e., $\mathrm{Tr}(A) = \sum_{i=1}^n a_{ii}$, a_{ii} is the element on the main diagonal of square matrix A.

Lemma 2.9 (Schur Complement, see [94]) *For a given symmetric matrix $S \in \mathbb{R}^{n\times n}$, and $S = \begin{bmatrix} S_{11} & S_{12} \\ S_{12}^T & S_{22} \end{bmatrix}$, where $S_{ij} \in \mathbb{R}^{n_i\times n_j}$ are matrix blocks with appropriate dimensions, the following statements are equivalent:(1) $S < 0$; (2) $S_{11} < 0$, $S_{22} - S_{12}^T S_{11}^{-1} S_{12} < 0$; (3) $S_{22} < 0$, $S_{11} - S_{12}S_{22}^{-1}S_{12}^T < 0$.*

Definition 2.10 (*Nonsingular M-matrix, see* [95]) An $n \times n$ matrix P with nonpositive offdiagonal elements is called a *nonsingular M-matrix* if all its principal minors are positive.

Definition 2.11 (*Nonsingular M-matrix, see* [95]) P is a nonsingular M-matrix if and only if there exists a positive diagonal matrix D such that PD is a diagonally dominant matrix.

Lemma 2.12 (see [95]) *Let D_0 be an $n \times n$ positive diagonal matrix and P be an $n \times n$ matrix with $P = (p_{ij})_{n\times n}$. If $D_0 - |P|$ is a nonsingular M-matrix with $|P| = (|p_{ij}|)_{n\times n}$, then $D_0 + P$ is nonsingular.*

In mathematics and, specifically, real analysis, the Dini derivatives are a class of generalizations of the derivative. They were introduced by Ulisse Dini (1845–1918), who was an Italian mathematician, and is known for his contribution to real analysis [96].

Definition 2.13 (*Dini derivative*)
(1) The upper *Dini derivative*, which is also called an upper right-hand derivative of a continuous function $f : \mathbb{R} \to \mathbb{R}$, is denoted by $f_+{}'$ and defined by

$$D^+ f(t) = f_+{}'(t) \triangleq \lim_{h\to 0_+} \sup \frac{f(t+h) - f(t)}{h},$$

where limsup is the supremum limit.

(2) The lower Dini derivative $f_-{}'$ is defined by

$$D_- f(t) = f_-{}'(t) \triangleq \lim_{h \to 0_+} \inf \frac{f(t+h) - f(t)}{h},$$

where liminf is the infimum limit.

(3) If $f(t)$ is defined on a vector space, then the upper Dini derivative at t in the direction d is defined by

$$D^+ f(t) = f_+{}'(t) \triangleq \lim_{h \to 0_+} \sup \frac{f(t+hd) - f(t)}{h},$$

where limsup is the supremum limit.

(4) If $f(t)$ is locally Lipschitz, then $f_+{}'(t)$ is finite. If $f(t)$ is differentiable at t, then the Dini derivative at t is the usual derivative at t.

Also,

$$D^- f(t) \triangleq \limsup_{h \to 0-} \frac{f(t+h) - f(t)}{h},$$

$$D_- f(t) \triangleq \liminf_{h \to 0-} \frac{f(t+h) - f(t)}{h}$$

are used to denote the upper Dini derivative and lower Dini derivative, respectively.

Therefore, when using the D notation of the Dini derivatives, the plus or minus sign indicates the left-hand or right-hand limit, and the placement of the sign indicates the infimum or supremum limit.

Lemma 2.14 (Schauder fixed-point theorem) *The Schauder fixed-point theorem is an extension of the Brouwer fixed-point theorem to topological vector spaces, which may be of infinite dimension. It asserts that if K is a convex subset of a topological vector space V and T is a continuous mapping of K into itself so that $T(K)$ is contained in a compact subset of K, then T has a fixed point.*

A consequence, called Schaefer's fixed-point theorem, is particularly useful for proving existence of solutions to nonlinear partial differential equations. Schaefer's theorem is in fact a special case of the far reaching Leray–Schauder theorem which was discovered earlier by Juliusz Schauder and Jean Leray. The statement is as follows: Let T be a continuous and compact mapping of a Banach space X into itself, such that the set

$$\{x \in X : x = \lambda T x \text{ for some } 0 \le \lambda \le 1\}$$

is bounded. Then T has a fixed point.

Lemma 2.15 (Brouwer's fixed-point theorem) *Brouwer's fixed-point theorem is a fixed-point theorem in topology, named after Luitzen Brouwer. It states that for any* *continuous function f mapping a compact convex set into itself there is a point x_0 such that $f(x_0) = x_0$. The simplest forms of Brouwer's theorem are for continuous functions f from a closed interval I in the real numbers to itself or from a closed disk D to itself. A more general form than the latter is for continuous functions from a convex compact subset K of Euclidean space to itself.*

Among hundreds of fixed-point theorems, Brouwer's is particularly well known, due in part to its use across numerous fields of mathematics. In its original field, this result is one of the key theorems characterizing the topology of Euclidean spaces, along with the Jordan curve theorem, the hairy ball theorem and the Borsuk–Ulam theorem. This gives it a place among the fundamental theorems of topology. The theorem is also used for proving deep results about differential equations and is covered in most introductory courses on differential geometry. It appears in unlikely fields such as game theory. In economics, Brouwer's fixed-point theorem and its extension, the Kakutani fixed-point theorem, plays a central role in the proof of existence of general equilibrium in market economies as developed in the 1950s by economics Nobel prize winners Kenneth Arrow and Grard Debreu.

Lemma 2.16 (Contraction mapping principle) *In mathematics, a contraction mapping, or contraction or contractor, on a metric space (M, d) is a function f from M to itself, with the property that there is some nonnegative real number $0 \le k < 1$ such that for all x and y in M,*

$$d(f(x), f(y)) \le kd(x, y).$$

The smallest such value of k is called the Lipschitz constant of f. Contractive maps are sometimes called Lipschitzian maps. If the above condition is instead satisfied for $k \le 1$, then the mapping is said to be a non-expansive map.

More generally, the idea of a contractive mapping can be defined for maps between metric spaces. Thus, if (M, d) and (N, d^1) are two metric spaces, and $f : M \to N$, then there is a constant $k < 1$ such that

$$d^1(f(x), f(y)) \le kd(x, y),$$

for all x and y in M. Every contraction mapping is Lipschitz continuous and hence uniformly continuous (for a Lipschitz continuous function, the constant k is no longer necessarily less than 1).

A contraction mapping has at most one fixed point. Moreover, the Banach fixed-point theorem states that every contraction mapping on a nonempty complete metric space has a unique fixed point, and that for any x in M the iterated function sequence $x, f(x), f(f(x)), f(f(f(x))), \ldots,$ converges to the fixed point. This concept is very useful for iterated function systems where contraction mappings are often used. Banach's fixed-point theorem is also applied in proving the existence of solutions of ordinary differential equations, and is used in one proof of the inverse function theorem.

In mathematics, in the field of differential equations, an initial value problem (also called the Cauchy problem by some authors) is an ordinary differential equation together with a specified value, called the initial condition, of the unknown function at a given point in the domain of the solution. In physics or other sciences, modeling a system frequently amounts to solving an initial value problem; in this context, the differential equation is an evolution equation specifying how, given initial conditions, the system will evolve with time.

Definition 2.17 (*Initial value problem*) An initial value problem is a differential equation $y^{'}(t) = f(t, y(t))$ with $f : \Omega \subset \mathbb{R} \times \mathbb{R}^n \to \mathbb{R}^n$, where Ω is an open set of $\mathbb{R} \times \mathbb{R}^n$, together with a point in the domain of $f(t_0, y_0) \in \Omega$ called the initial condition.

A solution to an initial value problem is a function y, that is, a solution to the differential equation and satisfies $y(t_0) = y_0$.

In higher dimensions, the differential equation is replaced with a family of equations $y_i^{'} = f_i(t, y_1(t), y_2(t), \ldots,)$ and $y(t)$ is viewed as the vector $(y_1(t), \ldots, y_n(t))$. More generally, the unknown function y can take values on infinite dimensional spaces, such as Banach spaces or spaces of distributions. Initial value problems are extended to higher orders by treating the derivatives in the same way as an independent function, e.g. $y^{''}(t) = f(t, y(t), y^{'}(t))$.

Brouwer's fixed-point theorem, Schauder fixed-point theorem, contraction mapping principle and initial value problem are from Wikipedia, the free encyclopedia on the internet.

Definition 2.18 (*Solution in the sense of Caratheodory, see* [10]) Consider the following ordinary differential equation (ODE) in $\mathbb{R}^n$,

$$\dot{x} = f(x, t), t \geq 0, x(0) = x_0. \tag{2.11}$$

By a solution of Eq. (2.11) we mean a continuously differentiable function of time $x(t)$ satisfying

$$x(t) = x_0 + \int_0^t f(x(s), s)\mathrm{d}s. \tag{2.12}$$

Such a solution to Eq. (2.11) is called a *solution in the sense of Caratheodory.*

Definition 2.19 (*Local existence and uniqueness, see* [10]) Consider the system (2.11). Assume that $f(x, t)$ is continuous in t and x, and that there exist T, r, k, h such that for all $t \in [0, T]$, we have

$$\begin{aligned} |f(x, t) - f(y, t)| &\leq k|x - y|, \forall x, y \in B(x_0, r), \\ |f(x_0, t)| &\leq h, \end{aligned} \tag{2.13}$$

with $B(x_0, r) = B_r = \{x \in \mathbb{R}^n : |x - x_0| \leq r\}$ is a ball of radius r centered at x_0. Then Eq. (2.11) has exactly one solution of the form of (2.12) on $[0, \delta]$ for δ sufficiently small.

Definition 2.20 (*Global existence and uniqueness, see* [10]) Consider the system (2.11) and assume that $f(x, t)$ is piecewise continuous with respect to t and for each $T \in [0, \infty)$ there exist finite constants k_T, h_T such that for all $t \in [0, T]$, we have

$$\begin{aligned} |f(x,t) - f(y,t)| \leq & k_T|x - y|, \forall x, y \in \mathbb{R}^n, \\ |f(x_0, t)| \leq & h_T. \end{aligned} \tag{2.14}$$

Then Eq. (2.11) has exactly one solution on $[0, T]$ for all $T < \infty$.

Definition 2.21 (*Continuous dependence on initial conditions, see* [10]) Consider the system (2.11) and let $f(x, t)$ satisfy the hypothesis (2.14). Let $x(\cdot)$, $y(\cdot)$ be two solutions of this system starting from x_0 and y_0 respectively. Then for given $\epsilon > 0$, there exists $\delta(\epsilon, T)$ such that

$$|x_0 - y_0| \leq \delta \Rightarrow |x(\cdot) - y(\cdot)| \leq \epsilon. \tag{2.15}$$

Definition 2.22 (*Lipschitz continuous, see* [10]) The function f is said to be locally Lipschitz continuous in x if for some $h > 0$ there exists $L \geq 0$ such that

$$|f(x_1, t) - f(x_2, t)| \leq L|x_1 - x_2|, \tag{2.16}$$

for all $x_1, x_2 \in B_h$, $t \geq 0$. The constant L is called the Lipschitz constant. A definition for globally Lipschitz continuous functions follows by requiring Eq. (2.16) to hold for $x_1, x_2 \in \mathbb{R}^n$. The definition of semi-globally Lipschitz continuous functions hold as well by requiring that Eq. (2.16) hold in B_h for arbitrary h but with L possibly a function of h. The Lipschitz property is by default assumed to be uniform in t.

If f is Lipschitz continuous in x, it is continuous in x. On the other hand, if f has bounded partial derivatives in x, then it is Lipschitz. Formally, if $D_1 f(x, t) \triangleq \left[\frac{\partial f_i}{\partial x_j}\right]$ denotes the partial derivative matrix of f with respect to x (the subscript 1 stands for the first argument of $f(x, t)$), then $|D_1 f(x, t)| \leq L$ implies that f is Lipschitz continuous with Lipschitz constant L (again locally, globally, or semi-globally depending on the region in x that the bound on $|D_2 f(x, t)|$ is valid).

Definition 2.5 provides a universal definition on the stability and asymptotic stability. However, to meet the needs of this book, we will use the following delayed neural networks (DNNs),

$$\dot{x}(t) = -Ax(t) + Bg(x(t)) + Cg(x(t - \tau(t))) + U, \tag{2.17}$$

to present some stability definitions for convenience, where $x(t) = (x_1(t), \ldots, x_n(t))^T \in \mathbb{R}^n$, $g(x(t)) = g_1(x_1(t)), \ldots, g_n(x_n(t))^T \in \mathbb{R}^n$, $g_i(x_i(t))$ is the neuronal activation function, $A = \text{diag}(a_1, \ldots, a_n)$, $a_i > 0$, B and C are connection matrices with appropriate dimensions, $\tau(t)$ is a time-varying delay, $0 \leq \tau(t) \leq \tau$, $\dot{\tau}(t) \leq \mu$, τ and μ are positive constants, $U = (U_1, \ldots, U_n)^T \in \mathbb{R}^n$ is the external constant input vector, $x_0(s) = \phi(s) \in \mathbb{C}$, $i = 1, \ldots, n$. As usual, the solution $x(t, x_0)$ is also called a trajectory of (2.17).

Let $\mathcal{D} \subset \mathbb{R}^n$ be a subset. $\mathcal{D}$ is said to be *invariant* under the system (2.17) if $x_0 \in \mathcal{D}$ implies $\Gamma_1(x_0) \subseteq \mathcal{D}$, where $\Gamma_1(x_0)$ is the trajectory of system (2.17) through x_0. A point x^* is called a *ω-limit point* of $\Gamma_1(x_0)$ if there is a subsequence $\{t_i\}$ such that $x^* = \lim_{i\to\infty} x(t_i, x_0)$. All the ω-limit points constitute the ω-limit set $\omega(\Gamma_1(x_0))$ of $\Gamma_1(x_0)$. The ω-limit set is invariant under the dynamics. Recall that a constant vector x^* is said to be an *equilibrium state* of the system (2.17) if x^* is a zero point of operator $F_1(x(t))$ defined by [97, 98],

$$F_1(x^*) = -Ax^* + Bg(x^*) + Cg(x^*) + U = 0. \tag{2.18}$$

The *equilibrium state* x^* is said to be stable if any trajectory of (2.17) can stay within a small neighborhood of x^* whenever the initial x_0 is close to x^*, and is said to be *attractive* if there is a neighborhood $\Xi(x^*)$, called the attraction basin of x^*, such that any trajectory of (2.17) initialized from a state in $\Xi(x^*)$ will approach to x^* as time goes to infinity. An equilibrium state x^* is said to be *asymptotically stable* if it is both stable and attractive, whilst the equilibrium state x^* is said to be *exponentially stable* if there exist a constant $\alpha > 0$ and a strictly increasing function $M : \mathbb{R} \to \mathbb{R}^+$ with $M(0) = 0$ such that the following inequality holds,

$$\|x(t, x_0) - x^*\| \leq M(\|x(t, x_0) - x^*\|)e^{-\alpha t}. \tag{2.19}$$

Further, x^* is said to be *globally asymptotically stable* if it is asymptotically stable, and $\Xi(x^*) = \mathbb{R}^n$. System (2.17) is said to be *globally convergent* if $x(t, x_0)$ converges to an equilibrium state of (2.17) for every initial point (the limit of $x(t, x_0)$ may not be the same for different x_0), whilst it is said to be *exponentially convergent* if it is globally convergent with $x(t, x_0)$, and its limit x^* satisfying (2.19) [97].

Note that, some exponential stability definitions are required to satisfy the form $\|x(t, x_0) - x^*\| \leq \bar{M}(\|x(t, x_0) - x^*\|)e^{-\alpha t}$ for some positive value $\bar{M} > 0$ and $\alpha > 0$. However, some scholars define $\bar{M} \geq 1$ instead of $\bar{M} > 0$. Although this is a trivial problem, when $0 < \bar{M} < 1$, the exponential characteristic of the dynamical behavior is difficult to achieve. According to the *contraction mapping principle*, the state may be convergent but may not be exponentially convergent. As far as the exponential form of stability is concerned, many stability definitions satisfy this form. Therefore, strictly speaking, to the authors' knowledge, it is better to define $\bar{M} \geq 1$ in the exponential stability definition.

2.6.2 Discussions on Some Stability Definitions

In this subsection, we will introduce the evolution of equilibrium from fixed, stable flow to invariant set, along with different stability understandings and stability definitions. This will help us to understand different kinds of stability definitions in depth.

Definition 2.23 (*Absolute stability, see* [22]) The neural network (2.17) is said to be *robustly absolutely stable* in the sector $[K_1, K_2]$ if the system is globally uniformly asymptotically stable for any nonlinear function $g(x)$ satisfying $(g(x(t)) - K_1x(t))^T(g(x(t)) - K_2x(t)) \leq 0$.

Definition 2.24 (*Absolute stability*) There exists a unique equilibrium point for DNNs (2.17) attracting all trajectories in phase space and, moreover, that this property is valid for all neuron activations within a specific class of nonlinear functions and for all constant input stimuli to the networks.

Definition 2.25 (*Delay-independent condition, see* [99]) The system (2.17) with $g(x(t)) = x(t)$ and $U = 0$ is said to be stable independent of delay if it is asymptotically stable for every $\tau(t) \in [0, \tau]$. In this case one says that the system (2.17) is *absolutely stable*.

Definition 2.26 The dynamical system defined by (2.17) is *globally asymptotically stable* (GAS) if there exits a unique equilibrium point, x^*, which is stable and to which every system trajectory converges.

The concept of absolute stability for DNNs (2.17) concerns the persistence of GAS when the input U is varied. (In general, one also varies the activation functions $g(\cdot)$, but the DNNs model assumes the activation functions belonging to a form of specific Class). When the types of activation function are given in (2.17), the following ABST definition is given in [100].

Definition 2.27 (*see* [100]) The dynamical system described by (2.17) with fixed activation function is said to be *absolutely stable* if it possesses a GAS equilibrium point for each input $U \in \mathbb{R}^n$.

Definition 2.28 (*Absolute stability, see* [101]) Consider the system (2.17), if we can find a Lyapunov function $V(x(t))$ such that $\dot{V}(x(t)) < 0$ for any initial condition $x(t_0) = \phi(t) \in \mathbb{C}$, then the system (2.17) is absolutely stable.

Neural network is called absolutely stable (ABST), i.e., that it possesses a globally asymptotically stable (GAS) equilibrium point for every neuron activation function and for every constant input vector. It is easily realized that the property of absolute stability is really desirable in view of solving many signal processing tasks. Consider, for example, self-organizing neural networks. Absolute stability guarantees that convergence is preserved even during the learning phase when parameters are slowly adjusted in an unpredictable way. Moreover, absolute stability ensures convergence of the neural network also when, for some parameter values, there are infinitely many *nonisolated equilibrium points* (e.g., a *manifold of equilibria*). This feature is useful in practical problems, requiring convergence even in the presence of nonisolated equilibria. For instance, concerning minimization of multilinear polynomials with nonisolated minima, it was noted that the natural choice of the neural network parameters leads to a network with nonisolated equilibria. Another significant case is related to the neural network for the solution of linear or quadratic programming problems with infinite solutions. Once more, the design procedure naturally leads to the presence of manifolds of equilibria. A third case is that of gradient systems, which possess manifolds of equilibria in the generic case. It is also worth to mention that there are interesting problems where it is explicitly required to implement an associative memory, which is able to store and retrieve some pattern within a set of infinitely many nonisolated equilibrium patterns.

Definition 2.29 (*Complete stability-I, see* [95]) DNNs (2.17) is said to be *completely stable* if for any initial continuous function $\phi(t)$, the solution $x(t, \phi(t))$ of (2.17) satisfies $\lim_{t\to\infty} x(t, \phi) =$ constant.

Definition 2.30 (*Complete stability-II, see* [95, 102]) DNNs (2.17) is said to be completely stable if for any initial value starting from x_0 at $t = t_0$, the trajectory $x(t; t_0, x_0)$ of (2.17) satisfies

$$\lim_{t\to\infty} \|x(t; t_0, x_0) - x^*\| = 0 \tag{2.20}$$

where x^* is an equilibrium point of neural networks (2.17).

Above definitions of complete stability mean that each trajectory converges toward an equilibrium point (a *stationary state*), possibly within a set of many equilibrium points. In [103], the property of complete stability is referred to as global pattern formation, in order to highlight the ability of the neural networks to produce a steady-state pattern, i.e., $\lim_{t\to\infty} x(t)$ in response to any input pattern U and initial activity pattern x_0.

It is shown that several classes of real-value neural networks (RVNNs) can be completely stable by combining energy minimization and the *LaSalle invariant principle*. The equilibrium points of such networks were required to be isolated. However, under the framework of combining the energy minimization method and the Cauchy convergence principle to study complete stability for RVNNs, the equilibrium points of such networks were no longer required to be isolated. Meanwhile, complete stability for continuous-time and discrete-time RVNNs was further considered by combining the energy minimization method and the Cauchy convergence principle, respectively. Furthermore, as the extensive versions of the existing complete stability results for RVNNs, complete stability for discrete-time complex-value neural networks (CVNNs) was investigated by the energy minimization method. Note that GAS implies complete stability, but not vice versa [95].

Definition 2.31 (*Strict Lyapunov function, see* [24]) An energy function V is said to be strict if and only if the set $\{x \in \mathbb{R}^n | \dot{V}(x) = 0\}$ coincides with the set of equilibrium points. This is equivalent to the fact that the energy is strictly decreasing along nonstationary solutions.

Remark 2.32 It is necessary to give some comparisons among absolute stability, complete stability, asymptotical stability, and global stability.

(1) As one of the basic enabling properties of multistable neural networks, complete stability, which allows each trajectory converges toward an equilibrium point, possibly within a set of many equilibrium points. In contrast, absolute stability has a unique equilibrium point.

(2) Complete stability of (2.17) can hold both in the generic case where (2.17) has finitely many equilibrium points, as well in degenerate situations where there are infinitely many *nonisolated equilibrium points* [24].

(3) Consider the neural networks (2.17) with symmetric interconnection matrices, and neuron activation is a continuous, nondecreasing, and bounded piecewise linear (PWL) function, such that $g(0) = 0$. Then, within this class global pattern formation is absolutely stable, i.e., complete stability holds for any choice of the parameters defining A, B, C, U and $g(\cdot)$ in neural network [24]. Few results are reported on absolute stability of global pattern formation [24]. In some constraints, complete stability and absolute stability can be equivalent.

(4) In the analysis of complete stability, the Lyapunov method and the classic LaSalle approach are no longer effective because of the multiplicity of attractors, some new method must be proposed, for example, a convergence theorem of Gauss–Seidel method [23–25]. While for the *absolute stability*, the Lyapunov method and the classic LaSalle approach are very effective.

(5) Both absolute stability and asymptotical stability are concerned with the neural networks with unique equilibrium point. That is, for a specified equilibrium point, how to determine the stability property of the unique/specific equilibrium point. By contrast, complete stability and global stability is about the total dynamics of the concerned networks, not for a specific equilibrium state. Therefore, for neural networks with given equilibrium point, the Lyapunov stability theory can be effective, while for the nonisolated equilibrium point, *Lyapunov stability theory* can not be applied.

(6) *Global stability* of neural network (2.17) has been extensively investigated in the context of absolute stability theory, i.e., global absolute stability. When the input nonlinearities satisfy the sector constraints for only some finite range of their arguments, the network can only be guaranteed to be locally asymptotically stable. By contrast, complete stability itself is the *global dynamics*. Therefore, there is no global/local complete stability concept.

Definition 2.33 (*Global dynamic behavior, see* [104]) A stable system equilibrium point of the neural network (2.17) is defined to be the state vector with all its components consisting of stable equilibrium states.

Definition 2.34 (*Global stability, see* [13]) If every nonequilibrium solution of neural network (2.17) converges to an equilibrium, then neural network (2.17) is said to be *global stable*. In order to ensure that neural network (2.17) is globally stable, one requires that the set of equilibria for the networks (2.17) is discrete set. Thus, any point in the limit set of state is an equilibrium of networks (2.17) as $t \to \infty$, and this point approaches to some equilibrium of networks (2.17).

It follows from the above definition that a cellular neural network is always at one of its stable system equilibrium point after the transient has decayed to zero. From the dynamical systems theory point of view, the transient of a cellular neural network is simply the trajectory starting from some initial state and ending at an equilibrium point of the system. Since any stable equilibrium point of a cellular neural network is a *limit point* of a set of trajectories of the corresponding differential equations (2.17), such an attracting limit point has a basin of attraction, namely, the union of all trajectories converging to this point. Therefore, the state space of a cellular neural network can be partitioned into a set of basins centered at the stable system equilibrium points.

Definition 2.35 Neural network (2.17) is said to be *bounded* if its each trajectory are bounded.

Definition 2.36 (*Globally attractive set, see* [98]) Let S be a compact subset of $\mathbb{R}^n$. Denote the ϵ-neighborhood of S by S_ϵ. A compact set S is called a *globally attractive set* of neural network (2.17) if, for any $\epsilon > 0$, all the trajectories of neural network (2.17) ultimately enter and remain in S_ϵ.

As a special class of RNNs, cellular neural networks have many outstanding features with stable dynamics when it is applied in pattern recognition and storage memory. Generally, cellular neural networks can be characterized by a large system of ordinary differential equations. Since all of the cells are arranged in a regular array, one can exploit many spatial properties, such as regularity, sparsity, and symmetry in studying the dynamics of cellular neural networks. There are two mathematical models which can characterize dynamical systems having these spatial properties. One is partial differential equation and the other is cellular automata. Partial differential equation, cellular automata, and cellular neural networks share a common property, namely, their dynamic behavior depend only on their spatial local interactions.

In general, the *limit set* of a complex nonlinear system is very difficult, if not impossible, to determine, either analytically or numerically. Although, for piecewise linear circuit, it is possible to find all dc solutions by using either a brute force algorithm or some more efficient ones, it is nevertheless very time consuming for large systems. For a cellular neural network, in view of the nearest neighbor interactive property, one can solve for all system equilibrium points by first determining the stable cell equilibrium states, and then using the neighbor interactive rules to find the corresponding system equilibrium. As presented above, the *dynamic behavior* of a cellular neural network with zero control operators and nonzero feedback operators is reminiscent of a two-dimensional cellular automaton. Both of them have the parallel signal processing capability and are based on the nearest neighbor interactive dynamic rules. The main difference between a cellular neural network and a cellular automata machine is in their dynamic behaviors. The former is a continuous time while the latter is a discrete-time dynamical system. Because the two systems have many similarities, one can use cellular automata theory to study the *steady-state behavior* of cellular neural networks. Another remarkable distinction between them is that while the cellular neural networks will always settle to stable equilibrium points in the steady state, a cellular automata machine is usually imbued with a much richer dynamical behavior, such as periodic, chaotic, and even more complex phenomena. Of course, one can train a cellular neural network by choosing a sigmoid nonlinearity. If one chooses some other nonlinearity for the nonlinear elements, many more complex phenomena will also occur in cellular neural networks.

Definition 2.37 (*$\mathcal{K}$-class function, see* [63]) A function $\Phi : [0, a] \to [0, +\infty)$ is said to be positive if $\Phi(s) > 0$ for all $s > 0$ and $\Phi(0) = 0$. A continuous function $\alpha : [0, a] \to [0, +\infty)$ is said to belong to class $\mathcal{K}$ if it is positive, strictly increasing and $\alpha(0) = 0$. It is said to belong to class $\mathcal{K}_\infty$ if $a = \infty$ and $\alpha(r) \to +\infty$ as $r \to \infty$. Similarly, the continuous function $\beta : [0, a] \times [0, \infty) \to [0, \infty)$ is said to belong to class $\mathcal{KL}$ if, for each fixed s, the mapping $\beta(r, s)$ belongs to class $\mathcal{K}$ with respect to r and, for each fixed r, the mapping $\beta(r, s)$ is decreasing with respect to s and $\beta(r, s) \to \infty$ as $s \to \infty$.

An example of a class $\mathcal{K}_\infty$ function is $\alpha(r) = r^c$ with $c > 0$. An example of a class $\mathcal{KL}$ function is $\beta(r, s) = r^c e^{-s}$ with $c > 0$.

The system

$$\dot{x}(t) = f(x(t), u(t)), \; y(t) = h(x(t)), \tag{2.21}$$

is said to be *forward complete* if for every initial state $x_0 = x(0) = \xi$ and for every input u defined on $\mathbb{R}^+$, $t_{\max} = +\infty$. The corresponding output is denoted by $y(t; \xi, u) = h(x(t; \xi, u))$ on the domain of definition of the solution, where the input means a measurable and locally essentially bounded function, $x(t; \xi, u)$ denotes the unique maximal solution of the *initial value problem* of (2.21) with $x_0 = x(0) = \xi$.

Definition 2.38 (*see* [105]) Let $u = 0$. For some region $\mathcal{D} \subseteq \mathbb{R}^n$, if for any $\epsilon > 0$, there exists a $\delta = \delta(\epsilon)$ such that when $x_0 \in \mathcal{D}$ and $\|x_0\| \le \delta$, the following inequality holds: $\|x(t, x_0)\| < \epsilon, \forall t \ge 0$. Then, system (2.21) is said to be *uniformly stable* (US) in $\mathcal{D}$ and $\mathcal{D}$ is called to be a stable region of system (2.21).

Definition 2.39 (*see* [105]) Let $u = 0$. system (2.21) is said to be *uniformly asymptotically stable* (UAS) in $\mathcal{D}$ if it is US in $\mathcal{D}$, and moreover the equality holds: $\lim_{t\to\infty} \|x(t, x_0)\| = 0, \forall t \ge 0$.

Definition 2.40 (*see* [105]) Let $u = 0$. If there exist positive constants $\alpha > 0$, $K \ge 1$ such that for any $x_0 \in \mathcal{D}$, $\|x(t, x_0)\| < K\|x_0\|e^{-\alpha t}, \forall t \ge 0$. Then system (2.21) is said to be *uniformly exponentially stable* (UES) in $\mathcal{D}$ and $\mathcal{D}$ is called to be an exponential stable region of system (2.21).

Definition 2.41 (*see* [60]) A forward complete system (2.21) is

(1) *input-to-output stable* (IOS) if there exist a $\mathcal{KL}$-function β and a $\mathcal{K}$-function γ such that

$$|y(t;\xi,u)| \le \beta(|\xi|,t) + \gamma(\|u\|), \forall t \ge 0; \tag{2.22}$$

(2) *output-Lagrange input-to-output stable* (OLIOS) if it is IOS and there exist some $\mathcal{K}$-functions σ_1, σ_2 such that

$$|y(t;\xi,u)| \le \max\{\sigma_1(|h(\xi)|), \sigma_2(\|u\|)\}, \forall t \ge 0; \tag{2.23}$$

(3) state-independent IOS (SIIOS) there exist some $\beta \in \mathcal{KL}$ and some $\gamma \in \mathcal{K}$ such that

$$|y(t;\xi,u)| \le \beta(|h(\xi)|,t) + \gamma(\|u\|), \forall t \ge 0. \tag{2.24}$$

Definition 2.42 (*see* [59]) The system (2.21) without $u(t)$ is *output-to-state stable* (OSS) if there exist some $\beta \in \mathcal{KL}$ and some $\gamma \in \mathcal{K}$ such that

$$|x(t,\xi)| \le \max\{\beta(|\xi|,t), \gamma(\|y_\xi|_{[0,t]}\|)\}, \tag{2.25}$$

for all $\xi \in \mathcal{X}$ and all $t \in [0, t_{\max}]$, where $|\xi|$ indicates the Euclidean norm, and $\|y_\xi|_{[0,t_0]}\|$ is the sup-norm of the restriction of y_ξ to real interval $[0, t_0]$, that is $\sup_{t\in[0,t_0]} y_\xi(t)$.

Definition 2.43 (*see* [110]) The system (2.21) is *globally asymptotically stable* (GAS) if there exists a function $\beta(s,t) \in \mathcal{KL}$, such that, with the control $u = 0$, given any initial state ξ, the solution exists for all $t > 0$ and it satisfies the estimate

$$|x(t)| \le \beta(|\xi|,t), \tag{2.26}$$

for all $t \ge 0$.

Definition 2.44 (*see* [110]) The system (2.21) is *input-to-state stable* (ISS) if there exist functions $\beta \in \mathcal{KL}$ and $\gamma \in \mathcal{K}_\infty$, such that for each measurable essentially bounded control $u(t)$ and each initial state ξ, the solution exists for each $t \ge 0$, and furthermore it satisfies

$$|x(t)| \le \beta(|\xi|,t) + \gamma(|u(t)|), \tag{2.27}$$

for all $t \ge 0$.

The above definition of GAS is of course equivalent to the usual one (stability plus attractivity) but it is much more elegant and easier to work with. The definition of an ISS system is a natural generalization of this. Since $\gamma(0) = 0$, an ISS system is necessarily GAS. For linear systems $\dot{x} = Ax + Bu$ with asymptotically stable matrix A, an estimate (2.27) is obtained from the variation of parameters formula, but in general, GAS does not imply ISS. The notion of ISS is somewhat related to the classical *total stability* notion, but total stability typically studies only the effect of small perturbations (or controls), while ISS concerns with the bounded behavior for arbitrary bounded controls [110].

Definition 2.45 (*Integral input-to-state stability (iISS), see* [49]) System (2.21) is iISS if there exist functions $\alpha \in \mathcal{K}_\infty$, $\beta \in \mathcal{KL}$ and $\gamma \in \mathcal{K}$, such that, for all $\xi \in \mathbb{R}^n$ and all u, the solution $x(t, \xi, u)$ is defined for all $t \geq 0$, and

$$\alpha(|x(t,\xi,u)|) \leq \beta(|\xi|, t) + \int_0^t \gamma(|u(s)|)\mathrm{d}s, \tag{2.28}$$

for all $t \geq 0$.

Observe that a system is iISS if and only if there exist functions $\beta \in \mathcal{KL}$ and $\gamma_1, \gamma_2 \in \mathcal{K}$ such that

$$\alpha(|x(t,\xi,u)|) \leq \beta(|\xi|, t) + \gamma_1\Big(\int_0^t \gamma_2(|u(s)|)\mathrm{d}s\Big), \tag{2.29}$$

for all $t \geq 0$, all $\xi \in \mathbb{R}^n$, and all u. Also note that if system (2.21) is iISS, then it is 0-GAS, that is, the 0-input system $\dot{x} = f(x, 0)$ is globally asymptotically stable (GAS). (That is, the *zero solution* of this system is globally asymptotically stable.)

Definition 2.46 (*see* [49]) A continuously differentiable function $V : \mathbb{R}^n \to \mathbb{R}$ is called an iISS-Lyapunov function for system (2.21) if there exist functions $\alpha_1, \alpha_2 \in \mathcal{K}_\infty$ and $\sigma \in \mathcal{K}$, and a continuous positive definite function α_3, such that

$$\alpha_1(|\xi|) \leq V(\xi) \leq \alpha_2(|\xi|) \tag{2.30}$$

for all $\xi \in \mathbb{R}^n$, and

$$DV(\xi)f(\xi, u) \leq -\alpha_3(|\xi|) + \sigma(|u|), \tag{2.31}$$

for all $\xi \in \mathbb{R}^n$, and all $u \in \mathbb{R}^m$.

Note that the estimate (2.30) amounts to the requirement that V must be *positive definite* (i.e., $V(x) > 0$ for all $x \neq 0$ and $V(0) = 0$), and *proper* (i.e., radially unbounded, namely, as $V(x) \to \infty$ as $|x| \to \infty$).

Stability so far is studied with respect to equilibrium points. Stability, however, can be studied with respect to invariant sets.

Definition 2.47 (*Stability of Sets, see* [87]) A set M is said to be stable for the system $\dot{x} = f(x)$ if for every $\epsilon > 0$, there is a $\delta = \delta(\epsilon) > 0$ such that $\mathrm{dist}_M(x) < \delta \Rightarrow \mathrm{dist}_M(x(t, x_0)) < \epsilon$ for all $t \geq 0$. Where $\mathrm{dist}_M(x)$ denotes the distance of x from the set M defined as $\mathrm{dist}_M(x) = \inf_{x \in M} \|x - z\|$.

Definition 2.48 (*Asymptotic stability of Sets, see* [87]). A set M is said to be locally asymptotically stable for the system $\dot{x} = f(x)$ if it is stable and additionally $\lim_{t\to\infty} \text{dist}_M(x(t, x_0)) = 0$ for all x such that $\text{dist}_M(x(t, x_0)) < \eta$ for some $\eta > 0$. If the limitation holds true for all $x \in \mathbb{R}^n$, then M is called a globally asymptotically stable set.

Definition 2.49 (*LaSalle's Principle, see* [87]). Let $\Omega \subset \mathcal{D}$ be a compact set that is positively invariant with respect to $\dot{x} = f(x)$. Let $V : D \to \mathbb{R}$ be a continuously differentiable and positive definite function such that the derivative of V, i.e., $L_f V$, is negative semidefinite. Define the set: $E \triangleq \{x \in \mathcal{D}, (L_f V)(x) = 0\}$. Let $\subset E$ be a maximal invariant set in E. Then M is globally asymptotically stable.

Aforementioned stability definitions are mainly for the nonlinear autonomous systems. In the following, we will consider a kind of systems that may be represented by equations of the form [84],

$$\frac{dx}{dt} = f(x, t), \tag{2.32}$$

where $x \in \mathbb{R}^n$, $x = (x_1, \ldots, x_n)^T$, $F : \mathbb{R}^n \times J \to \mathbb{R}$ are considered, $J = [t_0, \infty)$, t_0 is an finite initial time instant. It is assumed that f is continuous on $\mathbb{R}^n \times J$, which denote their Cartesian product. The solutions of (2.32) are denoted by $x(t; x_0, t_0)$ with $x(t_0; x_0, t_0) = x_0$. In general it is not required that $f(0, t) = 0$.

Let $S(t) \in \mathbb{R}^n$ for all $t \in J$. Assume that $S(t)$ is a connected open region. Let $\bar{S}(t)$ denote the closure of $S(t)$ and let $\partial S(t)$ denote the boundary of $S(t)$. Assume that $S(t)$ is bounded for all $t \in J$, and $\lim_{t\to t_a} S(t)$ exists for all $t_a \in J$, and that $\lim_{t\to t_a} S(t) = S(t_a)$. Henceforth, whenever the symbol S(t) (with appropriate subscripts) is used to denote a set, it is assumed that this set possesses the properties described above. Let$[S(t) - S_0(t)] = \{x \in \mathbb{R}^n : x \in S(t), x \notin S_0(t)\}$, $B(a) = \{x \in \mathbb{R}^n : \|x\| < a\}$, $\partial B(a) = \{x \in \mathbb{R}^n : \|x\| \le a\}$. Let t_i be any point (initial time) in J, and let $x_i = x(t_i; x_i, t_i)$. Then the following definitions are presented in [84].

Definition 2.50 (*Uniformly stable*) System (2.32) is *stable* with respect to $\{S_0(t), S(t), t_0\}$, if $x_0 \in S_0(t_0)$ implies $x(t; x_0, t_0) \in S(t)$ for all $t \in J$. System (2.32) is uniformly stable with respect to $(S_0(t), S(t))$, if for all $t_i \in J$, $x_i \in S_0(t_i)$ implies that $x(t; x_i, t_i) \in S(t)$ for all $t \in [t_i, \infty)$.

Definition 2.51 (*Unstable*) System (2.32) is *unstable* with respect to $\{S_0(t), S(t), t_0\}$, $S_0(t_0) \subseteq S(t_0)$, if $x_0 \in S_0(t_0)$ and a $t_c \in J$ such that $x(t_c; x_0, t_0) \in \partial S(t_c)$.

Definition 2.52 (*Uniformly asymptotically stable*) System (2.32) is asymptotically stable with respect to $\{S_0(t), S(t), S_f, t_0\}$, if it is stability respect to $\{S_0(t), S(t), t_0\}$, and if in addition, $x_0 \in S_0(t_0)$ implies $x(t; x_0, t_0) \to S_f$ as $t \to \infty$ ($d(x(t; x_0, t_0), S_f) \to 0$ as $t \to \infty$, where $d(x, S_f) = \inf_{y\in S_f} \|y - x\|$). System (2.32) is *uniformly asymptotically stable* with respect to $\{S_0(t), S(t), S_f\}$, if it is uniformly stable with respect to $\{S_0(t), S(t)\}$, and if in addition, for all $t_i \in J$, $x_i \in S_0(t_i)$ implies $x(t; x_i, t_i) \to S_f$, as $t \to \infty$.

Definition 2.53 (*Practically stable*) If in Definition 2.50, if $S(t) \equiv B(\beta) = \{x \in \mathbb{R}^n : \|x\| < \beta\}$, $S_0(t) \equiv S_0(t_0) = B(\alpha)$, $\alpha \leq \beta$, then system (2.32) is said to be *practically stable* with respect to $\alpha, \beta, t_0, \|\cdot\|$, and uniformly practically stable with respect to $\alpha, \beta, \|\cdot\|$. If in Definition 2.52, $S(t) \equiv B(\beta)$, $S_0(t) \equiv S_0(t_0) = B(\alpha)$, $S_f = B(0)$, $\alpha \leq \beta$, then system (2.32) is said to be practically asymptotically stable with respect to $(\alpha, \beta, t_0, \|\cdot\|)$, and uniformly practically asymptotically stable with respect to $(\alpha, \beta, \|\cdot\|)$. Finally, system (2.32) is said to be *practically exponentially stable* with respect to $(\alpha, \beta, \gamma, t_0, \|\cdot\|)$, $\alpha \leq \beta, \gamma > 0$, if $\|x_0\| < \alpha$ implies $\|x(t; x_0, t_0)\| \leq \beta e^{-\gamma(t-t_0)}$ for all $t \in J$.

In the following, suppose $f : \mathbb{R} \times \mathbb{C} \to \mathbb{R}^n$ is continuous and consider the following retarded functional differential equation or delayed nonlinear system,

$$\dot{x} = f(t, x_t), \tag{2.33}$$

where $x_t(\theta) = x(t+\theta)$ for $\theta \in [-r, 0]$. The function f will be supposed to be completely continuous and to satisfy enough additional smoothness conditions to ensure the solution $x(\sigma, \phi)(t)$ through (σ, ϕ) is continuous in (σ, ϕ, t) in the domain of definition of the function. A function x is said to be a *solution* of Eq. (2.33) on $[\sigma - r, \sigma + q]$ if there are $\sigma \in \mathbb{R}$ and $q > 0$ such that $x \in \mathbb{C}([\sigma - r, \sigma + q], \mathbb{R}^n)$, $(t, x_t) \in \mathbb{D}$, and $x(t)$ satisfies equation (2.33) for $t \in [\sigma, \sigma + q]$, where $\mathbb{C} = \mathbb{C}([-r, 0], \mathbb{R}^n)$, which is the Banach space of linear functions mapping the interval $[-r, 0]$ into $\mathbb{R}^n$ with the topology of uniform convergence. For given $\sigma \in \mathbb{R}$, $\phi \in \mathbb{C}$, we say $x(\sigma, \phi, f)$ is a solution of Eq. (2.33) with initial value ϕ at σ or simply a solution through (σ, ϕ). Finding a solution of Eq. (2.33) through (σ, ϕ) is equivalent to solving the integral equation

$$\begin{aligned} x_\sigma &= \phi \\ x(t) &= \phi(0) + \int_\sigma^t f(s, x_s) \mathrm{d}s, \, t \geq \sigma. \end{aligned} \tag{2.34}$$

Definition 2.54 (*Uniformly asymptotically stable, see* [91]) Suppose $f(t, 0) = 0$ for all $t \in \mathbb{R}$. The solution $x = 0$ of system (2.33) is said to be *stable* if for any $\sigma \in \mathbb{R}$, $\epsilon > 0$, there is a $\delta = \delta(\epsilon, \sigma)$ such that $\phi \in \mathcal{B}(0, \delta)$ implies $x_t(\sigma, \phi) \in \mathcal{B}(0, \epsilon)$ for $t \geq \sigma$. The solution $x = 0$ of system (2.33) is said to be *asymptotically stable* if it is stable and there is a $b_0 = b_0(\sigma) > 0$ such that $\phi \in \mathcal{B}(0, b_0)$ implies $x(\sigma, \phi)(t) \to 0$ as $t \to \infty$. The solution $x = 0$ of system (2.33) is said to be *uniformly stable* if the number δ in the definition is independent of σ. The solution $x = 0$ of system (2.33) is said to be *uniformly asymptotically stable* if it is uniformly stable and there is a $b_0 > 0$ such that, for every $\eta > 0$, there is a $t_0(\eta)$ such that $\phi \in \mathcal{B}(0, b_0)$ implies $x_t(\sigma, \phi) \in \mathcal{B}(0, \eta)$ for $t \geq \sigma + t_0(\eta)$ for every $\sigma \in \mathbb{R}$.

If $y(t)$ is any solution of system (2.33), then y is said to be stable if the solution $z = 0$ of the equation

$$\dot{z}(t) = f(t, z_t + y_t) - f(t, y_t), \tag{2.35}$$

is stable.

Definition 2.55 (*Uniformly ultimately bounded, see* [91]) A solution $x(\sigma, \phi)$ of system (2.33) is *bounded* if there is a $\beta(\sigma, \phi)$ such that $|x(\sigma, \phi)(t)| < \beta(\sigma, \phi)$ for $t \geq \sigma - r$, $r > 0$ is the maximal bound of time delay. The solutions are *uniformly bounded* if, for any $\alpha > 0$, there is a $\beta = \beta(\alpha) > 0$ such that, for all $\sigma \in \mathbb{R}$, $\phi \in \mathbb{C}$, and $|\phi| \leq \alpha$, we have $|x(\sigma, \phi)(t)| \leq \beta(\alpha)$ for all $t \geq \sigma$. The solutions are *ultimately bounded* if there is a constant β such that, for any $(\sigma, \phi) \in \mathbb{R} \times \mathbb{C}$, there is a constant $t_0(\sigma, \phi)$ such that $|x(\sigma, \phi)(t)| < \beta$ for $t \geq \sigma + t_0(\sigma, \phi)$. The solutions are *uniformly ultimately bounded* if there is a $\beta > 0$ such that, for any $\alpha > 0$, there is a constant $t_0(\alpha) > 0$ such that $|x(\sigma, \phi)(t)| \leq \beta$ for $t \geq \sigma + t_0(\alpha)$ for all $\sigma \in \mathbb{R}$, $\phi \in \mathbb{C}$, $|\phi| \leq \alpha$.

If $V : \mathbb{R}^n \to \mathbb{R}^n$ is a given positive definite continuously differentiable function, the derivative of V along a solution of system (2.33) is given by

$$\dot{V}(x(t)) = \frac{\partial V(x(t))}{\partial x} f(t, x_t). \tag{2.36}$$

In order for $\dot{V}$ to be nonpositive for all initial data, one would be forced to impose very severe restrictions on the function $f(\phi)$. In fact, the point $\phi(0)$ must play a dominant role and therefore, the results will apply only to equations which are very similar to ordinary differential equations.

A few moments of reflection in this proper direction indicates the tit is unnecessary to require that Eq. (2.36) be nonpositive for all initial data in order to have stability. In fact, if a solution of the Eq. (2.33) begins in a ball and is to leave this ball at some time t, then $|x_t| = |x(t)|$, that is, $|x(t+s)| \leq |x(t)|$ for all $[s \in [-r, 0]]$. Consequently, one need only consider initial data satisfying the latter property. This is the basic idea of *Razumikhin-type stability theorem.*

Lemma 2.56 (Razumikhin uniform stability theorem, see [91]) *Suppose* $f : \mathbb{R} \times \mathbb{C} \to \mathbb{R}^n$ *takes* $\mathbb{R} \times$ bounded sets of $\mathbb{C}$ *into bounded sets of* $\mathbb{R}^n$ *and consider the system (2.33). Suppose* $u, v, w : \mathbb{R}^+ \to \mathbb{R}^+$ *are continuous, nondecreasing functions,* $u(s)$, $v(s)$ *positive for* $s > 0$, $u(0) = v(0) = 0$. *If there is a continuous function* $V : \mathbb{R} \times \mathbb{R}^n \to \mathbb{R}$ *such that*

$$\begin{aligned} u(|x|) \leq & V(t, x) \leq v(|x|), t \in \mathbb{R}, x \in \mathbb{R}^n, \\ \dot{V}(t, \phi(0)) \leq & - w(|\phi(0)|) \text{ if } V(t+s, \phi(s)) \leq V(t, \phi(0)), s \in [-r, 0], \end{aligned} \tag{2.37}$$

then the solution $x = 0$ *of system (2.33) is uniformly stable.*

Lemma 2.57 (Razumikhin uniform asymptotical stability theorem, see [91]) *Suppose all of the conditions of Lemma 2.56 are satisfied and in addition* $w(s) > 0$ *if* $s > 0$. *If there is a continuous nondecreasing function* $p(s) > s$ *for* $s > 0$ *such that*

$$\begin{aligned} u(|x|) \leq & V(t, x) \leq v(|x|), t \in \mathbb{R}, x \in \mathbb{R}^n, \\ \dot{V}(t, \phi(0)) \leq & - w(|\phi(0)|) \text{ if } V(t+s, \phi(s)) \leq p(V(t, \phi(0))), s \in [-r, 0], \end{aligned} \tag{2.38}$$

then the solution $x = 0$ of system (2.33) is uniformly asymptotically stable. If $u(s) \to \infty$ as $s \to \infty$, then the solution $x = 0$ is also a global attractor for the system (2.33).

Lemma 2.58 (Razumikhin uniformly ultimately bounded theorem, see [91]) *Suppose $f : \mathbb{R} \times \mathbb{C} \to \mathbb{R}^n$ takes $\mathbb{R} \times$ bounded sets of $\mathbb{C}$ into bounded sets of $\mathbb{R}^n$ and consider the system (2.33). Suppose $u, v, w : \mathbb{R}^+ \to \mathbb{R}^+$ are continuous nondecreasing functions, $u(s) \to \infty$ as $s \to \infty$. If there is a continuous nondecreasing function $V : \mathbb{R} \times \mathbb{R}^n \to \mathbb{R}$, a continuous nondecreasing function $p : \mathbb{R}^+ \to \mathbb{R}^+$, $p(s) > s$ for $s > 0$, and a constant $H \geq 0$ such that*

$$\begin{aligned} u(|x|) \leq & V(t, x) \leq v(|x|), t \in \mathbb{R}, x \in \mathbb{R}^n, \\ \dot{V}(t, \phi(0)) \leq & -w(|\phi(0)|) \\ & \text{if } |\phi(s)| \geq H, V(t+s, \phi(s)) \leq p(V(t, \phi(0))), s \in [-r, 0], \end{aligned} \tag{2.39}$$

then the solutions of system (2.33) are uniformly ultimately bounded.

With the coupling of neural networks, complex neural networks (CNN) have become the hot topic in the scientific community. For this kind of CNN, new dynamics such as synchronization have been proposed and studied. For this purpose, *synchronization stability* is necessary to be introduced.

Colloquially, synchronization means correlated in-time behavior between different processes [106, 107]. Indeed, the Oxford Advanced dictionary, 12 defines synchronization as "to agree in time" and "to happen at the same time." From this intuitive definition we propose that synchronization requires the following four tasks [108, 109]: (1) Separating the dynamics of a large dynamical system into the dynamics of subsystems. (2) Measuring properties of the subsystems. (3) Comparing properties of the subsystems. (4) Determining whether the properties agree in time. If the properties agree then the systems are synchronized.

Consider the following dynamical system:

$$\begin{cases} \dot{x} = F_1(x, y, t), \\ \dot{y} = F_2(x, y, t), \end{cases} \tag{2.40}$$

where $x \in X \subset \mathbb{R}^{d_1}$, $y \in Y \subset \mathbb{R}^{d_2}$, and $t \in \mathbb{R}$. The space of all trajectories is defined by $Z = X \times Y$, and the global trajectory is denoted by $\Phi(z)$, $z = (x, y) \in \mathbb{R}^{d_1} otimes \mathbb{R}^{d_2}$, d_1 and d_2 are positive integers, respectively. The trajectory properties of each subsystem are defined by the functionals: $g_x : X \times \mathbb{R} \to \mathbb{R}^k$, $g_y : Y \times \mathbb{R} \to \mathbb{R}^k$. An example of functional commonly used is $g_x = x(t)$. The comparison of the functionals is made by the function $h : \mathbb{R}^k \times \mathbb{R}^k \to \mathbb{R}^k$, that is called comparison function. By using these functionals two synchronization definitions can be given.

Definition 2.59 (*Synchronization-1*) [108, 109]: The subsystems of equation (2.40) are synchronized on the trajectory $\Phi(z)$, with respect to the properties g_x, g_y and synchronization norm $\|\cdot\|_s$ if there is a time independent mapping h such that $\|h(g_x, g_y)\|_s = 0$.

Definition 2.60 (*Synchronization-2*) [108, 109]: The subsystems of equation (2.40) are synchronized with respect to the properties g_x, g_y and synchronization norm $\|\cdot\|_s$ if there is a time independent mapping h such that $\|h(g_x, g_y)\|_s = 0$.

The synchronization definition 2.60 is what many papers in the literature call synchronization. However, synchronization depends strongly on the trajectory. Two subsystems can be synchronized on some trajectories and not synchronized on other trajectories. Therefore, the trajectory dependence in synchronization definition 2.59 cannot be ignored.

The stability of synchronous motion is another issues raised by these two definitions. Specifically, stability is not required by the synchronization definition 2.59. This definition only requires that $\|h\| = 0$ exists for properties measured on the trajectory. If the trajectory of one of the sub-systems is perturbed then this condition may no longer hold. As a trivial example of this, consider two uncoupled identical Lorenz systems with parameter values that produce chaotic trajectories. If both systems have the same initial condition then they will follow the same trajectory and their motion will clearly be synchronous. However, because of chaos, this type of synchronization is very unstable. In contrast, the second synchronization definition 2.60 implicitly requires the notion of stability because it requires $\|h\| = 0$ for all trajectories.

A strength of the definition is that the properties and comparison functions are not specified, a priori. Due to different synchronous target, different applications require different properties and comparison functions. Those that are suitable for one application are often completely unsuitable for another. This also implies that synchronization stability is a relative stability with respect to specified or virtual target.

The major point about this definition is the appropriated choice of the functionals g_x, g_y and the synchronization norm. The appropriated choice of the functional and synchronization norm depends upon the required type of synchronization.

In the majority of chaotic applications the required synchronization is the one referred as identical. In this case the functional are given by $g_x = x(t)$ and $g_y = y(t)$. The synchronization stability of identical subsystems coupled in a master-slave configuration exhibiting chaotic behavior is now presented, based on a criterion that assures based on a criterion that assures the stability of the synchronous motion under small perturbations.

Definition 2.61 (*Synchronization stability*): The subsystems of equation (2.40) are synchronous stable with respect to the properties $g_x = x(t), g_y = y(t)$ and synchronization norm $\|\cdot\|_s$ if there is a time independent mapping h such that $\|h(g_x, g_y)\|_s = \|x(t) - y(t)\| = 0$.

When $\|x(t) - y(t)\| = 0$, it implies $x(t) = y(t) = s(t)$, where $s(t)$ is the synchronous state, and $s(t)$ can be an equilibrium point, a limit cycle, an aperiodic orbit, or a chaotic orbit.

Based on above different definitions of stability, we can present a brief review on the evolution process of *stability theory*.

(1) In the original stability definition, it is about a fixed equilibrium point of system $\dot{x}(t) = f(x(t))$ under some restrictions on the nonlinear function. In the phase, asymptotical stability, exponential stability, attractivity, and so on have been studied. For the multiple fixed or discrete equilibrium point, global stability, complete stability, convergence, etc., have been developed. When the solution of $\dot{x}(t) = f(x(t))$ is periodic or chaotic, such concepts as invariant sets and limit sets have been proposed. In general, stability in the sense of Lyapunov is about the nonlinear system without input, or the system with zero input and nonzero initial states.

(2) With the development of nonlinear system theory, some external actions can affect the dynamics of the autonomous system $\dot{x}(t) = f(x(t))$ In this case, such systems as $\dot{x}(t) = f(x(t), u(t))$ have been studied. In this phase, the external control input has direct influence on the nonlinear system. In order to more precisely describe the qualitative behavior of nonlinear system, such concepts as ISS, iISS, OSS, IOS, SIIOS, passive, dissipative, and so on have been proposed, and the relations among global asymptotical stability and ISS, OSS have been discussed. These concepts have generalized the stability conception in the sense of Lyapunov.

(3) Besides the internal states and certain input, a system always suffers from the external disturbance. Therefore, a more accurate description of the concerned nonlinear system is $x(t) = f(x(t), u(t), w(t))$, where $w(t)$ can represent uncertain input such as disturbance, noise, fault, and so on. In this phase, generalized input $(u(t), w(t))$ can be used to evaluate the stability in the sense of *boundedness*. Meanwhile, in the aspect of control performances, such concepts as $\mathcal{L}_2$ performance and $\mathcal{H}_\infty$ performance have been proposed. All above three-phased, the concerned nonlinear systems are isolated or are not connected to other systems (e.g., node system with respect to complex system with couplings).

(4) For the complex systems with couplings, such as *complex networks* and multi-agent systems, such concepts as *synchronization* used consensus have been proposed. Synchronization and consensus stresses the identical behavior among all the node dynamics. Even though the complex network is synchronous, the global dynamics of this complex networks can not be stable. This is a significant difference between synchronization and stability in the sense of Lyapunov (or called *stability of fixed point*). The *reference frame* of stability of fixed point is itself, or the origin of the system without any input. Stability of fixed point is the internal dynamical behavior of a system, while synchronization is the external dynamical behavior among different systems. The reference frame of synchronization can be the dynamics of any node systems or other external reference target. Therefore, synchronization is the upgrade of stability in the sense of Lyapunov. The essence of stability of fixed point is a *relative stability* with the inherent equilibrium of the concerned system. For the system $\dot{x}(t) = f(x(t))$, its inherent equilibrium is naturally the origin. This reflects that $f(0) = 0$ is the fundamental assumption on the nonlinear function. Or in another way, one can say that $\dot{x}(t) = f(x(t))$ with $f(0) = 0$ is just an *error system* required in the research of stability of fixed point, state estimation, system identification, regulation, tracking, and synchronization. Therefore, where there is the research of relative motion such as tracking and regulation, where there is the Lyapunov stability theory.

(5) The systems concerned in [84] is an interconnected large-scale systems, in which $f(0,t) = 0$ is not required. However, there is no any explanation on this features. Although some similar ideas to the present book may contain, to the best of the authors' knowledge, it is not explicitly noted in his series of studies on the stability of dynamical systems. Except the work in [84], above stability definitions are about the *total variables* of the systems $\dot{x}(t) = f(x(t))$ or $x(t) = f(x(t), u(t))$. That is to say, Lyapunov stability theory, synchronization, ISS, and its variants are all about the total variables or the whole systems. The global dynamics of the whole system ultimately have the same characteristics, for example, asymptotic stability, or synchronization, or chaos. These *qualitative concepts* are relative to the whole systems. In fact, there may coexist many different dynamics in a system, for example, periodic solution and stable equilibrium point. This phenomenon relates to the concept of *partial stability*. Partial stability is an important variants and complements to the original Lyapunov and Lagrange stability concepts. For a given motion of a dynamical system, say $x(t, x_0, t_0) = (y(t, x_0, t_0), z(t, x_0, t_0))$, partial stability concerns the qualitative behavior of the y-component of the motion, relative to disturbances in the entire initial vector $x(t_0, x_0, t_0) = x_0 = (y_0, z_0)$, or relative to disturbances in the initial component y_0. In the former case one speaks simply of y-stability, while in the latter case, one speaks more explicitly of y-stability under arbitrary z-perturbations [111–114]. We note in passing that problems concerning partial stability of dynamical systems are closely related to problems of stability with respect to two measures [115].

Thus, stability is *relative* to some reference frame. If the reference frame is the total variable, then the stability result is global; while if the reference frame is partial variables, then the stability result is partial.

2.7 Summary

This chapter is mainly concerned with the preliminaries of the dynamical systems and stability theory, since neural network is also a special kind of dynamical systems. For the stability analysis of dynamical systems, the famous Lyapunov stability concept is emphasized here. From the background of the different stability definitions, we can find that all the theory researches must conform to the contemporary requirement including the industry and information technology. Practical production demands need the scientific innovation and technical revolution. Therefore, by understanding the history of different stability concepts and analysis methods, one may have a deep insight into the research of science and technology, and arousing the study interests and enthusiasm of researchers. Note that, some contents about the research background and partial stability theory and definitions of this chapter are from the Wikipedia—the free encyclopedia on the internet, while some original comments for the development of stability theory and stability concepts are presented by the authors.

References

1. G. Birkhoff, Proof of poincare's geometric theorem. Trans. Amer. Math. Soc. **14**, 14–22 (1913)
2. A.M. Lyapunov, The general problem of the stability of motion (in Russian), Doctoral Dissertation, University of Kharkov, 1892 (English translation version: (1) *Stability of Motion*, (Academic Press, New York, 1966). (2) *The General Problem of the Stability of Motion* (Taylor and Francis, London, 1992))
3. A.M. Letov, *Stability of Nonlinear Control Systems* (Russian), Moscow 1955(English translation, Princeton, 1961)
4. R.E. Kalman, J.F. Bertram, Control system analysis and design via the second method of lyapunov: I-continuous-time systems. J. Fluids Eng. **82**(2), 371–394 (1960)
5. J.P. LaSalle, S. Lefschetz, *Stability by Lyapunov's Second Method with Applications* (Academic Press, New York, 1961)
6. P.C. Parks, Lyapunov's method in automatic control theory. Control **5**(3), 102–105 (1962)
7. R.E. Kalman, Lyapunov functions for the problem of Lurie in automatic control. Proc. Natl Acad. Sci. USA **49**(2), 201–207 (1963)
8. M.J. Smith, M.B. Wisten, A continuous day-to-day traffic assignment model and the existence of a continuous dynamic user equilibrium. Ann. Oper. Res. **60**(1), 59–79 (1995)
9. B. Liu, W. Lu, T. Chen, New criterion of asymptotic stability for delay systems with time-varying structures and delays. Neural Netw. **54**, 103–111 (2014)
10. S. Sastry, *Nonlinear System, Analysis, Stability and Control* (Springer, New York, 1999)
11. X. Liao, L. Wang, P. Yu, *Stability of Dynamical Systems* (Elsevier, The Netherlands, 2007)
12. C. Robinson, *Dynamical Systems: Stability, Symbolic Dynamics, and Chaos* (CRC Press, Boca Raton, 1998)
13. A.N. Michel, L. Hou, D. Liu, *Stability of Dynamical Systems: Continuous, Discontinuous, and Discrete Systems* (Birkhauser, Boston, 2008)
14. A.A. Andronov, L.S. Pontryagin, Systemes grossiers. Dokl. Akad. Nauk SSSR **14**(5), 247–250 (1937)
15. F. Diacu, P. Holmes, *Celestial Encounters* (Princeton University Press, Princeton, 1996)
16. G. Birkhoff, *Dynamical Systems* (Amer. Math. Soc., Providence, 1927)
17. S. Smale, Differentiable dynamical systems. Bull. Am. Math. Soc. **73**, 747–817 (1967)
18. A.I. Lur'e, V.N. Postnikov, On the theory of stability of control systems. Prikladnaya Matematika Mehkhanika **8**, 246–248 (1944)
19. H.K. Khalil, *Nonlinear Systems* (Prentice Hall, Upper Saddle River, 1996)
20. V. Kolmanovskii, A. Myshkis, *Applied Theory of Functional Differential Equations* (Kluwer Academic Press, Boston, 1992)
21. K. Gu, V.L. Kharitonov, J. Chen, *Stability of Time-Delay Systems* (Birkhauser, Boston, 2003)
22. Q. Han, Y. Tian, D. Han, Robust absolute stability criteria for a class of uncertain Lur'e systems of neutral type, in *Proceedings of the 6th World Congress on Intelligent Control and Automation*, Dalian, China, June 21–23, pp. 2339–2343 (2006)
23. Z. Zeng, J. Wang, Complete stability of cellular neural networks with time-varying delays. IEEE Trans. Circuits Syst.-I: Regul. Pap. **53**(4), 944–955 (2006)
24. M. Forti, Some extensions of a new method to analyze complete stability of neural networks. IEEE Trans. Neural Netw. **13**(5), 1230–1238 (2002)
25. N. Takahashi, L.O. Chua, On the complete stability of nonsymmetric cellular neural networks. IEEE Trans. Circuits Syst.-I: Fundam. Theory Appl. **45**(7), 754–758 (1998)
26. E.D. Sontag, Input to state stability: basic concepts and results. Lect. Notes Math. **1932**, 163–220 (2008)
27. Y. Lin, E.D. Sontag, Y. Wang, A smooth converse Lyapunov theorem for robust stability. SIAM J. Control Optim. **34**(1), 124–160 (1996)
28. Y. Lin, E. Sontag, Y. Wang, Various results concerning set input-to-state stability, in *Proceedings of the 34th IEEE Conference on Decision and Control*, vol. 2, pp. 1330–1335 (1995)

29. E.D. Sontag, Y. Lin, Stabilization with respect to noncompact sets: Lyapunov characterizations and effect of bounded inputs, in *Nonlinear Control Systems Design 1992*, IFAC Symposia Series, pp. 43–49 (1992)
30. E.D. Sontag, Smooth stabilization implies coprime factorization. IEEE Trans. Autom. Control **34**(4), 435–443 (1989)
31. E.D. Sontag, Y. Wang, On characterizations of input-to-state stability with respect to compact sets, in *Proceedings of IFAC Non-Linear Control Systems Design Symposium (NOLCOS'95)*, Tahoe City, CA, pp. 226–231 (1995)
32. E.D. Sontag, Smooth stabilization implies coprime factorization. IEEE Trans. Autom. Control **34**, 435–443 (1989)
33. A. Isidori, *Nonlinear Control Systems II* (Springer, London, 1999)
34. Z.-P. Jiang, L. Praly, Preliminary results about robust Lagrange stability in adaptive nonlinear regulation. Int. J. Adapt. Control Signal Process. **6**, 285–307 (1992)
35. Z.-P. Jiang, A. Teel, L. Praly, Small-gain theorem for ISS systems and applications. Math. Control Signals Syst. **7**, 95–120 (1994)
36. H.K. Khalil, *Nonlinear Systems*, 2nd edn. (Prentice-Hall, Upper Saddle River, 1996)
37. M. Krichman, E.D. Sontag, A version of a converse Lyapunov theorem for input output to state stability, in *Proceedings of IEEE Conference Decision and Control*, Tampa, Dec. 1998, IEEE Publications, pp. 4121–4126 (1998)
38. M. Krstic, H. Deng, *Stabilization of Uncertain Nonlinear Systems* (Springer, London, 1998)
39. M. Krstic, I. Kanellakopoulos, P.V. Kokotovic, *Nonlinear and Adaptive Control Design* (Wiley, New York, 1995)
40. D. Liberzon, E. Sontag, Y. Wang, On integral input-to-state stabilization, in *Proceedings of American Control Conference*, June, San Diego, pp. 1598–1602 (1999)
41. M. Arcak, A. Teel, Input-to-state stability for a class of Lurie systems. Automatica **38**, 1945–1949 (2002)
42. H. Edwards, Y. Lin, Y. Wang, On input-to-state stability for time varying nonlinear systems, in *Proceedings of 39th IEEE Conference Decision and Control*, Sydney, Australia, pp. 3501–3506 (2000)
43. N.C.S. Fah, Input-to-state stability with respect to measurement disturbances for one-dimensional systems. Control Optim. Calc. Var. **4**, 99–121 (1999)
44. L. Grune, Input-to-state dynamical stability and its Lyapunov function characterization. IEEE Trans. Autom. Control **47**, 1499–1504 (2002)
45. Z.-P. Jiang, Y. Wang, Input-to-state stability for discrete-time nonlinear systems. Automatica **37**, 857–869 (2001)
46. B. Ingalls, E.D. Sontag, A small-gain lemma with applications to input/output systems, incremental stability, detectability, and interconnections. J. Frankl. Inst. **339**, 211–229 (2002)
47. J.P. Hespanha, D. Liberzon, A.S. Morse, Supervision of integral-input-to-state stabilizing controllers. Automatica **38**, 1327–1335 (2002)
48. D. Angeli, Intrinsic robustness of global asymptotic stability. Syst. Control Lett. **38**, 297–307 (1999)
49. D. Angeli, E.D. Sontag, Y. Wang, A characterization of integral input to state stability. IEEE Trans. Autom. Control **45**, 1082–1097 (2000)
50. D. Angeli, An almost global notion of input-to-state stability. IEEE Trans. Autom. Control **49**, 866–874 (2004)
51. D. Angeli, B. Ingalls, E.D. Sontag, Y. Wang, Separation principles for input-output and integral-input to state stability. SIAM J. Control Optim. **43**, 256–276 (2004)
52. D. Angeli, B. Ingalls, E.D. Sontag, Y. Wang, Uniform global asymptotic stability of differential inclusions. J. Dyn. Control Syst. **10**, 391–412 (2004)
53. D. Angeli, E.D. Sontag, Y. Wang, Input-to-state stability with respect to in-puts and their derivatives. Int. J. Robust Nonlinear Control **13**, 1035–1056 (2003)
54. D. Angeli, E.D. Sontag, Y. Wang, Further equivalences and semiglobal versions of integral input to state stability. Dyn. Control **10**, 127–149 (2000)

55. M. Arcak, D. Angeli, E.D. Sontag, A unifying integral ISS framework for stability of nonlinear cascades. SIAM J. Control Optim. **40**, 1888–1904 (2002)
56. R.A. Freeman, Global internal stabilizability does not imply global external stabilizability for small sensor disturbances. IEEE Trans. Autom. Control **40**, 2119–2122 (1996)
57. E.D. Sontag, Y. Wang, Output-to-state stability and detectability of nonlinear systems. Syst. Control Lett. **29**, 279–290 (1997)
58. E.D. Sontag, Some connections between stabilization and factorization, in *Proceedings of IEEE Conference Decision and Control*, Tampa, pp. 990–995, Dec. (1989)
59. E.D. Sontag, Y. Wang, Output-to-state stability and detectability of nonlinear systems. Syst. Control Lett. **29**, 279–290 (1997)
60. E.D. Sontag, Y. Wang, Notions of input to output stability. Syst. Control Lett. **38**, 235–248 (1999)
61. E.D. Sontag, Y. Wang, On characterizations of the input-to-state stability property. Syst. Control Lett. **24**, 351–359 (1995)
62. P.P. Varaiya, R. Liu, Bounded-input bounded-output stability of nonlinear time varying differential systems. SIAM J. Control **4**, 698–704 (1966)
63. E.D. Sontag, On the input-to-state stability property. Eur. J. Control **1**(1), 24–36 (1995)
64. L.T. Grujic, On practical stability. Int. J. Control **17**(4), 881–887 (1973)
65. A.R. Teel, J. Peuteman, D. Aeyels, Semi-global practical asymptotic stability and averaging. Syst. Control Lett. **37**, 329–334 (1999)
66. L. Moreau, D. Aeyels, Practical stability and stabilization. IEEE Trans. Autom. Control **45**(8), 1554–1558 (2000)
67. X. Xu, G. Zhai, Practical stability and stabilization of hybrid and switched systems. IEEE Trans. Autom. Control **50**(11), 1897–1903 (2005)
68. L. Weiss, E.F. Infante, Finite time stability under perturbing forces and on product spaces. IEEE Trans. Autom. Control **12**(1), 54–59 (1967)
69. S. Bhat, D. Bernstein, Finite-time stability of continuous autonomous systems. SIAM J. Control Optim. **38**, 751–766 (2000)
70. Y. Hong, Z.-P. Jiang, Finite-time stabilization of nonlinear systems with parametric and dynamic uncertainties. IEEE Trans. Autom. Control **49**, 1950–1956 (2006)
71. Y.G. Hong, Z.-P. Jiang, G. Feng, Finite-time input-to-state stability and applications to finite-time control design. SIAM J. Control Optim. **48**(7), 4395–4418 (2010)
72. W.S. Chen, L.C. Jiao, Finite-time stability theorem of stochastic nonlinear systems. Automatica **46**(12), 2105–2108 (2010)
73. Y.Y. Cao, Z. Lin, D.G. Ward, An antiwindup approach to enlarging domain of attraction for linear systems subject to actuator saturation. IEEE Trans. Autom. Control **47**(1), 140–145 (2002)
74. A. Isidori, *Nonlinear Control Systems: A Introduction* (Springer, Berlin, 1989)
75. E.J. Davison, E.M. Kurak, A computational method for determining quadratic Lyapunov functions for nonlinear systems. Automatica **7**, 627–636 (1971)
76. R. Genesio, A. Vicino, New techniques for constructing asymptotic stability regions for nonlinear systems. IEEE Trans. Circuits Syst. **31**, 574–581 (1984)
77. A. Tesi, F. Villoresi, R. Genesio, On the stability domain estimation via a quadratic Lyapunov function: convexity and optimality properties for polynomial systems. IEEE Trans. Autom. Control **41**(1), 1650–1657 (1996)
78. A. Vannelli, M. Vidyasagar, Maximal Lyapunov function and domain of attraction for autonomous nonlinear systems. Automatica **21**, 69–80 (1985)
79. K. Ichikawa, R. Ortega, On stabilization of nonlinear systems with enlarged domain of attraction. Automatica **28**, 623–626 (1992)
80. D. Liaw, C. Lee, An approach to estimate domain of attraction for nonlinear control systems, in *Proceedings of the First International Conference on Innovative Computing, Information and Control (ICICIC'06)*, pp. 146–149 (2006)
81. S. Mayer, B. Tibken, S. Warthenpfuhl, Parallel computation of domains of attraction for nonlinear dynamic systems, in *Proceedings of 2011 Sixth International Symposium on Parallel Computing in Electrical Engineering*, pp. 87–92 (2011)

82. D. Jin, J. Peng, A new approach for estimating the attraction domain for Hopfield-type neural networks. Neural Comput. **21**, 101–120 (2009)
83. X. Liao, Stability of Hopfield-type neural networks: I. Sci. China: Ser. A **38**(4), 407–418 (1995)
84. A.N. Michel, Simple and interconnected systems: stability, boundedness and trajectory behavior. IEEE Trans. Circuits Syst. **17**(3), 292–301 (1970)
85. E. Thelen, E. Bates, Connectionism and dynamic systems: are they really different? Dev. Sci. **6**(4), 378–391 (2003)
86. P.H. Miller, *Theories of Developmental Psychology*, 4th edn. (Worth Publishers, New York, 2002)
87. P. Sopasakis, The Realm of Linear State Space Control Systems, http://users.ntua.gr/chvng/en (2011)
88. P.-A. Bliman, LMI characterization of the strong delay-independent stability of linear delay systems via quadratic Lyapunov-Krasovskii functionals. Syst. Control Lett. **43**, 263–274 (2001)
89. P.-A. Bliman, Lyapunov-Krasovskii method and strong delay-independent stability of linear delay systems, in *Proceedings of the Second IFAC Workshop on Linear Time Delay Systems*, Ancona, Italy, pp. 5–9 (2000)
90. N.N. Krasovskii, *Stability of Motion* (Stanford University Press, Stanford, 1963)
91. J.K. Hale, *Theory of Functional Differential Equations* (Springer, New York, 1977)
92. E.I. Verriest, A.F. Ivanov, Robust stabilization of systems with delayed feedback, in *Proceedings of the Second International Symposium on Implicit and Robust Systems*, Warszaw, Poland, pp. 190–193 (1991)
93. Q. Zhou, P. Shi, S. Xu, H. Li, Observer-based adaptive neural network control for nonlinear stochastic systems with time delay. IEEE Trans. Neural Netw. Learn. Syst. **24**(1), 71–80 (2013)
94. S. Boyd, L.E.I. Ghaoui, E. Feron, V. Balakrishnan, *Linear Matrix Inequality in System and Control Theory* (SIAM, Philadelphia, 1994)
95. W. Chen, W. Zheng, A new method for complete stability analysis of cellular neural networks with time delay. IEEE Trans. Neural Netw. **21**(7), 1126–1139 (2010)
96. W.B. Ford, *A brief account of the life and work of the late professor Ulisse Dini*. Bull. AM. Math. Soc. **26**(4), 173–177 (1920)
97. H. Qiao, J. Peng, Z. Xu, B. Zhang, A reference model approach to stability analysis of neural networks. IEEE Trans. Syst. Man Cybern.-Part B: Cybern. **33**(6), 925–936 (2003)
98. W. Zhou, J. Zurada, Discrete-time recurrent neural networks with complex-valued linear threshold neurons. IEEE Trans. Circuits Syst. II Exp. Br. **56**(8), 669–673 (2009)
99. J. Louisell, Absolute stability in linear delay-differential systems: ill-posedness and robustness. IEEE Trans. Autom. Control **40**(7), 1288–1291 (1995)
100. M.P. Joy, On the absolute stability of delayed cellular neural networks, in *1998 Fifth IEEE International Workshop on Cellular Neural Networks and their Applications*, London, England, 14–17, April, pp. 94–99 (1998)
101. R.Q. Lu, W.N. Zhou, H.Y. Su, J. Chu, A new absolute stability and stabilization conditions for a class of Lurie uncertain time-delay systems. in *IEEE International Conference on Systems, Man and Cybernetics*, vol. 2, pp. 1916–1921 (2003)
102. B. Zhou, Q. Song, Boundedness and complete stability of complex-valued neural networks with time delay. IEEE Trans. Neural Netw. Learn. Syst. **24**(8), 1227–1238 (2013)
103. M. Cohen, S. Grossberg, Absolute stability of global pattern formation and parallel memory storage by competitive neural networks, IEEE Trans. Syst. Man Cybern. **13**(5), 815–826 (1983)
104. L.O. Chua, L. Yang, Cellular neural networks: theory. IEEE Trans. Circuits Syst. **35**(10), 1257–1272 (1988)
105. B. Liu, D.J. Hill, Uniform stability and ISS of discrete-time impulsive hybrid systems. Nonlinear Anal. Hybrid Syst. **4**, 319–333 (2010)

106. H. Fujisaka, T. Yamada, Stability theory of synchronized motion in coupled oscillator systems. Progress Theoret Phys, **1**, 32–47 (1983)
107. L.O. Chua, M. Itoh, Chaos synchronization in Chua's circuit. J. Circuit Syst. Comp. **1**, 93–108 (1993)
108. R. Brown, A unifying definition of synchronization for dynamical systems, Chaos, **10**(2), 344–349 (2000)
109. U.F. Moreno, P.L. Peres, I.S. Bonatti, Synchronization stability analysis of chaotic oscillators, in *Proceedings of the IEEE Conference on Decision and Control*, vol. 5, pp. 4833–4838 (2000)
110. E.D. Sontag, Smooth stabilization implies coprime factorization. IEEE Trans. Autom. Control **34**, 435–443 (1989)
111. Y. Sun, A.P. Molchanov, A.N. Michel, Partial stability of dynamical systems, in *Proceedings of 15th International Symposium Mathematical Theory Networks Systems*, University of Notre Dame (2002)
112. A.N. Michel, A.P. Molchanov, Y. Sun, Partial stability and boundedness of discontinuous dynamical systems. Nonlinear Stud. **9**(3), 225–247 (2002)
113. A.N. Michel, A.P. Molchanov, Y. Sun, Partial stability and boundedness of general dynamical systems on metric spaces. Nonlinear Anal. TMA **52**(4), 1295–1316 (2003)
114. V.I. Vorotnikov, *Partial Stability and Control* (Birkhauser, Boston, 1998)
115. V. Lakshmikantham, X. Liu, *Stability Analysis in Terms of Two Measures* (World Scientific, Singapore, 1993)

Chapter 3
Survey of Dynamics of Cohen–Grossberg-Type RNNs

In Chaps. 1 and 2, we have introduced the history of artificial neural networks and the concepts of dynamical systems and stability, respectively, which are related to the research background of complex neural networks and the basis of qualitative stability analysis in the mathematical meaning. In this chapter, we will present a comprehensive review on the stability research of Cohen–Grossberg neural networks (a special kind of RNNs) in recent years. By analyzing the internal structures and external actions of RNNs, some elements that can be explored to study the stability of RNNs have been pointed out, which will provide some new insights and new alternatives to reduce the conservativeness of the stability criteria. The contents in this chapter are mainly from the result in [39], which forms the main fundamentals of complex neural networks in the whole book.

3.1 Introduction

The neural approach for solving optimization problems has attracted considerable attention in recent years, see [1–9] and references therein. However, some crucial drawbacks have seriously limited its applicability. One main drawback is that spurious suboptimal responses, due to the existence of many stable equilibrium points, are likely to be present [1, 4, 5, 9]. The main features that a *neural optimizer* of the Hopfield type (or the original Hopfield additive neural networks) should possess are as follows.

(1) The interconnection matrix should be symmetric. The property of symmetry is indeed inherently related to the optimization capabilities [1, 2, 7].

(2) There should be a unique equilibrium point which is *globally asymptotically stable* (GAS), i.e., locally stable and attracting all trajectories of motion. As discussed in [3–5], GAS is a necessary property to avoid the presence of spurious responses and to guarantee convergence towards the global optimal solution.

Z. Wang et al., *Qualitative Analysis and Control of Complex Neural Networks with Delays*, Studies in Systems, Decision and Control 34,
DOI 10.1007/978-3-662-47484-6_3

(3) The neural network should be *absolutely stable* (ABST). In accordance to [10, 11], by ABST of a neural network it means that there is a GAS equilibrium point for every neuron activation function belonging to the class $\mathcal{S}$ of sigmoidal (i.e., bounded increasing) functions and for every constant input vector to the neural network. In general, ABST is a relative concept to the total elements set of neural networks, for example, the entire set is composed of external input class, activation function class, and the time delay class. Occasionally, one may also define ABST with respect to some element. For example, the networks are ABST with respect to any forms of activation function belonging to a specified class, while other elements of networks are known and deterministic. ABST is important since in practical problems the neuron activation is known to belong to the class $\mathcal{S}$, but its shape is not specified exactly. A typical case is that of neural networks in the high-gain limit [1, 7], where GAS must hold for every sufficiently large value of the activation gain. Also, for neural networks running in real time, the input data are fed by using biasing input currents that are held constant in each given sampling interval and then vary at a clock rate in many applications [1]. Then, for each constant input vector there should be a unique GAS equilibrium. ABST neural networks are best suited for optimization problems, being devoid of spurious responses for every choice of the activation function and of the input vector.

Research on recurrently connected neural networks is an important topic in neural network theory. Among them, the *Cohen–Grossberg neural networks* were first proposed in the pioneer work of Cohen–Grossberg in 1983 [17], which includes the famous Hopfield neural networks as its special cases. Since Cohen–Grossberg neural networks, Hopfield neural networks and other recurrent neural networks have their promising potential for the tasks of classification, associative memory, parallel computations, and their ability to solve difficult optimization problems, they have greatly attracted the attentions of the scientific community. The success of most of these networks' applications relies heavily on understanding the underlying dynamic behavior of these networks. A thorough analysis of this dynamics is a necessary step toward a practical design of the recurrent neural networks. One of the most investigated problems is that the existence, uniqueness, and global asymptotic/exponential stability of the equilibrium point. The number of *equilibrium points* of the neural networks relate to their *storage capacity*. When designing an associative memory neural networks, we should make as many stable equilibrium states as possible to provide a memory system with large information capability, an attractive region of each stable equilibrium state as large as possible to provide the robustness and fault tolerance for information processing, and a convergence speed as high as possible to ensure the quick convergence of the neural network operation. Due to the properties of locally asymptotic stability, the associative memory network is used mainly for information retrieval, pattern recognition, and so on. On the other hand, to embed and solve many problems in applications of neural networks to parallel computations, signal processing, and other problems involving the optimization, the dynamic neural networks have to be designed to have only one equilibrium point which is global asymptotic stability (that is, the whole real space or the interested domain as its *domain of attraction*) to avoid the risk of spurious equilibrium point or the problem of local minima. In fact, the earlier applications of recurrent neural networks to optimization problems have suffered from the existence of a complicated set of equilibrium point. Thus, the global asymptotic/exponential stability of a unique equilibrium point for RNNs is of great importance from theoretical and applicable points of view.

Research on the dynamic behaviors of RNNs can date back to the early days of neural network science. For example, multistable and oscillatory behaviors were studied in [12–14] and chaotic behaviors are investigated in [15]. References [1, 7, 16] looked into the dynamic stability of symmetrically connected networks and showed their practical applicability to optimization problems. It should also be noted that Cohen and Grossberg [17] presented more rigorous analytical results on the global stability of the recurrent neural networks. In the early times, global stability of symmetrically connected networks has been widely studied, and by now most of the results have been well established [1, 4, 7, 16–18]. Because stability of symmetric RNNs usually ensures the local stability, and symmetry is too much ideal, local and global stability of asymmetrically connected neural networks begin to be widely studied since 1989 (Note that asymmetry is the source power of evolution). For example, some sufficient conditions for the local exponential stability and for the existence and the uniqueness of an equilibrium point have been obtained in [5, 19, 20]. However, they did not address the issue of global stability of the networks (Note that, global stability may have different definition. Here, it means the global stability of the unique equilibrium point). In practice, the topic of global stability is of more importance than that of local stability, and some sufficient conditions for global stability have been given in [21]. Reference [22] applies contraction mapping theory to obtain some sufficient conditions for global stability. Reference [23] generalizes some results in [21, 22] by use of a new Lyapunov function. Reference [24] proves that the diagonal stability of the interconnection matrix implies the existence and uniqueness of an equilibrium point and the global stability at the equilibrium point. References [4, 25, 26] propose three main features a neural network should possess. Among them, the important one is that the neural networks should be absolutely stable. They also point out that the negative semi-definiteness of the interconnection matrix guarantees the global stability of the Hopfield networks and prove the absolute global stability for a class of activation functions. Reference [27] applies the matrix measure theory to get some sufficient conditions for global and local stability. References [28, 29] discuss the stability of delayed neural networks by using Lyapunov functions and Lyapunov diagonal stability condition on interconnection matrix. References [31, 32] introduce a new method to address the global stability of Hopfield neural networks. Reference [33] proves the exponential stability under less restrictive conditions and also obtains an estimate of the accurate rate of convergence.

The analysis of dynamical behavior of Cohen–Grossberg neural networks was first discussed by Cohen and Grossberg in 1983 [17] and later studied further by Grossberg in 1988 [34], in which some global limit property with symmetric connection matrix was given. Since 2000, researches on Cohen–Grossberg neural networks with asymmetric connection matrices have become the main focus in the neural network community. Reference [35] presents some sufficient conditions for exponential stability of Cohen–Grossberg neural networks with asymmetric connection matrices. However, the signs of entries in the connection matrices are neglected in the conditions of theorems given in [35], and as a result the differences between the excitatory and inhibitory effects have been ignored. In addition, the activation functions are restricted to be bounded functions in [35]. Reference [36] gives a detailed description on the development history of Cohen–Grossberg neural networks, and some global asymptotic stability criteria have been established without requiring the boundedness on the activation function and the positive lower boundedness on the amplification function. For the case of bounded amplification function, some exponential stability criteria have been established. All the stability results in [36] are in the forms of Lyapunov diagonal stability and M-matrix, respectively.

Time delays are ubiquitous both in biological and artificial neural networks due to the processing of information [37–39]. For example, they may arise because of the finite propagation speed of signals along the axons in nervous systems or the finite switching speed of amplifier in neural circuits. Therefore, the study of effects of time delays on stability and convergent dynamics of neural networks has attracted considerable attention in recent years. Under certain symmetric connectivity assumptions, it is found that a neural network with time delays will be stable when the delays do not exceed certain (usually small) bounds [18, 40, 41]. For asymmetric neural networks with delays, sufficient stability conditions independent of or depending on the magnitude of delays are established [42–45]. These results are mostly based on *linearization analysis* and energy or Lyapunov functional method (Note that energy function is generally not the Lyapunov function. Energy function has a energy-like expression form, and it is a continuous and differentiable function. The foundation of energy function-based method is the LaSalle invariant set principle, while Lyapunov function method is on the basis of Lyapunov stability theory). Recently, almost all the stability results are for RNNs with delays, no matter discrete delay and distributed delays, and many different analysis methods are proposed and applied, and more and more elegant results have been reported and published.

Before stating the detailed review, we first give some explanations on some concepts, which will be used in this chapter. Regarding on the other conceptions, they are all formal and standard, which can be found in the literature or classical textbooks [39].

(1) On the conception of inhibitory and excitatory action. In fact, inhibitory and excitatory action corresponds to the competitive-cooperative connectivity of a network. By *competitive connection* it means the way in which a neuron's firing inhibits the firing of other neurons. Conversely, *cooperative connection* means the way in which a neuron's firing excites the firing of others. The competitive-cooperative connection pattern can thus be recognized by the sign of the weights: positive weights are due to excitatory coupling, negative weights are due to inhibitory coupling, while a zero weight indicates no interaction at all.

(2) Stability result in algebraic inequality form. Nowadays, there are many different expression forms of stability results, for example, in the form of M-matrix, H-matrix, matrix measure or norm, Lyapunov diagonal stability, LMI, matrix inequality, $\mathcal{P}$ or $\mathcal{P}_l$ Class, and so on. For a kind of stability criteria with scalar inequality form, which involves suitable parameter to be tuned, we call them as stability results

in algebraic inequality form. These kinds of stability results have such features: (a) Expressed in scalar inequality form and involving some parameters to be determined, (b) Derived mostly by algebraic inequalities, for example, Young inequality, Holder inequality and Minkowski inequality, and (c) Difficultly unified in a vector or matrix form due to different parameters being involved.

(3) Stability result in like-M-matrix form. These kind of stability results lie between M-matrix and algebraic inequality form. For some specific case, stability result in like-M-matrix form can be converted to stability result in M-matrix form. For general case, it is expressed in the algebraic inequality form. Stability result in algebraic inequality form generally cannot be expressed in any compact form.

3.2 Main Research Directions of Stability of RNNs

In a recurrent neural network, it is mainly composed of such components as self-feedback connection weights, activation functions, interconnection weights associated with activation functions, delays, delay interconnection weights associated with activation functions with state delays, and amplification function in Cohen–Grossberg-type neural networks. In a class of RNNs with Cohen–Grossberg type, which are mainly used in the optimization problem solving, associative memory, and pattern recognition, the self-feedback connection weight is always negative, and plays a stabilization role in the whole networks. For the interconnection weight and delay interconnection weight, they can be positive, negative, and zero, which correspond to the excitatory, inhibitory, and no action on each other. For the activation functions associated with state or delayed state, they may be the same or different in the whole networks. For time delay, it can be constant or time-varying, single or multiple delays, neutral, discrete and distributed delays, and so on. In order to improve the stability performance, one can stress any component in the RNNs, which will lead to better stability results. Along the aforementioned line, in this section, we will give detailed reviews on the development of stability theory of RNNs.

3.2.1 Development of Neuronal Activation Functions

Many existing results on the existence, uniqueness, and global asymptotic or exponential stability of the equilibrium point of RNNs concern the case that the activation functions are continuous, bounded, and strictly monotonically increasing. These assumptions make the results inapplicable to some important engineering problems. For example, when RNNs are designed for solving optimization problems in the presence of constraints (linear, quadratic, or more general programming problems), unbounded activation functions modeled by diode-like exponential-type functions are needed to impose the constraints. Because of the difference between the bounded activation function and unbounded activation function, which will affect the dynamics of RNNs, the extensions of stability results with bounded activation function to the unbounded case is not straightforward. Different from the bounded activation function where the existence of an equilibrium point is always guaranteed, for the case of *unbounded activation function*, it may happen that there are no equilibrium points [25, 26]. When considering the widely used piecewise linear neural networks (which is composed of the piecewise linear activation function) [3, 46], infinite interval with zero slope are present in the activation functions, making it of interest to drop the assumption of strictly monotonically increasing and continuous first-order derivative for the activation functions. Reference [26] studies a class of Hopfield neural network with unbounded monotonic activation function. References [49, 50] have shown that the absolute capacity of an associative memory model can be remarkably improved by replacing the usual Sigmoid activation functions with nonmonotonic activation functions. Therefore, it seems that for some purposes, nonmonotonic (and not necessarily smooth) functions might be better candidates for neuronal activation functions in designing RNNs. In many electronic circuits, amplifiers, which have neither monotonically increasing nor continuously differentiable input-output functions, are frequently adopted. For example, Ref. [1] designs a linear programming Hopfield network with piecewise linear (nonsmooth) activation function, and Ref. [44] studies the global attractivity of delayed Hopfield neural network models with bounded and nonmonotonic activation functions. Therefore, for different application fields, both RNNs with bounded and unbounded activation functions have been studied sufficiently. An apparent tendency is toward the unbounded activation function in theory. In practical applications, one can take specific kind of activation function to solve the corresponding problems, either bounded or unbounded activation functions.

The following *activation functions* have been used in the existing references.

(1) In the original papers [1, 7, 16, 18, 47], the activation function is a sigmoidal function such that

$$g^{'}(u_j) = \mathrm{d}g_j(u_j)/\mathrm{d}u_j > 0, \lim_{\zeta_i \to +\infty} g_i(\zeta_i) = 1,$$

$$\lim_{\zeta_i \to -\infty} g_i(\zeta_i) = -1, \lim_{|\zeta_i| \to \infty} g_i^{'}(\zeta_i) = 0, i, j = 1, \ldots, n. \tag{3.1}$$

Obviously, the activation function is continuous, differential, smooth, monotonic, and bounded. Also, a general Sigmoid function $g_i(u_i)$ is defined by the properties $|g_i(u_i)| \le M$ and $\frac{\mathrm{d}g_i(u_i)}{\mathrm{d}u_i} > 0$, where M is a constant [46, 48].

(2) Piecewise linear (PWL) functions [3, 46, 51–57]. A special case of PWL function has the following form,

$$g_i(s) = \frac{|s+1| - |s-1|}{2}. \tag{3.2}$$

A PWL multilevel neuron activation can be found in [57]. The use of PWL approximates in neural-network modeling features a number of advantages, among which we mention the capability to exactly locate the neural-network equilibrium points. This in turn has permitted to derive effective techniques to design neural networks for solving specific signal processing tasks, such as image processing, the implementation of content-addressable memories, and the solution of combinatorial optimization problems. More generally, PWL modeling and analysis have proven extremely useful to study circuits containing nonlinear resistors, also due to the fact that the theory of PWL functions is well established and their universal approximation properties of continuous functions are well understood.

(3) The following activation function has been widely used in the existing literature [58–63],

$$|g_i(\zeta) - g_i(\xi)| \le \delta_i|\zeta - \xi|, \tag{3.3}$$

no matter whether the activation function is bounded or not. As pointed out in [58], the type of activation functions in (3.3) is not necessarily monotonic and smooth.

(4) The following activation function has been used in the existing literatures [64–67],

$$0 < \frac{g_i(\zeta) - g_i(\xi)}{\zeta - \xi} \le \delta_i. \tag{3.4}$$

(5) The following activation function has been widely used in the existing literatures [60, 62, 68–71],

$$0 \le \frac{g_i(\zeta) - g_i(\xi)}{\zeta - \xi} \le \delta_i. \tag{3.5}$$

(6) The following activation function is developed in recent year [72–79],

$$\delta_i^- \le \frac{g_i(\zeta) - g_i(\xi)}{\zeta - \xi} \le \delta_i^+, \tag{3.6}$$

As pointed out in [75–78], δ_i^- and δ_i^+ may be positive, negative, or zero. Then, those previously used Lipschitz conditions (3.1), (3.4), and (3.5) are just the special cases of the condition (3.6). Note that for the well-known Lur'e system, the nonlinear function $g_i(s(t))$ is memoryless and possibly time-varying, which is piecewise continuous in t and globally Lipschitz in $s(t)$ and satisfies the sector condition

$$(g(u(t)) - K_1u(t))^T(g(u(t)) - K_2u(t)) \le 0,$$

where K_1 and K_2 are constant real matrices of appropriate dimensions and $K = K_1 - K_2$ is a symmetric positive definite matrix, $g(u(t)) = (g_1(u_1(t)), \ldots, g_n(u_n(t)))^T$. It is customary that such a nonlinear function $g(u)$ is said to belong to a sector $[K1, K2]$. Obviously, activation function (3.6) has the features of sector condition.

The above-listed activation functions belong to the class of continuous function. For more details on the relations of global Lipschitz continuous, partially Lipschitz continuous, and locally Lipschitz continuous, readers can refer to [80, 81]. In fact, some discontinuous activation functions also exist in practical applications. For example, in the classical Hopfield neural networks with graded response neurons [7], the standard assumption is that the activations are employed in the high-gain limit where they closely approximate a discontinuous hard comparator function. Another important example concerns the class of neural networks introduced by Kennedy and Chua [3] to solve linear and nonlinear programming problems, in which the constraint neurons are with a diode-like input-output activations. In order to guarantee the satisfaction of constraints, the diodes are required to possess a very high slope in the conducting region, i.e., they should approximate the discontinuous characteristic of an ideal diode. Therefore, the following activation function is for the discontinuous case.

(7) Discontinuous activation function [82–88]. Let $g_i(\cdot)$ be a continuous, nondecreasing function and in every compact set of $\mathcal{R}$, each $g_i(\cdot)$ has only finite discontinuous points. Therefore, in any compact set in $\mathcal{R}$, except some finite points ρ_k, there exist finite right and left limit $g_i(\rho^+)$ and $g_i(\rho_-)$ with $g_i(\rho^+) > g_i(\rho_-)$. In general, one assumes that $g_i(\cdot)$ is bounded, i.e., there exist a positive number $G > 0$, such that $g_i(\cdot) \leq G$. On the other hand, stability analysis of neural networks with *discontinuous activation functions* has drawn many scholars' attention and many related results have been published in the literature since the independent pioneering work of Forti and Nistri in [82, 83] and Lu and Chen in [84] (Note that, Ref. [84] was submitted in Oct. 2003, while the paper [82] was not published. Moreover, the model, method used and the activation functions (unbounded) in [84] are different from (bounded activation functions) those in [82]).

Therefore, activation functions have evolved from bounded cases to unbounded cases, from continuous to discontinuous, and from strictly monotonic case to nonmonotonic case. All these show the depth of stability research on the essence of RNNs.

3.2.2 Evolution of Uncertainties in Interconnection Matrix

For the deterministic and accurate connection weight matrix, lots of stability results have been published since 1980s. However, with the application and implementation of RNNs, the connection weight matrix can be disturbed or perturbed in the external environment. Therefore, robustness or ill-posedness of RNNs against such perturbations should be considered.

There are generally several kinds of expression forms of *uncertainty* in the literature.

(1) Uncertainty with match condition

$$\Delta A = MF(t)N,\ F^T(t)F(t) \leq I, \tag{3.7}$$

or

$$\Delta A = MF_0(t)N,\ F_0(t) = (I - F(t)J)^{-1}F(t),\ F^T(t)F(t) \leq I, \tag{3.8}$$

where M, N, J are all constant matrices, $J^T J \leq I$. This kind of uncertainty is very convenient in the stability proof based on LMI method. Robust stability for neural network with matched uncertainty (3.7) has been studied in [89].

(2) Interval uncertainty

$$A \in A_I = [\underline{A}, \overline{A}] = \{[a_{ij}] : \underline{a}_{ij} \leq a_{ij} \leq \overline{a}_{ij}\}. \tag{3.9}$$

This kind of uncertainty is usually used in the non-LMI method. However, if let $A_0 = (\overline{A}+\underline{A})/2$, $\Delta A = (\overline{A}-\underline{A})/2$, then interval uncertainty (3.9) can be expressed as the following form,

$$A_J = \{A = A_0 + \Delta A = A_0 + M_A F_A N_A | F_A^T F_A \leq I\}, \tag{3.10}$$

where M_A, N_A, F_A are well defined according to some arrangement of elements in $\underline{A}$ and $\overline{A}$. Obviously, interval uncertainty (3.9) has been changed into the form of uncertainty with match condition (3.7). It has been shown in [62, 90–92] that interval uncertainty (3.9) is equivalent to the uncertainty (3.10). Therefore, both interval uncertainty and matched uncertainty can be uniformly dealt with by LMI method. Robust stability for neural networks with interval uncertainty (3.9) has been studied in [62, 91, 93].

(3) Absolute value uncertainty or unmatched uncertainty

$$\Delta A = (\delta a_{ij}) \in \{|\delta a_{ij}| \leq \overline{a}_{ij}\}. \tag{3.11}$$

(4) Polytopic-type uncertainty:

$$A \in \Omega, \Omega = \left\{ A(\xi) = \sum_{k=1}^{p} \xi_k A_k, \sum_{k=1}^{p} \xi_k = 1, \xi_k \geq 0 \right\}, \tag{3.12}$$

where A_k are constant matrices with compatible dimensions, ξ_k are time-invariant uncertainties. Robust stability for system with this kind of uncertainty has been studied in [94–96].

Above uncertainties are for the case of constant and real-value interconnection matrices. Recently, some control system ideas have been incorporated into the stability research of RNNs. For example, similar to complex-value system and time-varying system, the interconnection matrix can be complex-value, time-varying, switching, jumping, or state-dependent [97–99]. In these cases, partial concepts of aforementioned uncertainty can be suitable, for example, the boundedness of uncertainty. However, the specific expression of uncertainty in these cases will be different from above uncertainty expressions (3.7)–(3.12). How to redefine the uncertainty in these complex cases is not an easy work.

3.2.3 Evolution of Time Delays

Time delays involved in RNNs have evolved from the single constant delay τ to the multiple delays τ_i and τ_{ij}. Corresponding, the time-varying cases evolves from single delay $\tau(t)$ to the multiple delays $\tau_i(t)$ and $\tau_{ij}(t)$, $i, j = 1, \ldots, n$. Note that all the delays such as τ, τ_i, τ_{ij}, $\tau_i(t)$, and $\tau_{ij}(t)$ are called discrete delays since they only have actions on the neural networks at some isolated or discrete point with respect to continuously distributed delays. For the continuously distributed delay $\int_{t-\tau(t)}^{t} u(s)\mathrm{d}s$, $\int_{t-\tau(t)}^{t} K_{ij}(s)u(t-s)\mathrm{d}s$, and $\int_{-\infty}^{t} K_{ij}(s)u(t-s)\mathrm{d}s$, or $\int_{0}^{\infty} K_{ij}(t-s)u(s)\mathrm{d}s$, where $K_{ij}(s)$ is a kernel function satisfying some constraint conditions. These kinds of delays may keep a short- or long-time interval. Therefore, the past history or information of system states may have direct influence on the present states [100, 101].

For the case of time-varying delay, the derivative of time-varying delay $\tau(t)$ (or $\tau_i(t)$ and $\tau_{ij}(t)$) has been relaxed from $\dot{\tau}(t) < 1$ (or $\dot{\tau}_i(t) < 1$ and $\dot{\tau}_{ij}(t) < 1$) to $\dot{\tau}(t) \geq 1$ (or $\dot{\tau}_i(t) \geq 1$ and $\dot{\tau}_{ij}(t) \geq 1$). That is, due to the contribution of free weight matrix method, delay partitioning method, and other methods, no matter slow time-varying delays or fast time-varying delays ($\dot{\tau}(t) \geq 1$) in RNNs can be dealt with easily at present.

It is common that *time-varying delay* $\tau(t)$ belongs to an interval $0 \leq \tau(t) \leq \overline{\tau}$ in the previous literature (date back to the end of 2006). Since 2007, the time-varying delay interval is expanded from $0 \leq \tau(t) \leq \overline{\tau}$ to $0 \leq \underline{\tau} \leq \tau(t) \leq \overline{\tau}$ [102]. The meaning of this expansion lies in the fact that the lower bound of time-varying delay in a practical system cannot be zero. On the other hand, the upper bound of time delay to be estimated in a real delayed system can be approximated to the real value if the nonzero lower bound of time delay is used. In addition, neutral-type delay is involved in RNNs due the implementation of electronic circuits in a real system.

Many other kinds of delays used in the control system and biological system have also been involved in RNNs, for example, stochastic time delay is introduced into the neural networks by [74, 79]. In the existing references for delayed RNNs, only the deterministic time delay is concerned, and the criteria are derived based only on the information of variation range of the time delay. Actually, the time-varying delay in RNNs often exists in a random fashion, and its probabilistic characteristic, such as Poisson distribution or normal distribution, can often be obtained by statistical methods. It often occurs in the real systems that some values of the delay are very large but the probabilities of the delay taking such large values are very small. In this case, if only the variation range of time delay is employed to derive the criteria, the results may be somewhat more conservative. Therefore, a challenging issue is how to derive some criteria for the uncertain stochastic RNNs which can exploit the available probability distribution of the delay and obtain a larger allowable variation range of the delay. As stated in [74], however, few results have been reported in the literature.

With the changes of different time delays, the proof procedure and the expression form of stability results are different, which will promote the development of neural network theory and the related topics in other disciplines.

3.2.4 Relations Between Equilibrium and Activation Functions

In general, there are two kinds of proof methods to the existence and uniqueness of the equilibrium point of RNNs in the literature. However, there is no clear explanation about the relations between these two proof methods yet. This problem is a fundamental topic in neural network stability theory, and may lead to misunderstanding for the readers. Based on this purpose, in this subsection, we will try our best to present an explanation on this problem.

For the bounded *activation function* $|g_i(x_i)| \leq M$ or the like-unbounded activation function $|g_i(x_i)| \leq \delta_i^0 |x_i| + \sigma_i^0$, where $M > 0$, $\delta_i^0 \geq 0$, and $\sigma_i^0 \geq 0$ are constants, the existence of equilibrium point is established mainly on the basis of *Brouwer's fixed point theorem*, Schauder fixed point theorem and *contraction mapping principle* [25, 26, 47, 103, 104]. Comparison principle, theory of monotone flow and monotone operator are also used to ensure the existence and uniqueness of the periodic solution of a class of RNNs [105].

In general, for the case of bounded/unbounded activation functions satisfying Lipschitz continuous conditions, the existence of the solution can be guaranteed by the existence theorem of ordinary differential equation [106], which is consistent with the results in [23, 25, 26, 36, 47, 103, 107, 108].

For the unbounded activation function (in general sense, no specific form of the activation function is given), the existence of equilibrium point in RNNs is established mainly on the basis of homeomorphism mapping [26, 108, 109], topological degree theory [110, 111], Leray–Schauder principle [112] and so on.

The existence of equilibrium point cannot guarantee the uniqueness of the fixed point. In general, there are two ways to deal with the uniqueness. One way is that the uniqueness of the equilibrium point can be followed directly by the global asymptotic/exponential stability of the equilibrium point. The other way to derive the sufficient conditions guaranteeing the uniqueness of equilibrium point is to use contraction mapping, homeomorphism mapping, contradiction method and comparison principle [103, 110, 111, 113].

Aforementioned statements have already concerned the following questions: Must the existence, uniqueness, and global asymptotic/exponential stability be simultaneously done in the proof of the stability of neural networks?

Obviously, for the case of bounded activation function, we can directly present the proof on the global asymptotic/exponential stability as it is well known that the bounded activation function always guarantees the existence of the equilibrium point. For the like-unbounded case, the existence of the equilibrium point is also guaranteed as the same as that bounded activation function. Therefore, it suffices to present the global asymptotic/exponential stability proof on the equilibrium point, and the uniqueness is directly followed from the global stability [114].

For the case of unbounded activation function (in general sense), one must provide the proof on the existence, uniqueness, and global asymptotic/exponential stability of the concerned neural networks, respectively. Otherwise, it is not rigorous in mathematical viewpoints. However, in the general case of bounded/unbounded activation functions satisfying Lipschitz continuous conditions, the existence of the solution can also be guaranteed by the existence theorem of ordinary differential equation [106]. On the other hand, the stability conditions guaranteeing the global asymptotic/exponential stability is always sufficient (at least at present), these stability conditions are also sufficient for the existence and uniqueness. In other word, an existence condition is usually a necessary condition for the global asymptotic/exponential stability. Therefore, for activation functions satisfying Lipschitz continuous conditions (no matter it is bounded or not), global asymptotic/exponential stability can be directly proved. For activation functions satisfying non-Lipschitz continuous conditions, the existence, uniqueness, and global asymptotic/exponential stability must be done in the stability proof of RNNs. The first step is to prove the existence, then the uniqueness and global asymptotical/exponential stability of the equilibrium point.

The above-introduced methods present two ways to prove the global stability of the *unique equilibrium*, which can derive the same conclusions with different orders. Different from above proof method, [31, 115] provide an effective approach, called direct method, to prove the stability, which prove the exponential convergence directly, and that the existence and uniqueness of the equilibrium point is a direct consequence of the exponential convergence. Furthermore, the boundedness of the activation function is not required and the derivatives of the state variables converge to zero exponentially in [31, 115]. The main idea of direct approach proposed in [31, 115] is differentiating the differentiable system with respect to time t directly, and obtaining a system with the derivatives of state variable. To the authors' knowledge, Ref. [31] is the first paper providing such a unified approach in the stability research of RNNs.

Note that in the existing literature, stability proof procedure for recurrent neural networks usually consists of two steps: First, prove the existence and uniqueness of the equilibrium point. Second, prove the stability of the equilibrium point.

3.2.5 Different Construction Methods of Lyapunov Functions

Generally, there are two concepts concerning the stability of systems with time delays. The first one is called as the delay-independent stability criteria which do not include any information about the size of the time delay and the change rate of time-varying delays [108, 112, 116–118]. For the systems with unknown delays, delay-independent stability criteria will play an important role in checking the stability problems. The second is called as the delay-dependent stability criteria, in which the sizc of the time delay is mainly taken explicitly in the formulation [102, 119]. Strictly speaking, if the information on the change rate of time-varying delays is involved in the stability conditions, this kind of stability result should be categorized into the delay-dependent one. That is to say, stability criteria associated with any information on time delays are all called delay-dependent criteria. As the information on the size of delays and the magnitude of change rate of time-varying delays are used in delay-dependent criteria, delay-dependent criteria are often be less conservative than delay-independent ones, especially when the size of time delay is small. Notice that most of the cited approaches for the stability analysis of RNNs with time delay are on the basis of Lyapunov–Krasovskii functional method.

It should be pointed out that, in the past few years, *linear matrix inequalities* (LMIs) have gained much attention for their computational tractability and usefulness in system engineering [120] because the so-called interior point method [121] has been proved to be numerically very efficient for solving the LMIs. The number of analysis and design problems that can be formulated as LMI problems is large and continues to grow.

LMI Control Toolbox implements the state-of-the-art of interior point LMI solvers. While these solvers are significantly faster than classical convex optimization algorithms, it should be kept in mind that the complexity of LMI computations remains higher than that of solving, say, a Riccati equation. For instance, problems with a thousand design variables typically take over an hour on today's workstations [122]. However, research on LMI optimization is a very active area in the applied math, optimization and the operation research community, and substantial speed-ups can be expected in the future.

In [123], the approach applied to *delay-dependent* stability analysis of RNNs with time delays is done by Razumikin-type techniques based on construction of a Lyapunov function. It is known, however, that there exists no general rule to guide how a proper Lyapunov function can be constructed for a given neural network. So, the construction of Lyapunov function frequently becomes very skillful, which makes the use of this approach somehow conservative.

In the early day of neural network stability theory (before 1990), most stability studies stem from the viewpoint of building the algebraic relations among the physical parameters in a neural network. Therefore, the stability criteria based on matrix measure, matrix norm, and M-matrix have been developed.

Since the physical parameters in a neural network may have some nonlinear relations or some constraint relations by some free variables, which will affect the stability criteria crucially, stability criteria based on algebraic inequality, e.g., Young inequality, Holder inequality, Poicare inequality, Hardy inequality, have become the hot topic. Although the stability criteria based on algebraic inequality method can be less conservative in theory, they are generally difficult to check due to more adjusting variables being involved while we have no prior information to know how to tune these variables. A compromise or balance between computational complexity and effectiveness should be a possible way to find some alternative methods to establish stability criteria. Since LMI is regarded as a powerful tool to deal with matrix operation, LMI-based stability criteria have received the attentions of researchers all over the world. LMI is a vector form to build the relation of the physical parameters in a neural network. Therefore, they have compact structure and elegant symmetry, which build a cube architecture to connect the physical parameters in a neural network. In this cube framework, different transformations and changes lead to the different stability criteria. Therefore, LMI-based stability analysis and synthesis method in RNNs has become one of main streams sine 2002.

In the stability analysis of RNNs based on Lyapunov theory, there are two keys to be stressed. One is how to construct effective *Lyapunov function*, the other is how to efficiently estimate the derivative of the Lyapunov function. The former is always concerned with the system thinking, the latter is often related to mathematical computational methods, especially the inequality techniques. If these two keys are harmony in the stability analysis, some effective results will be obtained. For the inequality methods, readers can refer to any textbook about inequality, e.g., [124].

In the subsection, we will first introduce some effective matrix analysis methods in the estimation of derivatives of Lyapunov function. These matrix analysis methods are not the matrix inequality methods, but they are useful to analyze the qualitative characteristics of RNNs. Then, we will introduce some decomposition methods, which will be helpful to constructing different Lyapunov functions.

The following three matrix analysis methods can be used in many given Lyapunov function, which will provide sufficient information to improve the effectiveness of the stability results.

(1) Free weight matrix method: This is a very powerful technique to deal with the case of fast time-varying delay, i.e., $\dot{\tau}(t) \geq 1$. Before the emergence of *free weight matrix method*, all the LMI-based stability results can only deal with the case of slow time-varying delay, i.e., $\dot{\tau}(t) < 1$. The assumption that $\dot{\tau}(t) < 1$ stems from the need to bound the growth variations in the delay factor as a time function. It may be considered restrictive but in some applications it is considered realistic and holds for a wide class of retarded functional differential equations. In most cases, the constraint condition $\dot{\tau}(t) < 1$ is from the requirement of mathematical analysis method instead of engineering application.

The essence of free weight matrix method is to add free variables/matrices in an identical equation, which will not effect the identical equation. For example, the following identical equation holds according to the Newton–Leibniz formula,

$$x(t) - x(t - \tau(t)) - \int_{t-\tau(t)}^{t} \dot{x}(t)\mathrm{d}t = 0, \tag{3.13}$$

or the following nonlinear system,

$$\dot{x}(t) - Ax(t) - Bx(t - \tau(t)) - Cf(x(t)) = 0. \tag{3.14}$$

If multiplying $x^T(t)Q$ or $x^T(t)Q + g^T(x(t))P$, $-x^T(t)Q$ or $x^T(t)Q - g^T(x(t))P$ on both sides of (3.13) and (3.14), respectively, the identical equations still hold. That is to say, element 0 is substituted by some implicit relations among the system parameters and redundant variables Q and P. In this case, we call Q and P the free weight matrix. Therefore, we can utilize the combination of $x(t)$ and $f(x(t))$, especially the combinations $x^T(t)Qx(t)$ or $-x^T(t)Qx(t)$, which will compensate the effects of fast/slow time-varying delays on the systems.

(2) Nondelay-matrix decomposition method: For the case of Hopfield and Cohen–Grossberg neural network without delay, some *matrix decomposition methods* are used [125–129]. In [125], the nondelayed connection matrix W satisfying $W^T W = WW^T$ is decomposed into the summation of its symmetric and skew-symmetric parts, $W = W_s + W_{ss}$, where $W_s = W_s^T$ is the symmetric part and $W_{ss} = -W_{ss}^T$ is the skew-symmetric part of W, respectively. Then based on matrix eigenvalue method, a necessary and sufficient condition is presented to guarantee the absolute stability of the concerned Hopfield neural network. Similarly, dropping out the restriction of $W^T W = WW^T$, a more general matrix decomposition is used in [126]. That is, for any matrix W, it can always be written as summation of its symmetric and skew-symmetric parts, i.e., $W = W^s + W^{ss}$, where $W^s = (W + W^T)/2$ and $W^{ss} = (W - W^T)/2$ are the symmetric and the skew-symmetric parts of W, respectively. Then based on the matrix eigenvalue method, a new necessary and sufficient condition is presented to guarantee the absolute stability of the concerned Hopfield neural network, which improves the result in [125]. This method is suitable for the results based on matrix eigenvalue method. Meanwhile, this method is mainly used in the Hopfield neural network without delay. In [127], the nondelayed matrix W is decomposed into the summation of n matrices W_i, where the ith column is composed by the ith column of W, and the other columns are all zeros. Similarly, in [128], the nondelayed matrix W is decomposed into the summation of n matrices W_i, where the ith row is composed by the ith row of W, and the other columns are all zeros. In general, the method used in [127, 128] will improve the stability result based on Lyapunov diagonal stability (LDS) as stated in [127, 128], while it will be conservative than the stability results based on LMI. For the Cohen–Grossberg neural network, the nondelayed matrix W is decomposed as the product of a symmetric matrix and the positive definite diagonal matrix, i.e., $W = DS$, where D is a positive definite diagonal matrix, and S is a symmetric matrix [129]. In general, $DS \neq SD$, therefore, the stability condition in [129] relaxed the condition in [17].

(3) Interval matrix partitioning method: This method is devoted to the robust stability analysis for neural networks with interval uncertainty, i.e., uncertainty connection matrix $A \in [\underline{A}, \overline{A}]$. Similar to delay partitioning method, the interval $A \in [\underline{A}, \overline{A}]$ is divided by $\tilde{A} = \frac{\overline{A} - \underline{A}}{m}$ or $\tilde{a}_{ij} = \frac{\overline{a}_{ij} - \underline{a}_{ij}}{m}$ where m is an integer greater than or equal to 2, or the split interval may be unequal. Then based on the LMI method, large sets of matrix inequalities need to be checked simultaneously. This *interval matrix partitioning/splitting method* has been proposed in [130].

Next, the following decomposition methods can provide four different ways to construct Lyapunov function. Along this way, it can present us some certifications to show the idea on how to construct the Lyapunov function.

(1) Delay-matrix partitioning/decomposition method: Since LMI is a very powerful tool to analyze the stability of many kinds of neural networks with different delays, it is natural to build some LMI-based stability criteria for the neural systems with different multiple delays $\tau_{ij}(t)$. For the case of $\tau(t)$ and $\tau_j(t)$, many LMI-based stability results have been established. In the case of τ_{ij}, *delay-matrix decomposition method* is proposed to specially deal with the term $x(t-\tau_{ij}(t))$ or $g(x(t-\tau_{ij}(t)))$ [116, 131–134]. For the neural systems with continuously distributed delay $\int_0^\infty K_{ij}(s)\mathrm{d}s$, the delay-matrix decomposition method is still valid [135–137]. The expression form of stability results based on delay-matrix decomposition method is a natural generalization of the expression form of those stability results for the case of $\tau(t)$ and $\tau_j(t)$. Delay-matrix decomposition method is a special method to analyze the system with multiple delays $\tau_{ij}(t)$ or distributed delays $\int_0^\infty K_{ij}(s)\mathrm{d}s$. It is the contribution of delay-matrix decomposition method that unifies many LMI-based stability results for neural networks with different kinds of delays into one framework. In such a framework, all the stability results can be compared and changed for neural networks with different delays. Note that in [68], a matrix decomposition method is proposed for a kind of pure delay neural networks. The delayed matrix B is decomposed into two parts: excitatory and inhibitory parts, i.e., $B = B^+ - B^-$, where $b_{ij}^+ = \max\{b_{ij}, 0\}$ signifying the excitatory weights and $b_{ij}^- = \max\{-b_{ij}, 0\}$ signifying the inhibitory weights. Obviously, the elements in $B^+ = (b_{ij}^+)$ and $B^- = (b_{ij}^-)$ are all nonnegative. Then, through a symmetric transformation, the network is embedded into an augmented cooperative dynamical system. Using the monotone dynamical system theory, such a system has a significant order-preserving or monotone property that is useful in the analysis of pure delay neural networks. Via this method, some detailed componentwise exponential convergence estimates have been established in the form of like-M-matrix. Obviously, the delay-matrix decomposition method in both [116, 131–137] and [68] is different in such aspects as analysis method, decomposition purpose, and the expression forms of stability results.

(2) Descriptor system method: This is a universal transformation method, which can transform an ordinary differential system into a like-descriptor system and use the analysis concept of descriptor system to study the normal differential system. Therefore, the number of vector dimension of the original differential system is enlarged from n-dimensions to $2n$-dimensions. With the augment of the dimensions, the number of adjustable matrices in the construction of Lyapunov functional will be increased. It is the essence of the descriptor system method that by increasing the state space and correspondingly the number of tuning matrices to decrease the conservativeness of the stability results. *Descriptor system method* can be used in many delayed systems in which the Newton–Leibniz formula holds.

(3) Delay partitioning method with fixed interval: Time delay is a very important parameter in RNNs with delays. Because interconnection weight coefficients or matrices have been sufficiently explored in the development of neural network stability theory, especially with the emergence of *free wight matrix method*, it seems that the stability criteria have reached the limit that no more space in connection weights can be used to further decrease the conservativeness of the stability results. Another important physical parameter time delay, which has always been ignored in the proof procedure of previous stability analysis methods, has been re-recognized and begins to be utilized in recent years, mainly in the aspect of *delay-dependent* stability results. For the case of discrete time delay, how to achieve the maximum upper bound of time delay has become one of important aspects in judging the conservativeness of stability criteria. In general, the larger the maximum value of time delay, the less conservative of the stability results. It is well known that time delay belongs to an fixed interval $[0, \tau_M]$, i.e., $0 \leq \tau(t) \leq \tau_M$. In the previous stability analysis methods, $\tau(t)$ is simply restricted to the interval $[0, \tau_M]$. However, according to the sample theory or the approximate theory, if the whole interval $[0, \tau_M]$ is divided into m subinterval and the subinterval distance or the sample frequency in the interval $[0, \tau_M]$ is suitable, then a new delay-dependent stability criteria can be obtained, which may decrease the conservativeness of the stability results. The subinterval distance or the sample frequency in the interval $[0, \tau_M]$ may be fixed or variable. This is the principle of *delay partitioning method*. The essence of delay partitioning method is to enlarge/augment the state space and involve too many adjustable variables, which has larger state space or system dimensions than that in descriptor system method. Meanwhile, delay partitioning approach can shorten the delay interval artificially, which may improve the integral accuracy of variables associated with time delay in the Lyapunov function. This characteristic is very similar with the approximation characteristic of neural networks in modeling and identifying. Delay partitioning method can be used to the system with discrete delays $\tau(t)$, $\tau_j(t)$, or $\tau_{ij}(t)$. For the case of $\tau_j(t)$ or $\tau_{ij}(t), i, j = 1, \ldots, n$, the expression form of stability results will be more complex. A challenging topic existing in the delay partitioning method is how to determine the number of subinterval and the subinterval distance to achieve the optimal upper bound value of time delay.

(4) Delay interval contraction method: *Delay partitioning approach* studied recently aims to divide a fixed delay interval $[\tau_m, \tau_M]$ into several subintervals $[\tau_{j-1}, \tau_j], j = 1, \ldots, l, (\tau_m = \tau_0 < \tau_1 \cdots < \tau_l = \tau_M)$, [138–140] by involving some adjusting parameters, which realize the contraction of subinterval. The delay partitioning approach has evolved from the equal interval division to nonequal interval division [141, 142] for the fixed delay interval. In theory, we can also achieve the same target as the delay partitioning method with fixed interval does to change the delay interval $[\tau_m, \tau_M]$ in another way by involving some adjusting parameters. For example, define $\tau_m(t) = \alpha\tau(t) + (1-\alpha)\tau_m, \tau_M(t) = \alpha\tau(t) + (1-\alpha)\tau_M$, where $\alpha \in [0, 1]$. In the case of $\alpha = 0$, the delay interval $[\tau_m(t), \tau_M(t)]$ is equivalent to $[\tau_m, \tau_M]$, where $\alpha = 1$, it is just the discrete delay $\tau(t)$. As a counterpart of delay partitioning method with fixed interval, this kind of method is regarded as *delay interval contraction method* or *dynamic delay interval method*. By continuously changing the size of α, the delay interval $[\tau_m(t), \tau_M(t)]$ varies from discrete delay $\tau(t)$ to the maximal delay interval $[\tau_m, \tau_M]$ dynamically, which realize the contraction of the whole delay interval. Similar idea behind this method has been shown in [142].

The principle of delay interval contraction method is on the basis of convex combination technique, which is similar to the essence of delay partitioning method with fixed interval, but the way to achieve is different.

After all, all the derived stability criteria based on aforementioned method are sufficient. How to get the sufficient and necessary stability condition for RNNs with delays still needs some suitable combinations of above methods and other effective methods.

3.2.6 Expression Forms of Stability Criteria

At present, there are many different methods to show the stability property of RNNs, such as Lyapunov stability theory [36, 143, 144], nonsmooth analysis [145–147], ordinary differential equation theory [111, 148–150], LaSalle invariant set theory [17, 129], nonlinear measure method [151], gradient-like system theory [9, 57, 152, 153], comparison principle of delay differential systems [45, 154], and so on. Correspondingly, the obtained stability criteria can be expressed in different forms due to different proof methods and using different stability theories, such as M-matrix form, algebraic inequality form, matrix norm form, matrix measure form, linear matrix inequality form, and the mixed form of the above forms. For example, linear matrix inequality [59–62, 69, 70, 72, 73, 75–79, 89, 102, 117, 131, 132, 134, 155], M-matrix results [36, 63, 67, 109, 112, 149, 150], Algebraic inequality results [64–66, 103, 111, 113, 118, 143, 144, 148, 149, 156–161], Matrix norm [18, 60, 93], additive diagonal stability [162–164], Lyapunov diagonal stability [24, 26, 29, 36, 128, 165–167], Matrix measure [27, 146] (Note that, compared with the matrix norm, the matrix measure can not only have positive values, but also negative values, whereas the matrix norm can only have nonnegative values. Therefore, the results expressed in the form of matrix measure are more precise than the ones expressed by matrix norms), Spectral radius [168, 169], and like-M-matrix result [68].

Since the expression forms of M-matrix, matrix measure, matrix norm, and spectral radius are only associated with the system parameters, and not any freedom or free variables can be used in the criteria, these kinds of stability criteria play important roles in the early days of recurrent neural network stability theory and now are becoming less and less. Similarly, *Lyapunov diagonal stability* result is generally simple and no complex variant forms can be used, this kind of stability result is also becoming more and more scare. Algebraic inequality results and LMI results have many different expression forms and represent the nature of dynamical system in the aspects of number and vector in real space, therefore, many different kinds of stability results based on algebraic inequality and LMI have been proposed. Generally speaking, their expression forms are becoming more and more complex, and involving more and more tuning variables and tuning matrices. The development of stability results are accompanied with the development of mathematical tools and conceptual innovation. After all, how to find simple and effective stability criteria is still a challenging topic.

With globally Lipschitz continuous and monotone nondecreasing activation functions, Lyapunov diagonal stability (LDS) results is reported in [26]. The LDS result for the global exponential stability (GES) is extended in [185]. It is shown in [165, 166] that the LDS result extends many existing conditions in the literature, such as M-matrix characteristic, lower triangular structure, negative semi-definiteness, diagonal stability, diagonal semistability, and other kinds of sufficient conditions [128].

3.2.7 Domain of Attraction

The *domain of attraction* of each equilibrium point coincides with the region that the network, starting from any initial guess in it, will evolve to the optimal solution. Therefore, identifying the domain of attraction is important in the application of neural networks [170–176]. The approaches extensively used in the existing investigation into this field of neural networks are mainly based on Lyapunov direct method and so depend on the construction of Lyapunov function. However, there is no general rule guiding us to construct an optimal Lyapunov function for a given system; that is, constructing a Lyapunov function requires skill. Between 1985 and 1995, L. Gruyitch obtained many interesting results concerning the determination of attraction domain [177]. However, in those results, the Lyapunov functions used to characterize the attraction domain are constructed by the method of characteristics, which strongly depends on the solutions of system.

Important work about the determination of attraction domain has been done by S. Balint and his colleagues since 1985 [178]. In 1985, S. Balint developed a novel Lyapunov function (named the optimal Lyapunov function in 1986) and proved that the determination of attraction domain can be reduced to the determination of the analytic domain of the vector field when the field is R-analytic. In this case, the optimal Lyapunov function can be found by solving linear systems of algebraic equations whose solutions are the coefficients of the expansion of the optimal Lyapunov function at the equilibrium point. In 1986, the special case when the Jacobian matrix is diagonalizable at the equilibrium point was considered in [179], where a recurrence formula for finding the coefficients of the expansion of the optimal Lyapunov function had been established. When this formula was used, the optimal Lyapunov functions and attraction domains were found for several two-dimensional systems. The hypothesis that the Jacobian matrix is diagonalizable is used only for finding the recurrence formula; otherwise, it is possible to find the coefficients as developed in [178] by solving some linear algebraic equations but impossible to find a recurrence formula. In 1987, a method for approximating the attraction domain by the region of convergence of the Taylor series of the optimal Lyapunov function was established in [180] under the assumption that the Jacobian matrix is diagonalizable at the equilibrium point. The region of convergence of the Taylor series, obtained by the recurrence formula given in [179], has been shown to be part of attraction domain, and its boundary in some directions touches the boundary of attraction domain. Hence, by using the recurrence formula, it is possible to find the radius of the convergence in any direction so that the region of convergence of the Taylor's series can be found. However, the weakness of the approximation method is that part of the attraction domain may be lost when the domain is not symmetric with respect to the equilibrium points.

In 2003, a new approach for approximating the attraction domain by gradually extending the Lyapunov function (called the "embryo" of the optimal Lyapunov function) obtained in a neighborhood of the equilibrium point was presented in [181], which shows that it is possible to improve the approximation obtained by the region of convergence of the Taylor series of the optimal Lyapunov function at the equilibrium point by expanding the obtained "embryo" of the optimal Lyapunov function at a point apart from the equilibrium point but in the region of convergence of the Taylor series and near the boundary of the region. In 2005, another important step in approximation of the attraction domain was made in [182], in the case when the Jacobian matrix is diagonalizable at the equilibrium point. That article discusses the capacity of the Taylor polynomials of the optimal Lyapunov function to provide good approximations of the attraction domain for autonomous and R-analytical systems and presents some very good approximations of the attraction domain for some systems. In 2006 and 2009, the method in [182] was applied to evaluate the attraction domains of Hopfield-type neural networks [184] where some good results were achieved.

The approximation methods developed in [182] are good approaches to determining the attraction domain and also provide a way to construct Lyapunov functions by using a practical recurrence formula. In the approximation methods presented in [182], the Taylor polynomials of the optimal Lyapunov function can approximate the optimal Lyapunov function with better accuracy as the order p of the polynomial increases, and so better approximations of the attraction domain can be achieved with greater recurrence, though only in the region of convergence of the Taylor series. The meaning is twofold. On one hand, if the region of convergence of the Taylor series is the whole attraction domain, the approximation accuracy to the attraction domain increases with the order p. On the other hand, if the region of convergence is not the whole attraction domain, the approximation accuracy may not increase monotonically with the order p. In this case, it cannot be expected that the approximation accuracy would be better after more recurrences. Although it would be possible to find a better approximation by lower order polynomials, how to choose the best finite order p that can give the best approximation is an unsolved theoretical problem.

3.2.8 Different Kinds of Neural Network Models

Before 2010, a lot of stability results are established for the real-valued neural networks. In this kind of network models, the connection coefficients are real-value constants, for example, the well-known Cohen–Grossberg neural networks, Hopfield networks, and other kind of recurrent neural networks. However, according to the different forms of the connection coefficients, different kinds of network models are proposed and studied. Since 2010, memristive neural networks and complex-valued recurrent neural networks are two main network models being widely studied.

Compared with the *real-valued neural networks*, the states, connection weights, and activation functions of the *complex-valued neural networks* are all complex valued. Therefore, there are many differences between the real-valued neural networks and the complex-valued ones [186–194]. In fact, the complex-valued neural networks have much more complicated properties than the real-valued ones in a lot of aspects and hence make it possible to solve many problems that cannot be solved by their real-valued counterparts. For example, both the XOR problem and the detection of symmetry problem cannot be solved with a single real-valued neuron, however, they can be solved by a single complex-valued neuron with the orthogonal decision boundaries. Therefore, it is very important to investigate the dynamical behaviors of complex-valued neural networks, especially the stability of complex-valued neural networks.

Memristive neural networks made of hybrid complementary metal-oxide-semiconductors have a very wide range of uses in bioinspired engineering [195–202]. Memristive neural networks are well suited to characterize the nonvolatile feature of the memory cell because of hysteresis effects. Analysis and synthesis of memristive neural networks are very attractive for neuromorphic systems in which the bionic memories are appropriate for innovative designs. The development of high-performance memristive neural networks would benefit a number of important applications in neural learning circuits, associative memories, new classes of artificial neural systems, and so forth. From a systems-theoretic point of view, a memristive neural network is a state-dependent nonlinear system family. Such system family can reveal coexisting solutions, jump, and transient chaos of rich and complex nonlinear behaviors. Over the years, a lot of pioneering works on nonlinear systems have been reported. With the development and application of memristors, the studies of such state-dependent nonlinear system family with its various generalizations may be an active area of research, to allow the memristors to be readily used in emerging technologies.

Similar studies have similar evolution process. In the aspects of activation function, time delay, uncertainties in connection coefficients, proof method, forms of stability results, and so on, along with the similar routines to real-valued neural networks, memristive neural networks and complex-valued recurrent neural networks will have more space to be developed. For example, it is well known that the activation functions are often chosen to be bounded and smooth in real-valued neural networks. However, in complex-valued neural networks, if the activation functions are chosen to be smooth and bounded, then according to Liouville's theorem, the activation functions will be a constant over the entire complex domain. Therefore, it is a big challenge to choose proper activation functions for the complex-valued neural networks.

3.3 Stability Analysis for Cohen–Grossberg-Type RNNs

3.3.1 Stability on Hopfield-Type RNNs

It has been recognized that the time delay, which is an inherent feature of signal transmission between neurons, is one of the main sources for causing instability and poor performances of neural networks [203–207]. Therefore, stability analysis for RNNs with constant or time-varying delays has been an attractive subject of research in the past few years. Various sufficient conditions, either delay-dependent or delay-independent, have been proposed to guarantee the global asymptotic or exponential stability for the RNNs [208–212], where only the discrete time delays have been handled.

For more details on stability of continuous time *Hopfield neural networks*, readers can refer to the books [211, 212]. In the following, we will stress the stability analysis of Hopfield networks with bounded activation function based on the finite length of the trajectory. The following contents come from the discussion and communication with Professor Tianping Chen from Fudan University of China.

In [153], the authors proposed the so-called finite length of the trajectory by proving Theorem 3 (more details see [153]): Suppose that $g \in \mathcal{G}_{\mathcal{A}}$, $D \in \mathcal{D}_{\mathcal{A}}$, and $U \in \mathcal{V}_{\mathcal{A}}$, Then, any trajectory of the following Hopfield neural networks

$$\dot{u}(t) = -Du(t) + Ag(u(t)) + U, \tag{3.15}$$

has finite length on $[0, \infty)$, i.e.,

$$\int_0^{\infty} \|\dot{u}(\sigma)\|_2 \mathrm{d}\sigma = \lim_{t\to\infty} \int_0^t \|\dot{u}(\sigma)\|_2 \mathrm{d}\sigma < \infty. \tag{3.16}$$

Once one has proved that the length of $u(t)$ on $[0, \infty)$ is finite, a standard mathematical argument permits to prove the existence of the limit of $u(t)$ as $t \to \infty$, i.e., convergence of toward an equilibrium point of (3.15), hence absolute stability for (3.15). The details are as follows. From Theorem 3 in [153], we have $\lim_{t\to\infty} \int_0^t \|\dot{u}(\sigma)\|_2 \mathrm{d}\sigma < \infty$. From Cauchy criterion on limit existence (necessary part), it follows that for any $\epsilon > 0$, there exists $T(\epsilon)$ such that when $s_2 > s_1 > T(\epsilon)$, it results $\int_{s_1}^{s_2} \|\dot{u}(\sigma)\|_2 \mathrm{d}\sigma < \epsilon$. Hence,

$$\epsilon > \int_{s_1}^{s_2} \|\dot{u}(\sigma)\|_2 \mathrm{d}\sigma \geq \| \int_{s_1}^{s_2} \dot{u}(\sigma) \mathrm{d}\sigma \|_2 = \|u(s_2) - u(s_1)\|_2 \tag{3.17}$$

for $s_2 > s_1 > T(\epsilon)$. On the basis of Cauchy criterion on limit existence (sufficient part), it follows that there exists the limitation $\lim_{t\to\infty} u(t) = u^* =$ constant, where u^* is an equilibrium point of (3.15).

On the other hand, in the Ref. [31], the following lemma is given.

Lemma: If $\|\frac{\mathrm{d}u(t)}{\mathrm{d}t}\|_2 \leq Ee^{-\eta t}$, where E is a constant. Then when $t \to \infty$, $u(t)$ has a limit u^* and

$$\|u(t) - u^*\|_2 \leq \frac{E}{\eta} e^{-\eta t}. \tag{3.18}$$

A brief proof can be done as follows,

$$\|u(t_2) - u(t_1)\|_2 \leq \int_{t_1}^{t_2} \|\frac{\mathrm{d}u(t)}{\mathrm{d}t}\|_2 \mathrm{d}t \leq \frac{E}{\eta}(e^{-\eta t_2} - e^{-\eta t_1}). \tag{3.19}$$

By the Cauchy convergence principle, $u(t)$ has a limit u^* and (3.18) holds. It is clear that the idea of finite length of the trajectory has been proposed and used in [31], which was published much earlier than that in [153]. Both contributions in [153] and [31] are novel and independent, which provide the new way to study the stability of neural networks out of the framework of Lyapunov stability theory.

3.3.2 Stability on Cohen–Grossberg-Type RNNs

Cohen and Grossberg [17] first proposed a kind of neural network model in 1983 described by the following equations (called *Cohen–Grossberg neural networks*),

$$\dot{u}_i(t) = -d_i(u_i(t))\Big[a_i(u_i(t)) - \sum_{j=1}^{n} w_{ij} g_j(u_j(t))\Big], \tag{3.20}$$

where $u_i(t)$ is the state variable of the ith neuron at time t, $d_i(u_i(t))$ is an *amplification function*, $a_i(u_i(t))$ is a well-defined function to guarantee the existence of the solution of system (3.20), $g_j(u_j(t))$ is an activation function describing the effects of input on the output of neuron, w_{ij} is the connection weight coefficient of the neurons, $i, j = 1, \ldots, n$. System (3.20) includes a number of models from neurobiology, population biology and evolution theory, as well as the following Hopfield neural network model [7] as a special case in mathematical description,

$$\dot{u}_i(t) = -\gamma_i u_i(t) + \sum_{j=1}^{n} w_{ij} g_j(u_j(s)) + U_i, \tag{3.21}$$

where U_i represents the external input source introduced from the outside of the network to the neuron. It is well known that if the connection matrix is symmetric, then every solution of systems (3.20) and (3.21) will always converge to an equilibrium point [7, 17, 46].

Note that the solution of system (3.21) depends upon the specification of an initial condition $u(\theta) = \phi(\theta)$. It is usually assumed that the given n-dimensional vector function $\phi(\theta)$ is continuous, though it should only be measurable for system (3.21) being well defined. Here, we will also assume a bounded and piecewise continuous initial function with finite discontinuity points. More specifically, we allow as an initial condition a piecewise constant function with a possible discontinuity at $\phi(\theta) = 0$. Such a function is clearly measurable. We note that piecewise constant type initial functions have also been used in determining the fundamental solution of linear delay systems.

One of the objectives of neural network theory is to study the qualitative behavior of the fixed point dynamics of the network (3.21). By definition, for a given constant input vector U, a fixed point or an *equilibrium* of system (3.21) is a point $u_e \in \mathbb{R}^n$ having the property that

$$0 = -\Gamma u_e + W g(u_e) + U, \tag{3.22}$$

where $\Gamma = \text{diag}(\gamma_1, \dots, \gamma_n)$, $W = (w_{ij})_{n \times n}$, $g(u) = (g_1(u_1), \dots, g_n(u_n))^T$, $U = (U_1, \dots, U_n)^T$, $u_e = (u_1^e, \dots, u_n^e)^T$. Since $g(u)$ is bounded and continuous, it follows readily from *Brouwer's fixed point theorem* that there is at least one solution u_e to the Eq. (3.20) for every constant vector U. The locations of such equilibria in $\mathbb{R}^n$ are determined by the connection pattern (i.e., the weight matrix W) of the neural network, by the relaxed matrix Γ, the nature of the nonlinearity $g(u)$, and the constant input U. If u_e is globally asymptotically stable, then it is the unique equilibrium that attracts all other trajectories. In this case, to each given input vector U the network associates a unique equilibrium to which it converges irrespective of the initial conditions. This establishes a one-to-one correspondence between the input space and the steady-state space, which is a desirable property in applications of neural network (3.21) to such problems as optimization and input patterns classification [26].

As is well known, both in biological and artificial neural networks, delays arises because of the processing of information. For example, in the electronic implementation of analog neural networks, delay occurs in the communication and response of amplifiers due to the finite switching speed of amplifiers [40]. On the other hand, the delayed cellular neural network can be used to solve some moving image processing problems [213, 214], and hence it is desirable to introduce delays into neural networks when dealing with problems associated with motions [146, 215–218]. So it is important and practical to incorporate delays into neural networks. Neural networks with *time delay* have much more complicated dynamics due to the incorporation of delay. Therefore, model (3.20) and its delayed version have attracted the attention of many researchers and have been extensively investigated due to their potential applications such as associative content-addressable memories, pattern recognition, and optimization. Such applications rely on the qualitative stability properties of the network. Thus, the qualitative analysis of the network's dynamic behavior is a prerequisite step for practical design and application of neural networks. Recently, some sufficient conditions for global asymptotic/exponential stability of Cohen–Grossberg neural networks have been studied in the literature, see, e.g., [18, 36, 103, 109, 114,

116, 117, 132, 143, 147, 149, 219–225]. When delays are incorporated into system (3.20), it is natural to expect that the property of global asymptotic stability remains if the delays are sufficiently small. This became a challenging topic at that times, which attracted many researchers to study this topic since 1983–2002. In fact, such an expectation is confirmed in [18] under a certain type of symmetry requirement. More precisely, Ref. [18] introduces constant discrete delays into (3.20), which yields the following form,

$$\dot{u}_i(t) = -\,d_i(u_i(t))\left[a_i(u_i(t)) - \sum_{k=0}^{N}\sum_{j=1}^{n} w_{ij}^{k} g_j(u_j(t-\tau_k))\right], \tag{3.23}$$

where $\tau_k \geq 0$ are bounded constant delays, w_{ij}^k are the connection weight coefficients, $0 = \tau_0 < \tau_1 < \tau_2 < \cdots < \tau_N$, and other notations are the same as those in system (3.20), $k = 0, \ldots, N$, $i, j = 1, \ldots, n$. Assume that matrix $W^e = (\sum_{k=1}^{N} w_{ij}^k) = \sum_{k=1}^{N} W_k$, $W_k = (w_{ij}^k)$, is symmetric and activation function $g_j(\cdot)$ is sigmoidal and also satisfies other conditions, system (3.23) is globally stable if the following condition holds,

$$\sum_{k=1}^{N} (\tau_k \beta \| W_k \|) < 1, \tag{3.24}$$

where $\beta = \max \|D(u(t))g^{'}(u(t))\| \leq \overline{d}\,\overline{\delta}$, $\overline{d} = \max\{\overline{d}_i\}$, $\overline{\delta} = \max\{\delta_i\}$, $\|A\| = [\lambda_{\max}(A^T A)]^{1/2}$.

Note that systems (3.20) and (3.21) are for the case of instantaneous transmission of signal, while the system (3.23) is for the case of pure delay transmission of signal. In biological system and practical implementation of neural network, both instantaneous transmission and pure delay transmission of signal often occurs simultaneously, even with more complicated phenomena. Therefore, many kinds of complex neural network models have been proposed and studied in the literature [58, 64, 77, 78, 107, 158–161, 226]. For example, the following Hopfield neural networks with delay,

$$\dot{u}(t) = -Cu(t) + W_0 g(u(t)) + W_1 g(u(t-\tau)). \tag{3.25}$$

For the purely delayed Hopfield networks,

$$\dot{u}(t) = -Cu(t) + W_1 g(u(t-\tau)), \tag{3.26}$$

from condition (3.24) we can derive that $\tau\overline{\beta}\|W_1\| < 1$ guarantees the global stability of (3.26), where $\beta = \max \|g^{'}(u)\|$ or maximum slope of activation function. It has been shown in [18] that if the condition (3.24) holds, then the asymptotic stability of an equilibrium point of (3.23) can be deduced from the asymptotic stability of the corresponding equilibrium point of system (3.20). In other words, if (3.24) holds, then system (3.20) and system (3.23) are both *globally stable*, both have similar local stability properties at an asymptotic stable equilibrium. This enables us to verify the asymptotic stability of the equilibrium point of system (3.23) by ascertaining the asymptotic stability of corresponding equilibrium point of system (3.20). Second contribution of [18] is that the locations of the (asymptotic) stable equilibria of system (3.23) will not depend on the delays τ_k. Third contribution of [18] is that the interconnection matrices $W_k, k = 1, \ldots, N$, is not required to be symmetric, while the summation $\sum_{k=1}^{N} W_k$ is required to be symmetric. This requirement significantly reduces the symmetric constraint condition on the interconnection conditions W_k. Therefore, if (3.24) holds, then system (3.23) and system (3.20) will have identical (asymptotic) stable equilibria.

The *symmetry* requirement is a very critical constraint because the neural networks should be robust against the asymmetric connection matrix. Much work has been done to relax this symmetric requirement [109, 225]. Although the symmetry of the connection matrix is canceled, the activation function is assumed to be globally Lipschitz and bounded. Without requiring the symmetry of connection matrix and the boundedness of the activation function, how to establish the stability criteria of RNNs has become a challenging topic since 2002.

The use of constant fixed time delays in the models of delayed feedback systems serves as a good approximation in simple circuits having a small number of neurons. However, neural networks usually have a spatial extent due to the presence of a multitude of parallel pathways with a variety of axon sizes and lengths, there will be a distribution of propagation delays. In this case, the signal propagation is no longer instantaneous and cannot be modeled with discrete delays. Reference [228] has proposed a neural circuit with *distributed delays|indexdistributed delay* that solves a general problem of recognizing patterns using a time-dependent signal. It is desired to model them by introducing continuously distributed delays [109, 169, 228–231]. Nowadays, there are generally two kinds of continuously distributed delays in the modeling of RNNs, i.e., finitely distributed delays and infinitely distributed delays. The following system with finitely distributed delays,

$$\dot{u}_i(t) = -a_i(u_i(t)) + \sum_{j=1}^{n} w_{ij} \int_{t-\tau(t)}^{t} g_j(u_j(s))\mathrm{d}s, \qquad (3.27)$$

has been investigated in the recent literature, where $\tau(t)$ is a time-varying delay, $i = 1, \ldots, n$. Model (3.27) and its variants have been studied in [76, 232–236] based on LMI method and other methods. Similarly, the following models with infinitely distributed delays,

$$\dot{u}_i(t) = -a_i(u_i(t)) + \sum_{j=1}^{n} w_{ij} \int_{-\infty}^{t} K_{ij}(t-s) g_j(u_j(s))\mathrm{d}s, \qquad (3.28)$$

and

$$\dot{u}_i(t) = -a_i(u_i(t)) + \sum_{j=1}^{n} w_{ij} g_j \left(\int_{-\infty}^{t} K_{ij}(t-s) u_j(s) \mathrm{d}s \right), \tag{3.29}$$

have also been investigated in the literature [135–137], where the delay kernel $K_{ij}(\cdot) : [0, \infty) \to [0, \infty)$ is a real-valued nonnegative continuous function, and the additional restrictions are as follows:

$$\int_{0}^{\infty} K_{ij}(s) \mathrm{d}s = 1, \int_{0}^{\infty} s K_{ij}(s) \mathrm{d}s < \infty, \tag{3.30}$$

and other notations are the same as those in system (3.20), $i, j = 1, \ldots, n$. If the delay kernel function $K_{ij}(s)$ is of the form $K_{ij}(s) = \delta(t - \tau_{ij})$, then model (3.28) and model (3.29) can be reduced to the following neural networks with discrete delays,

$$\dot{u}_i(t) = -a_i(u_i(t)) + \sum_{j=1}^{n} w_{ij} g_j(u_j(t - \tau_{ij})). \tag{3.31}$$

If the delay *kernel function* $K_{ij}(s)$ are of the form $K_{ij}(s) = L_{ij}(t)$ if $t \in [0, \tau_{ij}]$, otherwise, $K_{ij}(s) = 0$, then the duration intervals for time delays are finite, and thus, model (3.28) and model (3.29) can be reduced to the following neural networks with finite distributed delays,

$$\dot{u}_i(t) = -a_i(u_i(t)) + \sum_{j=1}^{n} w_{ij} \int_{t-\tau_{ij}}^{t} L_{ij}(t-s) g_j(u_j(s)) \mathrm{d}s, \tag{3.32}$$

and

$$\dot{u}_i(t) = -a_i(u_i(t)) + \sum_{j=1}^{n} w_{ij} g_j \left(\int_{t-\tau_{ij}}^{t} L_{ij}(t-s) u_j(s) \mathrm{d}s \right), \tag{3.33}$$

where $L_{ij}(t) \geq 0$ and $\int_0^{\tau_{ij}} L_{ij}(t) dt = 1$. If we further take a special form of the delay kernel function as $L_{ij}(t) = 1/\tau_{ij}$, $\tau_{ij} > 0$, then model (3.32) and model (3.33) can be reduced to the following form

$$\dot{u}_i(t) = -a_i(u_i(t)) + \sum_{j=1}^{n} w_{ij}/\tau_{ij} \int_{t-\tau_{ij}}^{t} g_j(u_j(s)) \mathrm{d}s, \tag{3.34}$$

and

$$\dot{u}_i(t) = -a_i(u_i(t)) + \sum_{j=1}^{n} w_{ij} g_j \left(\int_{t-\tau_{ij}}^{t} u_j(s)/\tau_{ij} \mathrm{d}s \right). \tag{3.35}$$

Therefore, the discrete delays and the finite distributed delays can be included in the model (3.28) and model (3.29) by choosing suitable kernel functions. Model (3.28) and its variants have been studied in [64, 158–160, 169, 229, 230, 237–248]. Model (3.29) and its variants have been studied in [42, 47, 109, 144, 249–253]. However, few LMI-based stability criteria have been established for model (3.28) and model (3.29) in the existing literature except [135].

Recently, a more general model, i.e., recurrent neural networks with a general continuously distributed delays, have been proposed and studied in the literature [254–265],

$$\begin{aligned}\dot{x}_i(t) = &- a_i x_i(t) + \sum_{j=1}^{n} \int_0^{\infty} g_j(x_j(t-s))\mathrm{d}J_{ij}(s) \\ &+ \sum_{j=1}^{n} \int_0^{\infty} g_j(x_j(t-\tau_{ij}(t)-s))\mathrm{d}K_{ij}(s) + U_i,\end{aligned} \tag{3.36}$$

where $x = (x_1, \ldots, x_n)^T$, $A = \mathrm{diag}(a_1, \ldots, a_n)$, $a_i > 0$, $g(x(t)) = (g_1(x_1(t)), \ldots, g_n(x_n(t)))^T$, $g_i(x_i(t))$ are the activation functions, $\tau_{ij}(t)$ are the time-varying delays with $0 < \tau_{ij}(t) \leq \tau_M$, and $\dot{\tau}_{ij}(t) \leq \mu_{ij}$, $\mu_{ij} > 0$ are positive constants, $\mathrm{d}J_{ij}(s)$ and $\mathrm{d}K_{ij}(s)$ are *Lebesgue–Stieltjes measures* for each $i, j = 1, \ldots, n$.

The model (3.36) is said to be general because it can involve a large kinds of different delays. In the existing references, different kinds of delays have been considered, for example, discrete or concentrated constant delays τ, τ_j, τ_{ij} and their time-varying counterparts; finitely distributed delays $\int_{t-\tau}^{t} g_j(x_j(s))\mathrm{d}s$, $\int_{t-\tau_j}^{t} g_j(x_j(s))\mathrm{d}s$, $\int_{t-\tau_{ij}}^{t} g_j(x_j(s))\mathrm{d}s$ and their time-varying counterparts; infinitely distributed delays $\int_{-\infty}^{t} k_{ij}\ (t-s) g_j(x_j(s))\mathrm{d}s$, where $g_j(\cdot)$ are the neuron activation functions, $k_{ij}(s)$ are some Kernel functions. All the above delays can be uniformly expressed in model (3.36). Therefore, it is more interesting to study the stability of model (3.36).

As pointed out in [255, 258], can we propose an effective approach to investigate them in a universal framework? An affirmative answer has been given in [255, 258] to integrate the different delays based on like-M-matrix framework, i.e.,

$$-\xi_i a_i + \sum_{j=1}^{n} \xi_j \delta_j \left\{ \int_0^{\infty} \left(|\mathrm{d}J_{ij}(s)| + |\mathrm{d}K_{ij}(s)| \right) \right\} < 0, \tag{3.37}$$

where $\xi_i > 0$ are some parameters to be determined. Obviously, when the Lebesgue–Stieltjes measures $\mathrm{d}J_{ij}(s)$ and $\mathrm{d}K_{ij}(s)$ take different form, stability result (3.37) can be expressed in M-matrix form. This is also the reason why we call the form of (3.37) as like-M-matrix.

Similar to model (3.28) and model (3.29), few LMI-based stability criteria for model (3.36) have been obtained except [265].

In the case that activation function is a bounded and Lipschitz continuous function, Ref. [47] presents some sufficient conditions for the existence, global asymptotic stability, and global exponential stability of the equilibrium point of model (3.29) using algebraic inequality method. Reference [47] is a very classical paper to discuss the dynamics of Cohen–Grossberg/Hopfield neural networks with infinite distributed delays using algebraic inequality method, including how to prove the existence of the equilibrium point, how to deal with the infinite distributed delay by constructing different Lyapunov functional, how to use the characteristic equation method to directly derive the stability criterion for Hopfield neural networks with infinite distributed delays, and so on.

Without requiring the boundedness, differentiability, and the monotonicity of the activation function, and the symmetry of the interconnection matrix, Ref. [109] has presented some sufficient conditions to guarantee the existence, uniqueness, and global asymptotic stability of the equilibrium point of neural networks (3.23) and (3.29) on the basis of M-matrix theory.

Although the global stability results for model (3.28) and its variants are expressed in different forms, such as M-matrix form and algebraic inequality form, all the results take the absolute value operation on the connection weight coefficients. Consequently, the sign of entries in connection matrix is ignored, which leads to the ignorance of the neuron's excitatory and inhibitory effects on the neural networks. As a result, the conservativeness of the stability criteria will be increased. Because LMI method can provide a rather well trade-off between conservativeness and the verification, many LMI-based stability results have been proposed for different kinds of neural networks in the literature [36, 116, 131, 132, 218, 232], which have considered the neuron's excitatory and inhibitory effects on the neural network and have reduced the conservativeness. Therefore, it is important to study the LMI-based stability criteria for model (3.28) and its variants as that did for model (3.27) and its modifications and other kinds of neural networks.

In the practical operations, diffusion effects cannot be avoided in the neural network and electric circuits when electrons are moving in a nonuniform electromagnetic field. Hence, it is essential to consider the state variables which vary in space as well as in time. Furthermore, modeling and analysis of the dynamics of biological populations by means of differential equations is one of the primary concerns in population growth problems. A well-known and extensively studied class of models in population dynamics is the Lotka–Volterra competition system which models the interaction among various competing species. In the case that the effect of dispersion of the population in a bounded habitat is taken into consideration, the governing equations for the population densities become a Lotka–Volterra competition system with reaction-diffusion terms. Therefore, it is also common to consider the reaction-diffusion effects in biological systems caused by the immigration of species [266–271]. On the other hand, a second-order cellular neural networks with reaction–diffusion term has been identified which is able to reproduce through parameter setting a rich variety of spatio-temporal behaviors, and be able to robustly reproduce the rich phenomenology associated with active wave propagation and pattern formation. These wave formation phenomena are exhibited by systems belonging to very different scientific disciplines, for example, in neurophysiology, the propagation of electrical impulses through the nervous system, or the propagation of the cardiac movement through the cardiac muscle [272]. Since Cohen–Grossberg neural networks (3.20) are a kind of competitive-cooperation networks, which can describe ecological systems and general neural networks, it is natural to study the effects of reaction diffusion in stability analysis. In [113, 148, 154, 156, 167, 239, 243, 250, 273–277], the authors have considered the stability problems of reaction–diffusion neural networks, which are usually expressed by partial differential equations.

For Hopfield-type recurrent neural networks with reaction-diffusion terms, [156] presents an algebraic inequality GES stability criterion and a like-LMI-based GES stability criterion. However, the like-LMI-based GES stability criterion is generally difficult to check due to some unknown parameters. For Hopfield-type recurrent neural networks with both reaction-diffusion terms and neutral type delay, [275] presents an LMI-based GES stability condition. If the boundary domain of the bounded compact set is known a prior, the main result in [275] is easy to check. Otherwise, it is not easy to be applicable. For Hopfield-type recurrent neural networks with reaction-diffusion terms and stochastic perturbation term, almost sure stability and moment exponential stability are established in [276] based on M-matrix theory, respectively.

For the *Cohen–Grossberg neural networks* with reaction diffusion terms, under the requirement of both bounded activation function and positive and bounded amplification function, [148] presents an algebraic inequality criterion guaranteeing the GES of the unique equilibrium point. The proof method in [148] is similar to those in [110, 154].

In the following subsections, in order to make comparisons with the existing stability results, the following assumptions are required for the Cohen–Grossberg neural networks (3.23).

Assumption 3.1 *Amplification function* $d_i(\zeta) \in C(\mathbb{R}, [0, \infty))$ and there exist constants $\underline{d}_i, \overline{d}_i$ such that $0 < \underline{d}_i \leq d_i(\zeta) \leq \overline{d}_i$ for $\zeta \in \mathbb{R}$.

Assumption 3.2 The behaved function $a_i(u_i(t)) \in C(\mathbb{R}, \mathbb{R})$ and there exists $\gamma_i > 0$ such that $\frac{a_i(\zeta)-a_i(\xi)}{\zeta-\xi} \geq \gamma_i$ for $\zeta, \xi \in \mathbb{R}$ with $\zeta \neq \xi$.

Assumption 3.3 $g_i(u_i(t)) \in C(\mathbb{R}, \mathbb{R})$ is globally Lipschitz with constant δ_i, i.e., $|g_i(\zeta) - g_i(\xi)| \leq \delta_i |\zeta - \xi|$ for $\zeta, \xi \in \mathbb{R}$.

Assumption 3.4 $g_i(u_i(t)) \in C(\mathbb{R}, \mathbb{R})$ is globally Lipschitz with constant δ_i, i.e., $0 \leq \frac{g_i(\zeta)-g_i(\xi)}{\zeta-\xi} \leq \delta_i$ for $\zeta \neq \xi, \zeta, \xi \in \mathbb{R}$.

Assumption 3.5 *Amplification function* $d_i(\zeta)$ is continuous with $d_i(0) = 0, d_i(\zeta) > 0$ for all $\zeta > 0$, and $\int_0^{\epsilon} \frac{ds}{d_i(s)} = +\infty$ for all $i = 1, \ldots, n$, $\epsilon > 0$ is a constant.

Note that in [18], the *behaved function* is assumed to be continuous, $\lim_{\zeta_i \to +\infty} a_i(\zeta_i) = +\infty$, $\lim_{\zeta_i \to -\infty} a_i(\zeta_i) = -\infty$. The activation function is a sigmoidal function such that $g'(u_j) = \frac{dg_j(u_j)}{du_j} > 0$, $\lim_{\zeta_i \to +\infty} g_i(\zeta_i) = 1$, $\lim_{\zeta_i \to -\infty} g_i(\zeta_i) = -1$, $\lim_{|\zeta_i| \to \infty} g_i'(\zeta_i) = 0$. Obviously, the activation in [18] is continuous, differential, smooth, monotonic, and bounded. Meanwhile, the concerned behaved function can guarantee the boundedness of system (3.23) with or without delays, which has been proved using the contradiction method. It can be seen that both Assumption 3.2 and Assumption 3.3 all include those hypothesis in [18] as special cases. The differentiable activation function can be relaxed to be continuous function, i.e., the existence of right and left derivatives of activation function, which includes the monotonically increasing piecewise linear function.

Denoting $\underline{D} = \text{diag}(\underline{d}_1, \ldots, \underline{d}_n)$, $\overline{D} = \text{diag}(\overline{d}_1, \ldots, \overline{d}_n)$, $\Delta = \text{diag}(\delta_1, \ldots, \delta_n)$ and $\Gamma = \text{diag}(\gamma_1, \ldots, \gamma_n)$, $\overline{d} = \max\{\overline{d}_i\}$, $\underline{d} = \min\{\underline{d}_i\}$, $\delta_M = \max\{\delta_i\}$, and $\gamma_m = \min\{\gamma_i\}$, which will be used in the sequel.

3.3.3 The Case with Nonnegative Equilibria

In this subsection, we will review the stability results on the Cohen–Grossberg neural networks (3.20) with *nonnegative equilibrium point*.

In the original paper of Cohen and Grossberg [17, 278], the concerned competitive neural network model is as follows:

$$\dot{u}_i(t) = d_i(u_i(t)) \left[a_i(u_i(t)) - \sum_{j=1}^{n} w_{ij} g_j(u_j(t)) \right], \tag{3.38}$$

is proposed as a kind of *competitive-cooperation dynamical system* for decision rules, pattern formation, and parallel memory storage. Hereby, each state of neuron might be the population size, activity, or concentration of the ith species in the system, which is always positive for all time. Based on such background, Cohen–Grossberg neural networks (3.38) has been studied under the following hypothesis [17]:

(a) Symmetry: matrix $W = (w_{ij})$ is a symmetric matrix of nonnegative constants;
(b) Continuity: function $d_i(\xi)$ is continuous for $\xi \geq 0$; function $a_i(\xi)$ is continuous for $\xi > 0$;
(c) Positivity: function $d_i(\xi) > 0$ for $\xi > 0$; function $g_i(\xi) \geq 0$ for $\xi \in (-\infty, \infty)$;
(d) Smoothness and monotonicity: function $g_i(\xi)$ is differentiable and monotone nondecreasing for $\xi \geq 0$.
(e) The *boundedness* of the trajectories are guaranteed by the following condition

$$\limsup_{\zeta \to +\infty} [a_i(\zeta) - w_{ii} g_i(\zeta)] < 0. \tag{3.39}$$

(f) For any positive initial data, the positivity of the trajectories are guaranteed by the following condition

$$\lim_{\zeta\to 0^+} a_i(\zeta) = +\infty, \tag{3.40}$$

or

$$\lim_{\zeta\to 0^+} a_i(\zeta) < +\infty, \quad \int_0^{\epsilon} \frac{ds}{d_i(s)} = +\infty \text{ for some } \epsilon > 0. \tag{3.41}$$

Under above hypotheses (a)–(d), (3.39), (3.40) or (3.41), all the trajectories of Cohen–Grossberg neural networks (3.38) are convergent (locally).

Note that for system (3.38) in [17], the relations between the activation function $g_i(u_i(t))$ and the behaved function $a_i(u_i(t))$ are constrained in (3.39), which is a very general restriction on the interaction of neural system. Obviously, the activation function $g_i(u_i(t))$ in (3.38) can be bounded or unbounded as $u_i(t) \to \infty$, which will affect the property of behaved function directly. For the behaved function $a_i(u_i(t))$, however, from the boundedness condition (3.39) and positivity conditions (3.40) or (3.41), we can conclude that if the behaved function $a_i(u_i(t))$ is monotonically decreasing in system (3.38), then both boundedness condition and positivity conditions naturally hold.

Note that the *Cohen–Grossberg neural network* (3.20) widely studied in the existing literature is just the opposite or counterpart of the original competitive system (3.38). Therefore, in order to keep the stability of the Cohen–Grossberg neural networks (3.20), corresponding to the constraint conditions for (3.38), the restriction conditions exerted on the Cohen–Grossberg neural networks (3.20) should be changed, which will not affect the qualitative stability property of the Cohen–Grossberg neural networks (3.20). The hypotheses (b)–(d) discussed above are not changed, while the boundedness condition and the positivity condition should be changed. Specifically,

(a1) Symmetry: matrix $W = (w_{ij})$ is a symmetric matrix of nonpositive constants;

(e1) The boundedness of the trajectories of Cohen–Grossberg neural network (3.20) is guaranteed by the following condition

$$\lim_{\zeta\to+\infty} \sup\, [-a_i(\zeta) + w_{ii} g_i(\zeta)] < 0. \tag{3.42}$$

(f1) For any positive initial data, the positivity of the trajectories for Cohen–Grossberg neural network (3.20) are guaranteed by the following condition

$$\lim_{\zeta\to 0^+} -a_i(\zeta) = +\infty, \tag{3.43}$$

or

$$\lim_{\zeta\to 0^+} -a_i(\zeta) < +\infty, \quad \int_0^{\epsilon} \frac{ds}{d_i(s)} = +\infty \text{ for some } \epsilon > 0. \tag{3.44}$$

Under above hypotheses (a1), (b)–(d), (3.42), (3.43) or (3.44), all the trajectories of Cohen–Grossberg neural networks (3.20) are convergent (locally). Similar to aforementioned discussions, we can conclude that if the behaved function $a_i(u_i(t))$ is monotonically increasing in system (3.20), then both boundedness condition and positive conditions naturally hold. This is the reason why the behaved function $a_i(u_i(t))$ in system (3.20) is usually assumed to be increasing in the existing literature, see Assumption 3.2.

In addition, another interesting question can be discussed as follows. Why does the so-called *Cohen–Grossberg-type neural network model* (3.20) studied widely in the existing literature overwhelms the original competitive neural network model (3.38)? In fact, these two models are counterparts to each other. It is well known that the competitive neural network (3.38) is mainly for the biological system, in which the states represent the population of some species and all the states are nonnegative. However, with the development and wide application of Hopfield neural networks in optimization problem and other fields [1, 7, 16, 228], the *Hopfield neural networks* have become a very hot topic in the engineering community. The standard Hopfield neural network model has the following structure,

$$\dot{u}_i(t) = -\gamma_i u_i(t) + \sum_{j=1}^{n} w_{ij} g_j(u_j(s)) + U_i. \tag{3.45}$$

Comparing the competitive neural network model (3.38), i.e.,

$$\dot{u}_i(t) = d_i(u_i(t)) \left[a_i(u_i(t)) - \sum_{j=1}^{n} w_{ij} g_j(u_j(t)) \right], \tag{3.46}$$

we can change the Hopfield neural network model (3.45) as follows:

$$\dot{u}_i(t) = -\left(\gamma_i u_i(t) - \sum_{j=1}^{n} w_{ij} g_j(u_j(s)) - U_i \right). \tag{3.47}$$

If we further consider the effects of *amplification function* on the neural network model (3.47) as that done on model (3.38), we then have the general Cohen–Grossberg-type neural network model (3.20), i.e.,

$$\dot{u}_i(t) = -d_i(u_i(t)) \left[a_i(u_i(t)) - \sum_{j=1}^{n} w_{ij} g_j(u_j(t)) \right]. \tag{3.48}$$

Obviously, in the aspect of model description, both models (3.48) and (3.46) are similar and have the same structure. However, in the aspect of physical essence, both models (3.48) and (3.46) represent different background and engineering meaning.

- Model (3.46) originates from the biological system, in which the elements w_{ij} and the states are all nonnegative. The amplification function $d_i(u_i(t))$ in (3.46) is nonnegative, i.e., $d_i(u_i(t)) > 0$ for $u_i(t) > 0$ and $d_i(u_i(t)) = 0$ for $u_i(t) = 0$, which represents the inhibitory and excitatory effects on the population of species.
- Model (3.48) originates from the Hopfield neural network model, which has the storage capability and computational capability and has very notable superiority in engineering application. Meanwhile, all the elements w_{ij} and the states in model (3.48) can be negative, positive, or mixture. The amplification function $d_i(u_i(t))$ in (3.48) is strictly positive, i.e., $d_i(u_i(t)) > 0$ for any $u_i(t) \in \mathbb{R}$, which represents the amplification capability on the magnitude of the state of the concerned engineering problems to be solved.

  In the viewpoint of mathematics, we can have the following statements to understand the relations between models (3.48) and (3.46).
- Model (3.46) and model (3.48) are counterparts in the description of general biological system, in which the states may be nonnegative and nonpositive. For example, if the *behaved function* in model (3.46) is decreasing, then the behaved function in model (3.48) is increasing; if the connection weight in model (3.46) is positive, then the connection weight in model (3.48) is negative.
- Model (3.46) and model (3.48) have many common features. For example, the activation function should be nondecreasing, the amplification function should be positive, and the connection weight matrices are required to be symmetric.
- Model (3.46) is widely used in the fields such as biological system, while model (3.48) is widely used in the fields such as industrial and engineering systems.

This is the reason why model (3.48) is called *Cohen–Grossberg-type neural network* model and why model (3.48) has been widely studied in engineering fields.

Since model (3.48) is also a kind of Cohen–Grossberg neural network structure, it can represent a class of biological system model if some assumption conditions in (3.46) are exerted on the model (3.48). Hence, it is also meaningful to consider the dynamics of (3.48) with positive trajectories. This is the reason why the dynamics of delayed Cohen–Grossberg neural network (3.48) with nonnegative amplification function and positive initial condition have received much attention in the neural network community, and the corresponding stability results have dropped out the requirement of symmetry on the connection weight matrix.

The main contribution of [17] is to discover the essence of the *symmetry* on the effects of dynamics of complex systems. Symmetry holds naturally and the nature in essence is symmetry and harmonic. The symmetric structure is so elegant that it attracted many scientists to devote themselves to the research on the nature of the human being and the natural science. In this aspect, Cohen and Grossberg set a good example for us. The contribution of [17] is important not only in natural science, but also in philosophy.

For Cohen–Grossberg neural networks (3.38), Ref. [129] proposes the following sufficient global stability condition based on LaSalle's invariance principle, if the nondelayed matrix W is decomposed as the product of a symmetric matrix and the positive definite diagonal matrix, i.e.,

$$W = DS, \tag{3.49}$$

where D is a positive definite diagonal matrix, and S is a symmetric matrix, then every bounded trajectories approaches to one of possibly large number of equilibrium points as $t \to \infty$. In general, $DS \neq SD$, therefore, the stability condition in [129] relaxed the condition in [17].

The following classical *Lotka–Volterra model* of competing species, which is a special case of Cohen–Grossberg neural networks (3.38), has been studied in [279],

$$\dot{u}_i(t) = G_i u_i \left(1 - \sum_{k=1}^{n} H_{ik} u_k(t) \right), \tag{3.50}$$

where $u_i(t)$ is the population of the ith species, $H_{ik} \leq 0$ for $i \neq k$ are the negative interaction parameters between different species, and $G_i > 0$ are constants. Model (3.50) is of great importance not only in modeling population dynamics but also in chemical kinetics, ecology, plasma physics, and neural network modeling in general.

Theorem 1 in [279] requires that if matrix $H = H_{ik}$ is symmetric and $H_{ii} > 0$, then each trajectory $u(t)$ starting at $u(0) \in O^+ = \{u \in \mathbb{R}^n : u_i(t) \geq 0, i = 1, \ldots, n\}$ is bounded for $t \geq 0$, and $\lim_{t \to +\infty} u(t) = u_e \in O^+$, where u_e is an equilibrium point of system (3.50).

Theorem 1 in [279] implies that any trajectory of (3.50) starting in O^+ converges toward a singleton, for any choice of the parameters $G_i > 0$, and for any choice of the symmetric neuron interconnection matrix H, including the situations where networks (3.50) possess infinitely nonisolated equilibrium points. According to the terminology used in [17], this means that the global pattern formation is *absolutely stable* within the class of neural network (3.50) with a symmetric interaction matrix H and positive parameters G_i. Theorem 1 in [279] extends the results in [17] as far as the convergence is concerned for model (3.50), because the result in [17] requires an additional assumption of isolated equilibrium point to ensure the convergence, while the result in [279] admits multiple and (possible) infinitely *nonisolated equilibrium points*.

Convergence of system (3.50) cannot be proved by means of the Lyapunov method and LaSalle's invariance principle. When system (3.50) possesses non-isolated equilibrium points, Lyapunov method only enables to show that system (3.50) is quasi-convergent. This means that for any trajectory $u(t)$ of (3.50), we have $\lim_{t \to +\infty} \dot{u}(t) = 0$, or equivalently, $u(t)$ approaches to the set of equilibrium points of (3.50) as $t \to +\infty$ (see [17]). However, in a quasi-convergent system, we cannot exclude that a trajectory $u(t)$ approaches to a manifold of equilibrium without converging to a singleton (see [279]). An example of this kind is given in [280], where a gradient system of a C^∞ function is constructed, such that all bounded trajectories indefinitely slide along a manifold of equilibria, and they display large size nonvanishing oscillations as $t \to +\infty$. The oscillatory behavior would be highly undesirable for a neural optimization solver or a neural associate memory. The crucial point is that $G_i u_i(t)$ will vanish at $u_i(t) = 0$. It can be seen that the term $G_i u_i(t)$

introduces singularities in the expression of the length of the trajectory $u(t)$, making it impossible to prove the finiteness of the trajectory length.

Therefore, despite that the description form of the consequences in [279] and [17] being same for the system (3.50), they represent different dynamics and require different proof methods. For the case of isolated equilibrium point, Ref. [17] presents the absolute stability property for system (3.50) via LaSalle's invariance principle, in which all the trajectories converge to the isolated equilibrium. While the consequence in [279] is to prove the convergence of the isolated or *nonisolated equilibrium point* for system (3.50) by exploiting the Lojasiewicz inequality for analytic-gradient systems, in which a suitable change of variable plays an important role in the application of Lojasiewicz inequality.

Moreover, for the system (3.50) with nonsymmetric matrix H, Ref. [279] also shows that if there exists two positive definite diagonal matrices D_a and D_b such that $H_a = D_a H D_b$ is a symmetric matrix, then the convergence of the trajectory in Theorem 1 in [279] still holds.

Although the generality of the research results in [17] is excellent, the specific application of this kind of research results is another question. Since the assumption conditions in [17] are too general, the applications of the results in [17] will be restricted in the practical systems. Especially, the requirement of the *symmetry* on the connection weight is too hard to implement in practice, which may lead to poor/no robustness in applications.

In the aspects of behaved function and symmetric connection weights, Ref. [112, 116, 117, 131, 167, 222] improved the conditions in [17], and the stability of the nonnegative/positive equilibrium point for corresponding Cohen–Grossberg neural networks with delays has been studied.

For the reaction-diffusion Cohen–Grossberg neural network,

$$\begin{aligned}\frac{\partial u_i(t)}{\partial t} &= \sum_{k=1}^{m} \frac{\partial}{\partial x_k}\left(D_{ik}\frac{\partial u_i(t,x)}{\partial x_k}\right)\\ &\quad - d_i(u_i(t,x))\left[a_i(u_i(t,x)) - \sum_{j=1}^{n} w_{ij} g_j(u_j(t,x))\right], \end{aligned} \tag{3.51}$$

under the following conditions on the *activation function* and *amplification function*:

$$(g_i(\zeta) - g_i(\xi))(\zeta - \xi) > 0, \ \text{for } \zeta \neq \xi, \tag{3.52}$$

$$(a_i(\zeta) - a_i(\xi))(\zeta - \xi) > 0 \text{ (or } \geq 0), \ \text{for } \zeta \neq \xi, \tag{3.53}$$

Amplification function $d_i(u_i(t,x))$ is nonnegative and continuous in $\mathbb{R}^n_+$,

$$d_i(u_i(t,x)) > 0 \text{ for } u_i(t,x) > 0 \tag{3.54}$$

$$\left(\frac{g_i(u_i(t,x)) - g_i(u_i^*)}{d_i(u_i(t,x))}\right)' \geq 0, \tag{3.55}$$

where $g_i(u_i(t,x))' = \frac{\mathrm{d}g_i(u_i(t,x))}{\mathrm{d}u_i(t,x)}$, Theorem 1 (or Theorem 2) in [167] requires $W = (w_{ij})$ to be *Lyapunov-Volterra quai-stable (or stable)*, i.e., there exists a positive diagonal matrix $P = \mathrm{diag}(p_1, \ldots, p_n)$ such that

$$-PW - (PW)^T \geq 0 \text{ (or } > 0), \tag{3.56}$$

then the nonnegative equilibrium point of system (3.51) is (locally) asymptotically stable. Further, if the following condition also holds (which guarantees the average *Lyapunov function* chosen in [167] to be positive infinity $+\infty$ as $u_i(t,x) \to +\infty$ or $u_i(t,x) \to 0^+$),

$$\begin{aligned} &\int_{u_i^*}^{+\infty} (g_i(u_i(t,x)) - g_i(u_i^*)) \frac{\mathrm{d}u_i(t,x)}{d_i(u_i(t,x))} \\ &= \int_{u_i^*}^{0} (g_i(u_i(t,x)) - g_i(u_i^*)) \frac{\mathrm{d}u_i(t,x)}{d_i(u_i(t,x))} = +\infty, \end{aligned} \tag{3.57}$$

then the nonnegative equilibrium point of system (3.51) is globally asymptotically stable. Some other different forms of stability criteria are also presented in [167]. From the stability results in [167], we can see the following interesting things.

(1) The stability condition (3.56) is a special case of stability condition (3.69) in [36] in the sequel, which means that whether the equilibrium point is nonnegative or not, the negative definite matrix W will guarantee the asymptotic stability of the equilibrium point of system (3.20). Under other additional assumption conditions, the negative definite matrix W will also guarantee the global asymptotic stability of the equilibrium point of system (3.51).

(2) The assumption conditions on the activation function, behaved function, and amplification function are different from those in Assumptions 3.1–3.3. On the one hand, the nonnegative property of the concerned equilibrium point in [167] is different from those equilibrium points in the existing references, in which the equilibrium points can be positive, negative, zero, and their combinations. On the other hand, any specific parameter on the activation function, behaved function, and amplification function is not involved in the stability criteria, which builds some general stability criteria for system (3.51) with the negative definiteness of matrix W. Because any information on the system parameters except W have not been used in the stability criteria, the conservativeness of the derived stability criteria will be increased.

(3) The existence condition on the nonnegative equilibrium points of system (3.51) is not discussed, which will effect the completeness of the studied results in [167].

(4) The stability criteria are only for the system without delays. For the delayed cases, it is not easy to establish such elegant stability criteria as stated in [167].

(5) The continuous dependence on the boundary data and initial condition is not discussed.

(6) Except above disadvantages, the main contribution of [167] is first introducing the reaction-diffusion term into the Cohen–Grossberg neural network, and based on some new assumptions on the system functions, some local and global asymptotic stability conditions have been established for the nonnegative equilibrium point of system (3.51), which are independent of reaction-diffusion term. All the assumptions on the system functions, e.g., activation function, behaved function, and amplification function, are rather general, which lead to the generalization of the stability criteria.

Note that, despite that the symmetry restriction on the matrix W is dropped out, Theorem 1 (or Theorem 2) in [167] will not always hold for any symmetric matrix W. If and only if symmetric matrix W is stable, e.g., Hurwitz stable, then Theorem 1 (or Theorem 2) in [167] always holds. Obviously, main results in [167] and [17] are different sufficient conditions to guarantee the (local) asymptotic stability for system (3.20) due to the different assumptions on the system functions and different construction of Lyapunov functional.

Therefore, we can draw the following conclusions for the comparisons of the different stability criteria. For the same system (3.20), under different assumptions on the physical parameters, it is natural to derive some different stability conditions. Also, with different Lyapunov functionals, the derived stability criteria are generally different. In particular, even we have derived some same stability conditions under different assumptions on the physical parameters of system (3.20) using same/different Lyapunov functionals, these stability conditions are generally different and cannot be compared if the assumptions on the system parameters are ignored. Therefore, before comparing one stability criterion with the other, it is very important to consider the restriction conditions or the assumptions on the system (3.20). Under different assumptions on the system parameters, the stability criteria cannot be compared in essence. Even the expression form of stability criteria are the same, they are not equivalent due to different assumptions on the system parameters.

In the following, we always suppose that the behaved function $a_i(u_i(t))$ satisfies Assumption 3.2 and amplification function $d_i(u_i(t))$ satisfies Assumption 3.5 if there are no other explanations.

When the activation function is a *quasi-Lipschitz condition* or *linear growth condition* $|g_i(s)| \leq \delta_i|s|+q_i$ instead of global Lipschitz Assumption 3.3 and the behaved function satisfies $a_i(s)\text{sgn}(s) \geq \gamma_i|s| - \beta_i$, bounded delays $\tau_{ij}(t)$ and positive initial data, Ref. [112] considers the following networks,

$$\dot{u}_i(t) = -d_i(u_i(t))\left[a_i(u_i(t)) - \sum_{j=1}^{n} w_{ij}g_j(u_j(t)) - \sum_{j=1}^{n} w_{ij}^1 g_j(u_j(t-\tau_{ij}(t)))\right]. \tag{3.58}$$

If the following matrix

$$\Gamma - (|W| + |W_1|)\Delta \tag{3.59}$$

is an M-matrix, then system (3.58) has a unique equilibrium point, which is globally asymptotically stable, where $|W| = (|w_{ij}|)$.

For system (3.58), Ref. [116] requires the following inequality to hold

$$2L_i\gamma_i - \sum_{j=1}^{n}(L_i w_{ij}\delta_j + L_j w_{ji}\delta_i) - \sum_{j=1}^{n}(L_i w_{ij}^1\delta_j + L_j w_{ji}^1\delta_i) > 0, \tag{3.60}$$

where $L_i > 0$, or the following matrix

$$2\Gamma - (|W| + |W_1|)\Delta - \Delta(|W| + |W_1|)^T \tag{3.61}$$

to be an M-matrix, which guarantees the unique equilibrium point of system (3.58), independent of time-varying delays. Obviously, the existence condition (3.60) in [116] improves the existence condition (3.59) in [112]. As same as that in (3.59), the global asymptotic stability condition in [116] is the same as condition (3.59).

Under Assumption 3.5, system (3.58) with $\tau_{ij}(t) = \tau$ has been studied in [117]. If $P[\Gamma|\Delta^{-1} - (W + W_1)]$ is Lyapunov diagonal stable, or equivalently, the following linear matrix inequality holds,

$$P[\Gamma\Delta^{-1} - W - W_1)] + [\Gamma\Delta^{-1} - W - W_1)]^T P > 0, \tag{3.62}$$

where P is a positive diagonal matrix, then there exists a unique nonnegative equilibrium point of system (3.58) with $\tau_{ij}(t) = \tau$. If there exists a positive diagonal matrix P and positive definite symmetric matrix Q such that the following linear matrix inequality holds,

$$\begin{bmatrix} 2P\Gamma\Delta^{-1} - PW - (PW)^T - Q & -PW_1 \\ -(PW_1)^T & Q \end{bmatrix} > 0, \tag{3.63}$$

then the unique nonnegative equilibrium point of system (3.58) with $\tau_{ij}(t) = \tau$ is globally asymptotically stable. If the unique equilibrium point is positive, then condition (3.63) can ensure the global exponential stability of system (3.58) with $\tau_{ij}(t) = \tau$.

Note that using *Schur Complement Lemma* [120], condition (3.63) is equivalent to the following form

$$2P\Gamma\Delta^{-1} - PW - (PW)^T - Q - PW_1Q^{-1}(PW_1)^T > 0. \tag{3.64}$$

Comparing the uniqueness condition (3.62) and the stability condition (3.64), we have the following relation

$$
\begin{aligned}
& -2P\Gamma\Delta^{-1} + PW + PW_1 + W^T P + W^T P \ \text{(uniqueness)} \\
& \leq -2P\Gamma\Delta^{-1} + PW + W^T P + Q + PW_1 Q^{-1}(PW_1)^T < 0. \ \text{(stability)} \quad (3.65)
\end{aligned}
$$

Or

$$
\begin{aligned}
& \text{Uniqueness conditions} \Leftarrow \text{stability conditions} \\
& \text{Uniqueness conditions} \not\Rightarrow \text{stability conditions} \quad (3.66)
\end{aligned}
$$

Obviously, the uniqueness and stability conditions are generally different and are not equivalent. As far as the existence of the *equilibrium* point is concerned, unique condition (3.62) is less conservative than condition (3.64), which can also guarantee the existence of the equilibrium point. Since the stability condition (3.64) ensures the uniqueness condition (3.62), then the proof on the existence and the proof of uniqueness in the existing stability are not necessary. It is well known that the existence condition or uniqueness condition cannot guarantee the stability of the concerned equilibrium point, since the interested unique equilibrium point may be unstable. Therefore, we can draw the following conclusion: almost all the existing stability conditions are sufficient and correspondingly guarantee the existence and uniqueness of the equilibrium points. If the concerned network models are well defined, in a non-strict sense of mathematical analytical viewpoint, the existence and uniqueness of the equilibrium point naturally holds if global stability conditions have been drawn. As scientific researchers, the existence of the equilibrium point must be guaranteed in advance; otherwise, the study will lose its meaning (on the contrary, engineers may be more interested in the implementation of a given performance for a practical system). This is one of fundamental differences between scholars and engineers in the aspects of considering an objective problem. After all, for rigorous consideration, all the preliminaries on the concerned problem must be stated clearly beforehand so as to ensure the well-posedness of the concerned problem.

For the following Cohen–Grossberg neural networks with finite distributed delays

$$
\begin{aligned}
\dot{u}_i(t) = & -d_i(u_i(t)) \Bigg[a_i(u_i(t)) - \sum_{j=1}^{n} w_{ij} g_j(u_j(t)) \\
& - \sum_{k=1}^{N} \sum_{j=1}^{n} w_{ij}^k g_j(u_j(t - \tau_{kj}(t))) \\
& - \sum_{l=1}^{r} \sum_{j=1}^{n} b_{ij}^l \int_{t-d_l}^{t} g_j(u_j(s)) \mathrm{d}s \Bigg], \quad (3.67)
\end{aligned}
$$

Ref. [131] studies the global asymptotic stability problem and establishes the following sufficient condition

$$-2P\Gamma\Delta^{-1} + PW + (PW)^T + \sum_{i=1}^{N}(PW_iQ_i^{-1}W_i^TP + Q_i)$$

$$+\sum_{l=1}^{r}(d_lY_l + d_iPB_lY_l^{-1}B_l^TP) < 0, \quad (3.68)$$

where P, Q_i, Y_l are positive diagonal matrices to be solved, $W = (w_{ij})$, $W_k = (w_{ij}^k)$, and $B_l = (b_{ij}^l)$. Obviously, condition (3.68) in [131] includes the condition (3.63) in [117] as a special case, and unifies many LMI-based stability results in the literatures.

3.3.4 Stability via M-Matrix or Algebraic Inequality Methods

In the aspects of amplification function and the behaved function, Ref. [18, 35, 36, 225] improves the conditions in [17], which has become the main tendency in the analysis of dynamics of Cohen–Grossberg neural networks, referring to Assumptions 3.2 (in which $a_i(u_i(t)) \propto u_i(t)$ instead of $a_i(u_i(t)) \propto 1/u_i(t)$) and Assumption 3.3. Therefore, a lot of different stability results on the equilibrium point of Cohen–Grossberg neural networks (3.20) and their variants have been established, and more and more less conservative and practical stability criteria have been proposed in different applications. This makes the research on recurrent neural network theory more and more fruitful and promising. It is the specific division, according to the principle from general to specialty, that makes the different disciplines develop quickly. A detailed global stability analysis for system (3.20) and its variants can be found in [64], in which all the results are based on algebraic inequality method.

Under Assumptions 3.2 and 3.3, Ref. [36] studies the global stability for system (3.20) without requiring the boundedness of activation function and positive lower bound of amplification function. If the following matrix

$$\Gamma\Delta^{-1} - W \quad (3.69)$$

is Lyapunov diagonal stable, i.e., there exists a positive diagonal matrix $P = \mathrm{diag}(p_1, \ldots, p_n)$ such that

$$P(\Gamma\Delta^{-1} - W) + (\Gamma\Delta^{-1} - W)^TP > 0 \quad (3.70)$$

is positive definite, then the unique equilibrium point of system (3.20) is the global asymptotic stability, which is independent of the amplification function [36]. More importantly, the relation between M-matrix and Lyapunov diagonal stability has been discussed in [36]. From (3.70) we have

$$p_i(\gamma_i\delta_i^{-1} - w_{ii}) - \frac{1}{2}\sum_{j=1, j\neq i}^{n}(p_iw_{ij} + p_jw_{ji})$$

$$\geq p_i(\gamma_i \delta_i^{-1} - w_{ii}) - \frac{1}{2} \sum_{j=1, j\neq i}^{n} (p_i |w_{ij}| + p_j |w_{ji}|). \tag{3.71}$$

or

$$\begin{aligned} & - p_i(\gamma_i \delta_i^{-1} - w_{ii}) + \frac{1}{2} \sum_{j=1, j\neq i}^{n} (p_i w_{ij} + p_j w_{ji}) \\ & \leq - p_i(\gamma_i \delta_i^{-1} - w_{ii}) + \frac{1}{2} \sum_{j=1, j\neq i}^{n} (p_i |w_{ij}| + p_j |w_{ji}|). \end{aligned} \tag{3.72}$$

If $\Gamma \Delta^{-1} - W^*$ is a nonsingular M-matrix, where $W^* = (w_{ij}^*)$, $w_{ij}^* = |w_{ij}|$ for $i \neq j$, $w_{ij}^* = w_{ij}$ for $i = j$, we can see that the right-hand side of (3.71) is equal and greater than 0, or the right-hand side of (3.72) is equal and less than 0. Therefore, from inequalities (3.71), (3.72) and M-matrix property, we can conclude that nonsingular M-matrix $\Gamma \Delta^{-1} - W^*$ also guarantees the global asymptotic stability of system (3.20) if $\Gamma \Delta^{-1} - W$ is Lyapunov diagonal stable. The relations are as follows

$$\begin{aligned} & \text{nonsingular M-matrix } \Gamma \Delta^{-1} - W^* \\ & \Rightarrow \Gamma \Delta^{-1} - W \text{ is Lyapunov diagonal stable} \\ & \Rightarrow \text{Global asymptotic stability}. \end{aligned} \tag{3.73}$$

Obviously, for the Cohen–Grossberg neural networks (3.20) without delays, we can see that LDS (Lyapunov diagonal stability) condition is less conservative than M-matrix stability condition.

For Cohen–Grossberg neural network (3.20) with positive interconnection coefficients, asymptotic stability criteria based on M-matrix and Lyapunov diagonal stable concept is equivalent. It is the M-matrix that builds a bridge between algebraic inequality and LMI method. However, for the Cohen–Grossberg neural network (3.20) with delays, asymptotic stability criteria based on M-matrix and LMI approach are not equivalent any more. Generally speaking, *M-matrix*-based stability results can have a unified expressions; see M-matrix-based stability results for Cohen–Grossberg neural network (3.20) with different kinds of delays in the sequel. However, we also see that LMI-based results often have different/various expressions for Cohen–Grossberg neural network (3.20) with different kinds of delays. Therefore, many different kinds of LMI-based stability results have been proposed in the literature for Cohen–Grossberg neural network (3.20) with different kinds of delays. If the lower bound of amplification function is given, then it is proved in [36] that the condition (3.69) or (3.70) also guarantees the global exponential stability. In contrast, exponential stability results in [35] requires the lower and upper bounds of amplification function in Cohen–Grossberg neural network (3.20).

When the activation function is a quasi-Lipschitz condition or *linear growth condition* $|g_i(s)| \leq \delta_i |s| + q_i$ instead of global Lipschitz Assumption 3.3 and the behaved function satisfies $a_i(s)\text{sgn}(s) \geq \gamma_i |s| - \beta_i$, where $q_i \geq 0$ and $\beta_i \geq 0$ are nonnegative constants, Ref. [112] studies the global stability for system (3.58) without requiring the boundedness of activation function and positive lower bound of amplification function. If the following matrix

$$\Gamma - |W + W_1|\Delta \tag{3.74}$$

is a nonsingular M-matrix, then system (3.58) has at least one equilibrium point. Since

$$\Gamma - |W + W_1|\Delta \geq \Gamma - (|W| + |W_1|)\Delta, \tag{3.75}$$

where $A = (a_{ij}) \geq B = (b_{ij})$ implies $a_{ij} \geq b_{ij}$, then if

$$\Gamma - (|W| + |W_1|)\Delta \tag{3.76}$$

is a nonsingular M-matrix, then system (3.58) also has at least one equilibrium point. Obviously, existence condition (3.74) is less conservative than condition (3.76). Further, under Assumptions 3.2 and 3.3, Theorem 2 in [112] states that a nonsingular M-matrix condition (3.76) also guarantees the global asymptotic stability of system (3.58).

For system (3.58) with bounded activation function, Ref. [118] establishes the following inequality to guarantee the global exponential stability,

$$\underline{d}_i\gamma_i - \sum_{j=1}^{n} \overline{d}_i(|w_{ij}| + |w_{ij}^1|)\delta_j > 0, \tag{3.77}$$

which is dependent on the bounds of amplification function. Condition (3.77) improves the stability results in [47].

Under Assumptions 3.2 and 3.3 and the positive lower bound of amplification, the following system

$$\dot{u}_i(t) = -d_i(u_i(t))\left[a_i(u_i(t)) - \sum_{j=1}^{n} w_{ij}^1 g_j(u_j(t - \tau_{ij}(t)))\right], \tag{3.78}$$

has been studied in [108]. It has been proved that if the following determinant holds,

$$\det(\Gamma - W_1 K) \neq 0 \tag{3.79}$$

for diagonal matrix K satisfying $-\Delta \leq K \leq \Delta$, then system (3.78) has a unique equilibrium point. Further, if the following matrix

$$\Gamma\Delta^{-1} - |W_1| \tag{3.80}$$

is a nonsingular M-matrix, then the equilibrium point of system (3.78) is globally exponentially stable. Obviously, from (3.80) we can deduce the condition (3.79).

Under Assumptions 3.1–3.3, [109] requires the following matrix

$$M_0 = \underline{D}\Gamma - \sum_{k=0}^{N} |W^k| \Delta \overline{D} \tag{3.81}$$

to be an nonsingular M-matrix, which guarantees the Cohen–Grossberg neural networks (3.23) to be global asymptotic stability, where $|W^k| = (|w_{ij}^k|)_{n\times n}$.

Note that the results in [109] can also be applied to the following networks,

$$\dot{u}_i(t) = -d_i(u_i(t)) \left[a_i(u_i(t)) - \sum_{k=0}^{N} \sum_{j=1}^{n} w_{ij} g_j(u_j(t - \tau_{ij}^k)) \right], \tag{3.82}$$

$$\dot{u}_i(t) = -d_i(u_i(t)) \left[a_i(u_i(t)) - \sum_{j=1}^{n} w_{ij} \int_{-\infty}^{t} K_{ij}(t-s) g_j(u_j(s)) \mathrm{d}s \right], \tag{3.83}$$

$$\dot{u}_i(t) = -d_i(u_i(t)) \left[a_i(u_i(t)) - \sum_{j=1}^{n} w_{ij} g_j \left(\int_{-\infty}^{t} K_{ij}(t-s) u_j(s) \mathrm{d}s \right) \right], \tag{3.84}$$

which includes the models (3.28), (3.29) and (3.31) as special cases. The unifies global asymptotic stability criterion for models (3.82), (3.83) and (3.84) requires

$$M_0^{'} = \underline{D}\Gamma - |W| \Delta \overline{D} \tag{3.85}$$

and

$$M_0^{''} = \Gamma - |W| \Delta \tag{3.86}$$

to be an nonsingular M-matrix, where $|W| = (|w_{ij}|)_{n\times n}$.

Obviously, stability results in the form of M-matrix in [109] can give a unified expression for Cohen–Grossberg neural networks with many different types of delays, and it is easy to check.

Under the same assumptions, some results in [103] have improved the main result in [109] for the model (3.82) with $N = 0$, which can be expressed as a set of algebraic inequalities independent of amplification functions (see Theorem 10 in [103]). For example, Theorem 10 in [103] includes the following M-matrix

$$M_0^{'''} = \Gamma - W^* \Delta \tag{3.87}$$

as a special case, where $W^* = (w_{ij}^*)$, $w_{ij}^* = |w_{ij}|$ for $i \neq j$, $w_{ij}^* = w_{ij}$ for $i = j$. It is clear that from (3.85), (3.86), and (3.87), we can have the following relations

$$\underline{D}\Gamma - |W|\Delta\overline{D} \Rightarrow \Gamma - |W|\Delta \Rightarrow \Gamma - W^*\Delta \Rightarrow \text{global asymptitic stability}, \tag{3.88}$$

and the reverse is generally false. Obviously, Theorem 10 in [103] is less conservative than conditions in [109].

Comparing the stability results in [109] with that in [103], we can find that the conservativeness of the results in [103] is decreased with involving many adjustable parameters at the expense of increasing the computational complexity. Altogether, the paper [103] presents a set of less conservative stability results in the algebraic inequality forms under different assumptions on amplification functions, activation functions, and the behaved functions for Cohen–Grossberg neural networks. Both [109] and [103] are excellent papers in briefly introducing the development and applications of recurrent neural networks. It should be emphasized that all the stability criteria can be compared only under the same prerequisite conditions. Otherwise, they cannot be compared.

Under Assumptions 3.1–3.3, and activation function being bounded, for the following Cohen–Grossberg neural network,

$$\dot{u}_i(t) = -d_i(u_i(t))\left[a_i(u_i(t)) - \sum_{j=1}^{n} w_{ij}g_j(u_j(t)) - \sum_{j=1}^{n} w_{ij}^1 g_j(u_j(t-\tau_{ij}))\right], \tag{3.89}$$

[114] requires the matrix $M_1 = (m_{ij}^1)_{n\times n}$ to be a nonsingular M-matrix, then the equilibrium point is unique and globally exponential stable, where $m_{ii}^1 = \gamma_i - w_{ii}\delta_i - |w_{ij}^1|\delta_i$, $M_{ij}^1 = -(|w_{ij}| + |w_{ij}^1|)\delta_j$ for $i \neq j$. Also, the main result in [114] is equivalent to the following nonsingular M-matrix

$$M_1 = \Gamma - W^*\Delta - |W_1|\Delta \tag{3.90}$$

or algebraic inequality

$$M_1^{'} = \zeta_i\gamma_i - \zeta_i w_{ii}\delta_i - \sum_{j=1,j\neq i}^{n} \zeta_j|w_{ji}|\delta_i - \sum_{j=1}^{n}\zeta_j|w_{ji}^1|\delta_i > 0, \tag{3.91}$$

$$M_1^{''} = \zeta_i\gamma_i - \zeta_i w_{ii}\delta_i - \sum_{j=1,j\neq i}^{n} \zeta_j|w_{ij}|\delta_j - \sum_{j=1}^{n}\zeta_j|w_{ij}^1|\delta_j > 0, \tag{3.92}$$

$$\begin{aligned} M_1^{'''} =& \zeta_i\gamma_i - \zeta_i w_{ii}\delta_i - \frac{\sum_{j=1,j\neq i}^{n}(\zeta_j|w_{ji}|\delta_i + \zeta_i|w_{ij}|\delta_j)}{2} \\ & - \frac{\sum_{j=1}^{n}(\zeta_j|w_{ji}^1|\delta_i + \zeta_i|w_{ij}^1|\delta_j)}{2} > 0, \end{aligned} \tag{3.93}$$

for positive constant $\zeta_i > 0$, where W^* is the same as that in (3.87), $|W_1| = (|w_{ij}^1|)_{n\times n}$.

We should notice that inequalities (3.91), (3.92) and (3.93) are equivalent to $\mu_1(\zeta M_1) < 0$ (strictly diagonally column-dominant), $\mu_\infty(M_1\zeta) < 0$ (strictly diagonally row-dominant) and $\mu_2(\zeta M_1) < 0$, respectively, where $\zeta = \mathrm{diag}(\zeta_1, \zeta_2, \ldots, \zeta_n)$ is a positive diagonal matrix, and for a matrix $M = (m_{ij})_{n\times n}$, the three matrix measures are defined by $\mu_1(M) = \max_i(m_{ii} + \sum_{j\neq i} m_{ji})$, $\mu_\infty(M) = \max_i(m_{ii} + \sum_{j\neq i} m_{ij})$, $\mu_2(M) = \lambda_{max}\{(M + M^T)/2\}$, $\lambda_{max}(\cdot)$ denotes the maximal eigenvalue of a symmetric square matrix. Therefore, the main result in [114] improved the results in [35, 47, 281–284].

When the *activation function* $g_i(u_i(t))$ is absolutely continuous, $0 \le g_i^{'}(u_i(t)) \le 1$ or $|f_i^{'}(u_i(t))| \le 1$, Assumption 3.1 and Assumption 3.2 hold, the following Cohen–Grossberg neural networks

$$\dot{u}_i(t) = -d_i(u_i(t))\left[a_i(u_i(t)) - \sum_{j=1}^{n} w_{ij}g_j(u_j(t)) - \sum_{j=1}^{n} w_{ij}^1 f_j(u_j(t-\tau_{ij}(t)))\right], \tag{3.94}$$

have been studied in [149], where $\dot{\tau}_{ij}(t) < 1$ and $\tau_{ij}(t) \le \tau_{ij}$. In the case of $|g_i^{'}(u_i(t))| \le 1$ and $|f_i^{'}(u_i(t))| \le 1$, Theorem 2 in [149] requires the following matrix

$$M_2 = \Gamma - |W| - |W_1| \tag{3.95}$$

to be a nonsingular M-matrix, or the following algebraic inequality to be true,

$$M_2^{'} = \zeta_j\gamma_j - \sum_{i=1}^{n} \zeta_i|w_{ij}| - \sum_{i=1}^{n} \zeta_i|w_{ij}^1| > 0, \tag{3.96}$$

where $\zeta_j > 0$, which all guarantee the global exponential stability of the equilibrium point for system (3.94). In the case of $0 \le g^{'}(u(t)) \le 1$ and $|f_i^{'}(u_i(t))| \le 1$, Theorem 1 in [149] requires the following algebraic inequality to be true,

$$M_2^{''} = \zeta_j(\gamma_j - \epsilon \underline{d}_j^{-1}) - \left[\zeta_j w_{jj} + \sum_{i=1, i\neq j}^{n} \zeta_i|w_{ij}|\right]^+ - \sum_{i=1}^{n} \zeta_i e^{\epsilon\tau_{ij}}|w_{ij}^1| \ge 0, \tag{3.97}$$

and Corollary 1.1 in [149] requires the following algebraic inequality to hold

$$M_2^{'''} = \zeta_j\gamma_j - \left[\zeta_j w_{jj} + \sum_{i=1, i\neq j}^{n} \zeta_i|w_{ij}|\right]^+ - \sum_{i=1}^{n} \zeta_i|w_{ij}^1| > 0, \tag{3.98}$$

which all guarantee the global exponential stability of the equilibrium point for system (3.94), where $\zeta_j > 0$ and $\epsilon > 0$, $x^+ = \max\{0, x\}$. According to [114, 149], (3.97) and (3.98) are equivalent. When $g_i(u_i(t)) = f_i(u_i(t))$ is bounded and $|g_i^{'}(u_i(t))| \leq 1$, for system (3.94) with constant delays, we can see that stability results (3.90) and (3.91) in [114] are less conservative than those stability results (3.95), (3.96), and (3.98) in [149], respectively, which all can be expressed in M-matrix form. As far as the time-varying delays are concerned, the results in [149] improves some existing results.

For the nonautonomous case of (3.94), i.e.,

$$\dot{u}_i(t) = -d_i(u_i(t))\Bigg[a_i(t, u_i(t)) - \sum_{j=1}^{n} w_{ij}(t)g_j(u_j(t)) - \sum_{j=1}^{n} w_{ij}^1(t)f_j(u_j(t-\tau_{ij}(t)))\Bigg], \tag{3.99}$$

Theorem 1 in [285] requires the following inequality to be true,

$$\max \sum_{j=1}^{n} \frac{\overline{d}_j\zeta_j\delta_j(w_{ij}^+(t) + |w_{ij}(t)|)}{\zeta_i\underline{d}_i\gamma_i(t)} < 1 \tag{3.100}$$

under the Assumption 3.4 and requirements of ω-periodic function on $\gamma_i(t) > 0$, $w_{ij}(t)$, $w_{ij}^1(t)$ and $\tau_{ij}(t)$ (see H_3 in [285]), then system (3.99) has an ω-periodic solution which is globally attractive, where $\zeta_i > 0$. Under the Assumption 3.3 and requirements of H_3 in [285], if the following inequality holds,

$$\max \sum_{j=1}^{n} \frac{\overline{d}_j\zeta_j\delta_j(|w_{ij}(t)| + |w_{ij}(t)|)}{\zeta_i\underline{d}_i\gamma_i(t)} < 1, \tag{3.101}$$

then system (3.99) has an ω-periodic solution which is globally attractive, where $\zeta_i > 0$. Obviously, for the case of constant coefficients, the results in [285] can be equivalent to some forms of M-matrix as those defined in (3.104) and (3.105) later.

For the Cohen–Grossberg neural network (3.94) with reaction-diffusion term,

$$\frac{\partial u_i(t)}{\partial t} = \sum_{k=1}^{m} \frac{\partial}{\partial x_k}\left(D_{ik}\frac{\partial u_i(t,x)}{\partial x_k}\right) - d_i(u_i(t,x))\Bigg[a_i(u_i(t,x)) - \sum_{j=1}^{n} w_{ij}g_j(u_j(t,x))$$

$$-\sum_{j=1}^{n} w_{ij}^{1} f_j(u_j(t-\tau_{ij}(t),x))\Bigg], \tag{3.102}$$

global exponential stability problem has been discussed in [148], where $x = (x_1, x_2, \ldots, x_m)^T \in \Omega \subset \mathbb{R}^m$, Ω is a bounded compact set with smooth boundary $\partial\Omega$ and measure mes$\Omega > 0$ in space $\mathbb{R}^m$, $u = (u_1(t,x), \ldots, u_n(t,x))^T$, $u_i(t,x)$ is the state of the ith unit at time t and in space x, $b_{ik} = b_{ik}(t,x,u) \geq 0$ denotes the transmission diffusion operator along the ith neuron. The boundary condition of (3.102) is given by

$$\begin{aligned}\frac{\partial u_i(t,x)}{\partial \bar{n}} &= \left(\frac{\partial u_i(t,x)}{\partial x_1}, \frac{\partial u_i(t,x)}{\partial x_2}, \ldots, \frac{\partial u_i(t,x)}{\partial x_m}\right)^T \\ &= 0, x \in \partial\Omega,\end{aligned}$$

$u_i(s,x) = \bar{\phi}_i(s,x)$, $\frac{\partial}{\partial t}u_i(s,x) = \frac{\partial \bar{\phi}_i(s,x)}{\partial t}$, in which $\bar{\phi}_i(s,x)$ are bounded and continuous differentiable functions, $s \in (-\infty, 0]$, $i = 1, 2, \ldots, n$, $\bar{n}$ is the outer normal vector of $\partial\Omega$.

Under Assumptions 3.1–3.3, and bounded activation functions are globally Lipschitz with positive constants δ_i, δ_i^0, i.e., $|g_i(\zeta) - g_i(\xi)| \leq \delta_i|\zeta - \xi|$, $|f_i(\zeta) - f_i(\xi)| \leq \delta_i^0|\zeta - \xi|$ for $\zeta, \xi \in \mathbb{R}$, Corollary 3.2 in [148] establishes a global exponential stability condition if the following inequality holds,

$$M_3 = \underline{d}_i\gamma_i - \sum_{j=1}^{n} \overline{d}_j|w_{ji}|\delta_i - \sum_{i=1}^{n} \overline{d}_j|w_{ji}^1|\delta_i^0 > 0. \tag{3.103}$$

Obviously, if

$$M_3^{'} = \underline{D}\Gamma - |W|\overline{D}\Delta - W_1\overline{D}\Delta^0 \tag{3.104}$$

is a nonsingular M-matrix, condition (3.103) is naturally satisfied, where $\underline{D} = \text{diag}(\underline{d}_1, \ldots, \underline{d}_n)$, $\overline{D} = \text{diag}(\overline{d}_1, \ldots, \overline{d}_n)$, and $\Delta^0 = \text{diag}(\delta_1^0, \ldots, \delta_n^0)$. For the Hopfield neural network with reaction-diffusion term (3.102), the following *M-matrix* has been derived to ensure the global exponential stability for system (3.102) in [154],

$$M_0^{''''} = \Gamma - W^+\Delta - |W_1|\Delta, \tag{3.105}$$

where $W^+ = (w_{ij}^+)$, $w_{ii}^+ = \max\{0, w_{ii}\}$, $w_{ij}^+ = |w_{ij}|$ for $i \neq j$. For the Hopfield neural network with reaction-diffusion term, which is a special case of (3.102) with reaction-diffusion term, in the case of $w_{ij} = 0$, main result in [110] requires

$$M_3^{''} = \Gamma - \Delta^0|W_1| \tag{3.106}$$

to be a nonsingular M-matrix, which guarantees the global exponential stability of the unique equilibrium point. It is clear that for Cohen–Grossberg neural network (3.102) with/without reaction-diffusion term, the stability conditions are the same in the expression form. However, the restriction conditions on reaction-diffusion term and delay terms are different in the prerequisite of the networks.

For stochastic Hopfield neural networks (3.102) with constant delays, i.e.,

$$\begin{aligned} \mathrm{d}y_i(t,x) = & \sum_{k=1}^{m} \frac{\partial}{\partial x_k}\left(D_{ik}\frac{\partial u_i(t,x)}{\partial x_k}\right) \\ & - \left[a_i(u_i(t,x)) - \sum_{j=1}^{n} w_{ij}g_j(u_j(t,x)) - \sum_{j=1}^{n} w_{ij}^1 f_j(u_j(t-\tau_{ij},x))\right]\mathrm{d}t \\ & + \sum_{j=1}^{n} \sigma_{ij}(y_j(t,x))\mathrm{d}\omega_j(t), \end{aligned} \tag{3.107}$$

global exponential stability problem has been discussed in [286], where $\omega(t) = (\omega_1(t),\ldots,\omega_n(t))^T$ is an n-dimensional Brownian motion defined on a complete probability space $(\Omega, , \mathcal{F}, \mathbb{P})$ with a natural filtration $\{\mathcal{F}_t\}_{t\geq 0}$ generated by $\{\omega(s) : 0 \leq s \leq t\}$, where we associate Ω with the canonical space generated by all $\{\omega_i(t)\}$, and denote by $\mathcal{F}$ the associated σ-algebra generated by $\{\omega(t)\}$ with the probability measure $\mathbb{P}$.

For the deterministic case of (3.107), condition (3.104) with $\underline{d}_i = \overline{d}_i = 1$ has been derived in [286] to guarantee the global exponential stability of the unique equilibrium point. For system (3.107), in the case of $\sigma_{ij}(y_j^*) = 0$ and $\sigma_{ij}(\cdot)$ being Lipschitz continuous with Lipschitz constant L_{ij}, the following M-matrices conditions have been derived,

$$M_4 = \Gamma - |W|\Delta - W_1\Delta^0 - \overline{C}, \tag{3.108}$$

$$M_4^{'} = \Gamma - |W|\Delta - W_1\Delta^0 - \tilde{C}, \tag{3.109}$$

which guarantee the almost sure exponential stability (or exponential stability in mean square) and mean-value exponential stability, respectively, where $\overline{C} = \mathrm{diag}(\overline{c}_1,\ldots,\overline{c}_n)$, $\overline{c}_i = -\gamma_i + \sum_{j=1}^{n} w_{ij}\delta_j + \sum_{j=1}^{n} w_{ij}^1\delta_j^0 + \sum_{j=1}^{n} L_{ij}^2 \geq 0$, $\tilde{C} = \mathrm{diag}(\tilde{c}_1,\ldots,\tilde{c}_n)$, $\tilde{c}_i = 0.5\sum_{j=1}^{n} L_{ij}^2 + K_1(\sum_{j=1}^{n} L_{ij}^2)^{1/2} \geq 0$, $K_1 > 0$ is a constant. Obviously, M-matrix (3.108) or (3.109) unifies many existing results as its special cases, for example, the stability results in [110, 114, 148, 149, 151].

For the reaction-diffusion Hopfield neural networks (3.102) with continuously distributed delays,

$$\frac{\partial u_i(t)}{\partial t} = \sum_{k=1}^{m} \frac{\partial}{\partial x_k}\left(D_{ik}\frac{\partial u_i(t,x)}{\partial x_k}\right) - \left[a_i(u_i(t,x)) - \sum_{j=1}^{n} w_{ij} g_j(u_j(t,x)) - \sum_{j=1}^{n} w_{ij}^1 \int_{-\infty}^{t} K_{ij}(t-s) g_j(u_j(s,x))\mathrm{d}s\right], \tag{3.110}$$

global exponential stability problem has been studied in [246, 287]. The following M-matrices have been derived to ensure the global exponential stability for system (3.110) in [246, 287], respectively,

$$M_0^{''''} = \Gamma - |W|\Delta - |W_1|\Delta, \tag{3.111}$$

$$\bar{M}_0^{''''} = \Gamma - W^+\Delta - |W_1|\Delta. \tag{3.112}$$

For the reaction-diffusion Hopfield neural networks (3.102) with continuously distributed delays,

$$\frac{\partial u_i(t)}{\partial t} = \sum_{k=1}^{m} \frac{\partial}{\partial x_k}\left(D_{ik}\frac{\partial u_i(t,x)}{\partial x_k}\right) - \left[a_i(u_i(t,x)) - \sum_{j=1}^{n} w_{ij}^1 f_j\left(\int_0^{\infty} K_{ij}(s) u_j(t-s,x)\mathrm{d}s\right)\right], \tag{3.113}$$

and

$$\int_0^{\infty} K_{ij}(s)\mathrm{d}s = k_{ij} > 0, \tag{3.114}$$

global exponential stability problem has been studied in [288]. The following M-matrix has been derived to ensure the global exponential stability for system (3.113) in [288],

$$M_0^{''''} = \Gamma - \Delta^0|W_1^a|, \tag{3.115}$$

where $W_1^a = (w_{ij}^1 k_{ij})$. If $k_{ij} = 1$ in (3.114), the global asymptotic stability problem for (3.113) has been studied in [289], which has the same result as that in (3.115).

For the following systems with distributed delays

$$\dot{u}_i(t) = -\left[a_i(u_i(t)) - \sum_{j=1}^{n} w_{ij}g_j(u_j(t)) - \sum_{j=1}^{n} w_{ij}^1 f_j(u_j(t-\tau_{ij}(t))) - \sum_{j=1}^{n} w_{ij}^2 \int_0^{\infty} K_{ij}(s)h_j(u_j(t-s))\mathrm{d}s\right], \tag{3.116}$$

and

$$\int_0^{\infty} e^{\lambda s} K_{ij}(s)\mathrm{d}s = k_{ij}(\lambda) > 0, \tag{3.117}$$

where $0 \le (h_i(\zeta) - h_i(\xi))/(\zeta - \xi) \le \delta_i^1$, $k_{ij}(0) = 1$, Ref. [150] establishes the following condition

$$\left[\lambda I - \varGamma + |W|\varDelta + e^{\lambda\tau}|W_1|\varDelta^0 + (\rho(\lambda) \otimes |W_2|\varDelta^1)\right]\zeta < 0, \tag{3.118}$$

or

$$\varGamma - |W|\varDelta - |W_1|\varDelta^0 - |W_2|\varDelta^1 \tag{3.119}$$

is a nonsingular M-matrix, to guarantee the global exponential stability of the unique equilibria for system (3.116), where $\lambda > 0$ is a positive number, I is an identity matrix with appropriate dimension, $0 \le \tau_{ij}(t) \le \tau$, $A \otimes B = (a_{ij}b_{ij})$, $\zeta = (\zeta_1, \ldots, \zeta_n)^T > 0$, $\zeta_i > 0$, $W_2 = (w_{ij}^2)$, $\varDelta^1 = \mathrm{diag}(\delta_1^1, \ldots, \delta_n^1)$, $\rho(\lambda) = (k_{ij}(\rho))$.

Note that systems (3.113) and (3.116) can be changed into the form of system (3.110) by coordinate transformation. More generally, for such kind of distributed delays

$$\frac{\partial u_i(t)}{\partial t} = \sum_{k=1}^{m} \frac{\partial}{\partial x_k}\left(D_{ik}\frac{\partial u_i(t,x)}{\partial x_k}\right) - \left[a_i(u_i(t,x)) - \sum_{j=1}^{n} w_{ij}^1 f_j\left(\int_0^{\mathrm{T}} K_{ij}(s)u_j(t-s,x)\mathrm{d}s\right)\right], \tag{3.120}$$

where $\mathrm{T} > 0$ is a positive constant, we can also use the same method as those in (3.113) and (3.110) to analyze the stability of (3.120), the asymptotic stability criteria can be expressed as (3.115) [290]. Therefore, it is the same to deal with the continuous distributed delays in (3.110), (3.113), and (3.120).

However, we can find that conditions (3.106) and (3.115) are different from (3.86), (3.90), (3.105), and (3.108) because of the different position of amplification function $\varDelta$. By carefully checking and comparing, we can find that the expression form of

M-matrix in [110, 288, 289] is not correct due to the misuse of M-matrix properties. Therefore, if Δ^0 in (3.106) and (3.115) were removed to the right side of $|W_1|$ and $|W_1^a|$, respectively, then it would lead to the correct consequences.

For the neutral-type Cohen–Grossberg neural networks with constant delays, i.e.,

$$\dot{u}_i(t) + \sum_{j=1}^{n} e_{ij}\dot{u}_j(t-\tau_j) = -d_i(u_i(t))\left[a_i(u_i(t)) - \sum_{j=1}^{n} w_{ij}g_j(u_j(t)) - \sum_{j=1}^{n} w_{ij}^1 g_j(u_j(t-\tau_j))\right], \tag{3.121}$$

global asymptotic stability problem has been discussed in [291], where $E = (e_{ij})$ shows how the derivatives of the neurons are delay feedforward connected in the network, i.e., the time delay occurs in the state velocity vector. If $E = 0$, then (3.121) describes a class of neural networks with retarded-type delays or discrete time delays. In the case of $a_i(u_i(t))$ and $a_i^{-1}(u_i(t))$ being continuously differentiable and $0 < \gamma_i \le a_i'(u_i(t)) \le \gamma_i^0 < \infty$, the following conditions are presented,

$$0 \le \|E\| < 1 \tag{3.122}$$

$$LP(1+\|E\|) + LR(1+\|E\|) + Q < \min_{1\le i\le n}\{\underline{d}_i\gamma_i\}, \tag{3.123}$$

which guarantees the global asymptotic stability of system (3.121), where $L = \max\{\delta_i\}$, $P = \max\{\overline{d}_i\}\|W\|$, $Q = \max\{\overline{d}_i\gamma_i^0\}\|E\|$ and $R = \max\{\overline{d}_i\}\|W_1\|$. When $E = 0$ in (3.121), condition (3.123) can be reduced to the following form,

$$L(\|W\| + \|W_1\|)\max_i\{\overline{d}_i\} \le \min_i\{\underline{d}_i\gamma_i\}. \tag{3.124}$$

Further, if $\overline{d}_i = \underline{d}_i = 1$, then (3.124) can be reduced to the following form

$$L(\|W\| + \|W_1\|) \le \min_i\{\gamma_i\} \tag{3.125}$$

where $\|B\|$ is the Euclidean norm induced from the vector norm.

The nonlinear measure proposed in [30] is only some rephrase of the part concerning with L_1 norm in Corollary 1 in [31]. In [27], by using matrix measure theory, the authors obtained some test conditions for global stability of Hopfield neural networks, where all the coefficients C_i assumed to be 1. Test conditions in Theorem 9 in [27] can be easily obtained from Proposition 4 in [31]. Moreover, by the *direct approach* in [31], the authors show that all the test conditions do not depend on the parameters C_i. Moreover, the direct method is more readable and applicable. It is also easy to see that the conditions in Proposition 4 in [31] are less restrictive than those in the Theorem 9 in [27].

3.3.5 Stability via Matrix Inequalities or Mixed Methods

In this subsection, the activation function is assumed to satisfy Assumption 3.4 if no other declaration is given.

For the following Cohen–Grossberg neural networks,

$$\dot{u}_i(t) = -d_i(u_i(t))\left[a_i(u_i(t)) - \sum_{j=1}^{n} w_{ij}g_j(u_j(t)) - \sum_{j=1}^{n} w_{ij}^1 f_j(u_j(t-\tau))\right], \tag{3.126}$$

which is a special case of system (3.94), global exponential stability has been studied in [60], and the following matrix inequality-based and matrix norm-based stability criteria have been presented, respectively,

$$2P\Gamma\Delta^{-1} - PW - (PW)^T - Q - PW_1Q^{-1}W_1^TP > 0 \tag{3.127}$$

or

$$\delta_M(\|W\| + \|W_1\|) < \gamma_m, \tag{3.128}$$

where P and Q are positive diagonal matrices, respectively, $\gamma_m = \min\{\gamma_i\}$, $\delta_M = \max\{\delta_i\}$, $i = 1, \ldots, n$. In fact, stability condition (3.127) is the same as stability condition (3.64) in [117]. In order to compare with the existing results, here we rewrite it again. It is easy to see that the stability condition (3.127) includes the stability conditions (3.56), Lyapunov diagonal stability condition (3.69) or (3.70) as special cases in expression forms. In the following, we will show that (3.128) can be recovered from (3.127). If $\|W_1\| = 0$, which implies $W_1 = 0$, we choose $P = I$ and $Q = (\gamma_m/\delta_M - \|W\|)I > 0$, respectively, then (3.127) becomes

$$2\Gamma\Delta^{-1} - W - W^T - (\gamma_m/\delta_M - \|W\|)I > 0. \tag{3.129}$$

If condition (3.129) holds, which implies that there exists any vector $x(t) \neq 0$ such that

$$x^T\Big(2\Gamma\Delta^{-1} - W - W^T - (\gamma_m/\delta_M - \|W\|)I\Big)x(t) > 0 \tag{3.130}$$

or

$$x^T\Big(2\gamma_m\delta_M^{-1} - 2\|W\| - (\gamma_m/\delta_M - \|W\|)\Big)x(t) > 0. \tag{3.131}$$

Obviously, condition (3.131) implies condition (3.128) with $\|W_1\| = 0$. This also means that for system (3.126) without delay, if the following conditions hold, which ensue the global exponential stability of the equilibrium point,

$$2P\Gamma\Delta^{-1} - PW - (PW)^T > 0, \tag{3.132}$$

then we can also derive a stability condition from (3.132) that if the following condition holds,

$$\delta_M \|W\| < \gamma_m, \tag{3.133}$$

then the equilibrium point of system (3.126) without delay is globally exponentially stable. For the case $\|W_1\| \neq 0$, we choose $P = I$ and $Q = \|W_1\| I > 0$, respectively, then (3.127) becomes

$$2\Gamma\Delta^{-1} - W - W^T - \|W_1\| I - \frac{1}{\|W_1\|} W_1 W_1^T > 0. \tag{3.134}$$

If condition (3.134) holds, which implies that there exists any vector $x(t) \neq 0$ such that

$$x^T \left(2\Gamma\Delta^{-1} - W - W^T - \|W_1\| I - \frac{1}{\|W_1\|} W_1 W_1^T\right) x(t) > 0 \tag{3.135}$$

or

$$x^T \left(2\gamma_m \delta_M^{-1} - 2\|W\| - 2\|W_1\|\right) x(t) > 0. \tag{3.136}$$

Obviously, condition (3.136) implies condition (3.128). Under Assumption 3.3, Ref. [224] presents the following global exponential stability criterion,

$$\overline{d}\delta_M(\|W\| + \|W_1\|) < \gamma_m \underline{d}, \tag{3.137}$$

and Ref. [322] presents the following global exponential stability criterion, respectively,

$$\overline{d}\delta_M(\|W\|_1 + \|W_1\|_1) < \gamma_m \underline{d}, \tag{3.138}$$

where $\|B\|_1 = \max_{1\leq i\leq n} \sum_{j=1}^{n} |b_{ij}|$ represents the first norm of matrix B, $B = (b_{ij})$. Since the Euclidean norm $\|B\|$ and the first norm $\|B\|_1$ are different in the measure space, they cannot be compared directly. Meanwhile, there exist some equivalent relations among different matrix norms, therefore, different matrix norm-based stability results have similar expression forms, see (3.137) and (3.138). Further, comparing (3.137) with (3.128), we can find the following relation,

$$(\|W\| + \|W_1\|) < \frac{\gamma_m \underline{d}}{\delta_M \overline{d}} \leq \frac{\gamma_m}{\delta_M} < 1. \tag{3.139}$$

Obviously, stability result (3.128) in [60] is conservative than result (3.137) in [224].

For system (3.83), or for system (3.126) with $W = 0$, under the assumption of bounded activation function, Ref. [168] requires the following condition to be true,

$$\rho(AM) < 1, A = \text{diag}\left(\frac{\overline{d}_1}{\underline{d}_1}, \ldots, \frac{\overline{d}_n}{\underline{d}_n}\right), M = (m_{ij}), m_{ij} = \frac{|w^1_{ij}|\delta_j}{\gamma_i}, \tag{3.140}$$

where $\rho(B) = \max_i |\lambda_i|$ is the spectral radius of a square matrix B, λ_i is the eigenvalue of B, then the uniqueness of system (3.83) or system (3.126) with $W = 0$ is uniformly stable, uniformly bounded, globally attractive and globally asymptotically stable. Obviously, from (3.140) we have

$$AM = \left(\frac{\overline{d}_i|w^1_{ij}|\delta_j}{\underline{d}_i\gamma_i}\right) = A\Gamma^{-1}W_1\Delta, \tag{3.141}$$

which implies

$$\rho(AM) \leq \|AM\| = \|A\Gamma^{-1}M\Delta\| \leq \frac{\overline{d}}{\underline{d}}\frac{\delta_M}{\gamma_m}\|W_1\|. \tag{3.142}$$

Comparing the conditions (3.142), (3.139), and (3.128), we can see that the stability result in [168] is less conservative.

For the following Cohen–Grossberg neural networks with continuously distributed delays,

$$\begin{aligned}\dot{u}_i(t) = &- d_i(u_i(t))\left[a_i(u_i(t)) - \sum_{j=1}^{n} w_{ij}g_j(u_j(t))\right.\\ &\left.- \sum_{j=1}^{n} w^1_{ij}g_j(u_j(t-\tau(t))) - \sum_{j=1}^{n} w^2_{ij}\int_{-\infty}^{t} K_j(t-s)g_j(u_j(s))\mathrm{d}s\right],\end{aligned} \tag{3.143}$$

LMI-based global exponential stability (GES) problem has been studied in [75], where

$$\int_0^{\infty} K_j(s)\mathrm{d}s = 1, \int_0^{\infty} sK_j(s)e^{2\lambda s}\mathrm{d}s = \pi_j(\lambda) < \infty, \lambda > 0. \tag{3.144}$$

System (3.143) can be written in a compact matrix-vector form,

$$\begin{aligned}\dot{u}(t) = &- D(u(t))\Big[A(u(t)) - Wg(u(t)) - W_1g(u(t-\tau(t)))\\ &- W_2\int_{-\infty}^{t} K(t-s)g(u(s))\mathrm{d}s\Big],\end{aligned} \tag{3.145}$$

Obviously, the distributed delay (3.144) in [75] is different from that in (3.30). Therefore, the analysis method in [75] cannot be applied to the neural networks with distributed delay (3.30). By using a descriptor system method [72, 76, 155, 292, 293], an LMI-based GES stability criterion has been derived, which has involved more tuning variables. Meanwhile, the results in [75] have no restrictions on the change rate of time-varying delays. When $d_i(u_i(t)) =$ constant in (3.144), this kind of model with constant delay is studied in [59], using Moon inequality [294] and the well-known Leibniz–Newton formula, an LMI-based global asymptotic stability criterion has been established, which is dependent on the magnitude of time delay.

For neural networks (3.94) with $f_i(u_i(t)) = g_i(u_i(t))$ and $\dot{\tau}_{ij}(t) \leq \eta < 1$, Theorem 4.1 in [237] requires the following matrix inequality to hold,

$$2P\Gamma\Delta^{-1} - PW - W^TP - (PQ^{-1}W_1)_\infty - \frac{1}{1-\eta}(PQW_1)_1 > 0, \qquad (3.146)$$

then the equilibrium point of (3.94) is globally exponentially stable, where P and Q are positive diagonal matrices to be determined, $B = (b_{ij})$, $B_1 = \text{diag}(\sum_{i=1}^n |b_{i1}|, \sum_{i=1}^n |b_{i2}|, \ldots, \sum_{i=1}^n |b_{in}|)$, $B_\infty = \text{diag}(\sum_{i=1}^n |b_{1i}|, \sum_{i=1}^n |b_{2i}|, \ldots, \sum_{i=1}^n |b_{ni}|)$.

For the following Cohen–Grossberg neural networks with continuously distributed delays,

$$\begin{aligned}\dot{u}_i(t) = &- d_i(u_i(t))\left[a_i(u_i(t)) - \sum_{j=1}^n w_{ij}g_j(u_j(t))\right.\\ &\left.- \sum_{j=1}^n w_{ij}^1 \int_{-\infty}^t K_{ij}(t-s)g_j(u_j(s))\text{d}s\right],\end{aligned} \qquad (3.147)$$

Theorem 5.2 in [237] requires the following matrix inequality to hold,

$$2P\Gamma\Delta^{-1} - PW - W^TP - (PQ^{-1}W_1)_\infty - (PQW_1)_1 > 0, \qquad (3.148)$$

then the equilibrium point of (3.147) is globally asymptotically stable. Further, if the following condition holds,

$$\int_0^\infty K_{ij}(s)e^{\delta_0 s}\text{d}s < \infty, \qquad (3.149)$$

where $\delta_0 > 0$ is a positive constant, then the equilibrium point of (3.147) is globally exponentially stable if (3.148) holds. Obviously, for the case of constant delays, condition (3.148) guarantees the global asymptotic stability for both (3.94) with $g(\cdot) = f(\cdot)$ and (3.147) with continuously distributed delays. The reason is that system (3.94) with $g(\cdot) = f(\cdot)$ can be deduced from system (3.147) if a suitable kernel function is selected (also see (3.30)–(3.31) for more details). Since the distributed delays are involved, under suitable assumption on kernel function, the equilibrium point of (3.147) is also globally exponentially stable if (3.148) holds. It is clear, on the one hand, that stability conditions both (3.146) and (3.148) only consider the inhibitory effect of W, and do not consider the inhibitory effect of W_1. On the other hand, both conditions (3.146) and (3.148) are not easy to check because unknown matrix Q is involved in computing the 1-norm and ∞-norm.

For the Cohen–Grossberg neural networks with finite distributed delays (3.67), Ref. [131] studies the global asymptotic stability problem and establishes the same sufficient condition as that in (3.68). Obviously, stability condition (3.68) in [131] unifies many LMI-based stability results in the literature.

For the Cohen–Grossberg neural networks (3.58), Ref. [116] establishes the following LMI-based global exponential stability conditions, independent of time-varying delays,

$$\begin{bmatrix} \Phi_1 & \Phi_2 & PB_1 & PB_2 & \dots & PB_n \\ * & \Phi_3 & QB_1 & QB_2 & \dots & QB_n \\ * & * & -H_1 & 0 & \dots & 0 \\ * & * & * & -H_2 & \dots & 0 \\ \vdots & \vdots & \vdots & \vdots & \ddots & \vdots \\ * & * & * & * & \dots & -H_n \end{bmatrix} > 0, \tag{3.150}$$

$$P/\bar{d} > \sum_{i=1}^{n} \Delta H_i \Delta, \tag{3.151}$$

where $\Phi_1 = -2P\Gamma + (P + Q\Delta)/\underline{d} + \theta I$, $\Phi_2 = R\Delta - Q\Gamma + PW$, $\Phi_3 = QW + (QW)^T - 2R$, P, Q, R and H_i are positive diagonal matrices, respectively. $\theta > 0$ is a positive constant, and $B_k = (b_{ij}^k)$, whose kth row is composed of the kth row of matrix $W_1 = (w_{ij}^1)$ and the other rows are all zeros.

For the Cohen–Grossberg neural networks (3.58) with $\tau_{ij}(t) = \tau(t)$, Ref. [116] establishes the following LMI conditions to guarantee the global exponential stability, independent of time-varying delays,

$$\begin{bmatrix} \Phi_1 & \Phi_2 & PW_1 \\ * & \Phi_3 & QW_1 \\ * & * & -H \end{bmatrix} > 0, \tag{3.152}$$

$$P/\bar{d} > \Delta H \Delta, \tag{3.153}$$

where H is a positive diagonal matrix, and the other notations are the same as those defined in (3.150).

Obviously, based on different assumptions on activation functions, amplification functions, and delay types, different LMI-based stability criteria have been established. With more matrix inequality techniques being involved in the proof procedure of stability analysis, these kinds of stability criteria begin to become more and more complex. Correspondingly, it is not an easy work to check the stability conditions and compare them with the other kind of stability criteria analytically. In the extreme case, how to evaluate the performance of an LMI-based stability criterion, it will be no other choice but to resort to the case study. Case study is plentiful, which will reflect the diversity of theory research. With the development of different theories on stability and computational mathematics, case study can also be deepened significantly, and some common understandings may be reached. In this way, theory research will become more and more refined, which gradually promotes the development of stability theory. LMI-based stability research is such a reflection on the understanding of the neural dynamical networks.

3.3.6 Topics on Robust Stability of RNNs

RNNs have been extensively studied in recent years, e.g., [80, 209, 281, 282, 295–302] and the references cited therein, because of their many important applications in various areas such as pattern recognition, associate memory, and combinatorial optimization. Since the existence of delays is frequently a source of instability for neural networks, the stability of neural networks with time delays has long been a focused topic of theoretical as well as practical importance. In the design and hardware implementation of neural networks, however, a common problem is that parameters acquired in neural networks are inaccurate. To design neural networks, vital data, such as the neurons fire rate, the synaptic interconnection weight and the signal transmission delays, etc., usually need to be measured, acquired, and processed by means of statistical estimation which definitely leads to estimation errors. Moreover, parameter fluctuation in neural network implementation on very large scale integration (VLSI) chips is also unavoidable. In practice, it is possible to explore the range of the above-mentioned vital data as well as the bounds of circuit parameters by engineering experience even from incomplete information. This fact implies that a good neural network should have certain robustness which paves the way for introducing the theory of interval matrices and interval dynamics to investigate the global stability of interval neural networks. Otherwise, the neural network is not meaningful in the practical applications. For example, as using an interval neural network having certain robustness to solve optimization problems, we need not consider spurious suboptimal responses for each parameter value of the neural network, which is of great importance. Therefore, besides global exponential stability, complete stability, asymptotic stability, and periodic oscillation of neural networks with time delays have been extensively investigated, e.g., [67, 71, 80, 118, 205, 208, 209, 232, 282, 296–298, 300, 301, 303]. As pointed out in [71], *robust stability* is seldom considered for neural networks with and without delay. There exist several related results on robust stability, see, [123, 304–307]. In [305, 306], global robust stability of delayed interval Hopfield neural networks are investigated with respect to the bounded and strictly increasing activation functions. Several M-matrix conditions to ensure the robust stability are given for delayed interval Hopfield neural networks. In [123], the authors view the uncertain parameters as perturbations and give some testable results for robust stability of continuous-time Hopfield neural networks without time delay. In [71], the authors view the interval uncertain parameters as the matched uncertainty and give some LMI-based robust testable results.

In general, there are several kinds of expression forms for *uncertainty* of interconnection matrices, which have been introduced in Sect. 3.2.2. In the following, we will mainly consider the interval uncertainty.

Interval uncertainty can be expressed in the following form [71, 304],

$$A_I = [\underline{A}, \overline{A}] = \{A = (a_{ij})_{n\times n} : \underline{A} \le A \le \overline{A},\ \text{i.e.,}\ \underline{a}_{ij} \le a_{ij} \le \overline{a}_{ij}\}. \tag{3.154}$$

In the proof of the robust stability of this kind of interval system, some inequalities are evolved gradually. In [304], the following inequality is used,

$$\|A\|_2^2 \le \max\{\|\underline{A}\|_2^2, \|\overline{A}\|_2^2\}, \tag{3.155}$$

where $\|A\|_2 = \sqrt{\lambda_{\max}(A^T A)}$ is the normal Euclidean norm. As pointed out in [71], inequality (3.155) does not always hold, for example, take

$$\underline{A} = \begin{bmatrix} 0 & -16 \\ -16 & 0 \end{bmatrix}, \overline{A} = \begin{bmatrix} 16 & 0 \\ 0 & 16 \end{bmatrix}, A = \begin{bmatrix} 16 & -16 \\ -16 & 16 \end{bmatrix}, \tag{3.156}$$

clearly, $\underline{A} \le A \le \overline{A}$, and $\|\underline{A}\|_2 = \|\overline{A}\|_2 = 16$, $\|A\|_2 = 32$. Therefore, for this case the inequality (3.155) does not hold. In [71], the following inequality is presented,

$$\|A\|_2 \le \|A_*\|_2 + \|A^*\|_2, \tag{3.157}$$

where $A^* = (\overline{A} + \underline{A})/2$, $A_* = (\overline{A} - \underline{A})/2$.

As a commonly used concept, we give the following robust definition.

Definition 3.6 (*Robust stability, see* [208, 298, 308–310]) The neural network model given by (3.89) with the parameter ranges defined by (3.154) is globally *robust stable* if the unique equilibrium point $u^* = (u_1^*, u_2^*, \ldots, u_n^*)$ of the model is globally asymptotically stable for all $W \in W_I$, $W_1 \in W_{1I}$.

For the *interval uncertainty* (3.154), a splitting or partitioning interval matrix method has been proposed in [130] to prove the robust stability. Specifically, the interval $A \in [\underline{A}, \overline{A}]$ is divided by $\tilde{A} = \frac{\overline{A}-\underline{A}}{m}$ or $\tilde{a}_{ij} = \frac{\overline{a}_{ij}-\underline{a}_{ij}}{m}$ where m is an integer greater than or equal to 2. The disadvantage of the partitioning interval matrix method will produce a large set of criteria to be checked, which leads to the complexity and difficult check of the robust result. The advantage is that it presents another method to study the robust stability of RNNs with interval parameter uncertainties.

Similar to the delay partitioning method, in which time delay varies in a fixed interval or bounded range, interval matrix can also be partitioned by as many subinterval matrices as possible. In this partitioning way, the total uncertainties are divided by many almost deterministic subinterval matrices. Then based on the similar principle of neural network approximator, some weighted combination methods are applied to the subinterval matrix, which forms the fundamental principle of partitioning interval matrix method. It is interesting that the universal approximation capability of neural networks have two main applications: one is to model and identify the external controlled systems, the other is to analyze the interval dynamics by partitioning the interval number. It is the internal partitioning/decomposition approximation and external synthesis approximation that make neural networks more powerful. Stability analysis of neural networks to some degree is a optimal approximated process with the combinations of interconnection matrix, activation functions, time delay, amplifications, and external inputs.

By the way, as an important parameter of RNNs, *time delay* cannot be avoidable [311]. Nowadays, both delay-dependent and delay-independent stability criteria have been investigated. Delay-independent stability criteria are robust to variations of time delay, while delay-dependent stability criteria are sensitive to the variations of time delay. At present, more emphasis is placed on the robust stability with respect to the variations of interconnection coefficients, while delay-dependent robust stability conditions is sensitive to time delay, no matter the change rate of time-varying delay and the size of time delay. This is a strange phenomenon in the robust stability research of RNNs. From the other viewpoint of understanding, robust stability is also a *relative conception*, who is robust to some variables while is sensitive to other variables. *Robust stability*, similar to stability definition, is also defined subjectively. Based on this kind of understanding, we can know that the current robust stability is not with respect to the total perturbations of the networks, on the contrary, it is relative to the partial uncertainties or the interested uncertainties. Therefore, for the variation of time delay, how to establish the corresponding robust stability criteria has not gained the attention of researchers.

3.3.7 Other Topics on Stability Results of RNNs

In the aforementioned subsections, the focus is on the stability in Lyapunov sense and global asymptotical stability of a kind of continuous-time real-value additive recurrent neural networks with *discrete delays* (or called isolated delays or concentrated delays with respect to the distributed delays). These kinds of stability criteria are the main streams in the current researches of neural network theory. However, there are many qualitative characteristics of RNNs, such as absolute stability, complete stability, and stability in Lagrange sense. Other topics include the periodic solution, $\omega-$limit set, aperiodic solution, local stability, partial stability, practical stability. In these cases, the stability problems may not be concerned with the fixed point, but they may be concerned with the invariant set or $\omega-$limit set, even the boundedness of the state or output trajectories.

Besides the above qualitative features, there are some other directions to be further investigated for Cohen–Grossberg neural networks(CGNN). (1) One is in the aspect of neural network models. For example, the CGNN can be extended to the cases: the connection weight coefficients can be complex valued, which forms the complex-valued CGNN; the connection weight coefficients can be state-dependent, which forms the memristive neural networks; the neuronal activation function can be discontinuous, which forms the discontinuous neural networks; the time delay can be distributed or stochastic, which forms the neural networks with distributed delays and stochastic neural networks; when the number of variables in the derivative of differential equations exceeds to one, which is more suitable to be expressed by partial differential equations, some neural networks expressed in the partial differential equation are necessary; when the variables are involved in different structures, bidirectional memory neural networks, competing neural networks, and gene regulation networks are formed, to name a few. For these neural network models, the introduced contents in above subsections can be continued to study. (2) The other is in the aspect of analysis methods. Different stability property or qualitative characteristic may require different analysis method, especially for further improving the effectiveness of the obtained stability criteria. In this way, such mathematical methods as algebraic method, geometric method, set method, manifold method, and so on can be used. Except the mathematical analytical methods, maybe some other nonanalytical methods are still more effective in the qualitative analysis of complex networks. (3) Stability problem is doomed to be connected with some practical applications. Different applications may require different stability concepts and stability analysis methods. Different neural models may reflect different description of engineering applications. Different qualitative characteristics may represent different requirements of the designers. Different level or degree of stability may indicate different response of the process systems. Therefore, it is better to connect some stability analysis to some specific engineering or social applications, which will make the stability research more meaningful and will provide a unified framework for the qualitative analysis and quantitative control/synthesis.

3.3.8 Qualitative Evaluation on the Stability Results of RNNs

From above subsections, we can draw the following conclusions.

(1) At present, there are two kinds of stability results for the neural networks with delays, one is the delay-dependent one, the other is *delay-independent* one. Specially, if any delay-related information is involved in the stability criteria, all these kinds of stability results are called *delay-dependent* stability results. For example, such information as the size of time delay, the change rate of time-varying delay, the kernel function of the distributed delay, the density function of the stochastic delay, and so on, is involved in the stability criteria, the corresponding stability criteria are all called delay-dependent ones. If no information about delay is involved in the stability criteria, this kind of stability result is called as delay-independent one.

In fact, delay-independent stability criterion is a relative concept. Although not any information about delay is involved in the stability criteria, some information about delay is usually involved in the assumption conditions. For example, such information as the boundedness of time delay, slowly time-varying delay (e.g., $\dot{\tau}(t) < 1$), partially known probability transition density, known kernel function and so on, are always assumed in the prerequisite or assumption. Therefore, strictly speaking, no delay-independent stability condition exists in the literature. This is the reason why the prerequisite must be considered before a stability criterion formula is used to judge whether a neural network is stable or not. Ignoring the prerequisite and directly verifying the criterion, there will not be a correct understanding on the concerned problem. In general, if partial information of time delay can be relaxed in the stability criteria, for example, the size of time delay, we then call this criterion as independent of the size of time delay (although the bounded delay is unknown). Similarly, stability criterion independent of probability transition density, independent of change rate of time delay, etc., are all meaningful results for the stability research of delayed neural networks, despite some other information about delay is required in the prerequisite. This is the relativity of partial-delay-information independent stability result.

(2) When discussing the delay-dependent stability criterion, researchers usually say that delay-dependent criteria are better than delay-independent criteria, especially when the size of time delay is very small. This assertion seems odd. According to [2, 18], when the size of time delay does not exceed a limit (for the case of constant delay), the neural networks with delay have the same dynamical property as the neural networks without delay. Therefore, when the size of time delay is too small, it is not necessary to establish the delay-dependent stability criterion. If the size of time delay exceeds some limit, it is necessary to discuss the stability of RNNs with large delay. Naturally, a question may arise, is it necessary to spend too much effort finding some techniques to establish the delay-dependent stability criteria? In this respect, there is a misleading understanding on reducing the conservativeness of the stability results. Except the size of time delay, there are many delay information effecting the stability of the delayed neural networks, such as the change rate of time-varying delay, the delayed kernel function and probability transition density. Therefore, the traditional delay-dependent and delay-independent concept is only concerned with the size of time delay. With the development of neural network theory, such concept should be updated. Except the case of discrete or concentrated delay (with respect to distributed delay), the meaning of delay-dependent and delay-independent concept seems no longer meaningful. In this case, discussing the effects of other delay-related information (no limited to the size of time delay) on the stability of delayed neural networks looks meaningful, which forms the initial value and boundary value problems of the concerned neural networks. The meaning of delay-dependent stability criteria lies in the fact that it can present a more close relationship among the total information of RNNs and the qualitative stability property of RNNs, which can provide a theoretical foundation for the synthesis of RNNs. If the stability of fixed equilibrium point is further upgraded, qualitative characteristics such as invariant set, passivity, dissipativity, and synchronization of the complex neural networks will play more important roles. After all, neural networks fall into the interdisciplinary fields,

both mathematical methods and engineering applications are two main streams in the development of stability theory. Stability problem is the original and fundamental investigation of the concerned problems, and other researches such as control schemes, invariant sets, and passivity are all on the basis of the stability theory of fixed equilibrium point.

Above two points are for the discussion on the delay-dependent/independent stability criteria form. The following points are for the specific methods of the stability research on RNNs.

(3) With the increase of additive terms in (3.82), the negative term in the M-matrix condition is also increasing. This means that the stability criteria based on M-matrix will become more and more conservative with the increase of additive terms (such as different types of delays) in the network structure, which will affect the accurate judgement of stability for the parallel computational neural network models. With the increase of the complexity, the conservativeness of stability criteria will be increased. This assertion is also suitable for the case of LMI-based stability criteria, matrix norm-based stability results and algebraic inequality-based results. Meanwhile, all the stability results based on M-matrix neglect the signs of the connection weight coefficients, which lead to the ignorance of the inhibitory effect of neurons on the neural networks. It is well known that the inhibitory action of neurons can stabilize the neural network while excitatory action will destabilize the neural networks. The balance of a system dynamics is kept by the interactions of inhibitory and excitatory effect. In the applications such as optimization problems and image processing, the inhibitory action of neurons plays an important role in practice. Therefore, the simple M-matrix-based stability criteria are derived at the expense of ignoring the inhibitory action of neurons.

(4) M-matrix-based method builds a bridge among algebraic inequality methods, matrix measure methods, spectral norm methods and Lyapunov diagonal stability methods. Especially, matrix-based method will build the connection with linear matrix inequality method through the Lyapunov diagonal stability methods. Furthermore, M-matrix-based stability result is a very flexible, effective, and simple criterion to determine the dynamic properties of a complex recurrent neural network. In the development of stability theory of RNNs, M-matrix method plays an important role in promoting the stability research both in applications and theory. Interestingly, some global asymptotic stability results and global exponential stability results for RNNs can be expressed in the same form of M-matrix.

3.4 Necessary and Sufficient Conditions for RNNs

Nowadays, most stability results for Hopfield-type and Cohen–Grossberg-type neural networks are sufficient. Few necessary and sufficient conditions for asymptotic/exponential stability of RNNs with delay has been published. However, there exist some necessary and sufficient conditions for different dynamics of recurrent neural networks with/without delay. Especially, for the delayed case, some necessary and sufficient conditions for the attractiveness of a class of recurrent neural networks with delay have been published [312].

Note that the sufficient asymptotic/exponential stability criteria in the existing literature are all established on the basis of *strict inequality* (i.e., > 0 or < 0). It is natural to ask: what will happen if the strict inequalities are placed by the nonstrict inequalities (i.e., ≥ 0 or ≤ 0)? For the case of necessary and sufficient conditions, one must consider the critical case, i.e., the nonstrict inequalities. For the case of nonstrict inequalities, Ref. [313] establishes a global convergence condition based on M-matrix-like (or algebraic inequality) method for the following neural network

$$\dot{u}_i(t) = -\gamma_i u_i(t) + \sum_{j=1}^{n} w_{ij} g_j(u_j(t)) + \sum_{j=1}^{n} w_{ij}^1 f_j(u_j(t-\tau_{ij})), \tag{3.158}$$

where $g_i(u_i(t)) = \tanh(\alpha_i u_i(t))$ and $f_i(u_i(t)) = \tanh(\beta_i u_i(t))$, i.e., $|g_i(u_i(t))| \leq \alpha_i |u_i(t)|$ and $|f_i(u_i(t))| \leq \beta_i |u_i(t)|$, $\alpha_i > 0, \beta_i > 0, i = 1, \ldots, n$. Obviously, model (3.158) is a special case of model (3.94). If

$$M_c = -\gamma_j + \alpha_j \left(w_{jj} + \sum_{i=1, i\neq j}^{n} |w_{ij}| \right)^{+} + \beta_j \sum_{i=1}^{n} |w_{ij}^1| \leq 0, \tag{3.159}$$

then there is a unique equilibrium point such that every solution of (3.158) satisfies $\lim_{t\to\infty} u(t) = u^*$. It is clear that the following matrix

$$\Gamma - W^+\alpha - |W_1|\beta \in \mathcal{P}_0, \tag{3.160}$$

is equivalent to the condition (3.159), where $\mathcal{P}_0$ will be defined in (3.164). Comparing the condition (3.160) in [313] with (3.95) in [149], we can see that the result in [313] further relaxes the restrictive condition on the stability/convergence, which is close to the necessary condition on the stability/convergence.

For the following pure delay Hopfield neural network,

$$\dot{u}(t) = -\Gamma u(t) + W_1 f(u(t-\tau)), \tag{3.161}$$

Theorem 3 in [68] presents the following necessary and sufficient condition

$$\left(\sigma I - \begin{bmatrix} \Gamma & 0 \\ 0 & \Gamma \end{bmatrix} + e^{\sigma\tau} \begin{bmatrix} W_1^+ & W_1^- \\ W_1^- & W_1^+ \end{bmatrix} \begin{bmatrix} \Delta & 0 \\ 0 & \Delta \end{bmatrix} \right) \eta \leq 0, \tag{3.162}$$

then system (3.161) is guaranteed componentwise exponential convergence, where $\eta = [\alpha^T, \beta^T]^T$, $\alpha > 0$ and $\beta > 0$ are two constant vectors with appropriate dimensions, $\sigma > 0$ is a scalar, I is an identity matrix with appropriate dimension, $(w_{ij}^1)^+ =$

$\max\{w_{ij}^1, 0\}$ signifying the excitatory weights and $(w_{ij}^1)^- = \max\{-w_{ij}^1, 0\}$ signifying the inhibitory weights. Obviously, the elements in W_1^+ and W_1^- are all nonnegative.

In this subsection, we will introduce some recent development in the aspects of necessary and sufficient condition on the dynamics of RNNs.

For the case of *Hopfield neural networks* without any delay, under some assumptions on the activation and connection weight matrix, some sufficient and necessary stability and attractiveness conditions have been established in the existing literature. For the following Hopfield neural network, absolute stability (ABST) has been studied in [125, 126, 314, 315],

$$\dot{u}_i(t) = -\gamma_i u_i(t) + \sum_{j=1}^{n} w_{ij} g_j(u_j(t)) + U_i, \tag{3.163}$$

where $W = (w_{ij})_{n\times n}$ is the connection matrix, and the activation function $g_i(u_i(t)))$ is a class of sigmoid function that consists of smooth, strictly monotone, increasing functions which are saturated as $u_i(t) \to \pm\infty$ (e.g., $\tanh(u_i(t)))$, $i = 1\ldots, n$. In Ref. [314], for the nonsymmetric case of connection matrix W in (3.163), a necessary and sufficient condition as follows

$$-W \in \mathcal{P}_0, \tag{3.164}$$

is presented to guarantee the uniqueness of the equilibrium point for any bounded activation function Class g, where $\mathcal{P}_0$ denotes the class of square matrices A defined by one of the following equivalent properties [314]: (i) All principal minors of A are nonnegative; (ii) Every real eigenvalue of A as well as of each principal submatrix of A is nonnegative; (iii) $\det(K + A) \neq 0$ for every diagonal matrix $K = \text{diag}(K_1, \ldots, K_n)$ with $K > 0, i = 1, \ldots, n$. That is to say, the negative semi-definite matrix W is a necessary and sufficient condition to guarantee the uniqueness of the equilibrium point for system (3.163) with asymmetric connection matrix. However, condition (3.164) is not in general sufficient for the absolute stability of system (3.163) with asymmetric connection matrix. On the contrary, for the system (3.163) with symmetric connection matrix, it has been shown that condition (3.164), or the negative semi-definite matrix W, is a necessary and sufficient condition to guarantee the absolute stability of the unique equilibrium point. This is consistent with the result in [315]. The ABST result was also extended to the absolute exponential stability (AEST) in [316, 317]. At this point, we make some comments on the necessary and sufficient stability result. It is well known that a symmetric Hopfield neural network always exhibits convergent dynamics, but it may run into local minimum when applied to optimization problems [16]. Condition (3.164) shows that the largest subclass of symmetric Hopfield networks that can guarantee achievement of global minimization without spurious response are those with negative semi-definite weight matrices.

In [318], a conjecture is raised: the necessary and sufficient condition for *ABST* of the neural network is that its connection matrix W belongs to the Class of matrices W such that all eigenvalues of matrix $(W - D_1)D_2$ has negative real parts for arbitrary positive diagonal matrices D_1 and D_2. This condition is proven to be a necessary and sufficient condition for ABST of the neural network with two neurons [319]. The necessity of such a condition for ABST is proven in [318] and implies that all existing sufficient conditions for ABST in the literature are special cases of that in [318]. However, whether or not such a condition is sufficient for ABST of a general neural network remains unknown in the case of more than two neurons. Within the class of partially Lipschitz continuous and monotone nondecreasing activation functions (this class includes the sigmoidal activation as a special case), a recent AEST result is given in [163] under a mild condition that the connection weight matrix W belongs to the Class of additively diagonally stable matrix introduced in [162], i.e., for any positive diagonal matrix D_1, there exists a positive diagonal matrix D_2 such that $D_2(W - D_1) + (W - D_1)^T D_2 < 0$. This condition extends the condition in [320] that the connection weight matrix W is an H-matrix with nonpositive diagonal elements.

As a comparison, the following two conditions have been proposed in [21], respectively,

$$\delta_M < \frac{\gamma_i}{w_{ii} + 0.5\sum_{j\neq i, j=1}^{n}(|w_{ij}| + |w_{ji}|)}, \tag{3.165}$$

and

$$w_{ii} \leq -0.5 \sum_{j\neq i, j=1}^{n} (|w_{ij}| + |w_{ji}|), \tag{3.166}$$

which guarantee the system (3.163) with symmetric connection matrix to have a unique global asymptotic stable equilibrium point.

In the GAS condition (3.166), it is independent of the maximum slope δ_M. Clearly, (3.166) assures GAS for all functions g in the Class S (i.e., bounded, monotonic, strictly increasing activation function) and hence represents a sufficient condition for ABST. On the contrary, (3.165) does not represent a condition for ABST, since (3.165) requires a limitation on the maximum slope δ_M of the sigmoidal function g.

For a symmetric W, (3.166) can be rewritten as $w_{ii} \leq -\sum_{j\neq i, j=1}^{n} |w_{ij}|$, which means that W is weakly row sum dominant and hence is negative semi-definite. Therefore, (3.166) implies the condition of (3.164). However, the class of negative semi-definite matrices is wider than that of matrices satisfying a row dominance condition, which implies that condition (3.166) is more restrictive than the condition (3.164).

In [125], W is a normal matrix in (3.163), i.e., $W^T W = W W^T$, and system (3.163) is called normal Hopfield neural network, which is a special case of original Hopfield neural network (3.21). The following sufficient and necessary condition is established in [125],

$$\max_i Re\ \lambda_i(W) \leq 0, \tag{3.167}$$

or

$$\max_i \lambda_i \left(\frac{W + W^T}{2}\right) \leq 0, \tag{3.168}$$

then the normal neural network (3.163) is absolute stability, where $\lambda_i(B)$ represents the i-th eigenvalue of matrix B. $Re\ \lambda_i(B)$ represents the real part of eigenvalue $\lambda_i(B)$. Since a symmetric matrix is normal, then for a symmetric neural network, the negative semi-definiteness result in [314] is obviously a special case of the result in [125]. Furthermore, a sufficient global asymptotic stability condition is also presented in [125],

$$\delta_M \max_i Re\ \lambda_i(W) \leq \gamma_m. \tag{3.169}$$

Condition (3.169) is clearly weaker than condition (3.167). In particular, the condition (3.169) even possibly allows an unstable connection matrix W, i.e., $Re\ \lambda_i(W) > 0$ for some i.

In [126], the same Hopfield neural network (3.163) has been further discussed. By removing the assumption of normal matrix on W, a more general matrix decomposition method is proposed. That is $W = W^s + W^{ss}$, where $W^s = (W + W^T)/2$ and $W^{ss} = (W - W^T)/2$ are the symmetric and the skew-symmetric parts of W, respectively. Then based on the matrix eigenvalue method and a solvable Lie algebra condition, a new necessary and sufficient condition is presented to guarantee the absolute stability of the concerned Hopfield neural network. Specifically, suppose that $\{W^s, W^{ss}\}$ generates a solvable Lie algebra, if and only if the following condition holds,

$$\max_i Re\ \lambda_i(W) \leq 0, \tag{3.170}$$

or

the symmetric part W^s of the weight matrix W is negative semi-definite, (3.171)

then the system (3.163) is absolutely stable, which includes the results in [125, 314, 315] as special cases. In the case of smooth activation function for Hopfield neural network (3.163), condition (3.170) also ensures the exponential convergence of the equilibrium point. It is clear that the exponential convergence result stated in [126] is also valid for neural networks under the conditions of stability results in [125, 314, 315]. However, such an exponential convergence result has not been obtained in [125, 314, 315]. Therefore, the results in [126] improves the results in [125, 314, 315].

The importance of *absolute stability* is as follows. In general, absolute stability of a neural network implies existence and global attractiveness of a unique equilibrium of the system for every neuron activation function of sigmoid type and for every constant input to the neural network. The insensitivity of this property to model details is of important physical significance, since in most cases the neuron activation is known to belong to the sigmoid class but its shape is not exactly specified. A typical example of this case is a neural network in the high-gain limit. Moreover, the global attractiveness inherent in absolute stability ensures a neural network running in real time without the need of resetting the activations when changing inputs. It also precludes spurious responses for every choice of the activation function and of the external input. This is particularly desirable for neural optimization and classification problems.

For (3.163) with global Lipschitz condition satisfying Assumption 3.4, the following necessary and sufficient stability condition has been derived in [127, 321]

$$-\Gamma + W\Delta \text{ is nonsingular or } \det(-\Gamma + W\Delta) \neq 0, \tag{3.172}$$

which ensures that system (3.163) has a unique equilibrium point. According to the discussion, the absolute stability discussed in [125] is the attractiveness of the equilibrium point.

For the case of Hopfield neural network with delays,

$$\dot{u}_i(t) = -\gamma_i u_i(t) + \sum_{j=1}^{n} w_{ij}^1 g_j(u_j(t - \tau_{ij})), \tag{3.173}$$

global attractivity of system (3.173) was studied in [312] based on M-matrix structure, where $W_1 = (w_{ij}^1)$ is the connection matrix, and the activation function $g_i(0) = 0$, $g_i(u_i(t))$ saturates at ± 1 for any $u_i(t) \in \mathbb{R}$, i.e., $\lim_{u_i(t)\to\pm\infty} = \pm 1$, $g_i^{'}(u_i(t))$ is continuous such that $g_i^{'}(s) = \frac{\mathrm{d}g_i(s)}{\mathrm{d}s} > 0$ for any $s \in \mathbb{R}$, $g^{'}(0) = 1$, and $0 < \bar{g}_i(u_i(t)) < m_b$ for any $m_b > 0$, where $\bar{g}_i(u_i(t)) = \max\{g_i(u_i(t)),\ -g_i(-u_i(t))\}$, $i = 1, \ldots, n$, $\max\{\tau_{ij}\} = \tau_M \geq 0$, the initial conditions are continuous on $[-\tau_M,\ 0]$. That is to say, the activation function is a Sigmoid-type and positive saturation function. The following necessary and sufficient condition has been obtained to guarantee the global attractivity of the origin of system (3.173),

$$\det(-\Gamma + W_1) \neq 0, \text{ and } \Gamma - |W_1| \text{ is an } \mathcal{P}_0\text{-matrix}, \tag{3.174}$$

where $\mathcal{P}_0$-matrix is defined in (3.164). Comparing the result (3.174) in [312] with those results (3.79) and (3.80) in [108], we can see the theorem conditions in [312] improve the results in [108] for system (3.173), which is a special case of the model (3.78) studied in [108]. Thus, generally speaking, the results in [108] have wider application ranges than that in [312].

Also, the following result is presented in [312] if

$$\Gamma - |W_1| \text{ is a nonsingular M-matrix,} \tag{3.175}$$

then the origin of system (3.173) is globally exponential stable for any time delay $\tau_{ij} \geq 0$, which is the same as that result (3.80) in [108]. As far as the model (3.173) is concerned, the result (3.175) in [312] improves the results in [108].

3.5 Summary

In summary, stability studies for recurrent neural networks with or without delays have achieved great development in the last three decades, no matter in the wideness and the deepness. However, there are still many new problems to be proposed and solved. All these changes accompany the development of mathematical theory, especially the applied mathematics and computational mathematics. Keeping in mind, any forms of stability criteria have their own feasible ranges, and one cannot expect that only one stability result can tackle all the stability problems existed in the recurrent neural networks. Every kind of stability result, for example, in the forms of algebraic inequality, M-matrix, and LMI, has its own advantages, which has considered different tradeoff between computational complexity and efficiency of the stability results. One form of stability result is not absolutely superior to the other form of stability result, it only reflects different aspects of the investigated recurrent neural networks. Therefore, it is the different forms of stability results that promote the development of the stability theory of recurrent neural networks.

References

1. D.W. Tank, J.J. Hopfield, Simple neural optimization networks: an A/D converter, signal decision circuit, and a linear programming circuit. IEEE Trans. Circuits Syst. **33**(5), 533–541 (1986)
2. J.H. Li, A.N. Michel, W. Porod, Qualitative analysis and synthesis of a class of neural networks. IEEE Trans. Circuits Syst. **35**, 976–986 (1988)
3. M.P. Kennedy, L.O. Chua, Neural networks for nonlinear programming. IEEE Trans. Circuits Syst. **35**(5), 554–562 (1988)
4. M. Forti, S. Manetti, M. Marini, A condition for global convergence of a class of symmetric neural networks. IEEE Trans. Circuits Syst. **39**, 480–483 (1992)
5. A.N. Michel, D.L. Gray, Analysis and synthesis of neural networks with lower block triangular interconnecting structure. IEEE Trans. Circuits Syst. **37**(2), 1267–1283 (1990)
6. A. Rodriguez-Vazquez, R. Dominguez-Castro, A. Rueda, E. Sanchez-Sinencio, Nonlinear switched-capacitor "neural" networks for optimization problems. IEEE Trans. Circuits Syst. **37**(3), 384–398 (1990)
7. J.J. Hopfield, Neurons with graded response have collective computational properties like those of two-state neurons. Proc. Natl. Acad. Sci. **81**, 3088–3092 (1984)

8. J.J. Hopfield, Neural networks and physical systems with emergent collective computational abilities. Proc. Natl. Acad. Sci. USA **79**, 2554–2558 (1982)
9. M. Vidyasagar, Location and stability of high-gain equilibria of nonlinear neural networks. IEEE Trans. Neural Netw. **4**, 660–672 (1993)
10. C. Corduneanu, *Principles of Differential and Integral Equations* (Allyn and Bacon, Boston, 1971)
11. W. Hahn, *Theory and Application of Lyapunov's Direct Method* (Prentice Hall, Englewood Cliffs, 1963)
12. S. Amari, Characteristics of randomly connected threshold element networks and neural systems. Proc. IEEE **59**, 35–47 (1971)
13. S. Amari, Characteristics of random nets of analog neuron-like elements. Proc. IEEE **2**, 643–657 (1972)
14. H. Wilson, A. Cowan, Excitatory and inhibitory interconnections in localized populations of model neurons. Biophys. J. **12**, 1–24 (1972)
15. H. Sompolinsky, A. Crisanti, Chaos in random neural networks. Phys. Rev. Lett. **61**, 259–262 (1988)
16. J.J. Hopfield, D.W. Tank, Computing with neural circuits: a model. Science **233**, 625–633 (1986)
17. M.A. Cohen, S. Grossberg, Absolute stability and global pattern formation and parallel memory storage by competitive neural networks. IEEE Trans. Syst. Man Cybern. **13**(5), 815–826 (1983)
18. H. Ye, A.N. Michel, K. Wang, Qualitative analysis of Cohen-Grossberg neural networks with multiple delays. Phys. Rev. E **51**(3), 2611–2618 (1995)
19. A.N. Michel, J.A. Farrel, W. Porod, Qualitative analysis of neural networks. IEEE Trans. Circuits Syst. **36**(2), 229–243 (1989)
20. H. Yang, T. Dillon, Exponential stability and oscillation of Hopfield graded response neural networks. IEEE Trans. Neural Netw. **5**, 719–729 (1994)
21. M. Hirsch, Convergent activation dynamics in continuous time networks. Neural Netw. **2**, 331–349 (1989)
22. D.G. Kelly, Stability in contractive nonlinear neural networks. IEEE Trans. Biomed. Eng. **3**, 231–242 (1990)
23. K. Matsuoka, Stability condition for nonlinear continuous neural networks with asymmetric connection weight. Neural Netw. **5**, 495–500 (1992)
24. E. Kaszkuurewicz, A. Bhaya, On a class of globally stable neural circuits. IEEE Trans. Circuits Syst.-I: Fundam. Theory Appl. **41**, 171–174 (1990)
25. M. Forti, On global asymptotic stability of a class of nonlinear systems arising in neural networks theory. J. Differ. Equ. **113**, 246–264 (1994)
26. M. Forti, A. Tesi, New conditions for global stability of neural networks with applications to linear and quadratic programming problems. IEEE Trans. Circuits Syst.-I **42**(7), 354–366 (1995)
27. Y. Fang, T. Kincaid, Stability analysis of dynamical neural networks. IEEE Trans. Neural Netw. **7**(4), 996–1006 (1996)
28. M.P. Joy, On the global convergence of a class of functional differential equations with applications in neural network theory. J. Math. Anal. Appl. **232**, 61–81 (1999)
29. M.P. Joy, Results concerning the absolute stability of delayed neural networks. Neural Netw. **13**, 613–616 (2000)
30. H. Qiao, J.G. Peng, Z.B. Xu, Nonlinear measure: a new approach to exponential stability analysis for Hopfield-Type neural networks. IEEE Trans. Neural Netw. **12**(2), 360–370 (2001)
31. T. Chen, S. Amari, Stability of asymmetric Hopfield neural networks. IEEE Trans. Neural Netw. **12**(1), 159–163 (2001)
32. T. Chen, S. Amari, New theorems on global convergence of some dynamical systems. Neural Netw. **14**, 251–255 (2001)
33. T. Chen, W. Lu, S. Amari, Global convergence rate of recurrently connected neural networks. Neural Comput. **14**, 2947–2958 (2002)

34. S. Grossberg, Nonlinear neural networks: principles, mechanisms, and architectures. Neural Netw. **1**, 17–61 (1988)
35. L. Wang, X.F. Zou, Exponential stability of Cohen-Grossberg neural networks. Neural Netw. **15**, 415–422 (2002)
36. W. Lu, T. Chen, New conditions on global stability of Cohen-Grossberg neural networks. Neural Comput. **15**, 1173–1189 (2003)
37. J. Jian, B. Wang, Stability analysis in Lagrange sense for a class of BAM neural networks of neutral type with multiple time-varying delays. Neurocomputing **149**, 930–939 (2015)
38. H. Zhang, Z. Wang, D. Liu, Global asymptotic stability of recurrent neural networks with multiple time-varying delays. IEEE Trans. Neural Netw. **19**(5), 855–873 (2008)
39. H. Zhang, Z. Wang, D. Liu, A Comprehensive review of stability analysis of continuous-time recurrent neural networks. IEEE Trans. Neural Netw. Learn. Syst. **25**(7), 1229–1262 (2014)
40. C.M. Marcus, R.M. Westervelt, Stability of analog neural networks with delay. Phys. Rev. A **39**(1), 347–359 (1989)
41. T.A. Burton, Averaged neural networks. Neural Netw. **6**, 677–680 (1993)
42. K. Gopalsamy, X. He, Stability in asymmetric Hopfield nets with transmission delays. Phys. D: Nonlinear Phenom. **76**(4), 344–358 (1994)
43. Y. Zhang, Global exponential stability and periodic solutions of delay Hopfield neural networks. Int. J. Syst. Sci. **27**, 227–231 (1996)
44. P. Van Den Driessche, X.F. Zou, Global attractivity in delayed Hopfield neural networks models. SIAM J. Appl. Math. **58**(6), 1878–1890 (1998)
45. T. Chu, An exponential convergence estimate for analog neural networks with delay. Phys. Lett. A **283**, 113–118 (2001)
46. L.O. Chua, L. Yang, Cellular neural networks: theory. IEEE Trans. Circuits Syst. **35**(10), 1257–1272 (1988)
47. L. Wang, Stability of Cohen-Grossberg neural networks with distributed delays. Appl. Math. Comput. **160**(1), 93–110 (2005)
48. X. Liang, L. Wu, New sufficient conditions for absolute stability of neural networks. IEEE Trans. Circuits Syst.-I: Fundam. Theory Appl. **45**(5), 584–586 (1998)
49. M. Morita, S. Yoshizawa, K. Nakano, Analysis and improvement of the dynamics of autocorelation associative memory. Trans. Inst. Electron. Inf. Commun. Eng. **J73-D-III**(2), 232–242 (1990)
50. M. Morita, S. Yoshizawa, K. Nakano, Memory of correlated patterns by associative memory neural networks with improved dynamics, in *Proceedings of the INNC*, Paris, France, pp. 868–871 (1990)
51. J. Li, A.N. Michel, W. Porod, Analysis and synthesis of a class of neural networks: linear systems operating on a closed hypercube. IEEE Trans. Circuits Syst. **36**, 1405–1422 (1989)
52. B.D. Calvert, Neural nets modeled by an O.D.E. on a cube. Nonlinear Anal. **47**, 2929–2937 (2001)
53. C.W. Shih, Complete stability for a class of cellular neural networks. Int. J. Bifurc. Chaos **11**, 169–177 (2001)
54. L.O. Chua, C.A. Desoer, E.S. Kuh, *Linear and Nonlinear Circuits* (McGraw-Hill, New York, 1987)
55. L.O. Chua, S.M. Kang, Section-wise piecewise-linear functions: canonical representation, properties, and applications. Proc. IEEE **65**, 915–929 (1977)
56. J.-N. Lin, R. Unbehauen, Canonical piecewise-linear networks. IEEE Trans. Neural Netw. **6**, 43–50 (1995)
57. M. Forti, Some extensions of a new method to analyze complete stability of neural networks. IEEE Trans. Neural Netw. **13**(5), 1230–1238 (2002)
58. Z. Wang, Y. Liu, X. Liu, On global asymptotic stability of neural networks with discrete and distributed delays. Phys. Lett. A **345**(6), 299–308 (2005)
59. J.H. Park, On global stability criterion of neural networks with continuously distributed delays. Chaos Solitons Fractals **37**, 444–449 (2008)

60. W. Zhao, Global exponential stability analysis of Cohen-Grossberg neural networks with delays. Commun. Nonlinear Sci. Numer. Simul. **13**, 847–56 (2008)
61. H. Huang, G. Feng, J. Cao, An LMI approach to delay-dependent state estimation for delayed neural networks. Neurocomputing **71**(15), 2857–2867 (2008)
62. C. Li, J. Chen, T. Huang, A new criterion for global robust stability of interval neural networks with discrete time delays. Chaos Solitons Fractals **31**, 561–570 (2007)
63. J. Zhang, Globally exponential stability of neural networks with variable delays. IEEE Trans. Circuits Syst.-I: Fundam. Theory Appl. **50**(2), 288–291 (2003)
64. X.F. Liao, C.D. Li, K. Wong, Criteria for exponential stability of Cohen-Grossberg neural networks. Neural Netw. **17**, 1401–1414 (2004)
65. X. Liao, K.-W. Wong, Global exponential stability for a class of retarded functional differential equations with applications in neural networks. J. Math. Anal. Appl. **293**, 125–148 (2004)
66. F. Tu, X. Liao, Harmless delays for global asymptotic stability of Cohen-Grossberg neural networks. Chaos Solitons Fractals **26**, 927–933 (2005)
67. J. Cao, T. Chen, Global exponential robust stability and periodicity of delayed neural networks. Chaos Solitons Fractals **22**, 957–963 (2004)
68. T. Chu, Z. Zhang, Z. Wang, A decomposition approach to analysis of competitive-cooperative neural networks with delays. Phys. Lett. A **312**, 339–347 (2003)
69. J. Sun, G.P. Liu, J. Chen, D. Rees, Improved stability criteria for neural networks with time-varying delay. Phys. Lett. A **373**, 342–348 (2009)
70. S. Xu, J. Lam, D.W.C. Ho, Y. Zou, Global robust exponential stability analysis for interval recurrent neural networks. Phys. Lett. A **325**, 124–133 (2004)
71. J. Cao, H. Li, L. Han, Novel results concerning global robust stability of delayed neural networks. Nonlinear Anal.: Real World Appl. **7**, 458–469 (2006)
72. W. Chen, X. Lu, Z. Guan, W. Zheng, Delay-dependent exponential stability of neural networks with variable delay: an LMI approach. IEEE Trans. Circuits Syst.-II: Express Br. **53**(9), 837–842 (2006)
73. W. Chen, W. Zheng, Global exponential stability of impulsive neural networks with variable delay: an LMI approach. IEEE Trans. Circuits Syst.-I: Regul. Pap. **56**(6), 1248–1259 (2009)
74. Y. Zhang, D. Yue, E. Tian, Robust delay-distribution-dependent stability of discrete-time stochastic neural networks with time-varying delay. Neurocomputing **72**(4), 1265–1273 (2009)
75. T. Li, M. Fei, Y. Guo, Q. Zhu, Stability analysis on Cohen-Grossberg neural networks with both time-varying and continuously distributed delays. Nonlinear Anal.: Real World Appl. **10**, 2600–2612 (2009)
76. T. Li, M. Fei, Y. Guo, Q. Zhu, Exponential state estimation for recurrent neural networks with distributed delays. Neurocomputing **71**, 428–438 (2007)
77. Z. Wang, H. Shu, Y. Liu, Robust stability analysis of generalized neural networks with discrete and distributed delays. Chaos Solitons Fractals **30**(4), 886–896 (2006)
78. Y. Liu, Z. Wang, Y. Liu, Global exponential stability of generalized neural networks with discrete and distributed delays. Neural Netw. **19**(5), 667–675 (2006)
79. D. Yue, Y. Zhang, E. Tian, C. Peng, Delay-distribution-dependent exponential stability criteria for discrete-time recurrent neural networks with stochastic delay. IEEE Trans. Neural Netw. **19**(7), 1299–1306 (2008)
80. J. Cao, J. Wang, Absolute exponential stability of recurrent neural networks with Lipschitz continuous activation functions and time delays. Neural Netw. **17**, 379–390 (2004)
81. J. Xu, Y.-Y. Cao, Y. Sun, J. Tang, Absolute exponential stability of recurrent neural networks with generalized activation function. IEEE Trans. Neural Netw. **19**(6), 1075–1089 (2008)
82. M. Forti, P. Nistri, Global convergence of neural networks with discontinuous neuron activations. IEEE Trans. Circuits Syst.-I **50**(11), 1421–1435 (2003)
83. M. Forti, P. Nistri, D. Papini, Global exponential stability and global convergence in finite time of delayed neural networks with infinite gain. IEEE Trans. Neural Netw. **16**(6), 1449–1463 (2005)
84. W. Lu, T.P. Chen, Dynamical behaviour of Cohen-Grossberg neural networks with discontinuous activation functions. Neural Netw. **18**, 231–242 (2005)

85. W. Lu, T.P. Chen, Dynamical behaviors of delayed neural network systems discontinuous activation functions. Neural Comput. **18**, 683–708 (2006)
86. D. Papini, V. Taddei, Global exponential stability of the periodic solution of the delayed neural networks with discontinuous activations. Phys. Lett. A **343**, 117–128 (2005)
87. W. Lu, T.P. Chen, Almost periodic dynamics of a class of delayed neural networks with discontinuous activations. Neural Comput. **20**, 1065–1090 (2008)
88. X. He, W. Lu, T. Chen, Nonnegative periodic dynamics of Cohen-Grossberg neural networks with discontinuous activations and discrete time Delays. Lect. Notes Comput. Sci. **5551**, 579–588 (2009)
89. V. Singh, Robust stability of cellular neural networks with delay: linear matrix inequality approach. IEE Proc.-Control Theory Appl. **151**(1), 125–129 (2004)
90. J. Lu, G. Chen, Robust stability and stabilization of fractional-order interval systems: an LMI approach. IEEE Trans. Autom. Control **54**(6), 1294–1299 (2009)
91. C. Li, X. Liao, R. Zhang, Global robust asymptotical stability of multiple-delayed interval neural networks: an LMI approach. Phys. Lett. A **328**, 452–462 (2004)
92. L.S. Wang, *Recurrent neural networks with delays* (Science Press, Beijing, China, 2008) (in Chinese)
93. H. Qi, New sufficient conditions for global robust stability of delayed neural networks. IEEE Trans. Circuits Syst.-I: Regul. Pap. **54**(5), 1131–1141 (2007)
94. Y. He, Q. Wang, W. Zheng, Global robust stability for delayed neural networks with polytopic type uncertainties. Chaos Solitons Fractals **26**, 1349–1354 (2005)
95. Y. He, M. Wu, J. She, G. Liu, Parameter-dependent Lyapunov functional for stability of time-delay systems with polytopic-type uncertainties. IEEE Trans. Autom. Control **49**(5), 828–832 (2004)
96. J. Geromel, P. Colaneri, Robust stability of time varying polytopic systems. Syst. Control Lett. **55**, 81–85 (2006)
97. A. Wu, Z. Zeng, An improved criterion for stability and attractability of memristive neural networks with time-varying delays. Neurocomputing **145**, 316–323 (2014)
98. L. Wei, H. Chen, Global exponential stability of a class of impulsive neural networks with unstable continuous and discrete dynamics. Neurocomputing **147**, 225–234 (2015)
99. B. Xu, D. Xu, Attraction of stochastic neutral delay differential equations with Markovian switching. IMA J. Math. Control Inf. **31**(1), 15–31 (2014)
100. H. Chen, S. Zhong, J. Shao, Exponential stability criterion for interval neural networks with discrete and distributed delays. Appl. Math. Comput. **250**, 121–130 (2015)
101. Z. Zhao, J. Jian, B. Wang, Global attracting sets for neutral-type BAM neural networks with time-varying and infinite distributed delays. Nonlinear Anal.-Hybrid Syst. **15**, 63–73 (2015)
102. Y. He, G. Liu, D. Rees, M. Wu, Stability analysis for neural networks with time-varying interval delay. IEEE Trans. Neural Netw. **18**(6), 1850–1854 (2007)
103. S. Guo, L. Huang, Stability analysis of Cohen-Grossberg neural networks. IEEE Trans. Neural Netw. **17**(1), 106–116 (2006)
104. B. Wang, Y. Zeng, J. Cheng, Further improvement in delay-dependent stability criteria for continuous-time systems with time-varying delays. Neurocomputing **147**, 324–329 (2015)
105. B. Chen, J. Wang, Global exponential periodicity and global exponential stability of a class of recurrent neural networks. Phys. Lett. A **329**(2), 36–48 (2004)
106. R.K. Miller, A.N. Michel, *Ordinary Differential Equations* (Academic Press, New York, 1982)
107. Y. Shen, J. Wang, Noise-induced stabilization of the recurrent neural networks with mixed time-varying delays and Markovian-switching parameters. IEEE Trans. Neural Netw. **18**(6), 1857–1862 (2007)
108. J.Y. Zhang, Y. Suda, H. Komine, Global exponential stability of Cohen-Grossberg neural networks with variable delays. Phys. Lett. A **338**, 44–50 (2005)
109. Y. Chen, Global asymptotic stability of delayed Cohen-Grossberg neural networks. IEEE Trans. Circuits Syst.-I: Regul. Pap. **53**(2), 351–357 (2006)
110. L. Wang, D. Xu, Global exponential stability of Hopfield reaction-diffusion neural networks with time-varying delays. Sci. China (Series F) **46**(6), 466–474 (2003)

111. B. Liu, L. Huang, Existence and exponential stability of periodic solutions for a class of Cohen-Grossberg neural networks with time-varying delays. Chaos Solitons Fractals **32**, 617–627 (2007)
112. K. Lu, D. Xu, Z. Yang, Global attraction and stability for Cohen-Grossberg neural network system with delays. Neural Netw. **19**, 1538–1549 (2006)
113. J. Lu, Robust global exponential stability for interval reaction-diffusion Hopfield neural networks with distributed delays. IEEE Trans. Circuits Syst.-II: Express Br. **54**(12), 1115–1119 (2007)
114. H. Lu, Global exponential stability analysis of Cohen-Grossberg neural networks. IEEE Trans. Circuits Syst.-II: Express Br. **52**(9), 476–479 (2005)
115. T. Chen, Global exponential stability of delayed Hopfield neural networks. Neural Netw. **14**, 977–980 (2001)
116. H. Zhang, Z. Wang, D. Liu, Robust stability analysis for interval Cohen-Grossberg neural networks with unknown time varying delays. IEEE Trans. Neural Netw. **19**(11), 1942–1955 (2008)
117. W. Lu, T. Chen, $\mathbb{R}^n_+$-global stability of a Cohen-Grossberg neural network system with non-negative equilibria. Neural Netw. **20**, 714–722 (2007)
118. J. Cao, J. Liang, Boundedness and stability for Cohen-Grossberg neural network with time-varying delays. J. Math. Anal. Appl. **296**, 665–685 (2004)
119. Y. He, M. Wu, J. She, Delay-dependent exponential stability of delayed neural networks with time-varying delay. IEEE Trans. Circuits Syst.-II: Express Br. **53**(7), 553–557 (2006)
120. S. Boyd, L.E.I. Ghaoui, E. Feron, *Linear Matrix Inequality in System and Control Theory* (SIAM, Philadelphia, 1994)
121. Y. Nesterov, A. Nemirovsky, *Interior Point Polynomial Methods in Convex Programming: Theory and Applications* (SIAM, Philadelphia, 1993)
122. P. Gahinet, A. Nemirovsky, A. Laub, M. Chilali, *LMI Control Toolbox: for Use with Matlab* (The MATH Works Inc., Natick, 1995)
123. H. Ye, A.N. Michel, K.N. Wang, Robust stability of nonlinear time-delay system with applications to neural networks. IEEE Trans. Circuits Syst.-I: Fundam. Theory Appl. **43**(7), 532–543 (1996)
124. G. Hardy, J.E. Littlewood, G. Polya, *Inequality*, 2nd edn. (Cambridge University Press, Cambridge, 1954)
125. T. Chu, C. Zhang, Z. Zhang, Necessary and sufficient condition for absolute stability of normal neural networks. Neural Netw. **16**, 1223–1227 (2003)
126. T. Chu, C. Zhang, New necessary and sufficient conditions for absolute stability of neural networks. Neural Netw. **20**, 94–101 (2007)
127. S. Hu, J. Wang, Global asymptotic stability and global exponential stability of continuous-time recurrent neural networks. IEEE Trans. Autom. Control **47**(5), 802–807 (2002)
128. S. Hu, J. Wang, Global stability of a class of continuous-time recurrent neural networks. IEEE Trans. Circuits Syst.-I: Fundam. Theory Appl. **49**(9), 1334–1337 (2002)
129. J. Ruan, A global stable analysis for CGNN and CNN with asymmetric weights, in *Proceedings of 1993 International Joint Conference on Neural Networks*, pp. 2327–2330 (1993)
130. V. Singh, Improved global robust stability of interval delayed neural networks via split interval: generalizations. Appl. Math. Comput. **206**, 290–297 (2008)
131. H. Zhang, Z. Wang, D. Liu, Global asymptotic stability and robust stability of a class of Cohen-Grossberg neural networks with mixed delays. IEEE Trans. Circuits Syst.-I: Regul. Pap. **56**(3), 616–629 (2009)
132. H. Zhang, Y. Wang, Stability analysis of Markovian jumping stochastic Cohen-Grossberg neural networks with mixed time delays. IEEE Trans. Neural Netw. **19**(2), 366–370 (2008)
133. H. Zhang, Z. Wang, Global asymptotic stability of delayed cellular neural networks. IEEE Trans. Neural Netw. **18**(3), 947–950 (2007)
134. H. Zhang, Z. Wang, D. Liu, Robust exponential stability of recurrent neural networks with multiple time-varying delays. IEEE Trans. Circuits Syst.-II: Express Br. **54**(8), 730–735 (2007)

135. Z. Wang, H. Zhang, B. Jiang, LMI-based approach for global asymptotic stability analysis of recurrent neural networks with various delays and structures. IEEE Trans. Neural Netw. **22**(7), 1032–1045 (2011)
136. Z. Wang, H. Zhang, P. Li, An LMI approach to stability analysis of reaction-diffusion Cohen-Grossberg neural networks concerning Dirichlet boundary conditions and distributed delays. IEEE Trans. Syst. Man Cybern. Part B-Cybern. **40**(6), 1596–1606 (2010)
137. Z. Wang, H. Zhang, Global asymptotic stability of reaction-diffusion Cohen-Grossberg neural networks with continuously distributed delays. IEEE Trans. Neural Netw. **21**(1), 39–49 (2010)
138. C. Ge, C. Hua, X. Guan, New delay-dependent stability criteria for neural networks with time-varying delay using delay-decomposition approach. IEEE Trans. Neural Netw. Learn. Syst. **25**(7), 1378–1383 (2014)
139. S. Mou, H. Gao, W. Qiang, K. Chen, New delay-dependent exponential stability for neural networks with time delay. IEEE Trans. Syst. Man Cybern. B Cybern. **38**(2), 571–576 (2008)
140. H. Zeng, Y. He, M. Wu, C. Zhang, Complete delay-decomposing approach to asymptotic stability for neural networks with time-varying delays. IEEE Trans. Neural Netw. **22**(5), 806–812 (2011)
141. H. Zhang, Z. Liu, G.-B. Huang, Z. Wang, Novel weighting delay-based stability criteria for recurrent neural networks with time varying delay. IEEE Trans. Neural Netw. **21**(1), 91–106 (2010)
142. Z. Wang, L. Liu, Q. Shan, H. Zhang, Stability criteria for recurrent neural networks with time-varying delay based on secondary delay partitioning method. IEEE Trans. Neural Netw. Learn. Syst. (2015). doi:10.1109/TNNLS.2014.2387434. (to appear in 2005)
143. H. Lu, R. Shen, F.-L. Chung, Global exponential convergence of Cohen-Grossberg neural networks with time delays. IEEE Trans. Neural Netw. **16**(6), 1694–1696 (2005)
144. Y. Meng, S. Guo, L. Huang, Convergence dynamics of Cohen-Grossberg neural networks with continuously distributed delays. Appl. Math. Comput. **202**, 188–199 (2008)
145. H. Qi, L. Qi, Deriving sufficient conditions for global asymptotic stability of delayed neural networks via non-smooth analysis. IEEE Trans. Neural Netw. **15**(1), 99–109 (2004)
146. H. Qi, L. Qi, X. Yang, Deriving sufficient conditions for global asymptotic stability of delayed neural networks via non-smooth analysis-II. IEEE Trans. Neural Netw. **16**(6), 1701–1706 (2005)
147. K. Yuan, J. Cao, An analysis of global asymptotic stability of delayed Cohen-Grossberg neural networks via nonsmooth analysis. IEEE Trans. Circuits Syst.-I **52**(9), 1854–1861 (2005)
148. Q. Zhou, L. Wan, J. Sun, Exponential stability of reaction-diffusion generalized Cohen-Grossberg neural networks with time-varying delays. Chaos Solitons Fractals **32**, 1713–1719 (2007)
149. T. Chen, L. Rong, Robust global exponential stability of Cohen-Grossberg neural networks with time delays. IEEE Trans. Neural Netw. **15**(1), 203–206 (2004)
150. K. Li, X. Zhang, Z. Li, Global exponential stability of impulsive cellular neural networks with time-varying and distributed delay. Chaos Solitons Fractals **41**, 1427–1434 (2009)
151. A. Wan, M. Wang, J. Peng, H. Qiao, Exponential stability of Cohen-Grossberg neural networks with a general class of activation functions. Phys. Lett. A **350**, 96–102 (2006)
152. M. Forti, A. Tesi, A new method to analyze complete stability of PWL cellular neural networks. Int. J. Bifurc. Chaos **11**(3), 655–676 (2001)
153. M. Forti, A. Tesi, Absolute stability of analytic neural networks: an approach based on finite trajectory length. IEEE Trans. Circuits Syst.-I **51**(12), 2460–2469 (2004)
154. J. Liang, J. Cao, Global exponential stability of reaction-diffusion recurrent neural networks with time-varying delays. Phys. Lett. A **314**, 434–442 (2003)
155. X. Liao, Y. Liu, S. Guo, H. Mai, Asymptotic stability of delayed neural networks: a descriptor system approach. Commun. Nonlinear Sci. Numer. Simul. **14**, 3120–3133 (2009)
156. J. Qiu, Exponential stability of impulse neural networks with time-varying delays and reaction-diffusion terms. Neurocomputing **70**, 1102–1108 (2007)
157. J. Liang, J. Cao, Boundedness and stability for recurrent neural networks with variable coefficients and time-varying delays. Phys. Lett. A **318**, 53–64 (2003)

158. X. Liao, K.-W. Wong, S. Yang, Convergence dynamics of hybrid bidirectional associative memory neural networks with distributed delays. Phys. Lett. A **316**, 55–64 (2003)
159. X. Liao, Q. Liu, W. Zhang, Delay-dependent asymptotic stability for neural networks with distributed delays. Nonlinear Anal.: Real World Appl. **7**, 1178–1192 (2006)
160. X. Liao, K.-W. Wong, C. Li, Global exponential stability for a class of generalized neural networks with distributed delays. Nonlinear Anal.: Real World Appl. **5**, 527–547 (2004)
161. C. Li, X. Liao, New algebraic conditions for global exponential stability of delayed recurrent neural networks. Neurocomputing **64**, 319–333 (2005)
162. S. Arik, V. Tavsanoglu, A comment on "comments on 'necessary and sufficient condition for absolute stability of neural networks"'. IEEE Trans. Circuits Syst. I **45**, 595–596 (1998)
163. X.B. Liang, J. Wang, An additive diagonal-stability condition absolute exponential stability of a general class of neural network. IEEE Trans. Circuits Syst. **I** (48), 1308–1317 (2001)
164. S. Hu, J. Wang, Absolute exponential stability of a class of continuous-time recurrent neural networks. IEEE Trans. Neural Netw. **14**(1), 35–46 (2003)
165. X.B. Liang, L.D. Wu, Global exponential stability of a class of neural circuits. IEEE Trans. Circuits Syst. **I** (46), 748–751 (1999)
166. X.B. Liang, A comment on equilibria, stability, and instability of Hopfield neural networks. IEEE Trans. Neural Netw. **11**, 1506–1507 (2000)
167. X. Liao, S. Yang, S. Cheng, Y. Shen, Stability of generalized networks with reaction-diffusion terms. Sci. China (Series F) **44**(5), 389–395 (2001)
168. L. Wan, J.H. Sun, Global asymptotic stability of Cohen-Grossberg neural network with continuously distributed delays. Phys. Lett. A **342**, 331–340 (2005)
169. J. Zhang, X. Jin, Global stability analysis in delayed Hopfield neural network models. Neural Netw. **13**, 745–753 (2000)
170. D. Burshtein, Long-term attraction in higher order neural networks. IEEE Trans. Neural Netw. **9**(1), 42–50 (1998)
171. G. Chesi, Computing output feedback controllers to enlarge the attraction domain in polynomial systems. IEEE Trans. Autom. Control **49**(10), 1846–1853 (2004)
172. P. Driessche, X. Zou, Global attractivity in delayed Hopfield neural networks. SIAM J. Appl. Math. **58**(6), 1878–1890 (1998)
173. D. Zhou, J. Shen, X. Ren, Estimation of attraction domain and exponential convergence rate of dynamic feedback neural nets, in *Proceedings of ICSP2000*, pp. 1598–1601 (2000)
174. K.L. Moore, D.S. Naidu, maximal domains of attraction in a Hopfield neural network with learning, in *Proceeding of American Control Conference*, pp. 2894–2896 (1993)
175. J. Cao, On the domain of attraction and convergence rate of Hopfield continuous feedback neural networks, in *Proceedings of the 3rd World Congress on Intelligent Control and Automation*, pp. 3277–3280 (2000)
176. D. Zhou, L. Zhang, D. Zhao, Study of attraction domain and exponential convergence rate of Hopfield continuous dynamic feedback neural nets, in *Proceedings of the 4th World Congress on Intelligent Control and Automation*, pp. 2035–2038 (2002)
177. L. Gruyitch, J. Richard, P. Borne, J. Gentina, *Stability Domains, Nonlinear Systems in Aviation, Aerospace, Aeronautics, Astronautics* (Chapman and Hall, London, 2004)
178. S. Balint, Considerations concerning the maneuvering of some physical systems. An. Univ. Timisoara **23**, 8–16 (1985)
179. S. Balint, A. Balint, V. Negru, The optimal Lyapunov function in diagonalizable case. An. Univ. Timisoara **24**, 1–7 (1986)
180. S. Balint, V. Negru, A. Balint, T. Simiantu, An approach of the region of attraction by the region of convergence of the series of the optimal Lyapunov function. An. Univ. Timisoara **25**, 15–30 (1987)
181. E. Kaslik, A. Balint, S. Balint, Gradual approximation of the domain of attraction by gradual extension of the "embryo" of the transformed optimal Lyapunov function. Nonlinear Stud. **10**(1), 67–78 (2003)
182. E. Kaslik, A. Balint, S. Balint, Methods for determination and approximation of the domain of attraction. Nonlinear Anal.: Real World Appl. **60**(4), 703–717 (2005)

183. E. Kaslik, S. Balint, Configurations of steady states for Hopfield-type neural networks. Appl. Math. Comput. **182**(1), 934–946 (2006)
184. D. Jin, J. Peng, A new approach for estimating the attraction domain for Hopfield-type neural networks. Neural Comput. **21**, 101–120 (2009)
185. Y. Zhang, P. Peng, A. Fu, Estimate of exponential convergence rate and exponential stability for neural networks. IEEE Trans. Neural Netw. **10**(6), 1487–1493 (1999)
186. N. Ozdemir, B. Iskender, N. Qzgur, Complex valued neural network with Mobius activation function. Commun. Nonlinear Sci. Numer. Simul. **16**(12), 4698–703 (2011)
187. N. Hashimoto, Y. Kuroe, T. Mori, An energy function for complex-valued neural networks and its applications. Proc. Annu. Conf. Inst. Syst. Control Inf. Eng. **43**, 391–392 (1999)
188. Y. Kuroe, M. Yoshid, T. Mori, On activation functions for complex-valued neural networks: existence of energy functions, in *Artificial Neural Networks and Neural Information Processing*, pp. 174–175 (2003)
189. T. Kim, T. Adali, Fully complex multi-layer perceptron network for nonlinear signal processing. J. VLSI Signal Process. **32**(2), 29–43 (2002)
190. B. Zhou, Q. Song, Boundedness and complete stability of complex-valued neural networks with time delay. IEEE Trans. Neural Netw. Learn. Syst. **24**(8), 1227–1238 (2013)
191. X. Liu, K. Fang, B. Liu, A synthesis method based on stability analysis for complex-valued Hopfield neural network, in *Proceedings 7th Asian Control Conference*, pp. 1245–1250 (2009)
192. A. Hirose, *Complex-Valued Neural Networks: Theories and Applications* (World Scientific, Singapore, 2003)
193. W. Zhou, J. Zurada, Discrete-time recurrent neural networks with complex-valued linear threshold neurons. IEEE Trans. Circuits Syst. II Express Br. **56**(8), 669–673 (2009)
194. C. Duan, Q. Song, Boundedness and stability for discrete-time delayed neural network with complex-valued linear threshold neurons. Discret. Dyn. Nature Soc. **2010**, 1–19 (2010)
195. L. Chua, Memristor-the missing circuit element. IEEE Trans. Circuit Theory **18**, 507–519 (1971)
196. D. Strukov, G. Snider, G. Stewart, R. Williams, The missing memristor found. Nature **453**, 80–83 (2008)
197. M. Ventra, Y. Pershin, L. Chua, Circuit elements with memory: memristors, memcapacitors, and meminductors. Proc. IEEE **97**, 1717–1724 (2009)
198. M. Itoh, L.O. Chua, Memristor cellular automata and memristor discrete-time cellular neural networks. Int. J. Bifurcat. Chaos **19**(11), 3605–3656 (2009)
199. Y.V. Pershin, M. Di Ventra, Experimental demonstration of associative memory with memristive neural networks. Neural Netw. **23**(7), 881–886 (2010)
200. F.Z. Wang, N. Helian, S.N. Wu, X. Yang, Y.K. Guo, G. Lim, M.M. Rashid, Delayed switching applied to memristor neural networks. J. Appl. Phys. **111**(7), 07E317-1–07E317-3 (2012)
201. A. Chandrasekar, R. Rakkiyappan, J. Cao, S. Lakshmanand, Synchronization of memristor-based recurrent neural networks with two delay components based on second-order reciprocally convex approach. Neural Netw. **57**, 79–93 (2014)
202. A. Wu, Z. Zeng, Lagrange stability of memristive neural networks with discrete and distributed delays. IEEE Trans. Neural Netw. Learn. Syst. **25**(4), 690–703 (2014)
203. S. Arik, Stability analysis of delayed neural networks. IEEE Trans. Circuits Syst.-I: Fundam. Theory **47**(7), 1089–1092 (2000)
204. P. Baldi, A.F. Atiya, How delays affect neural dynamics and learning. IEEE Trans. Neural Netw. **5**(10), 612–621 (1994)
205. J. Cao, Periodic oscillation and exponential stability of delayed CNNs. Phys. Lett. A **270**, 157–163 (2000)
206. X. Liao, G. Chen, E.N. Sanchez, LMI-based approach for asymptotically stability analysis of delayed neural networks. IEEE Trans. Circuits Syst.-I: Fundam. Theory Appl. **49**, 1033–1039 (2002)
207. Z. Wang, D.W.C. Ho, X. Liu, State estimation for delayed neural networks. IEEE Trans. Neural Netw. **16**(1), 279–284 (2005)

208. J. Cao, D. Huang, Y. Qu, Global robust stability of delayed recurrent neural networks. Chaos Solitons Fractals **23**, 221–229 (2005)
209. J. Cao, J. Wang, Global asymptotic stability of a general class of recurrent neural networks with time-varying delays. IEEE Trans. Circuits Syst.-I: Fundam. Theory **50**(1), 34–44 (2003)
210. Z.G. Zeng, J. Wang, X.X. Liao, Global exponential stability of a general class of neural networks with time-varying delays. IEEE Trans. Circuits Syst.-I: Regul. Pap. **50**(10), 1353–1358 (2003)
211. Z. Wang, *Stability of Continuous-time Recurrent Neural Networks with Delays* (Northeastern University Press, Shenyang, 2007) (in Chinese)
212. Z. Wang, *Stability and Synchronization of Complex Neural Dynamical Networks* (Science Press, Beijing, 2014) (in Chinese)
213. L.O. Chua, T. Roska, CNN: a new paradigm of nonlinear dynamics in space, in *World Congress of Nonlinear Analysis*, pp. 2979–2990 (1992)
214. P. Civalleri, M. Gilli, L. Pabdolfi, On stability of cellular neural networks with delay. IEEE Trans. Circuits Syst.-I **40**(3), 157–165 (1993)
215. T. Roska, C.W. Wu, M. Balsi, L.O. Chua, Stability and dynamics of delay-type general and cellular neural networks. IEEE Trans. Circuits. Syst.-I: Fundam. Theory Appl. **39**(6), 487–490 (1992)
216. T. Roska, C.W. Wu, L.O. Chua, Stability of cellular neural networks with dominant nonlinear and delay-type template. IEEE Trans. Circuits Syst.-I: Fundam. Theory Appl. **40**(4), 270–272 (1993)
217. X.X. Liao, J. Wang, Algebraic criteria for global exponential stability of cellular neural networks with multiple time delays. IEEE Trans. Circuits Syst.-I: Fundam. Theory Appl. **50**(2), 268–275 (2003)
218. Z. Wang, H. Zhang, W. Yu, Robust exponential stability analysis of neural networks with multiple time varying delays. Neurocomputing **70**, 2534–2543 (2007)
219. C. Huang, L. Huang, Dynamics of a class of Cohen-Grossberg neural networks with time-varying delays. Nonlinear Anal.: Real World Appl. **8**, 40–52 (2007)
220. T. Huang, A. Chan, Y. Huang, J. Cao, Stability of Cohen-Grossberg neural networks with time-varying delays. Neural Netw. **20**(8), 868–873 (2007)
221. T. Huang, C. Li, G. Chen, Stability of Cohen-Grossberg neural networks with unbounded distributed delays. Chaos Solitons Fractals **34**(3), 992–996 (2007)
222. W. Lu, T. Chen, Dynamical behaviors of Cohen-Grossberg neural networks with discontinuous activation functions. Neural Netw. **18**, 231–242 (2005)
223. Z. Wang, H. Zhang, W. Yu, Robust stability of Cohen-Grossberg neural networks via state transmission matrix. IEEE Trans. Neural Netw. **20**(1), 169–174 (2009)
224. C.C. Huang, C.J. Chang, T.L. Liao, Globally exponential stability of generalized Cohen-Grossberg neural networks with delays. Phys. Lett. A **319**(2), 157–166 (2003)
225. L. Wang, X. Zou, Harmless delays in Cohen-Grossberg neural network. Phys. D: Nonlinear Phenom. **170**(2), 162–173 (2002)
226. Y. Shen, J. Wang, Almost sure exponential stability of recurrent neural networks with Markovian switching. IEEE Trans. Neural Netw. **20**(5), 840–855 (2009)
227. A. Bouzerdoum, T.R. Pattison, Neural network for quadratic optimization with bound constraints. IEEE Trans. Neural Netw. **4**(3), 293–304 (1993)
228. D.W. Tank, J.J. Hopfield, Neural computation by concentrating information in time. Proc. Nat. Acad. Sci. USA **84**, 1896–1991 (1987)
229. W. Chen, W. Zheng, Global asymptotic stability of a class of neural networks with distributed delays. IEEE Trans. Circuits Syst.-I: Regul. Pap. **53**(3), 644–652 (2006)
230. K. Gopalsamy, X. He, Delay-independent stability in bidirectional associative memory networks. IEEE Trans. Neural Netw. **5**(6), 998–1002 (1994)
231. V.S.H. Rao, B.R.M. Phaneendra, V. Prameela, Global dynamics of bidirectional associative memory networks with transmission delays. Differ. Equ. Dyn. Syst. **4**, 453–471 (1996)
232. J. Cao, K. Yuan, H.X. Li, Global asymptotic stability of recurrent neural networks with multiple discrete delays and distributed delays. IEEE Trans. Neural Netw. **17**(6), 1646–1651 (2006)

233. H. Li, B. Chen, Q. Zhou, S. Fang, Robust exponential stability for uncertain stochastic neural networks with discrete and distributed time-varying delays. Phys. Lett. A **372**, 3385–3394 (2008)
234. R. Rakkiyappan, P. Balasubramaniam, S. Lakshmanan, Robust stability results for uncertain stochastic neural networks with discrete interval and distributed time-varying delays. Phys. Lett. A **372**, 5290–5298 (2008)
235. Q. Song, Z. Wang, Neural networks with discrete and distributed time-varying delays: a general stability analysis. Chaos Solitons Fractals **37**, 1538–1547 (2008)
236. J. Yu, K. Zhang, S. Fei, T. Li, Simplified exponential stability analysis for recurrent neural networks with discrete and distributed time-varying delays. Appl. Math. Comput. **205**, 465–474 (2008)
237. S. Chen, W. Zhao, Y. Xu, New criteria for globally exponential stability of delayed Cohen-Grossberg neural network. Math. Comput. Simul. **79**(5), 1527–1543 (2009)
238. Z. Huang, X. Wang, Y. Xia, Exponential stability of impulsive Cohen-Grossberg networks with distributed delays. Int. J. Circuit Theory Appl. **36**, 345–365 (2008)
239. Z. Li, K. Li, Stability analysis of impulsive Cohen-Grossberg neural networks with distributed delays and reaction-diffusion terms. Appl. Math. Model. **33**, 1337–1348 (2009)
240. J. Liang, J. Cao, Global asymptotic stability of bi-directional associative memory networks with distributed delays. Appl. Math. Comput. **152**, 415–424 (2004)
241. J. Liang, J. Cao, Global output convergence of recurrent neural networks with distributed delays. Nonlinear Anal.: Real World Appl. **8**, 187–197 (2007)
242. Y. Liu, Z. You, L. Cao, On the almost periodic solution of cellular neural networks with distributed delays. IEEE Trans. Neural Netw. **18**(1), 295–300 (2007)
243. X. Lou, B. Cui, W. Wu, On global exponential stability and existence of periodic solutions for BAM neural networks with distributed delays and reaction-diffusion terms. Chaos Solitons Fractals **36**, 1044–1054 (2008)
244. X. Nie, J. Cao, Multistability of competitive neural networks with time-varying and distributed delays. Nonlinear Anal.: Real World Appl. **10**, 928–942 (2009)
245. J. Zhang, Absolute stability analysis in cellular neural networks with variable delays and unbounded delay. Comput. Math. Appl. **47**, 183–194 (2004)
246. Z. Zhao, Q. Song, J. Zhang, Exponential periodicity and stability of neural networks with reaction-diffusion terms and both variable and bounded delays. Comput. Math. Appl. **51**, 475–486 (2006)
247. J. Shao, Global exponential convergence for delayed cellular neural networks with a class of general activation functions. Nonlinear Anal.: Real World Appl. **10**, 1816–1821 (2009)
248. L. Zhou, G. Hu, Global exponential periodicity and stability of cellular neural networks with variable and distributed delays. Appl. Math. Comput. **195**, 402–411 (2008)
249. B. Cui, W. Wu, Global exponential stability of Cohen-Grossberg neural networks with distributed delays. Neurocomputing **72**, 386–391 (2008)
250. P. Liu, F. Yi, Q. Guo, J. Yang, W. Wu, Analysis on global exponential robust stability of reaction-diffusion neural networks with S-type distributed delays. Phys. D **237**, 475–485 (2008)
251. S. Mohamad, K. Gopalsamy, Dynamics of a class of discrete-time neural networks and their continuous-time counter parts. Math. Comput. Simul. **53**, 1–39 (2000)
252. S. Mohamad, Exponential stability preservation in discrete-time analogues of artificial neural networks with distributed delays. J. Comput. Appl. Math. **215**, 270–287 (2008)
253. W. Wu, B. Cui, X. Lou, Global exponential stability of Cohen-Grossberg neural networks with distributed delays. Math. Comput. Model. **47**, 868–873 (2008)
254. T.P. Chen, W.L. Lu, Stability analysis of dynamical neural networks, in *IEEE International Conference Neural Networks and Signal Processing*, Nanjing, China, December, pp. 14–17 (2003)
255. T.P. Chen, Universal approach to study delay dynamical systems. Lect. Notes Comput. Sci. **3610**, 245–253 (2005)

256. T.P. Chen, W.L. Lu, G.R. Chen, Dynamical behaviors of a large class of general delayed neural networks. Neural Comput. **17**, 949–968 (2005)
257. T.P. Chen, W.L. Lu, Global asymptotic stability of Cohen-Grossberg neural networks with time-varying and distributed Delays. Lect. Notes Comput. Sci. **3971**, 192–197 (2006)
258. T.P. Chen, Universal approach to study delayed dynamical systems. Stud. Comput. Intell. **35**, 85–110 (2007)
259. W.L. Lu, T.P. Chen, On periodic dynamical systems. Chin. Ann. Math. Ser. B **25**, 455–462 (2004)
260. W.L. Lu, T.P. Chen, Global exponential stability of almost periodic solution for a large class of delayed dynamical systems. Sci. China Ser. A-Math. **48**, 1015–1026 (2005)
261. W.L. Lu, T.P. Chen, Almost periodic dynamics of a class of delayed neural networks with discontinuous activations. Neural Comput. **20**, 1065–1090 (2008)
262. X.F. Liao, C.D. Li, Global attractivity of Cohen-Grossberg model with finite and infinite delays. J. Math. Anal. Appl. **315**, 244–262 (2006)
263. P.C. Liu, F.Q. Yi, Q. Guo, J. Yang, W. Wu, Analysis on global exponential robust stability of reaction-diffusion neural networks with S-type distributed delays. Phys. D **237**, 475–485 (2008)
264. L.S. Wang, R.J. Zhang, Y.F. Wang, Global exponential stability of Reaction-Diffusion cellular neural networks with S-type distributed time delays. Nonlinear Anal.: Real World Appl. **10**, 1101–1113 (2009)
265. Z. Wang, H. Zhang, J. Feng, Stability analysis of recurrent neural networks with distributed delays satisfying Lebesgue-Stieljies measures, in *7th International Symposium on Neural Networks*, pp. 504–511 (2010)
266. V. Capasso, A. Di Liddo, Asymptotic behaviour of reaction-diffusion systems in population and epidemic models: the role of cross diffusion. J. Math. Biol. **32**, 453–463 (1994)
267. X. Liao, J. Li, Stability in Gilpin-Ayala competition models with diffusion. Nonlinear Anal. **28**, 1751–1758 (1997)
268. C.V. Pao, Global asymptotic stability of Lotka-Volterra competition systems with division and time delays. Nonlinear Anal.: Real World Appl. **5**, 91–104 (2004)
269. S. Raychaudhuri, D.K. Sinha, J. Chattopadhyay, Effect of time-varying cross-diffusivity in a two-species Lotka-Volterra competitive system. Ecol. Model. **92**, 55–64 (1996)
270. F. Rothe, Convergence to the equilibrium state in the Volterra-Lotka diffusion equations. J. Math. Biol. **3**, 319–324 (1976)
271. X. Zhao, Permanence and positive periodic solutions of n-species competition reaction-diffusion systems with spatial inhomogeneity. J. Math. Anal. Appl. **197**, 363–378 (1996)
272. T. Serrano-Gotarredona, A. Rodriguez-Vazquez, On the design of second order dynamics reaction-diffusion CNNs. J. VLSI Signal Process. **23**, 351–371 (1999)
273. Y. Lv, W. Lv, J. Sun, Convergence dynamics of stochastic reaction-diffusion recurrent neural networks with continuously distributed delays. Nonlinear Anal.: Real World Appl. **9**, 1590–1606 (2008)
274. L. Wang, Y. Gao, Global exponential robust stability of reaction-diffusion interval neural networks with time-varying delays. Phys. Lett. A **350**, 342–348 (2006)
275. J.L. Qiu, J.D. Cao, Delay-dependent exponential stability for a class of neural networks with time delays and reaction-diffusion terms. J. Frankl. Inst. **346**, 301–304 (2009)
276. J.H. Sun, L. Wan, Convergence dynamics of stochastic reaction-diffusion recurrent neural networks with delay. Int. J. Bifurc. Chaos **15**(7), 2131–2144 (2005)
277. H. Zhao, K. Wang, Dynamical behaviors of Cohen-Grossberg neural networks with delays and reaction-diffusion terms. Neurocomputing **70**, 536–543 (2006)
278. S. Grossberg, Biological competition: decision rules, pattern formation, and oscillations. Proc. Natl. Acad. Sci. USA **77**(4), 2338–2342 (1980)
279. M. Forti, Convergence of a subclass of Cohen-Grossberg neural networks via the Lojasiewicz inequality. IEEE Trans. Syst. Man Cybern.-Part B: Cybern. **38**(1), 252–257 (2008)
280. J. Palis, W. De Melo, *Geometric Theory of Dynamical Systems* (Springer, Berlin, 1982)

281. S. Arik, An improved global stability result for delayed cellular neural networks. IEEE Trans. Circuits Syst.-I: Fundam. Theory **49**(8), 1211–1214 (2002)
282. J. Cao, Global stability conditions for delayed CNNs. IEEE Trans. Circuits Syst.-I: Fundam. Theory **48**(11), 1330–1333 (2001)
283. T. Chen, L. Rong, Delay-independent stability analysis of Cohen-Grossberg neural networks. Phys. Lett. A **317**(6), 436–449 (2003)
284. H. Lu, On stability of nonlinear continuous-time neural networks with delays. Neural Netw. **13**, 1135–1143 (2000)
285. Z. Yuan, L. Huang, D. Hu, B. Liu, Convergence of nonautonomous Cohen-Grossberg-type neural networks with variable delays. IEEE Trans. Neural Netw. **19**(1), 140–147 (2008)
286. J. Sun, L. Wan, Convergence dynamics of stochastic reaction-diffusion recurrent neural networks with delays. Int. J. Bifurc. Chaos **15**(7), 2131–2144 (2005)
287. Q. Song, J. Cao, Z. Zhao, Periodic solutions and its exponential stability of reaction-diffusion recurrent neural networks with continuously distributed delays. Nonlinear Anal.: Real World Appl. **7**, 65–80 (2006)
288. Z. Tang, Y. Luo, F. Deng, Global exponential stability of reaction-diffusion Hopfield neural networks with distributed delays. Lect. Notes Comput. Sci. **3496**, 174–180 (2005)
289. Y. Luo, F. Deng, B. Zhao, Global asymptotic stability of reaction-diffusion neural networks with distributed delays. Acta Electronica Sinica **33**(2), 218–221 (2005)
290. Y. Luo, F. Deng, Z. Tang, F. Yang, Global exponential stability of distributed delays neural networks with reaction-diffusion terms. Control Theory Appl. **24**(2), 210–214 (2007)
291. C. Cheng, T. Liao, J. Yan, C. Hwang, Globally asymptotic stability of a class of neutral-type neural networks with delays. IEEE Trans. Syst. Man Cybern.-Part B: Cybern. **36**(5), 1191–1195 (2006)
292. E. Fridman, New Lyapunov-Krasovskii functionals for stability of linear retard and neutral type systems. Syst. Control Lett. **43**, 309–319 (2001)
293. R. Krishnasamy, P. Balasubramaniam, A descriptor system approach to the delay-dependent exponential stability analysis for switched neutral systems with nonlinear perturbations. Nonlinear Anal.-Hybrid Syst. **15**, 23–36 (2015)
294. Y. Moon, P. Park, W. Kwon, Y. Lee, Delay-dependent robust stabilization of uncertain state-delayed systems. Int. J. Control **74**(14), 1447–1455 (2001)
295. S. Arik, Global asymptotic stability of a larger class of neural networks with constant time delay. Phys. Lett. A **311**, 504–511 (2003)
296. J. Cao, A set of stability for delayed cellular neural networks. IEEE Trans. Circuits Syst.-I: Fundam. Theory **48**(4), 494–498 (2001)
297. J. Cao, L. Wang, Exponential stability and periodic oscillatory solution in BAM networks with delays. IEEE Trans. Neural Netw. **13**(2), 457–463 (2002)
298. J. Cao, J. Wang, Global asymptotic and robust stability of recurrent neural networks with time delays. IEEE Trans. Circuits Syst.-I **52**(3), 417–426 (2005)
299. J. Cao, J. Wang, X. Liao, Novel stability criteria of delayed cellular neural networks. Int. J. Neural Syst. **13**(5), 367–375 (2003)
300. J. Cao, D. Zhou, Stability analysis of delayed cellular neural networks. Neural Netw. **11**, 1601–1605 (1998)
301. J.A. Farrell, A.N. Michel, A synthesis procedure for Hopfield's continuous-time associative memory. IEEE Trans. Circuits Syst. **37**(7), 877–884 (1990)
302. S. Hu, J. Wang, Global exponential stability of continuous-time interval neural networks. Phys. Rev. E **65**, 036133 (2002)
303. T. Liao, F. Wang, Global stability for cellular neural networks with time delay. IEEE Trans. Neural Netw. **11**(6), 1481–1484 (2000)
304. S. Arik, Global robust stability of delayed neural networks. IEEE Trans. Circuits Syst.-I: Fundam. Theory **50**(1), 156–160 (2003)
305. X.F. Liao, J.B. Yu, Robust stability for interval Hopfield neural networks with time delay. IEEE Trans. Neural Netw. **9**, 1042–1046 (1998)

306. X.F. Liao, K. Wong, Z.F. Wu, G.R. Chen, Novel robust stability criteria for interval delayed Hopfield neural networks. IEEE Trans. Circuits Syst.-I **48**, 1355–1358 (2001)
307. V. Singh, A novel global robust stability criterion for neural networks with delay. Phys. Lett. A **337**, 369–373 (2005)
308. C.Y. Sun, C.B. Feng, Global robust exponential stability of interval neural networks with delays. Neural Proc. Lett. **17**, 107–115 (2003)
309. C.Y. Sun, C.B. Feng, On robust exponential periodicity of interval neural networks with delays. Neural Proc. Lett. **20**, 53–61 (2004)
310. H. Huang, J. Cao, Y. Qu, Global robust stability of delayed neural networks with a class of general activation functions. J. Comput. Syst. Sci. **69**, 688–700 (2004)
311. S.I. Niculescu, *Delay Effects on Stability: A Robust Control Approach*, Lecture Notes in Control and Information Sciences (LNCIS) (Springer, London, 2001)
312. W. Ma, Y. Saito, Y. Takeuchi, M-matrix structure and harmless delays in a Hopfield-type neural network. Appl. Math. Lett. **22**, 1066–1070 (2009)
313. T. Chen, Global Convergence of delayed dynamical systems. IEEE Trans. Neural Netw. **12**(6), 1532–1536 (2001)
314. M. Forti, S. Manetsi, M. Marini, Necessary and sufficient condition for absolute stability of neural networks. IEEE Trans. Circuits Syst.-I: Fundam. Theory Appl. **41**(7), 491–494 (1994)
315. X. Liang, L. Wu, A simple proof of a necessary and sufficient condition for absolute stability of symmetric neural networks. IEEE Trans. Circuits Syst.-I: Fundam. Theory Appl. **45**(9), 1010–1011 (1998)
316. X.B. Liang, Y. Yamaguchi, Necessary and sufficient conditions for absolute exponential stability of Hopfield-type neural networks. IEICE Trans. Inf. Syst. **E79-D**, 990–993 (1996)
317. X.B. Liang, T. Yamaguchi, Necessary and sufficient conditions for absolute exponential stability of a class of nonsymmetric neural networks. IEICE Trans. Inf. Syst. **E80-D**, 802–807 (1997)
318. E. Kaszkurewicz, A. Bhaya, Comments on 'Necessary and sufficient condition for absolute stability of neural networks'. IEEE Trans. Circuits Syst. I **42**, 497–499 (1995)
319. X.B. Liang, J. Wang, A proof of Kaszkurewicz and Bhaya's conjecture on absolute stability of neural networks in two-neuron case. IEEE Trans. Circuits Syst. **I** (47), 609–611 (2000)
320. X.B. Liang, J. Wang, Absolute exponential stability of neural networks with a general class of activation functions. IEEE Trans. Circuits Syst. I **47**, 1258–1263 (2000)
321. J.C. Juang, Stability analysis of Hopfield-type neural network. IEEE Trans. Neural Netw. **10**, 1366–1374 (1999)
322. S. Arik, Z. Orman, Global stability analysis of Cohen-Grossberg neural networks with time varying delays. Phys. Lett. A **341**, 410–421 (2005)
323. C. Enache, Spatial decay bounds and continuous dependence on the data for a class of parabolic initial-boundary value problems. J. Math. Anal. Appl. **323**, 993–1000 (2006)
324. R. Haberman, *Applied Partial Differential Equations with Fourier Series and Boundary Value Problems*, 4th edn. (Addison Wesley, Reading, 2004)
325. K. Li, Q. Song, Exponential stability of impulsive Cohen-Grossberg neural networks with time-varying delays and reaction-diffusion terms. Neurocomputing **72**, 231–240 (2008)
326. P. Van Den Driessche, J.H. Wu, X.F. Zou, Stability role of inhibitory self-connections in a delayed neural network. Phys. D **150**, 84–90 (2001)
327. L. Wang, X.F. Zou, Harmless delays in Cohen-Grossberg neural networks. Phys. D **170**, 162–173 (2002)

Chapter 4
Delay-Partitioning-Method Based Stability Results for RNNs

Chapter 3 has presented many ways on how to use the time delay to establish the effective stability criteria for RNNs with delays. In this chapter, inspired by the discussion in Chap. 3, a new delay splitting method is proposed. By nonuniformly splitting the interval of time delay through involving many adjustable parameters, along with the construction of a new Lyapunov function, some delay-dependent stability results are established. By some comments and comparisons with the existing results, the effectiveness and novelty of the obtained criteria are verified. The contents in this chapter are from the results in [28].

4.1 Introduction

As a special class of nonlinear dynamical systems, recurrent neural networks (RNNs) have been paid much attention in the past decades due to their wide range of applications, such as signal processing [1, 2], combinatorial optimization [3–5], associative memories [6–8], pattern recognition [9, 10], etc. Since the RNNs are usually implemented by VLSI or digital circuit, etc., they are not much associated with training of ensembles. Meanwhile, time delay is also commonly encountered in the implementation of RNNs due to the finite switching speed of amplifier, and it is frequently a source of oscillation in neural networks. Therefore, the stability of RNNs with delay has become a topic of great theoretical and practical importance. Generally, when a neural network is applied to solve an optimization problem, it needs to have a unique and globally stable equilibrium point. Thus, it is of great interest to establish some sufficient conditions that ensure the global asymptotic stability of a unique equilibrium point of RNNs with delay [11–28].

So far, the stability criteria of RNNs with time delay are classified into two categories, i.e., delay-independent [11–19] and delay-dependent [20–23]. Since delay-dependent criteria tend to be less conservative, especially when the size of time delay is small, much attention has been paid to the delay-dependent category. For

Z. Wang et al., *Qualitative Analysis and Control of Complex Neural Networks with Delays*, Studies in Systems, Decision and Control 34,
DOI 10.1007/978-3-662-47484-6_4

the delay-dependent case considered in this current stability study, less conservative criteria means that the larger allowable upper bound of delay can be calculated for a certain system, or the larger allowable ranges of parameters in system are obtained for a fixed delay.

Nowadays, there are two kinds of useful methods dealing with problems associated with time delay: free-weighting matrix approach [29–31] and augmented Lyapunov functional method [32–35]. Reference [30] considers the term $-\int_{t-h}^{t-d(t)} \dot{x}^T(s)Z\dot{x}(s)$ ds in Lyapunov functional, which is usually neglected in previous literature, where $d(t)$ denotes a time-varying delay and h denotes the upper bound of $d(t)$, i.e., $d(t) \in [0,\ h]$. Reference [34] uses a new augmented *Lyapunov function*, which is similar to the functions applied to the descriptor system with delay in [36]. We can see that the free-weighting matrix approach is used as a main tool to make the criteria less conservative in literature, and only the lower and upper bound of delay function $d(t)$ are considered. Recently, a novel method is proposed for *Hopfield neural networks* with constant delay in [37], which is carried out by dividing the constant time delay interval $[0,\ h]$ into m subintervals with the same size. This method utilizes the information in the interval $[0,\ h]$ to achieve the aim of reducing the conservativeness of stability criterion.

It is worth noticing that the delay interval $[0, d(t)]$ is usually considered as a single interval in literature for RNNs with time-varying delay $d(t)$. Thus, how to utilize the information in $[0, d(t)]$ in order to further obtain less conservative stability for RNNs with time-varying delay $d(t)$ motivates our present study. Our proposed approach is to divide the delay interval $[0,\ d(t)]$ into smaller variable subintervals, and study the stability based on these subintervals. As a result, we will introduce the idea of dividing delay interval $[0,\ d(t)]$ from the viewpoint of mathematics.

The idea of dividing the delay interval $[0, d(t)]$ is as follows: We determine a point α (or multiple points α_i, $i = 1, 2, \ldots, K$), where $\alpha \in (0, d(t))$ (or $\alpha_i \in (0, d(t))$), i.e., $0 < \alpha < d(t)$ (or $0 < \alpha_i < d(t)$), such that interval $[0, d(t)]$ can be divided into two subintervals $[0, \alpha]$ and $[\alpha, d(t)]$ (or $K + 1$ subintervals $[0, \alpha_1]$, $[\alpha_1, \alpha_2], \ldots, [\alpha_K, d(t)]$).

Different from [37] which divides the delay interval into the fixed subintervals with the same size, this chapter provides a dynamic mode to divide delay interval since the point α or points α_i can be chosen arbitrarily in the interval $[0, d(t)]$. For the sake of convenience, in this chapter, we let $\alpha = \rho d(t)$ or $\alpha_i = \rho_i d(t)$, where $\rho \in (0, 1)$ or $\rho_i \in (0, 1)$, and $(\rho_1, \rho_2, \ldots, \rho_K)$ is a parameter sequence satisfying $0 < \rho_1 < \rho_2 < \cdots < \rho_K < 1$. Therefore, $\rho d(t)$ or $\rho_i d(t)$ are nominated as weighting-delays, parameters ρ or ρ_i are weighting-delay parameters, and $(\rho_1 d(t), \rho_2 d(t), \ldots, \rho_K d(t))$ is weighting-delay sequence. Meanwhile, $[0, \rho d(t)]$ and $[\rho d(t), d(t)]$ or $[0, \rho_1 d(t)]$, $[\rho_1 d(t), \rho_2 d(t)], \ldots, [\rho_K d(t), d(t)]$ are called two or $K + 1$ variable subintervals, respectively.

Therefore, this chapter proposes the new method referred to the *weighting-delay* to deal with the stability of RNNs with time-varying delay, so that the larger allowable upper bound can be obtained by introducing these variable subintervals. Different from previous studies, the weighting-delay method has the following features:

(1) The idea of dividing delay interval will be used in this method. Unlike the previous works [29–35] which treat the delay interval $[0,\ d(t)]$ as one single interval, $[0,\ d(t)]$ will be divided into several subintervals in this chapter. It means that much more information in the interval $[0,\ d(t)]$ can be utilized.

(2) Different from the fixed subintervals with the same size mode in [37], the *weighting-delay method* is the one with dynamic subintervals. That is to say, the delay interval is divided into variable subintervals in this method. It implies that the points $\rho_i d(t)$ are reconfigurable at delay interval $[0,\ d(t)]$. Compared to the fixed subintervals, it has inherent flexibility, and should be more suitable to deal with time-varying delay $d(t)$.

(3) Since the weighting-delays are introduced in the delay interval, it is clear that the stability results based on the weighting-delay method is related to the number of subintervals, and the size of the variable subintervals or the position of the variable points (the values of parameters ρ_i). When the positions of weighting-delays vary, the stability results of proposed criteria are also different. In order to obtain the optimal weighting-delay sequence, we proposed an implementation algorithm based on an optimization method.

Based on these features, some novel weighting-delay-based stability criteria for RNNs with time-varying delay are developed by linear matrix inequality (LMI) technique. Four illustrative examples show that our results are less conservative than those in previous literature.

4.2 Problem Formulation

Consider the following RNNs with time-varying delay $d(t)$:

$$\begin{aligned} \dot{z}(t) &= -Dz(t) + Af(z(t)) + Bf(z(t - d(t))) + U, \\ z(t) &= \phi(t), \quad \forall t \in [-h, 0], \end{aligned} \tag{4.1}$$

where $z(\cdot) = \left(z_1(\cdot), z_2(\cdot), \ldots, z_n(\cdot)\right)^T$ is the neuron state vector, $f(z(\cdot)) = \left(f_1(z_1(\cdot)), f_2(z_2(\cdot)), \ldots, f_n(z_n(\cdot))\right)^T$ denotes the neuron activation function, and $U = (U_1, U_2, \ldots, U_n)^T$ is a bias value vector. $D = \text{diag}(d_1, d_2, \ldots, d_n)$ is a diagonal matrix with $d_i > 0, i = 1, 2, \ldots, n$. A and B are the connection weight matrix and the delay connection weight matrix, respectively. The initial condition $\phi(t)$ is a continuous and differentiable vector-valued function, where $t \in [-h,\ 0]$. The time delay $d(t)$ is a differentiable function that satisfies

$$0 \leqslant d(t) \leqslant h, \tag{4.2}$$

and

$$0 \leqslant \dot{d}(t) \leqslant \mu, \tag{4.3}$$

where $h > 0$ and $\mu \geqslant 0$. Obviously, $d(t)$ is a time-varying continuous function with the upper bound h.

For systems (4.1), $f(z(t))$ and $f(z(t - d(t)))$ are two different signals, i.e., it is different for dynamical characteristics of RNNs only with $f(z(t))$ connection and with two connections. Similarly, RNNs with time-varying delay $d(t)$ or with constant delay have different dynamical characteristics, either. Compared with Hopfield neural networks only with the delayed connection and constant delay in [37], we studied the stability of a class of neural networks with two connections and time-varying delay $d(t)$. Since delay $d(t)$ is a time-varying function, it will be described using the parameters h and μ. Meanwhile, system (4.1) can be reduced to the neural networks in [37] if $A = 0$ and $\dot{d}(t) = 0$ (i.e., $d(t) = h$). Thus, it means that system (4.1) with two connections studied in this chapter is more general than that in [37].

In addition, it is assumed that each neuron *activation function* in system (4.1), $f_i(\cdot), i = 1,\ 2, \ldots, n$, satisfies the following condition:

$$0 \leqslant \underline{l}_i \leqslant \frac{f_i(u) - f_i(v)}{u - v} \leqslant \bar{l}_i \tag{4.4}$$

where $\forall u, v \in \mathbb{R}, u \neq v, \underline{l}_i$ are some nonnegative constants, and $\bar{l}_i$ are some positive constants, $i = 1, 2, \ldots, n$.

Assume the equilibrium point of system (4.1) be denoted by $z^* = (z_1^*, z_2^*, \ldots, z_n^*)^T$. Define $x_i(\cdot) = z_i(\cdot) - z_i^*$, system (4.1) can then be transformed into the following *error system*:

$$\begin{aligned} \dot{x}(t) &= -Dx(t) + Ag(x(t)) + Bg(x(t - d(t))), \\ x(t) &= \varphi(t), \forall t \in [-h, 0], \end{aligned} \tag{4.5}$$

where $x(\cdot) = \big(x_1(\cdot), x_2(\cdot), \ldots, x_n(\cdot)\big)^T$ is the state vector of the transformed system, the initial condition $\varphi(t) = \phi(t) - z^*$. $g(x(t)) = \big(g_1(x_1(t)), g_2(x_2(t)), \ldots, g_n(x_n(t))\big)^T$, and $g_i(x_i(t)) = f_i(x_i(t) + z_i^*) - f_i(z_i^*), i = 1,\ 2, \ldots, n$. Functions $g_i(\cdot), i = 1, 2, \ldots, n$, satisfy the following condition:

$$\begin{cases} 0 \leqslant \underline{l}_i \leqslant \frac{g_i(x_i)}{x_i} \leqslant \bar{l}_i, & \text{if } x_i \neq 0, \\ g_i(x_i) = 0, & \text{if } x_i = 0. \end{cases} \tag{4.6}$$

The following lemmas will be used to prove the results of this chapter.

Lemma 4.1 (Jensen's inequality) (see [38]) *For any constant matrix* $\Omega > 0$, *vector function* $\chi(t)$ *with appropriate dimensions, and function* $\sigma(t) \in \mathbb{R}$ *satisfying* $0 < \sigma(t) \leqslant \delta$ *we have*

$$
\begin{aligned}
&\left[\int_{t-\sigma(t)}^{t} \chi(s)\mathrm{d}s\right]^T \Omega \left[\int_{t-\sigma(t)}^{t} \chi(s)\mathrm{d}s\right] \\
&\leqslant \sigma(t) \int_{t-\sigma(t)}^{t} \chi^T(s)\Omega\chi(s)\mathrm{d}s \\
&\leqslant \delta \int_{t-\delta}^{t} \chi^T(s)\Omega\chi(s)\mathrm{d}s.
\end{aligned}
$$

Lemma 4.2 *The following inequalities*

$$
\begin{cases} \Delta + \alpha X_1 < 0, \\ \Delta + \alpha X_2 < 0, \end{cases} \tag{4.7}
$$

are equivalent to the following condition:

$$
\Delta + \tau X_1 + (\alpha - \tau) X_2 < 0, \tag{4.8}
$$

where X_1, X_2, Δ *are constant matrices with appropriate dimensions,* $\tau \in [0,\ \alpha]$, *and* $\alpha > 0$.

Proof The proof is on the basis of the idea of *convex combination* in [42].

(1) (4.7)⇒(4.8)

Let matrices Δ_1 and Δ_2 satisfy the following condition

$$
\begin{aligned}
\Delta_1 &= \Delta + \alpha X_1 < 0, \\
\Delta_2 &= \Delta + \alpha X_2 < 0.
\end{aligned}
$$

We can get

$$
\tau\Delta_1 + (\alpha - \tau)\Delta_2 < 0,
$$

i.e.,

$$
\alpha(\Delta + \tau X_1 + (\alpha - \tau) X_2) < 0.
$$

Since $\alpha > 0$, the following inequality holds

$$
\Delta + \tau X_1 + (\alpha - \tau) X_2 < 0.
$$

(2) (4.8)⇒(4.7)

Since variable τ satisfies the following condition at interval $[0,\ \alpha]$

$$
\Delta + \tau X_1 + (\alpha - \tau) X_2 < 0.
$$

Then, for variable $\tau = \alpha$ and $\tau = 0$, respectively, the following inequalities hold

$$\varDelta + \alpha X_1 < 0 \text{ and } \varDelta + \alpha X_2 < 0.$$

4.3 GAS Criteria with Single Weighting-Delay

In this section, we will establish weighting-delay-independent stability criterion and weighting-delay-dependent stability criterion, respectively.

4.3.1 Weighting-Delay-Independent Stability Criterion

In this subsection, we will consider the delay interval $[0, h]$, which is divided into two subintervals $[0, \rho h]$ and $[\rho h, h]$, where ρh is a *weighting-delay*, $\rho \in (0, 1)$ and $0 < \rho h < h$. For this case, a weighting-delay-independent stability criterion can be proposed.

Theorem 4.3 *The equilibrium point of system (4.5) with time-varying delay $d(t)$ satisfying (4.2) and (4.3) is unique and globally asymptotically stable, if there exist matrices:*

$$\begin{aligned}
P = P^T = \begin{bmatrix} P_{11} & P_{12} \\ P_{12}^T & P_{22} \end{bmatrix} > 0,\ R = R^T > 0,\\
Q = Q^T > 0,\ Y_1 = Y_1^T > 0,\ Y_2 = Y_2^T > 0,\\
Z = Z^T = \begin{bmatrix} Z_{11} & Z_{12} \\ Z_{12}^T & Z_{22} \end{bmatrix} > 0,\ G = G^T > 0,\\
\varLambda = \mathrm{diag}(\lambda_1,\ \lambda_2, \ldots, \lambda_n) > 0,\\
W_1 = \mathrm{diag}(w_{11},\ w_{12}, \ldots, w_{1n}) > 0,\\
W_2 = \mathrm{diag}(w_{21},\ w_{22}, \ldots, w_{2n}) > 0,
\end{aligned}$$

such that the following LMI holds:

$$\begin{bmatrix} \varTheta & \varGamma_1 & \varGamma_2 & h\bar{A}^T (G + Z_{22}) \\ * & -Z_{11} & 0 & 0 \\ * & * & -Z_{11} & 0 \\ * & * & * & -(G + Z_{22}) \end{bmatrix} < 0, \tag{4.9}$$

where

$$\Theta = \begin{bmatrix} \Theta_{11} & Z_{22} & G & -P_{12} & \Theta_{15} & \Theta_{16} \\ * & -Q-2Z_{22} & 0 & Z_{22} & 0 & 0 \\ * & * & \Theta_{33} & G & 0 & \Theta_{36} \\ * & * & * & \Theta_{44} & 0 & 0 \\ * & * & * & * & \Theta_{55} & \Lambda B \\ * & * & * & * & * & \Theta_{66} \end{bmatrix},$$

$$\begin{aligned}
\Theta_{11} &= -P_{11}D - DP_{11} + Q + R + Y_1 + P_{12} + P_{12}^T - G \\
&\quad + h^2 Z_{11} - h^2 Z_{12} D - h^2 D Z_{12}^T - Z_{22} - 2\underline{L} W_1 \overline{L}, \\
\Theta_{15} &= P_{11}A - D\Lambda + \left(\underline{L} + \overline{L}\right) W_1 + h^2 Z_{12} A, \\
\Theta_{16} &= P_{11}B + h^2 Z_{12} B, \\
\Theta_{33} &= -(1-\mu)Y_1 - 2G - 2\underline{L} W_2 \overline{L}, \\
\Theta_{36} &= \left(\underline{L} + \overline{L}\right) W_2, \\
\Theta_{44} &= -R - G - Z_{22}, \\
\Theta_{55} &= \Lambda A + A^T \Lambda + Y_2 - 2W_1, \\
\Theta_{66} &= -(1-\mu)Y_2 - 2W_2, \\
\Gamma_1 &= \begin{bmatrix} P_{22}^T - DP_{12} - Z_{12}^T \\ Z_{12}^T \\ 0 \\ -P_{22}^T \\ A^T P_{12} \\ B^T P_{12} \end{bmatrix}, \Gamma_2 = \begin{bmatrix} P_{22}^T - DP_{12} \\ -Z_{12}^T \\ 0 \\ Z_{12}^T - P_{22}^T \\ A^T P_{12} \\ B^T P_{12} \end{bmatrix},
\end{aligned}$$

$\bar{A} = [-D, 0,\ 0,\ 0,\ A,\ B]$, $\overline{L} = \text{diag}(\overline{l}_i)$, $\underline{L} = \text{diag}(\underline{l}_i)$, $i = 1, 2, \ldots, n$, *and* $*$ *denotes the symmetric terms in a symmetric matrix.*

Proof The proof will include two steps.

(1) *The uniqueness of the equilibrium point*

Assuming $x^\star$ is a nonzero equilibrium point, which satisfies the equilibrium equation,

$$\dot{x}^\star = -Dx^\star + Ag(x^\star) + Bg(x^\star) = 0. \tag{4.10}$$

Thus, we can get the following equations,

$$\begin{aligned}
&2(x^\star)^T [P_{11}\ \ P_{12}] \begin{bmatrix} -Dx^\star + Ag(x^\star) + Bg(x^\star) \\ x^\star - x^\star \end{bmatrix} = 0, \\
&2g^T(x^\star)\Lambda\left(-Dx^\star + Ag(x^\star) + Bg(x^\star)\right) = 0, \\
&h^2 \begin{bmatrix} x^\star \\ \dot{x}^\star \end{bmatrix}^T Z \begin{bmatrix} x^\star \\ \dot{x}^\star \end{bmatrix} \geqslant 0,
\end{aligned}$$

where P_{11}, P_{12}, Z, and Λ are defined at Theorem 4.3.

Then, according to $\mu \geqslant 0$ and condition (4.6), there exist the following equations,

$$\begin{aligned}(x^\star)^T Y_1 x^\star - (1-\mu)(x^\star)^T Y_1 x^\star \geqslant 0,\\ g^T(x^\star) Y_2 g(x^\star) - (1-\mu) g^T(x^\star) Y_2 g(x^\star) \geqslant 0,\\ -(W_1 + W_2)\left(g(x^\star) - \underline{L}x^\star\right)\left(g(x^\star) - \overline{L}x^\star\right) \geqslant 0,\end{aligned}$$

where Y_1, Y_2, W_1, and W_2 are defined at Theorem 4.3.

Let $\zeta^\star = \left[(x^\star)^T, (x^\star)^T, (x^\star)^T, (x^\star)^T, g^T(x^\star), g^T(x^\star)\right]^T$, applying the equations above and proper deviation, we can get

$$(\zeta^\star)^T \left(\Theta + h^2 \bar{A} T (G + Z_{22}) \bar{A}\right) \zeta^\star \geqslant 0, \text{ for } \zeta^\star \neq 0. \tag{4.11}$$

However, applying *Schur complements* for (4.9), we can get

$$(\zeta^\star)^T \left(\Theta + h^2 \bar{A} T (G + Z_{22}) \bar{A}\right) \zeta^\star < 0, \text{ for } \zeta^\star \neq 0. \tag{4.12}$$

Thus, there exists a contradiction between (4.11) and (4.12), which implies that the origin should be the unique equilibrium point under condition (4.9), i.e., $x^\star = 0$.

(2) *The stability of the equilibrium point*

Construct the following augmented *Lyapunov–Krasovskii functional*:

$$V(x(t)) = \sum_{i=1}^{6} V_i(x(t)), \tag{4.13}$$

where

$$\begin{aligned}
V_1(x(t)) &= \eta_1^T(t) P \eta_1(t),\\
V_2(x(t)) &= 2\sum_{i=1}^{n} \lambda_i \int_0^{x_i(t)} g_i(s)\mathrm{d}s,\\
V_3(x(t)) &= \int_{t-d(t)}^{t} \left(x^T(s) Y_1 x(s) + g^T(x(s)) Y_2 g(x(s))\right)\mathrm{d}s,\\
V_4(x(t)) &= \int_{t-\rho h}^{t} x^T(s) Q x(s)\mathrm{d}s + \int_{t-h}^{t} x^T(s) R x(s)\mathrm{d}s,\\
V_5(x(t)) &= h\int_{-h}^{0}\int_{t+\theta}^{t} \eta_2^T(s) Z \eta_2(s)\mathrm{d}s\mathrm{d}\theta,\\
V_6(x(t)) &= h\int_{-h}^{0}\int_{t+\theta}^{t} \dot{x}^T(s) G \dot{x}(s)\mathrm{d}s\mathrm{d}\theta,
\end{aligned}$$

where $\eta_1^T(t) = \left[x^T(t), \int_{t-h}^{t} x^T(s)\mathrm{d}s\right]$, $\eta_2^T(t) = \left[x^T(t), \dot{x}^T(t)\right]$, $P = P^T = \begin{bmatrix} P_{11} & P_{12} \\ P_{12}^T & P_{22} \end{bmatrix} > 0$, $\Lambda = \mathrm{diag}(\lambda_1, \lambda_2, \ldots, \lambda_n) > 0$, $Y_1 = Y_1^T > 0$, $Y_2 = Y_2^T > 0$, $R = R^T > 0$, $Q = Q^T > 0$, $Z = Z^T = \begin{bmatrix} Z_{11} & Z_{12} \\ Z_{12}^T & Z_{22} \end{bmatrix} > 0$, and $G = G^T > 0$.

For functional $V(x(t))$, we can verify that it satisfies the following condition

$$\beta_1 \|x(t)\|^2 \leqslant V(x(t)) \leqslant \beta_2 \|x(t)\|_c^2, \tag{4.14}$$

where $\|x(t)\|_c := \sup_{-h \leqslant \theta \leqslant 0} \|x(t+\theta)\|$, $\beta_1 = \lambda_m(P)$, and $\beta_2 = (1+h)\lambda_M(P) + 2\lambda_M(\Lambda\overline{L}) + h\lambda_M(Y_1) + h\lambda_M(Y_2)\lambda_M(\overline{L}^2) + \rho h \lambda_M(Q) + h\lambda_M(R) + 0.5h^3\lambda_M(Z)$ $[1 + 3\lambda_M(D^T D) + 3\lambda_M(A^T A)\lambda_M(\overline{L}^2) + 3\lambda_M(B^T B)\lambda_M(\overline{L}^2)] + 1.5h^3\lambda_M(G)$ $[\lambda_M(D^T D) + \lambda_M(A^T A)\lambda_M(\overline{L}^2) + \lambda_M(B^T B)\lambda_M(\overline{L}^2)]$.

Then, calculating the time derivatives of $V_1(x(t))$, $V_2(x(t))$, $V_3(x(t))$, and $V_4(x(t))$, respectively, along the trajectories of system (4.5), they yield

$$\begin{aligned} \dot{V}_1(x(t)) = {} & 2 \begin{bmatrix} x(t) \\ \int_{t-h}^{t} x(s)\mathrm{d}s \end{bmatrix}^T \begin{bmatrix} P_{11} & P_{12} \\ P_{12}^T & P_{22} \end{bmatrix} \\ & \times \begin{bmatrix} -Dx(t) + Ag(x(t)) + Bg(x(t-d(t))) \\ x(t) - x(t-h) \end{bmatrix}, \end{aligned} \tag{4.15}$$

$$\begin{aligned} \dot{V}_2(x(t)) = {} & -2g^T(x(t))\Lambda D x(t) + 2g^T(x(t))\Lambda A g(x(t)) \\ & + 2g^T(x(t))\Lambda B g(x(t-d(t))), \end{aligned} \tag{4.16}$$

$$\begin{aligned} \dot{V}_3(x(t)) \leqslant {} & x^T(t)Y_1 x(t) + g^T(x(t))Y_2 g(x(t)) \\ & - (1-\mu)x^T(t-d(t))Y_1 x(t-d(t)) \\ & - (1-\mu)g^T(x(t-d(t)))Y_2 g(x(t-d(t))), \end{aligned} \tag{4.17}$$

$$\begin{aligned} \dot{V}_4(x(t)) = {} & x^T(t)Qx(t) - x^T(t-\rho h)Qx(t-\rho h) \\ & + x^T(t)Rx(t) - x^T(t-h)Rx(t-h). \end{aligned} \tag{4.18}$$

Then, using Lemma 4.1 and the Leibniz–Newton formula, the time derivatives of $V_5(x(t))$ and $V_6(x(t))$ can be obtained as follows:

$$\begin{aligned} \dot{V}_5(x(t)) = {} & h^2 \begin{bmatrix} x(t) \\ \dot{x}(t) \end{bmatrix}^T \begin{bmatrix} Z_{11} & Z_{12} \\ Z_{12}^T & Z_{22} \end{bmatrix} \begin{bmatrix} x(t) \\ \dot{x}(t) \end{bmatrix} \\ & - h \int_{t-\rho h}^{t} \begin{bmatrix} x(s) \\ \dot{x}(s) \end{bmatrix}^T \begin{bmatrix} Z_{11} & Z_{12} \\ Z_{12}^T & Z_{22} \end{bmatrix} \begin{bmatrix} x(s) \\ \dot{x}(s) \end{bmatrix} \mathrm{d}s \\ & - h \int_{t-h}^{t-\rho h} \begin{bmatrix} x(s) \\ \dot{x}(s) \end{bmatrix}^T \begin{bmatrix} Z_{11} & Z_{12} \\ Z_{12}^T & Z_{22} \end{bmatrix} \begin{bmatrix} x(s) \\ \dot{x}(s) \end{bmatrix} \mathrm{d}s \end{aligned}$$

$$
\begin{aligned}
&\leqslant h^2 \begin{bmatrix} x(t) \\ \dot{x}(t) \end{bmatrix}^T \begin{bmatrix} Z_{11} & Z_{12} \\ Z_{12}^T & Z_{22} \end{bmatrix} \begin{bmatrix} x(t) \\ \dot{x}(t) \end{bmatrix} \\
&\quad - \int_{t-\rho h}^{t} \begin{bmatrix} x(s) \\ \dot{x}(s) \end{bmatrix}^T \mathrm{d}s \begin{bmatrix} Z_{11} & Z_{12} \\ Z_{12}^T & Z_{22} \end{bmatrix} \int_{t-\rho h}^{t} \begin{bmatrix} x(s) \\ \dot{x}(s) \end{bmatrix} \mathrm{d}s \\
&\quad - \int_{t-h}^{t-\rho h} \begin{bmatrix} x(s) \\ \dot{x}(s) \end{bmatrix}^T \mathrm{d}s \begin{bmatrix} Z_{11} & Z_{12} \\ Z_{12}^T & Z_{22} \end{bmatrix} \int_{t-h}^{t-\rho h} \begin{bmatrix} x(s) \\ \dot{x}(s) \end{bmatrix} \mathrm{d}s
\end{aligned}
$$

$$
\begin{aligned}
&= h^2 \begin{bmatrix} x(t) \\ \dot{x}(t) \end{bmatrix}^T \begin{bmatrix} Z_{11} & Z_{12} \\ Z_{12}^T & Z_{22} \end{bmatrix} \begin{bmatrix} x(t) \\ \dot{x}(t) \end{bmatrix} \\
&\quad - \begin{bmatrix} \int_{t-\rho h}^{t} x^T(s)\mathrm{d}s \\ \int_{t-\rho h}^{t} \dot{x}^T(s)\mathrm{d}s \end{bmatrix} \begin{bmatrix} Z_{11} & Z_{12} \\ Z_{12}^T & Z_{22} \end{bmatrix} \begin{bmatrix} \int_{t-\rho h}^{t} x(s)\mathrm{d}s \\ \int_{t-\rho h}^{t} \dot{x}(s)\mathrm{d}s \end{bmatrix} \\
&\quad - \begin{bmatrix} \int_{t-h}^{t-\rho h} x^T(s)\mathrm{d}s \\ \int_{t-h}^{t-\rho h} \dot{x}^T(s)\mathrm{d}s \end{bmatrix} \begin{bmatrix} Z_{11} & Z_{12} \\ Z_{12}^T & Z_{22} \end{bmatrix} \begin{bmatrix} \int_{t-h}^{t-\rho h} x(s)\mathrm{d}s \\ \int_{t-h}^{t-\rho h} \dot{x}(s)\mathrm{d}s \end{bmatrix} \\
&= h^2 \begin{bmatrix} x(t) \\ \dot{x}(t) \end{bmatrix}^T \begin{bmatrix} Z_{11} & Z_{12} \\ Z_{12}^T & Z_{22} \end{bmatrix} \begin{bmatrix} x(t) \\ \dot{x}(t) \end{bmatrix} \\
&\quad - \int_{t-\rho h}^{t} x^T(s)\mathrm{d}s Z_{11} \int_{t-\rho h}^{t} x(s)\mathrm{d}s \\
&\quad - \int_{t-h}^{t-\rho h} x^T(s)\mathrm{d}s Z_{11} \int_{t-h}^{t-\rho h} x(s)\mathrm{d}s \\
&\quad - 2\int_{t-\rho h}^{t} x^T(s)\mathrm{d}s Z_{12}(x(t) - x(t-\rho h)) \\
&\quad - 2\int_{t-h}^{t-\rho h} x^T(s)\mathrm{d}s Z_{12}\,(x(t-\rho h) \\
&\quad - x(t-h)) - \upsilon^T(t)\tilde{Z}\upsilon(t),
\end{aligned} \tag{4.19}
$$

$$
\begin{aligned}
\dot{V}_6(x(t)) &\leqslant h^2\dot{x}^T(t)G\dot{x}(t) - \int_{t-d(t)}^{t} \dot{x}^T(s)\mathrm{d}s G \int_{t-d(t)}^{t} \dot{x}(s)\mathrm{d}s \\
&\quad - \int_{t-h}^{t-d(t)} \dot{x}^T(s)\mathrm{d}s G \int_{t-h}^{t-d(t)} \dot{x}(s)\mathrm{d}s \\
&= h^2\dot{x}^T(t)G\dot{x}(t) - \upsilon^T(t)\tilde{G}\upsilon(t),
\end{aligned} \tag{4.20}
$$

where $\dot{V}_6(x(t))$ is obtained using the derivation similar to $\dot{V}_5(x(t))$, $\upsilon^T(t) = [x^T(t), x^T(t-\rho h), x^T(t-d(t)), x^T(t-h)]$, and

$$\tilde{Z} = \begin{bmatrix} 1 & -1 & 0 & 0 \\ -1 & 2 & 0 & -1 \\ 0 & 0 & 0 & 0 \\ 0 & -1 & 0 & 1 \end{bmatrix} Z_{22}, \quad \tilde{G} = \begin{bmatrix} 1 & 0 & -1 & 0 \\ 0 & 0 & 0 & 0 \\ -1 & 0 & 2 & -1 \\ 0 & 0 & -1 & 1 \end{bmatrix} G.$$

On the other hand, according to (4.6), we can obtain that

$$\left(g_i(x_i(t)) - \underline{l}_i x_i(t)\right) \left(g_i(x_i(t)) - \overline{l}_i x_i(t)\right) \leqslant 0, \tag{4.21}$$

and

$$\begin{aligned} &(g_i(x_i(t-d(t))) - \underline{l}_i x_i(t-d(t))) \\ &\qquad \times \left(g_i(x_i(t-d(t))) - \overline{l}_i x_i(t-d(t))\right) \leqslant 0, \end{aligned} \tag{4.22}$$

where $i = 1, 2, \ldots, n$.

For any $W_1 = \mathrm{diag}\,(w_{11}, w_{12}, \ldots, w_{1n}) > 0$, $W_2 = \mathrm{diag}\,(w_{21}, w_{22}, \ldots, w_{2n}) > 0$, by (4.21) and (4.22), we have

$$\begin{aligned} 0 \leqslant & -2g^T(x(t))W_1 g(x(t)) + 2g^T(x(t))W_1 \overline{L}x(t) \\ & +2x^T(t)\underline{L}W_1 g(x(t)) - 2x^T(t)\underline{L}W_1\overline{L}x(t) \\ & -2g^T(x(t-d(t)))W_2 g(x(t-d(t))) \\ & +2g^T(x(t-d(t)))W_2\overline{L}x(t-d(t)) \\ & +2x^T(t-d(t))\underline{L}W_2 g(x(t-d(t))) \\ & -2x^T(t-d(t))\underline{L}W_2\overline{L}x(t-d(t)), \end{aligned} \tag{4.23}$$

where $\overline{L} = \mathrm{diag}(\overline{l}_i)$ and $\underline{L} = \mathrm{diag}(\underline{l}_i)$, $i = 1, 2, \ldots, n$.

Thus, combining (4.15)–(4.20) and adding (4.23), the derivative of $V(x(t))$ is obtained as follows:

$$\begin{aligned} \dot{V}(x(t)) \leqslant & \ \zeta^T(t)\Big(\Theta + \Gamma_1 Z_{11}^{-1}\Gamma_1^T + \Gamma_2 Z_{11}^{-1}\Gamma_2^T \\ & + h^2\bar{A}^T(Z_{22} + G)\bar{A}\Big)\zeta(t), \end{aligned} \tag{4.24}$$

where $\zeta^T(t) = \left(\upsilon^T(t), g^T(x(t)), g^T(x(t-d(t)))\right)$.

Obviously, if $\Theta + \Gamma_1 Z_{11}^{-1}\Gamma_1^T + \Gamma_2 Z_{11}^{-1}\Gamma_2^T + h^2\bar{A}^T(Z_{22} + G)\bar{A} < 0$, then $\dot{V}(x(t)) < -\varepsilon\|x(t)\|^2$ for a small $\varepsilon > 0$. Using Schur complements, we can know that the above inequality is equivalent to (4.9). As a result, according to Lyapunov stability theory and inequality (4.14), the equilibrium point of the system (4.5) is globally asymptotically stable.

According to (1) and (2), if (4.9) is satisfied, equilibrium point of the system (4.5) is unique and globally asymptotically stable.

When $\mu \geqslant 1$ or μ is unknown, by letting $Y_1 = 0$ and $Y_2 = 0$ in Theorem 4.3, the stability result independent of $\dot{\tau}(t)$ can be obtained. Meanwhile, as a universal tool, LMI has been widely applied in many fields, such as dynamical systems theory, control systems, and neural networks and its applications, etc. In this chapter, we have obtained the stability criteria for RNNs with time-varying delay using Lyapunov stability theory. Here, the LMI can be just used to express and solve the proposed stability criteria. On the other hand, there are three points to show the advantages of LMI in stability analysis of RNNs.

(1) Since the LMI-based stability results consider the sign difference of the elements in connection matrices, neuron's excitatory and inhibitory effects on the neural network have been considered, which overcome the shortcomings of the results based on other methods (such as M-matrix and algebraic inequality, etc.).

(2) Currently, the stability problem based on *Lyapunov stability theory* can be considered as an optimization problem with multiple matrix variables and many constraints. Thus, LMI is the best and only tool to solve this kind of stability problem till now.

(3) Along with the development of computer and the related software, LMI-based stability criteria can be solved easily using the interior point algorithm.

Remark 4.4 In Theorem 4.3, the delay interval $[0,\ h]$ is divided into two subintervals $[0,\ \rho h]$ and $[\rho h,\ h]$. For two variable subintervals, the weighting-delay-independent stability criterion (i.e., Theorem 4.3) is developed, i.e., stability results will not be affected by the value of parameter ρ. Since the terms with weighting-delay ρh do not appear in LMI (4.9), the computation burden of Theorem 4.3 is not increased while less conservative stability results are obtained.

4.3.2 Weighting-Delay-Dependent Stability Criterion

From Sect. 4.3.1, we can know that Theorem 4.3 is a criterion independent of weighting-delay, i.e., the effect of weighting-delay parameter ρ is neglected. Then, when the effect of ρ is considered, what about the stability results? According to this idea, the weighting-delay-dependent stability criterion can be proposed. In this subsection, the delay interval $[0,\ d(t)]$ will be divided into two subintervals $[0, \rho d(t)]$ and $[\rho d(t), d(t)]$, i.e., the corresponding weighing-delay term $\rho d(t)$ is introduced. It is clear that $\rho d(t)$ satisfies the following conditions:

$$0 < \rho d(t) < d(t) \leqslant h,\ \ 0 \leqslant \rho \dot{d}(t) \leqslant \rho\mu < \mu,$$

where $\rho \in (0, 1)$.

Theorem 4.5 *The equilibrium point of system (4.5) with time-varying delay $d(t)$ satisfying (4.2) and (4.3) is unique and globally asymptotically stable, for a given parameter ρ satisfying $0 < \rho < 1$, if there exist matrices*

$$
\begin{aligned}
&P = P^T = \begin{bmatrix} P_{11} & P_{12} \\ P_{12}^T & P_{22} \end{bmatrix} > 0,\ R = R^T > 0,\\
&Z = Z^T = \begin{bmatrix} Z_{11} & Z_{12} \\ Z_{12}^T & Z_{22} \end{bmatrix} > 0,\ Q = Q^T > 0,\\
&R = R^T > 0,\ Y_1 = Y_1^T > 0,\ Y_2 = Y_2^T > 0,\\
&\Lambda = \mathrm{diag}(\lambda_1,\ \lambda_2, \ldots, \lambda_n) > 0,\\
&W_1 = \mathrm{diag}(w_{11},\ w_{12}, \ldots, w_{1n}) > 0,\\
&W_2 = \mathrm{diag}(w_{21},\ w_{22}, \ldots, w_{2n}) > 0,\\
&M^T = \begin{bmatrix} M_1^T & M_2^T & M_3^T & 0 & 0 & 0 \end{bmatrix},\\
&N^T = \begin{bmatrix} N_1^T & N_2^T & N_3^T & 0 & 0 & 0 \end{bmatrix},\\
&S^T = \begin{bmatrix} S_1^T & S_2^T & S_3^T & 0 & 0 & 0 \end{bmatrix},
\end{aligned}
$$

such that the following LMIs hold:

$$
\Phi + h\bar{A}^T Z_{22}\bar{A} + \rho h\bar{M}Z^{-1}\bar{M}^T + (1-\rho)h\bar{N}Z^{-1}\bar{N}^T < 0, \tag{4.25}
$$

and

$$
\Phi + h\bar{A}^T Z_{22}\bar{A} + h\bar{S}Z^{-1}\bar{S}^T < 0, \tag{4.26}
$$

where

$$
\Phi = \Omega + \bar{\Omega} + \bar{\Omega}^T,
$$

$$
\Omega = \begin{bmatrix}
\Omega_1 & 0 & 0 & -P_{12} & \Theta_{15} & \Theta_{16} \\
* & -(1-\rho\mu)Q & 0 & 0 & 0 & 0 \\
* & * & \Omega_3 & 0 & 0 & \Theta_{36} \\
* & * & * & -R & 0 & 0 \\
* & * & * & * & \Theta_{55} & \Lambda B \\
* & * & * & * & * & \Theta_{66}
\end{bmatrix},
$$

$$
\begin{aligned}
\Omega_1 =& -P_{11}D - DP_{11} + Q + R + Y_1 + hZ_{11} + P_{12} + P_{12}^T\\
&- hZ_{12}D - hDZ_{12}^T - 2\underline{L}W_1\overline{L},\\
\Omega_3 =& -(1-\mu)Y_1 - 2\underline{L}W_2\overline{L},\\
\bar{\Omega} =& [M \quad -M+N \quad -N+S \quad -S \quad 0 \quad 0],\\
\bar{M} =& [\Gamma \quad -M],\ \bar{N} = [\Gamma \quad -N],\ \bar{S} = [\Gamma \quad -S],\\
\Gamma^T =& \begin{bmatrix} -P_{12}^T D + P_{22} & 0 & 0 & -P_{22} & P_{12}^T A & P_{12}^T B \end{bmatrix},
\end{aligned}
$$

and the other parameters are the same as those defined in Theorem 4.3.

Proof Similar with Theorem 4.3, there are two steps in the Proof of Theorem 4.5:

(1) *The uniqueness of the equilibrium point*

The proof is similar to that of Theorem 4.3, and thus it is omitted here.

(2) *The stability of the equilibrium point*

Construct the following augmented *Lyapunov–Krasovskii functional*:

$$\bar{V}(x(t)) = V_1(x(t)) + V_2(x(t)) + V_3(x(t)) + \bar{V}_4(x(t)) + \bar{V}_5(x(t)), \tag{4.27}$$

where $V_1(x(t))$, $V_2(x(t))$, and $V_3(x(t))$ are the same as those in Theorem 4.3, and

$$\bar{V}_4(x(t)) = \int_{t-\rho d(t)}^{t} x^T(s)Qx(s)\mathrm{d}s + \int_{t-h}^{t} x^T(s)Rx(s)\mathrm{d}s,$$
$$\bar{V}_5(x(t)) = \int_{-h}^{0}\int_{t+\theta}^{t} \eta_2^T(s)Z\eta_2(s)\mathrm{d}s\mathrm{d}\theta,$$

where the parameters are defined in Theorem 4.5.

Similar to Theorem 4.3, we can verify that $\bar{V}(x(t))$ satisfies the following condition

$$\beta_1\|x(t)\|^2 \leqslant \bar{V}(x(t)) \leqslant \bar{\beta}_2\|x(t)\|_c^2, \tag{4.28}$$

where $\beta_1 = \lambda_m(P)$ and $\bar{\beta}_2 = (1+h)\lambda_M(P) + 2\lambda_M(\Lambda\overline{L}) + h\lambda_M(Y_1) + h\lambda_M(Y_2)\lambda_M(\overline{L}^2) + \rho h\lambda_M(Q) + h\lambda_M(R) + 0.5h^3\lambda_M(Z)[1 + 3\lambda_M(D^TD) + 3\lambda_M(A^TA)\lambda_M(\overline{L}^2) + 3\lambda_M(B^TB)\lambda_M(\overline{L}^2)]$.

Calculating the time derivatives of $\bar{V}_4(x(t))$ and $\bar{V}_5(x(t))$ along the trajectories of system (4.5), respectively, they yield

$$\begin{aligned}\dot{\bar{V}}_4(x(t)) \leqslant{}& x^T(t)Qx(t) + x^T(t)Rx(t)\\ &-(1-\rho\mu)x^T(t-\rho d(t))Qx(t-\rho d(t))\\ &-x^T(t-h)Rx(t-h),\end{aligned} \tag{4.29}$$

$$\begin{aligned}\dot{\bar{V}}_5(x(t)) ={}& h\begin{bmatrix} x(t)\\ \dot{x}(t)\end{bmatrix}^T \begin{bmatrix} Z_{11} & Z_{12}\\ Z_{12}^T & Z_{22}\end{bmatrix}\begin{bmatrix} x(t)\\ \dot{x}(t)\end{bmatrix}\\ &-\int_{t-h}^{t}\eta_2^T(s)Z\eta_2(s)\mathrm{d}s.\end{aligned} \tag{4.30}$$

where $\dot{V}_1(x(t))$, $\dot{V}_2(x(t))$, and $\dot{V}_3(x(t))$ are given in (4.15)–(4.17).

According to the *Leibniz–Newton formula*, the following equations hold for any matrices M, N, and S with appropriate dimensions,

$$2\zeta^T(t)M\left[x(t) - x(t-\rho d(t)) - \int_{t-\rho d(t)}^{t}\dot{x}(s)\mathrm{d}s\right] = 0, \tag{4.31}$$

$$2\zeta^T(t)N\left[x(t-\rho d(t))-x(t-d(t))-\int_{t-d(t)}^{t-\rho d(t)}\dot{x}(s)\mathrm{d}s\right]=0, \tag{4.32}$$

$$2\zeta^T(t)S\left[x(t-d(t))-x(t-h)-\int_{t-h}^{t-d(t)}\dot{x}(s)\mathrm{d}s\right]=0, \tag{4.33}$$

where $\zeta^T(t)=[x^T(t), x^T(t-\rho d(t)), x^T(t-d(t)), x^T(t-h), g^T(x(t)), g^T(x(t-d(t)))]$.

Thus, combining (4.15)–(4.17), (4.29), and (4.30), adding (4.23) and (4.31)–(4.33), the derivative of $\bar{V}(x(t))$ is obtained as follows:

$$\begin{aligned}\dot{\bar{V}}(x(t)) \leqslant\ & \zeta^T(t)\left[\Phi+h\bar{A}^T Z_{22}\bar{A}+\bar{\Phi}\right]\zeta(t)\\ &-\int_{t-\rho d(t)}^{t}\left[Z\eta_2(s)-\bar{M}^T\zeta(t)\right]^T Z^{-1}\left[Z\eta_2(s)-\bar{M}^T\zeta(t)\right]\mathrm{d}s\\ &-\int_{t-d(t)}^{t-\rho d(t)}\left[Z\eta_2(s)-\bar{N}^T\zeta(t)\right]^T Z^{-1}\left[Z\eta_2(s)-\bar{N}^T\zeta(t)\right]\mathrm{d}s\\ &-\int_{t-h}^{t-d(t)}\left[Z\eta_2(s)-\bar{S}^T\zeta(t)\right]^T Z^{-1}\left[Z\eta_2(s)-\bar{S}^T\zeta(t)\right]\mathrm{d}s,\end{aligned}$$

where $\bar{\Phi}=\rho d(t)\bar{M}Z^{-1}\bar{M}^T+(1-\rho)d(t)\bar{N}Z^{-1}\bar{N}^T+(h-d(t))\bar{S}Z^{-1}\bar{S}^T$.

Since $Z>0$, then the last three terms in (4.34) are all less than zero. Thus, if $\Phi+h\bar{A}^T Z_{22}\bar{A}+\bar{\Phi}<0$, we have $\dot{\bar{V}}(x(t))<0$.

Using Lemma 4.2, let $\alpha=h, \tau=d(t), \Delta=\Phi+h\bar{A}^T Z_{22}\bar{A}, X_1=\rho\bar{M}Z^{-1}\bar{M}^T+(1-\rho)\bar{N}Z^{-1}\bar{N}^T$ and $X_2=\bar{S}Z^{-1}\bar{S}^T$, then $\Phi+h\bar{A}^T Z_{22}\bar{A}+\Xi<0$ is equivalent to the inequalities (4.25) and (4.26) at $d(t)=h$ and $d(t)=0$. Thus, if (4.25) and (4.26) are satisfied, then $\dot{\bar{V}}(x(t))<-\varepsilon\|x(t)\|^2$ for some $\varepsilon>0$, i.e., the equilibrium point of the system (4.5) is globally asymptotically stable.

Remark 4.6 Obviously, when weighting-delay parameter ρ is set with different values, the stability results in Theorem 4.5 are also changed. Compared with previous results, the negative definite terms with parameters $(1-\rho)$, $\frac{1-\rho}{\rho}$, and $\frac{\rho}{1-\rho}$ are added to some main diagonal elements of LMIs (4.25) and (4.26). Thus, by properly choosing the value of parameter ρ, it can lead to less conservative stability results. Meanwhile, Theorem 4.5 is studied based on two subintervals $[0, \rho d(t)]$ and $[\rho d(t), d(t)]$. Thus, the criteria based on such subintervals with time-varying delay $d(t)$ should be more suitable to deal with; some examples are provided in Sect. 4.6 to prove this result.

Remark 4.7 For Theorem 4.5, when $\mu\geqslant 1$, there are two cases: (1) When $1\leqslant\mu<\frac{1}{\rho}$, stability result can be obtained by setting matrices $Y_1=0$ and $Y_2=0$; (2) when $\mu\geqslant\frac{1}{\rho}$, let $Q=0$, $Y_1=0$ and $Y_2=0$ in Theorem 4.5, the criterion independent of $\dot{\tau}(t)$ can be derived. Namely, for the case 2), we can obtain the criterion with unknown μ.

In Theorem 4.5, the weighting-delay parameter ρ belongs to $(0,\ 1)$. Then, when $\rho \in [0,\ 1]$, the following corollary independent of weighting-delay can be developed.

Corollary 4.8 *The equilibrium point of system (4.5) with time-varying delay $d(t)$ satisfying (4.2) and (4.3) is unique and globally asymptotically stable if there exist matrices*

$$
\begin{aligned}
&P = P^T = \begin{bmatrix} P_{11} & P_{12} \\ P_{12}^T & P_{22} \end{bmatrix} > 0,\ R = R^T > 0,\\
&Z = Z^T = \begin{bmatrix} Z_{11} & Z_{12} \\ Z_{12}^T & Z_{22} \end{bmatrix} > 0,\ Q = Q^T > 0,\\
&Y_1 = Y_1^T > 0,\ Y_2 = Y_2^T > 0,\\
&\Lambda = \mathrm{diag}(\lambda_1,\ \lambda_2, \ldots, \lambda_n) > 0,\\
&W_1 = \mathrm{diag}(w_{11},\ w_{12}, \ldots, w_{1n}) > 0,\\
&W_2 = \mathrm{diag}(w_{21},\ w_{22}, \ldots, w_{2n}) > 0,\\
&M^T = \begin{bmatrix} M_1^T & M_2^T & M_3^T & 0 & 0 & 0 \end{bmatrix},\\
&N^T = \begin{bmatrix} N_1^T & N_2^T & N_3^T & 0 & 0 & 0 \end{bmatrix},\\
&S^T = \begin{bmatrix} S_1^T & S_2^T & S_3^T & 0 & 0 & 0 \end{bmatrix},
\end{aligned}
$$

such that the following LMIs hold:

$$\Phi_1 + h\bar{A}^T Z_{22}\bar{A} + h\bar{M}Z^{-1}\bar{M}^T < 0, \tag{4.34}$$

$$\Phi_0 + h\bar{A}^T Z_{22}\bar{A} + h\bar{N}Z^{-1}\bar{N}^T < 0, \tag{4.35}$$

$$\Phi_1 + h\bar{A}^T Z_{22}\bar{A} + h\bar{S}Z^{-1}\bar{S}^T < 0, \tag{4.36}$$

$$\Phi_0 + h\bar{A}^T Z_{22}\bar{A} + h\bar{S}Z^{-1}\bar{S}^T < 0, \tag{4.37}$$

where

$$\Phi_j = \Omega_j + \bar{\Omega} + \bar{\Omega}^T, \quad j = 0, 1,$$

$$\Omega_0 = \begin{bmatrix} \Omega_{11} & 0 & 0 & -P_{12} & \Theta_{15} & \Theta_{16} \\ * & -Q & 0 & 0 & 0 & 0 \\ * & * & \Omega_{33} & 0 & 0 & \Theta_{36} \\ * & * & * & -R & 0 & 0 \\ * & * & * & * & \Omega_{55} & \Lambda B \\ * & * & * & * & * & \Omega_{66} \end{bmatrix},$$

$$\Omega_1 = \begin{bmatrix} \Omega_{11} & 0 & 0 & -P_{12} & \Theta_{15} & \Theta_{16} \\ * & -(1-\mu)Q & 0 & 0 & 0 & 0 \\ * & * & \Omega_{33} & 0 & 0 & \Theta_{36} \\ * & * & * & -R & 0 & 0 \\ * & * & * & * & \Omega_{55} & \Lambda B \\ * & * & * & * & * & \Omega_{66} \end{bmatrix},$$

and the other parameters are the same as those defined in Theorems 4.3 and 4.5.

Proof Similar to the proof of Theorem 4.5, constructing the same augmented Lyapunov–Krasovskii functional as that in Theorem 4.5, we can get $\dot{V}(x(t))$ as same as (4.34). By omitting the integral terms which are less than zero, $\dot{V}(x(t))$ can be rewritten as:

$$\dot{V}(x(t)) \leqslant \zeta^T(t)\big[\Phi + h\bar{A}^T Z_{22}\bar{A} + (1-\rho)d(t)\bar{N}Z^{-1}\bar{N}^T + \rho d(t)\bar{M}Z^{-1}\bar{M}^T + (h-d(t))\bar{S}Z^{-1}\bar{S}^T\big]\zeta(t).$$

If the following inequality holds

$$\Phi + h\bar{A}^T Z_{22}\bar{A} + \rho d(t)\bar{M}Z^{-1}\bar{M}^T + (1-\rho)d(t)\bar{N}Z^{-1}\bar{N}^T + (h-d(t))\bar{S}Z^{-1}\bar{S}^T < 0, \tag{4.38}$$

then $\dot{V}(x(t)) < -\varepsilon\|x(t)\|^2$ for some $\varepsilon > 0$. In addition, because of $0 \leqslant \rho \leqslant 1$, let $\alpha = 1$, $\tau = \rho$, $\Delta = \Phi + h\bar{A}^T Z_{22}\bar{A} + (h-d(t))\bar{S}Z^{-1}\bar{S}^T$, $X_1 = d(t)\bar{M}Z^{-1}\bar{M}^T$ and $X_2 = d(t)\bar{N}Z^{-1}\bar{N}^T$, using Lemma 4.2, inequality (4.38) is equivalent to the following two inequalities at $\rho = 1$ and $\rho = 0$.

$$\Phi_1 + h\bar{A}^T Z_{22}\bar{A} + d(t)\bar{M}Z^{-1}\bar{M}^T + (h-d(t))\bar{S}Z^{-1}\bar{S}^T < 0, \tag{4.39}$$

$$\Phi_0 + h\bar{A}^T Z_{22}\bar{A} + d(t)\bar{N}Z^{-1}\bar{N}^T + (h-d(t))\bar{S}Z^{-1}\bar{S}^T < 0. \tag{4.40}$$

Similar to Theorem 4.5, inequalities (4.39) and (4.40) are equivalent to the inequalities (4.34)–(4.37), which are derived using Lemma 4.2 again. Then, the criterion will be derived.

4.4 GAS Criteria with Multiple Weighting-Delays

In previous section, the stability of RNNs with time-varying delay was studied based on single weighting-delay. Naturally, the idea of introducing multiple weighting-delays into the delay interval will be considered. In this section, the delay interval $[0, d(t)]$ is divided into $K+1$ dynamical subintervals, i.e., $[0, \rho_1 d(t)]$, $[\rho_1 d(t), \rho_2 d(t)], \ldots, [\rho_K d(t), h]$, where $\rho_1 < \rho_2 < \cdots < \rho_K$. That is to say, there is a parameter sequence $(\rho_1, \rho_2, \ldots, \rho_K)$ satisfying the following conditions,

$$0 < \rho_1 d(t) < \rho_2 d(t) < \cdots < \rho_K d(t) < d(t),$$
$$0 \leqslant \rho_i \dot{d}(t) \leqslant \rho_i \mu,$$

where $\rho_i \in (0,\ 1)$, $i = 1, 2, \ldots, K$, K is a positive integer and $K \geqslant 2$. Then, a stability criterion with multiple weighting-delays can be proposed.

Theorem 4.9 *The equilibrium point of system (4.5) with time-varying delay $d(t)$ satisfying (4.2) and (4.3) is unique and globally asymptotically stable, if there exist parameters ρ_i satisfying $0 < \rho_1 < \rho_2 < \cdots < \rho_K < 1$, symmetric matrices $P = P^T > 0$, $R = R^T > 0$, $Y = Y^T > 0$, $Z = Z^T > 0$, $Q_j = Q_j^T > 0$, and diagonal matrices $\Lambda = \mathrm{diag}(\lambda_1,\ \lambda_2, \ldots, \lambda_n) > 0$, $W_1 = \mathrm{diag}(w_{11},\ w_{12}, \ldots, w_{1n}) > 0$, $W_2 = \mathrm{diag}(w_{21},\ w_{22}, \ldots, w_{2n}) > 0$, where $i = 1, 2, \ldots, K$, $j = 1, 2, \ldots, K+1$, and K is a positive integer, such that the following matrix inequality holds:*

$$\Upsilon = \begin{bmatrix} \Upsilon_1 & \frac{1}{\rho_1}Z & 0 & \cdots & 0 & 0 & 0 & \Upsilon_a & PB \\ * & \Upsilon_2 & \frac{1}{\rho_2-\rho_1}Z & \cdots & 0 & 0 & 0 & 0 & 0 \\ * & * & \Upsilon_3 & \cdots & 0 & 0 & 0 & 0 & 0 \\ * & * & * & \ddots & \vdots & \vdots & \vdots & \vdots & \vdots \\ * & * & * & * & \Upsilon_{K+1} & \frac{1}{1-\rho_K}Z & 0 & 0 & 0 \\ * & * & * & * & * & \Upsilon_{K+2} & Z & 0 & \left(\underline{L}+\overline{L}\right)W_2 \\ * & * & * & * & * & * & -R-Z & 0 & 0 \\ * & * & * & * & * & * & * & \Upsilon_{K+4} & \Lambda B \\ * & * & * & * & * & * & * & * & \Upsilon_{K+5} \end{bmatrix} \tag{4.41}$$

$$\Upsilon + h^2 \tilde{A}^T Z \tilde{A} < 0 \tag{4.42}$$

where Υ is shown in (4.41), $\Upsilon_a = PA - D\Lambda + \left(\underline{L}+\overline{L}\right)W_1$,

$$\begin{aligned}
\Upsilon_1 &= -PD - DP + Q_1 + R - \frac{1}{\rho_1}Z - 2\underline{L}W_1\overline{L},\\
\Upsilon_2 &= -(1-\rho_1\mu)Q_1 + Q_2 - \frac{1}{\rho_1}Z - \frac{1}{\rho_2-\rho_1}Z,\\
\Upsilon_3 &= -(1-\rho_2\mu)Q_2 + Q_3 - \frac{1}{\rho_2-\rho_1}Z - \frac{1}{\rho_3-\rho_2}Z,\\
&\ \ \vdots\\
\Upsilon_{K+1} &= -(1-\rho_K\mu)Q_K + Q_{K+1} - \frac{1}{\rho_K-\rho_{K-1}}Z\\
&\quad - \frac{1}{1-\rho_K}Z,\\
\Upsilon_{K+2} &= -(1-\mu)Q_{K+1} - \frac{1}{1-\rho_K}Z - Z - 2\underline{L}W_2\overline{L},\\
\Upsilon_{K+4} &= \Lambda A + A^T\Lambda + Y - 2W_1,\\
\Upsilon_{K+5} &= -(1-\mu)Y - 2W_2,\\
\tilde{A} &= \begin{bmatrix} -D\ 0 \cdots 0\ A\ B \end{bmatrix}_{(K+5)n\times n},
\end{aligned}$$

and the other parameters are the same as those defined in Theorem 4.3.

Proof Similar with Theorems 4.3 and 4.5, the proof still includes two steps.

(1) *The uniqueness of the equilibrium point*

Assume there is a nonzero equilibrium point $x^\star$. Similar to Theorem 4.3, according to $\mu \geqslant 0$, $0 < \rho_i < 1$ $(i = 1, 2, \ldots, n)$, (4.6), and the equilibrium equation (4.10), we can obtain the following equations

$$\begin{aligned}
&2(x^\star)^T P\left[-Dx^\star + Ag(x^\star) + Bg(x^\star)\right] = 0,\\
&2g^T(x^\star)\Lambda\left[-Dx^\star + Ag(x^\star) + Bg(x^\star)\right] = 0,\\
&\sum_{i=0}^{K-1}\left[(x^\star)^T Q_{i+1}x^\star - (1-\rho_{i+1}\mu)(x^\star)^T Q_{i+1}x^\star\right] \geqslant 0,\\
&(x^\star)^T Q_{K+1}x^\star - (1-\mu)(x^\star)^T Q_{K+1}x^\star \geqslant 0,\\
&g^T(x^\star)Yg(x^\star) - (1-\mu)g^T(x^\star)Yg(x^\star) \geqslant 0,\\
&-(W_1 + W_2)\left(g(x^\star) - \underline{L}x^\star\right)\left(g(x^\star) - \overline{L}x^\star\right) \geqslant 0,
\end{aligned}$$

where the parameters are defined in Theorem 4.9.

Let $\varsigma^\star = \left((x^\star)^T, \ldots, (x^\star)^T, g^T(x^\star), g^T(x^\star)\right)^T$, applying the equations above and proper deviation, we can get

$$(\varsigma^\star)^T \Upsilon \varsigma^\star \geqslant 0, \tag{4.43}$$

for $\varsigma^\star \neq 0$.

However, applying *Schur complements* for (4.42), we can get

$$(\varsigma^\star)^T \Upsilon \varsigma^\star < 0, \tag{4.44}$$

for $\varsigma^\star \neq 0$.

Thus, there is a contradiction between (4.43) and (4.44), which implies that the origin is the unique equilibrium point of system (4.5) under the condition (4.42), i.e., $x^\star = 0$.

(2) *The stability of the equilibrium point*

Construct the following Lyapunov–Krosovskii functional:

$$\begin{aligned}
\tilde{V}(x(t)) = {} & x^T(t)Px(t) + 2\sum_{i=1}^{n}\lambda_i\int_0^{x_i(t)} g_i(s)\mathrm{d}s\\
& + \sum_{i=0}^{K}\int_{t-\rho_{i+1}d(t)}^{t-\rho_i d(t)} x^T(s)Q_{i+1}x(s)\mathrm{d}s\\
& + \int_{t-h}^{t} x^T(s)Rx(s)\mathrm{d}s
\end{aligned}$$

$$
\begin{aligned}
&+\int_{t-d(t)}^{t} g^T(x(s))Yg(x(s))\mathrm{d}s \\
&+h\int_{-h}^{0}\int_{t+\theta}^{t} \dot{x}^T(s)Z\dot{x}(s)\mathrm{d}s\mathrm{d}\theta,
\end{aligned} \tag{4.45}
$$

where $P = P^T > 0$, $\Lambda = \mathrm{diag}(\lambda_1, \lambda_2, \ldots, \lambda_n) > 0$, $Y = Y^T > 0$, $R = R^T > 0$, $Z = Z^T > 0$, $Q_i = Q_i^T > 0$, $i = 1, 2, 3, \ldots, K + 1$ and parameters $\rho_0 = 0$, $\rho_{K+1} = 1$.

According to $\dot{d}(t) \geqslant 0$ from (4.3), Jensen's inequality and Leibniz–Newton formula, calculating the time derivatives of $\tilde{V}(x(t))$ along the trajectories of system (4.5) yields

$$
\begin{aligned}
\dot{\tilde{V}}(x(t)) \leqslant &-2x^T(t)PDx(t) + 2x^T(t)PAg(x(t)) \\
&+2x^T(t)PBg(x(t-d(t)))) \\
&-2g^T(x(t))\Lambda Dx(t) + 2g^T(x(t))\Lambda Ag(x(t)) \\
&+2g^T(x(t))\Lambda Bg(x(t-d(t)))) \\
&+\sum_{i=0}^{K} x^T(t-\rho_i d(t))Q_{i+1}x(t-\rho_i d(t)) \\
&-\sum_{i=0}^{K}(1-\rho_{i+1}\dot{d}(t))x^T(t-\rho_{i+1}d(t))Q_{i+1}x(t-\rho_{i+1}d(t)) \\
&+x^T(t)Rx(t) - x^T(t-h)Rx(t-h) + g^T(x(t))Yg(x(t)) \\
&-(1-\mu)g^T(x(t-d(t)))Yg(x(t-d(t))) \\
&+h^2\dot{x}^T(t)Z\dot{x}(t) - \varpi^T(t)\hat{Z}\varpi(t),
\end{aligned} \tag{4.46}
$$

where $\varpi^T(t) = \big(x^T(t), x^T(t-\rho_1 d(t)), x^T(t-\rho_2 d(t)), \ldots, x^T(t-\rho_K d(t)), x^T(t-d(t)), x^T(t-h)\big)$, $\hat{Z}$ is shown in (4.48).

Thus, adding (4.23) into $\dot{\tilde{V}}(x(t))$, we have

$$
\dot{\tilde{V}}(x(t)) \leqslant \varsigma^T(t)\left[\Upsilon + h^2\tilde{A}^T Z\tilde{A}\right]\varsigma(t), \tag{4.47}
$$

where $\varsigma^T(t) = [\varpi^T(t), g^T(x(t)), g^T(x(t-d(t)))]$.

Thus, if $\Upsilon + h^2\tilde{A}^T Z\tilde{A} < 0$, then $\dot{V}(x(t)) < -\varepsilon\|x(t)\|^2$ for some small $\varepsilon > 0$, i.e., the equilibrium point of the system (4.5) is *globally asymptotically stable*.

Remark 4.10 Obviously, Theorem 4.9 is the weighting-delay sequence-dependent criterion, and it depends on not only the value of parameter sequence $(\rho_1, \rho_2, \ldots, \rho_K)$, but also the number of subintervals, i.e., the number K of weighting-delays.

Thus, by properly choosing the number of weighting-delay and the value of weighting-delay parameter sequence, the less conservative stability results can be obtained.

Remark 4.11 Theorem 4.9 is a typical trade-off between conservativeness and complexity. Our results are less conservative by involving more parameters. That is to say, by increasing the useful parameters, the aim of reducing conservativeness will be achieved. Meanwhile, the global stability conditions are expressed in the form of LMI, which can be checked easily using the interior point algorithm. Thus, the overall computation complexity looks similar to the previous methods. And then, some studies on simultaneously reducing the complexity and conservativeness still need to be further carried out for the weighting-delay method.

In the following, a corollary based on augmented Lyapunov–Krasovskii functional can be proposed.

Corollary 4.12 *The equilibrium point of system (4.5) with time-varying delay $d(t)$ satisfying (4.2) and (4.3) is unique and globally asymptotically stable, if there exist parameters ρ_i satisfying $0 < \rho_1 < \rho_2 < \cdots < \rho_K < 1$, matrices*

$$\hat{Z} = \begin{bmatrix} -\frac{1}{\rho_1} & \frac{1}{\rho_1} & 0 & \cdots & 0 & 0 & 0 \\ * & -\frac{1}{\rho_1} - \frac{1}{\rho_2 - \rho_1} & \frac{1}{\rho_2 - \rho_1} & \cdots & 0 & 0 & 0 \\ * & * & -\frac{1}{\rho_2 - \rho_1} - \frac{1}{\rho_3 - \rho_2} & \cdots & 0 & 0 & 0 \\ * & * & * & \ddots & \vdots & \vdots & \vdots \\ * & * & * & * & \hat{Z}_K & \frac{1}{1-\rho_K} & 0 \\ * & * & * & * & * & -\frac{1}{1-\rho_K} - 1 & 1 \\ * & * & * & * & * & * & -1 \end{bmatrix} Z, \tag{4.48}$$

$$\hat{Z}_K = -\frac{1}{\rho_K - \rho_{K-1}} - \frac{1}{1-\rho_K}.$$

$$P = P^T = \begin{bmatrix} P_{11} & P_{12} \\ P_{12}^T & P_{22} \end{bmatrix} > 0, \; Y = Y^T > 0,$$

$$Z = Z^T = \begin{bmatrix} Z_{11} & Z_{12} \\ Z_{12}^T & Z_{22} \end{bmatrix} > 0, \; R = R^T > 0,$$

$$\Lambda = \mathrm{diag}(\lambda_1, \lambda_2, \ldots, \lambda_n) > 0, \; Q_j = Q_j^T > 0,$$

$$W_1 = \mathrm{diag}(w_{11}, w_{12}, \ldots, w_{1n}) > 0,$$

$$W_2 = \mathrm{diag}(w_{21}, w_{22}, \ldots, w_{2n}) > 0,$$

where $i = 1, 2, \ldots, K$, *and* $j = 1, 2, \ldots, K+1$, *such that the following LMI holds:*

$$\begin{bmatrix} \Xi & \Pi_1 & \cdots & \Pi_{K+1} & \Pi_{K+2} & h\tilde{A}^T Z_{22} \\ * & -\frac{1}{\rho_1} Z_{11} & \cdots & 0 & 0 & 0 \\ * & * & \ddots & \vdots & \vdots & \vdots \\ * & * & * & -\frac{1}{1-\rho_K} Z_{11} & 0 & 0 \\ * & * & * & * & -Z_{11} & 0 \\ * & * & * & * & * & -Z_{22} \end{bmatrix} < 0, \tag{4.49}$$

where Ξ, $\Pi_1, \ldots, \Pi_{K+2}$ *are shown in (4.50) and (4.51),*

$$\Xi = \begin{bmatrix} \Xi_1 & \frac{1}{\rho_1} Z_{22} & 0 & \cdots & 0 & 0 & -P_{12} & \Theta_{15} & \Theta_{16} \\ * & \Xi_2 & \frac{1}{\rho_2-\rho_1} Z_{22} & \cdots & 0 & 0 & 0 & 0 & 0 \\ * & * & \Xi_3 & \cdots & 0 & 0 & 0 & 0 & 0 \\ * & * & * & \ddots & \vdots & \vdots & \vdots & \vdots & \vdots \\ * & * & * & * & \Xi_{K+1} & \frac{1}{1-\rho_K} Z_{22} & 0 & 0 & 0 \\ * & * & * & * & * & \Xi_{K+2} & Z_{22} & 0 & \Theta_{36} \\ * & * & * & * & * & * & -R-Z_{22} & 0 & 0 \\ * & * & * & * & * & * & * & \Upsilon_{K+4} & \Lambda B \\ * & * & * & * & * & * & * & * & \Upsilon_{K+5} \end{bmatrix} \tag{4.50}$$

$$\begin{cases} \Pi_1^T = \left[-P_{12}^T D + P_{22} - \frac{1}{\rho_1} Z_{12} \ \frac{1}{\rho_1} Z_{12} \ 0 \ 0 \cdots 0 \ -P_{22} \ P_{12}^T A \ P_{12}^T B \right]_{n \times (K+5)n}, \\ \Pi_2^T = \left[-P_{12}^T D + P_{22} - \frac{1}{\rho_2-\rho_1} Z_{12} \ \frac{1}{\rho_2-\rho_1} Z_{12} \ 0 \cdots 0 \ -P_{22} \ P_{12}^T A \ P_{12}^T B \right]_{n \times (K+5)n}, \\ \quad \vdots \\ \Pi_{K+1}^T = \left[-P_{12}^T D + P_{22} \ 0 \cdots 0 \ -\frac{1}{1-\rho_K} Z_{12} \ \frac{1}{1-\rho_K} Z_{12} \ -P_{22} \ P_{12}^T A \ P_{12}^T B \right]_{n \times (K+5)n}, \\ \Pi_{K+2}^T = \left[-P_{12}^T D + P_{22} \ 0 \cdots 0 \ 0 \ -Z_{12} \ Z_{12} - P_{22} \ P_{12}^T A \ P_{12}^T B \right]_{n \times (K+5)n} \end{cases} \tag{4.51}$$

$$\begin{aligned} \Xi_1 = & -P_{11} D - D P_{11} + Q_1 + R + P_{12} + P_{12}^T + h^2 Z_{11} \\ & - h^2 Z_{12} D - h^2 D Z_{12}^T - \frac{1}{\rho_1} Z_{22} - 2\underline{L} W_1 \overline{L}, \\ \Xi_2 = & -(1 - \rho_1 \mu) Q_1 + Q_2 - \frac{1}{\rho_1} Z_{22} - \frac{1}{\rho_2 - \rho_1} Z_{22}, \\ \Xi_3 = & -(1 - \rho_2 \mu) Q_2 + Q_3 - \frac{1}{\rho_2 - \rho_1} Z_{22} - \frac{1}{\rho_3 - \rho_2} Z_{22}, \\ & \vdots \end{aligned}$$

$$\Xi_{K+1} = -(1-\rho_K\mu)Q_K + Q_{K+1} - \frac{1}{\rho_K - \rho_{K-1}} Z_{22} - \frac{1}{1-\rho_K} Z_{22},$$

$$\Xi_{K+2} = -(1-\mu)Q_{K+1} - \frac{1}{1-\rho_K} Z_{22} - Z_{22} - 2\underline{L}W_2\overline{L},$$

and the other parameters are the same as those defined in Theorems 4.3 and 4.9.

Proof The proof is similar with Theorem 4.9 using the augmented Lyapunov–Krasovskii functional V_1 and V_5 in (4.13) to replace the terms $x^T(t)Px(t)$ and $h\int_{-h}^{0}\int_{t+\theta}^{t}\dot{x}^T(s)Z\dot{x}(s)dsd\theta$ in (4.45). Here, the detailed proof is omitted.

Remark 4.13 For stability criteria to deal with Hopfield neural networks with constant delay in [37], the delay interval $[0,\ h]$ was divided into m fixed subintervals $[0, h/m]$, $[2h/m, 3h/m], \ldots, [(m-1)h/m, h]$, where m is an integer and $m \geqslant 1$. Different from that, the weighting-delay method proposed in this chapter is one with dynamic subintervals to handle RNNs with time-varying delay, where the subintervals are $[0,\ \rho_1 d(t)]$, $[\rho_2 d(t), \rho_2 d(t)], \ldots, [\rho_K d(t),\ d(t)]$, respectively. The sizes of subintervals are variable by altering parameter sequence $(\rho_1, \rho_2, \ldots, \rho_K)$ satisfying $0 < \rho_1 < \rho_2 < \cdots < \rho_K < 1$, where $i = 1, 2, \ldots, K$. That is to say, sequence $(h/m, 2h/m, \ldots, (m-1)h/m)$ is obtained in [37], while weighting-delay sequence $(\rho_1 d(t), \rho_2 d(t), \ldots, \rho_K d(t))$ is derived in this chapter. Obviously, when $d(t)$ is constant delay, parameter sequence $(\rho_1, \rho_2, \ldots, \rho_K)$ is more general and flexile than parameter sequence $(1/m, 2/m, \ldots, (m-1)/m)$.

4.5 Implementation of Optimal Weighting-Delay Parameters

Compared with other results in previous literature, the key problem is how to choose the values of weighting-delay parameters ρ or ρ_i, where $i = 1, 2, \ldots, K$. Since Theorem 4.3 and Corollary 4.8 are weighting-delay-independent stability criteria, they can be directly obtained using LMI ToolBox of MATLAB. The solutions to obtaining the values of weighting-delay parameters will be proposed for Theorems 4.5, 4.9, and Corollary 4.12 as follows:

4.5.1 The Single Weighting-Delay Case

Since only a weighting-delay $\rho d(t)$ is used in Theorem 4.5, the value of the *weighting-delay* parameter ρ can be chosen by trial and error. Then the stability results can be verified by MATLAB. Thus, a satisfactory value ρ can be obtained.

4.5.2 *The Multiple Weighting-Delays Case*

Obviously, the method in Sect. 4.5.1 is not suitable for Theorem 4.9 and Corollary 4.12. Thus, an optimization process is presented to simultaneously solve the stability results and obtain the value of the weighting-delay parameter sequence $(\rho_1, \rho_2, \ldots, \rho_K)$. On the other hand, for the number of weighting-delays, K, it is chosen by trial and error.

Take Theorem 4.9 for example, the optimization process is listed as follows to find the optimal weighting-delay parameter sequence. It consists of three steps.

1. Define $\delta_1 = \rho_1, \delta_2 = \rho_2 - \rho_1, \ldots, \delta_K = \rho_K - \rho_{K-1}$. According to the range of ρ_i, we can know that δ_i satisfies the following conditions

$$\sum_{i=1}^{K} \delta_i < 1, \tag{4.52}$$

$$\delta_i > 0. \tag{4.53}$$

2. Given μ, K, and $(\delta_1, \delta_2, \ldots, \delta_K)$, we can obtain maximum allowable h based on LMI ToolBox of MATLAB, which subjects to matrix inequality (4.42) and other restrictions in Theorem 4.9.
3. Choose the $K + 1$ groups of parameter sequence $\Delta^{(j)} = (\delta_1^{(j)}, \delta_2^{(j)}, \ldots, \delta_K^{(j)})$, where $\delta_i^{(j)}$ satisfying (4.52) and (4.53), $i = 1, 2, \ldots, K$ and $j = 1, 2, \ldots, K+1$. Based on these data and step (2), applying N-M simplex method proposed in [41], the optimal weighting-delay parameter sequence and the corresponding maximum allowable h can be obtained for Theorem 4.9.

4.6 Illustrative Examples

In this section, three numerical examples and an application example are given to verify the effectiveness of the criteria proposed in this chapter.

1. Numerical Examples

Example 4.14 Consider RNNs (4.5) with the following parameters [21],

$$D = \mathrm{diag}(1.2769,\ 0.6231,\ 0.9230,\ 0.4480),$$

$$A = \begin{bmatrix} -0.0373 & 0.4852 & -0.3351 & 0.2336 \\ -1.6033 & 0.5988 & -0.3224 & 1.2352 \\ 0.3394 & -0.0860 & -0.3824 & -0.5785 \\ -0.1311 & 0.3253 & -0.9534 & -0.5015 \end{bmatrix},$$

Table 4.1 Allowable upper bound of h for different μ and ρ in Example 4.14

Weighting-delay parameter	Methods	$\mu = 0.1$	$\mu = 0.5$	$\mu = 0.9$	Unknown μ
–	[22] and [29]	3.2775	2.1502	1.3164	1.2598
–	[30]	3.2793	2.2245	1.5847	1.5444
–	[34]	3.2819	2.2261	1.6035	1.5593
–	[31]	3.3039	2.5376	2.0853	2.0389
–	Theorem 4.3	3.2844	2.2376	1.6272	1.5777
–	Corollary 4.8	3.3574	2.5912	2.1303	2.0779
$\rho = 0.6$	Theorem 4.5	3.3574	2.5915	2.1306	2.0779

$$B = \begin{bmatrix} 0.8674 & -1.2405 & -0.5325 & 0.0220 \\ 0.0474 & -0.9164 & 0.0360 & 0.9816 \\ 1.8495 & 2.6117 & -0.3788 & 0.8428 \\ -2.0413 & 0.5179 & 1.1734 & -0.2775 \end{bmatrix},$$

$$\bar{l}_1 = 0.1137,\ \bar{l}_2 = 0.1279,\ \bar{l}_3 = 0.7994,\ \bar{l}_4 = 0.2368,$$

$$\underline{l}_1 = \underline{l}_2 = \underline{l}_3 = \underline{l}_4 = 0,$$

and time-varying delay $d(t)$ satisfies (4.2) and (4.3).

By setting different μ and parameter ρ (or ρ_1, ρ_2), the upper bound h of time delay $d(t)$ has been studied in [22, 29–31, 34], respectively. The corresponding results are shown in Table 4.1, where "–" means the results are not applicable to the corresponding cases. Specially, if $\mu = 0$, i.e., $d(t)$ is constant, the upper bound h is 1.4224 in [21], 1.9321 in [23], 3.5841 in [30, 31], respectively. And then, applying Theorem 4.3 and Theorem 4.5 in this chapter, h is 3.5869 and 3.6156. From Table 4.1, it is clear that the results in this chapter improve upon the existing delay-dependent results. Figure 4.1 shows the state response of Example 4.14 with constant delay $h = 3.6156$, when the initial value is $[12.2,\ 13.6,\ 11.4,\ 10.7]^T$.

Example 4.15 Consider the RNNs (4.5) with the following parameters [22],

$$D = \begin{bmatrix} 2 & 0 \\ 0 & 2 \end{bmatrix},\quad A = \begin{bmatrix} 1 & 1 \\ -1 & -1 \end{bmatrix},\quad B = \begin{bmatrix} 0.88 & 1 \\ 1 & 1 \end{bmatrix},$$

$$\bar{l}_1 = 0.4,\ \bar{l}_2 = 0.8,\ \underline{l}_1 = \underline{l}_2 = 0,$$

and time-varying delay $d(t)$ satisfies (4.2) and (4.3).

The corresponding results are given in Table 4.2 by using Theorems 4.3, 4.5 and Corollary 4.8 of this chapter and methods in [22, 29–31], when different μ and parameter ρ are set, respectively. Figure 4.2 shows the state response of Example 4.15 with $h = 1.5571$ and $\mu = 1$, when the initial value is $[8,\ -10]^T$. Obviously, when ρ

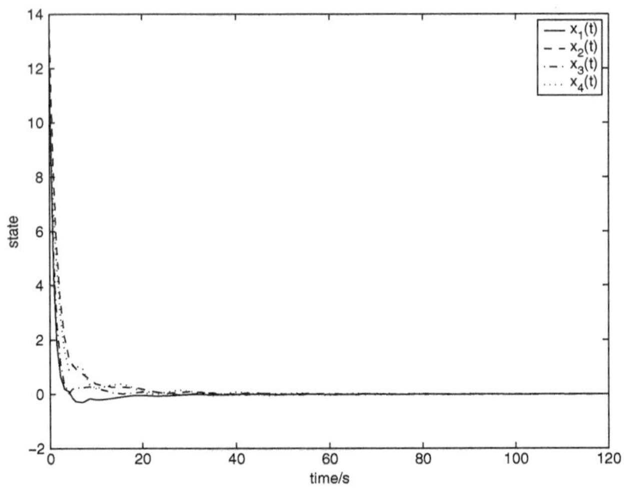

Fig. 4.1 State response *curves* with $\mu = 0$ and $h = 3.6156$ for Example 4.14, when the initial value is $[12.2,\ 13.6,\ 11.4,\ 10.7]^T$

Table 4.2 Allowable upper bound of h for different μ and ρ in Example 4.15

Parameter	Methods	$\mu = 0.8$	$\mu = 0.9$	$\mu = 1$	$\mu = 1.1$
–	[22] and [29]	1.2281	0.8636	0.8298	0.8298
–	[30]	1.6831	1.1493	1.0880	1.0880
–	[31]	2.3534	1.6050	1.5103	1.5103
–	Theorem 4.3	1.6831	1.1494	1.0880	1.0880
–	Corollary 4.8	2.3534	1.6050	1.5103	1.5103
$\rho = 0.1$	Theorem 4.5	2.3734	1.6133	1.5136	1.5130
$\rho = 0.2$	Theorem 4.5	2.3961	1.6270	1.5209	1.5190
$\rho = 0.3$	Theorem 4.5	2.4215	1.6466	1.5320	1.5276
$\rho = 0.4$	Theorem 4.5	2.4490	1.6715	1.5449	1.5358
$\rho = 0.5$	Theorem 4.5	2.4779	1.7003	1.5556	1.5387
$\rho = 0.6$	Theorem 4.5	2.5065	1.7286	1.5571	1.5303
$\rho = 0.7$	Theorem 4.5	2.5308	1.7457	1.5416	1.5139
$\rho = 0.8$	Theorem 4.5	2.5406	1.7273	1.5161	1.5103
$\rho = 0.9$	Theorem 4.5	2.5053	1.6583	1.5103	1.5103

varies in this example, the results of Theorem 4.5 are different, too. Based on the data of Table 4.2, the results of Theorem 4.5 are not monotonically increasing. Therefore, the satisfactory result of criterion for $\mu = 0.8$ will be obtained at $\rho \in (0.7,\ 0.9)$, i.e., $(0.7,\ 0.9)$ is the better interval to design the value of ρ for $\mu = 0.8$. Similarly, for $\mu = 0.9$, $\mu = 1$, and $\mu = 1.1$, the corresponding intervals of parameter ρ are $(0.6,\ 0.8)$, $(0.5,\ 0.7)$, and $(0.4,\ 0.6)$, respectively.

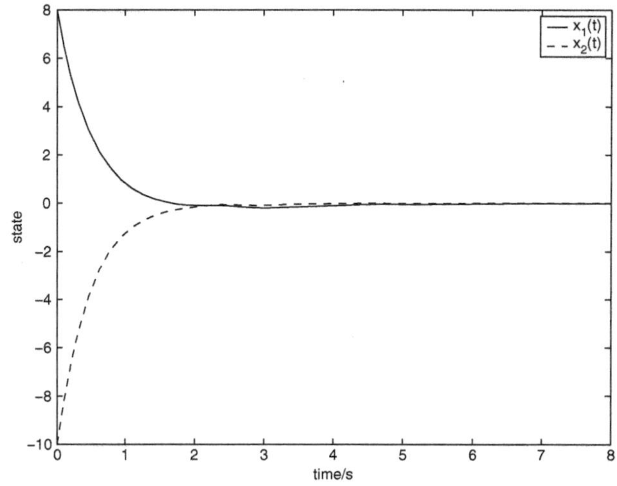

Fig. 4.2 State response *curves* with $\mu = 1$ and $h = 1.5571$ for Example 4.15, when the initial value is $[8, \ -10]^T$

Example 4.16 Consider RNNs (4.5) with the following parameters [37],

$$D = \begin{bmatrix} 4.1989 & 0 & 0 \\ 0 & 0.7160 & 0 \\ 0 & 0 & 1.9985 \end{bmatrix},$$

$$B = \begin{bmatrix} -0.1052 & -0.5069 & -0.1121 \\ -0.0257 & -0.2808 & 0.0212 \\ 0.1205 & -0.2153 & 0.1315 \end{bmatrix},$$

$$\bar{l}_1 = 0.4129, \ \bar{l}_2 = 3.8993, \ \bar{l}_3 = 1.0160,$$

$$\underline{l}_1 = \underline{l}_2 = \underline{l}_3 = 0,$$

$A = 0$, and time-varying delay $d(t)$ satisfies (4.2) and (4.3).

First, the upper bound h for this example with constant delay has been studied between [37] and this chapter, where the optimal parameter value ρ_i are obtained by applying the method proposed in Sect. 4.5.2, $m = K + 1$, and $i = 1, \ldots, K$. The corresponding results are shown in Table 4.3. For this example, the stability results is subequal between Theorem 4.9 and [37] when constant delay is employed. Then, the upper bound h for this example with time-varying delay $d(t)$ is given in Table 4.4 by using fixed subintervals method like [37] and the method proposed in Sect. 4.5.2 for Theorem 4.9, where the optimal weighting-delay parameters are given in brackets. From Table 4.4, it is clear that weighting-delay method dividing

Table 4.3 Allowable upper bound of h when $d(t)$ is constant delay using optimization method in Example 4.16, where $i = 1, \ldots, K$ and $K = m - 1$

Results in this chapter		Results in [37]	
Optimal parameters	Theorem 4.9	Parameters	[37]
$\rho_i = i \times 0.3333, K = 2$	2.5492	$m = 3$	2.54
$\rho_i = i \times 0.2500, K = 3$	2.5720	$m = 4$	2.57
$\rho_i = i \times 0.2000, K = 4$	2.5825	$m = 5$	2.581
$\rho_i = i \times 0.1000, K = 9$	2.5966	$m = 10$	2.596
$\rho_i = i \times 0.0667, K = 14$	2.5992	$m = 15$	2.597

Table 4.4 Allowable upper bound of h for time-varying delay $d(t)$ with different μ using optimization method at $K = 2$ for Theorem 4.9 in Example 4.16

μ	h ($\rho_1 = 1/3, \rho_2 = 2/3$)	h (optimal value of parameters)
0.1	2.1573	2.1927 ($\rho_1 = 0.2051, \rho_2 = 0.4935$)
0.5	1.4276	1.4948 ($\rho_1 = 0.1000, \rho_2 = 0.2187$)
0.9	1.1870	1.2030 ($\rho_1 = 0.7613, \rho_2 = 0.8819$)
1	1.1790	1.1887 ($\rho_1 = 0.6483, \rho_2 = 0.8817$)

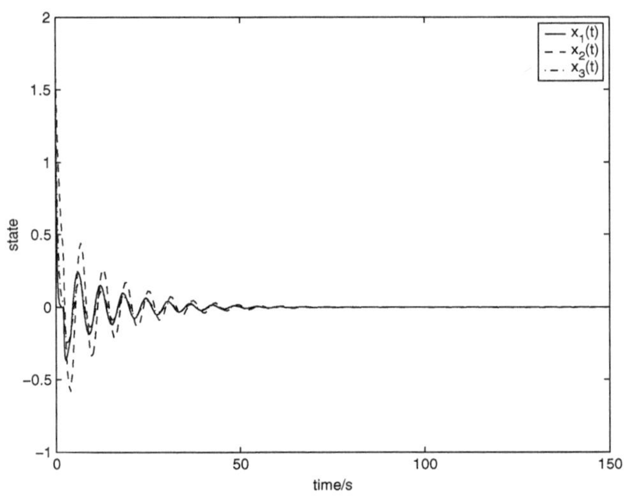

Fig. 4.3 State response *curves* with $\mu = 0$ and $h = 2.5992$ for Example 4.16, when the initial value is $[1.2,\ 1.6,\ 1.4]^T$

delay interval $[0, d(t)]$ is more suitable for dealing with stability problem with time-varying delay. Figure 4.3 shows the state response of Example 4.16 with constant delay $h = 2.5992$, when the initial value is $[1.2,\ 1.6,\ 1.4]^T$.

2. An Application Example

Example 4.17 Consider the continuous PH neutralization of an acid stream by a highly concentrated basic stream, which can be expressed in the following form [18, 39]:

$$v\dot{y}(t) = -fy(t) - u(t), \quad PH = w_2 \tanh w_1 y(t), \tag{4.54}$$

where v is the volume of the mixing tank, $u(t)$ is the manipulated variable representing the base flow rate, f is the acid flow rate, w_1 and w_2 are some constants, $y(t)$ is the strong acid, and PH is the measured output signal.

In fact, time delay is always inevitable in this control process. Therefore, we slightly modify model (4.54) as follows:

$$\begin{cases} v\dot{y}(t) = -fy(t) - u(t), \\ PH = w_2 \tanh(w_1 y(t)) + w_2 w_3 \tanh(w_1 y(t-\tau)). \end{cases} \tag{4.55}$$

The purpose of this application is to find the maximum allowable upper bound of delay τ for a feedback gain K by adopting output feedback controller $u = -K \times PH$ such that the closed-loop system is asymptotically stable. The closed-loop system can be expressed in the following form:

$$\begin{aligned} \dot{y}(t) = -\frac{f}{v}y(t) + \frac{Kw_2}{v}\tanh(w_1 y(t)) \\ + \frac{Kw_2w_3}{v}\tanh(w_1 y(t-\tau)), \end{aligned} \tag{4.56}$$

Let $x(t) = w_1 y(t)$. Then, system (4.56) is changed into the following form:

$$\dot{x}(t) = -Dx(t) + A\tanh(x(t)) + B\tanh(x(t-\tau)), \tag{4.57}$$

where $D = \frac{f}{v}$, $A = \frac{Kw_1w_2}{v}$, and $B = \frac{Kw_1w_2w_3}{v}$.

Then, we take $f = 5.8154$, $v = 1500.3732$, $w_1 = 28.9860$, $w_2 = -3.8500$, and $w_3 = 2.56$. Meanwhile, based on the results of [18], we choose the feedback gain $K = 0.5022$. Thus, we can obtain $D = 0.0039$, $A = -0.0374$, and $B = -0.0956$. Using criteria in [30, 31] and [34], Theorems 4.3 and 4.5 in this chapter, the maximum allowable upper bound τ of system (4.57) is 17.4956. Meanwhile, the maximum allowable $\tau = 18.2871$ for Theorem 4.9, when $\rho_1 = 0.3333$ and $\rho_2 = 0.6667$. Namely, the better result can be obtained using our criteria. Thus, the origin is the equilibrium point of system (4.57) based on the feedback gain $K = 0.5022$.

4.7 Summary

In this chapter, the delay-dependent stability problem for RNNs with time-varying delay is studied based on weighting-delay method. By introducing the new variables nominated as weighting-delays, delay interval $[0,\ d(t)]$ is divided into several variable subintervals, i.e., several dynamic subintervals. It implies that the stability results depend on the positions of weighting-delays in delay interval, which can be denoted as the form of parameter value ρ (or ρ_i, $i = 1, 2, \ldots, K$). Compared with previous results, several negative definite terms with weighting-delay parameters will be added to our proposed criteria, which leads to less conservative stability results. Specially, when weighting-delay sequence $(\rho_1 d(t), \rho_2 d(t), \ldots, \rho_K d(t))$ is applied, the optimal parameter sequence $(\rho_1, \rho_2, \ldots, \rho_K)$ is obtained by an optimization method. Compared with the fixed subintervals method used in [37], it is clear that weighting-delay method by applying the variable subintervals with delay $d(t)$ has inherent flexibility, and is more suitable for dealing with stability problem with time-varying delay. As a result, in both theory and practice, the criteria based on weighting-delay method are less conservative in dealing with the stability of RNNs with time-varying delay.

References

1. L.O. Chua, L. Yang, Cellular neural networks: applications. IEEE Trans. Circuits Syst. **35**(10), 1273–1290 (1988)
2. T. Roska, L.O. Chua, Cellular neural networks with non-linear and delay-type template elements and non-uniform grids. Int. J. Circuit Theory Appl. **20**(5), 469–481 (1992)
3. Y.H. Chen, S.C. Fang, Neurocomputing with time delay analysis for solving convex quadratic programming problems. IEEE Trans. Neural Netw. **11**(1), 230–240 (2000)
4. Y. Xia, J. Wang, A general projection neural network for solving monotone variational inequalities and related optimization problems. IEEE Trans. Neural Netw. **15**(2), 318–328 (2004)
5. R. Fantacci, M. Forti, M. Marini, L. Pancani, Cellular neural networks approach to a class of communication problems. IEEE Trans. Circuits Syst. I, Fund. Theory Appl. **46**(12), 1457–1467 (1999)
6. J.A. Farrell, A.N. Michel, A synthesis procedure for Hopfield's continuous-time associative memory. IEEE Trans. Circuits Syst. **37**(7), 877–884 (1990)
7. A.N. Michel, J.A. Farrell, H.F. Sun, Analysis and synthesis techniques for Hopfield type synchronous discrete-time neural networks with applications to associative memory. IEEE Trans. Circuits Syst. **37**(11), 1356–1366 (1990)
8. T. Nishikawa, Y.C. Lai, F.C. Hoppensteadt, Capacity of oscillatory associative-memory networks with error-free retrieval. Phys. Rev. Lett. **92**(10), 101–108 (2004)
9. D.L. Wang, Emergent synchrony in locally coupled neural oscillators. IEEE Trans. Neural Netw. **6**(4), 941–948 (1995)
10. S.S. Young, P.D. Scott, N.M. Nasrabadi, Object recognition using multilayer Hopfield neural network. IEEE Trans. Image Process. **6**(3), 357–372 (1997)
11. T. Liao, F. Wang, Global stability for cellular neural networks with time delay. IEEE Trans. Neural Netw. **11**(6), 1481–1484 (2000)
12. S. Arik, An analysis of global asymptotic stability of delayed neural networks. IEEE Trans. Neural Netw. **13**(5), 1239–1242 (2002)

13. S. Arik, Global aymptotic stability of a larger class of neural networks with constant time delay. Phys. Lett. A **311**, 504–511 (2003)
14. V. Singh, A generalized LMI-based approach to the global asymptotic stability of delayed cellular neural networks. IEEE Trans. Neural Netw. **15**(1), 223–225 (2004)
15. J. Cao, J. Wang, Global asymptotic stability of a class of recurrent neural networks with time-varying delays. IEEE Trans. Circuits Syst. I **50**(1), 34–44 (2003)
16. J. Cao, J. Wang, Global asymptotic stability and robust stability of recurrent neural networks with time delays. IEEE Trans. Circuits Syst. I **52**(2), 417–426 (2005)
17. J. Cao, J. Wang, Global exponential stability and periodicity of recurrent neural networks with time delays. IEEE Trans. Circuits Syst. I **52**(5), 920–931 (2005)
18. H. Zhang, Z. Wang, D. Liu, Global asymptotic stability of recurrent neural networks with multiple time-varying delays. IEEE Trans. Neural Netw. **19**(5), 855–873 (2008)
19. H. Zhang, Z. Wang, Global asymptotic stability of delayed cellular neural networks. IEEE Trans. Neural Netw. **18**(3), 947–950 (2007)
20. X. Liao, G. Chen, E. Sanchez, Delay-dependent exponential stability analysis of delay neural networks: An LMI approch. Neural Netw. **15**, 855–866 (2002)
21. S. Xu, J. Lam, D. Wu, Y. Zou, New global asymptotic stability criteria for delayed cellular neural networks. IEEE Trans. Circuits Syst. II, Exp. Br. **52**(6), 349–353 (2005)
22. C. Hua, C. Long, X. Guan, New results on stability analysis of neural networks with time-varying delays. Phys. Lett. A **352**, 335–340 (2006)
23. H. Cho, J. Park, Novel delay-dependent robust stability criterion of delayed cellular neural networks. Chaos Solitons Fractals **32**, 1194–1200 (2007)
24. R. Sathy, P. Balasubramaniam, R. Chandran, Robust stability of fuzzy Markov type Cohen-Grossberg neural networks by delay decomposition approach. Iranian J. Fuzzy Syst. **11**(2), 1–16 (2014)
25. X. Zhou, J. Tian, H. Ma, Improved delay-dependent stability criteria for recurrent neural networks with time-varying delays. Neurocomputing **129**, 401–408 (2014)
26. X. Liao, Y. Liu, H. Wang, Exponential estimates and exponential stability for neutral-type neural networks with multiple delays. Neurocomputing **14**, 868–883 (2015)
27. W. Feng, S. Yang, H. Wu, Further results on robust stability of bidirectional associative memory neural networks with norm-bounded uncertainties. Neurocomputing **148**, 535–543 (2015)
28. H. Zhang, Z. Liu, G. Huang, Z. Wang, Novel weighting delay-based stability criteria for recurrent neural networks with time varying delay. IEEE Trans. Neural Netw. **21**(1), 91–106 (2010)
29. Y. He, M. Wu, J. She, Delay-dependent exponential stability of delayed neural networks with time-varying delay. IEEE Trans. Circuits Syst. II, Exp. Br. **53**(7), 553–557 (2006)
30. Y. He, G. Liu, D. Rees, New delay-dependent stability criteria for neural networks with time-varying delay. IEEE Trans. Neural Netw. **18**(1), 310–314 (2007)
31. Y. He, G. Liu, D. Rees, M. Wu, Stability Analysis for neural networks with time-varying interval delay. IEEE Trans. Neural Netw. **18**(6), 1850–1854 (2007)
32. M. Parlakçi, Robust stability of uncertain neutral systems: a novel augmented Lyapunov functional appr̃oach. IET Control Theory Appl. **1**(3), 802–809 (2007)
33. C. Peng, Y. Tian, Delay-dependent robust stability criteria for uncertain systems with interval time-varying delay. J. Comput. Appl. Math. **214**(2), 480–494 (2008)
34. T. Li, L. Guo, C. Sun, C. Lin, Further result on delay-dependent stability criterion of neural networks with time-varying delays. IEEE Trans. Neural Netw. **19**(4), 726–730 (2008)
35. X.G. Li, X.J. Zhu, Stability analysis of neutral systems with distributed delays. Automatica **44**(8), 2197–2201 (2008)
36. E. Fridman, U. Shaked, A descriptor system approach to H_∞ control of linear time-delay systems. IEEE Trans. Autom. Control **47**(2), 253–270 (2002)
37. S. Mou, H. Gao, J. Lam, W. Qiang, A new criterion of delay-dependent asymptotic stability for Hopfield neural networks with time delay. IEEE Trans. Neural Netw. **19**(3), 532–535 (2008)
38. K. Gu, V. Kharitonov, J. Chen, *Stability of Time-Delay Systems*, (Cambridge University Press, Cambridge, 2003)

39. M. Liu, Delayed standard neural network model and its application. ACTA Automatica Sinica **31**(5), 750–758 (2005). (in Chinese)
40. J. Hale, *Theory of Functional Differential Equations* (Springer, New York, 1977)
41. J.A. Nelder, R. Mead, A simplex method for function minimization. Comput. J. **7**, 308–313 (1965)
42. P. Park, J. Ko, Stability and robust stability for systems with a time-varying delay. Automatica **43**(10), 1855–1858 (2007)

Chapter 5
Stability Criteria for RNNs Based on Secondary Delay Partitioning

Chapter 4 presented a new way to establish the delay-dependent stability results for RNNs with delay. The main feature of the method in Chap. 4 is to split the time delay with fixed interval by inserting some virtual sampling points or weighting coefficients, which leads to the nonuniformly changeable subintervals. In this chapter, we will present another method to decompose the interval of time delay and change the sizes of the subintervals. This new method is called the secondary delay partitioning method, and the effectiveness of the established stability result is verified by numerical simulation. The contents of this chapter are mainly from the result in [26].

5.1 Introduction

Delay-dependent stability criteria for delayed recurrent neural networks (RNN) have received considerable attention in recent years [1–27], because they are less conservative than the delay-independent stability results by incorporating such information as magnitude of time delay, change rate of time-varying delay, and connection matrices, especially when the sizes of time-delays are small. Till date, there are many delay-dependent stability results for RNN with different kinds of delays published.

In the early research on RNN, the concerned delay was often constant $\tau > 0$ or time-varying $\tau(t)$, which belonged to an interval $0 < \tau(t) \leq \tau_M$. In practice, the lower bound of time-varying delay may not be zero, then the time-varying delay was extended to the interval $0 < \tau_m \leq \tau(t) \leq \tau_M$ [4]. Based on this requirement, many new *delay-dependent* stability criteria have been established, and the conservativeness of the stability results is further reduced. Among these stability results, delay partitioning approach played an important role [2, 7, 11–13, 16–21]. For the case of constant delay τ, the interval $[0, \tau]$ is divided into l subinterval uniformly, $l \geq 1$, i.e., $[0, \frac{\tau}{l}]$, $[\frac{\tau}{l}, \frac{2\tau}{l}], \ldots, [\frac{(l-1)}{l}\tau, \tau]$ [11, 12]. By constructing corresponding Lyapunov functional, the LMI-based global asymptotical stability criterion had been derived. For the case of time-varying delay, the delay partitioning

Z. Wang et al., *Qualitative Analysis and Control of Complex Neural Networks with Delays*, Studies in Systems, Decision and Control 34,
DOI 10.1007/978-3-662-47484-6_5

approach was studied in [2, 14, 18], and the interval $\tau(t) \in [0, \tau_M]$ was divided into subinterval $[0, \tau_M] = \bigcup_{j=1}^{l}[\tau_{j-1}, \tau_j]$ with $0 = \tau_0 < \tau_1 < \tau_2, \ldots, < \tau_l = \tau_M$. In [16], the authors also divided the interval $\tau(t) \in [0, \tau_M]$ into subinterval $[0, \tau_M] = \bigcup_{j=1}^{l}[\tau_{j-1}, \tau_j]$ with $0 = \tau_0 < \tau_1 < \tau_2, \ldots, < \tau_l = \tau_M$, and constructed some corresponding Lyapunov functions to derive the LMI-based stability criteria. In [14], $\tau(t) \in [0, \tau_M]$ is divided into two parts $[0, \tau_m]$ and $[\tau_m, \tau_M]$, and the interval $[0, \tau_m]$ is divided into l subinterval $[0, \frac{\tau_m}{l}], [\frac{\tau_m}{l}, \frac{2\tau_m}{l}], \ldots, [\frac{(l-1)}{l}\tau_m, \tau_m], l \geq 1$. By constructing corresponding *Lyapunov function*, the LMI-based global asymptotical stability criterion was derived. As pointed out in [18], the constant delay partitioning approach is not suitable for the case of time-varying delay. Thus, in order to deal with the relation between the states with partitioned delays and the state with time-varying delay $\tau(t)$, a value-set method, that is, there exists a j such that $\tau(t) \in [\tau_{j-1}, \tau_j], j = 1, \ldots, l$, was proposed in [18], and later it was combined with the reciprocal *convex combination* (RCC) approach [28] to study the stability problem of delayed RNN [2, 14]. In [9], by dividing the interval $[\tau_m, \tau_M]$ into $[\tau_m, \tau(t)] \bigcup [\tau(t), \tau_M]$, some stability criteria based on RCC approach was obtained for delayed RNN, where convex combination information on $\tau_M - \tau(t)$ and $\tau(t) - \tau_m$ was used. Different from the methods in [9, 11, 12, 14, 16, 18], the authors in [2] used the delay partitioning approach and reciprocal convex combination approach by dividing the interval $\tau(t) \in [0, \tau_M]$ into subintervals $[0, \tau_M] = \bigcup_{j=1}^{l}[\tau_{j-1}, \tau_j]$ with $0 = \tau_0 < \tau_1 < \tau_2, \ldots, < \tau_l = \tau_M$, and some stability criteria were established. The *delay partitioning approach* was also applied to discrete-time RNN [29, 30].

Summarizing the above main methods, we find that the delay intervals $[0, \tau_M]$ and $[\tau_m, \tau_M]$ are usually divided into $\bigcup_{j=1}^{l}[\tau_{j-1}, \tau_j]$, and $[\tau_m, \tau(t)] \bigcup [\tau(t), \tau_M]$, respectively. This can be named as first delay partitioning. If $[\tau_m, \tau(t)]$ and $[\tau(t), \tau_M]$ are further divided into several subintervals, which is named as secondary delay partitioning, and some corresponding Lyapunov functionals are constructed, how about the obtained stability result? For example, considering a time-varying delay $\tau(t) = \sin(\frac{t}{3}) + \cos(\frac{t}{2})$, the relation between $\tau(t)$ and its interval bounds $[\tau_m, \tau_M]$ is depicted in Fig. 5.1, where the dotted line is the fixed boundary of the interval of the time-varying delay and the thick line is $\tau(t)$, respectively.

Motivated by the above argument, we present a novel delay-dependent stability criterion for RNN with time-varying delay in this paper. In this aspect, we nonuniformly decompose the delay intervals $[\tau_M - \tau(t)]$ and $[\tau(t) - \tau_m]$ into multiple subintervals, and construct a new *Lyapunov–Krasovskii functional* by choosing different weighting matrices on different subintervals. Then, we employ the new Lyapunov–Krasovskii functional and extend reciprocal convex combination approach to formulate a new delay-dependent stability criterion for a class of RNNs with time-varying delays. The main contributions of the chapter are as follows. (1) Both delay intervals $[\tau_m, \tau(t)]$ and $[\tau(t), \tau_M]$ are further divided into many subintervals by involving dynamic weighting parameters. Comparing with the methods in [2, 14, 18], the proposed method may build a closer relation among the states $x(t - \tau(t))$, $x(t - \tau_M)$, $x(t - \tau_m)$ and other states associated with the different subintervals. (2) An extended RCC is established and a double integral term with variable upper and lower limits

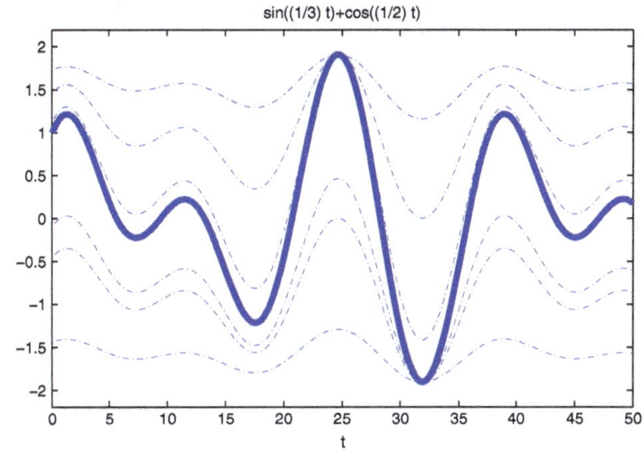

Fig. 5.1 The segmentation of time-varying delay $\tau(t)$

as a Lyapunov functional is constructed, simultaneously, which are used to tackle the cross terms among the states associated with the subintervals.

5.2 Problem Formulation and Preliminaries

We consider the following RNNs with time-varying delay,

$$\dot{u}(t) = Au(t) + B\hat{f}(u(t)) + C\hat{f}(u(t - \tau(t))) + \hat{I}, \tag{5.1}$$

where $u(t) = (u_1(t), \ldots, u_n(t))^T$ is the state of neuron, $A = -\text{diag}(a_1, \ldots, a_n)$ with $a_i > 0$, B and C are the interconnected matrices with appropriate dimensions, $\hat{f}(u) = (\hat{f}_1(u_1), \hat{f}_2(u_2), \ldots, \hat{f}_n(u_n))^T$ is the activation function. $\hat{I} = (\hat{I}_1, \hat{I}_2, \ldots, \hat{I}_n)^T$ is a constant input vector, $\tau(t)$ is the time-varying delay satisfying

$$0 \le \tau_m \le \tau(t) \le \tau_M, \quad \rho_m \le \dot{\tau}_i(t) \le \rho_M,$$

where τ_m, τ_M, ρ_m and ρ_M are known constants, $i = 1, \ldots, n$.

Assumption 5.1 The bounded *activation functions* $\hat{f}_i(\cdot)$ satisfy the following condition for $\forall s_1, s_2 \in \mathbb{R}, s_1 \ne s_2$,

$$\delta_i^- \le \frac{\hat{f}_i(s_1) - \hat{f}_i(s_2)}{s_1 - s_2} \le \delta_i^+, \tag{5.2}$$

where δ_i^- and δ_i^+ are known constants, which can be positive, zero and negative, $i = 1, \ldots, n$.

According to [7, 10], Assumption 5.1 always ensures the existence of an *equilibrium* point u^*. Denote $\Sigma = \mathrm{diag}(\delta_i^-)$ and $\Gamma = \mathrm{diag}(\delta_i^+)$, $i = 1, \ldots, n$.

We shift the equilibrium point u^* of the system (5.1) to the origin by the transformation $x(t) = u(t) - u^*$, it yields the following *error system*,

$$\dot{x}(t) = Ax(t) + Bf(x(t)) + Cf(x(t - \tau(t))), \tag{5.3}$$

where $f(x(t)) = \hat{f}(x(t) + u^*) - \hat{f}(u^*)$, $\phi(t)$ is a continuously real-valued function on $[-\tau_M, 0]$.

In order to derive our main result, the following lemmas are introduced.

Lemma 5.2 (Jensen integral inequality, see [17]) *For any symmetric positive definite constant matrix $Q > 0$, any scalars a and b with $a < b$, and a vector function $\varrho(t) : [a, b] \to \mathbb{R}$ such that the integrals concerned are well-defined, then the following inequality holds:*

$$\Big(\int_a^b \varrho(s)\mathrm{d}s\Big)^T Q \Big(\int_a^b \varrho(s)\mathrm{d}s\Big) \le (b-a)\int_a^b \varrho(s)^T Q \varrho(s)\mathrm{d}s.$$

The RCC approach in [28] is a useful approach to deal with the convex combination problem. However, the RCC approach in [28] is a basic expression, it may not be easy to be used directly, especially for the complex case of *convex combination.* Now we will present an extended RCC.

Lemma 5.3 *For any vectors $h_1, \ldots, h_N$ with appropriate dimensions, positive scalars $\alpha_i > 0$, $\sum_{i=1}^N \alpha_i = 1$, $R > 0$, if there exist appropriately dimensioned matrices $S_1, \ldots, S_{\frac{N(N-1)}{2}}$ satisfying the following conditions:*

$$\begin{bmatrix} R & S_j \\ * & R \end{bmatrix} \ge 0,\ j = 1, \ldots, \frac{N(N-1)}{2},$$

then the following inequality holds:

$$-\sum_{i=1}^N \frac{1}{\alpha_i} h_i^T R h_i \le -\begin{bmatrix} h_1 \\ h_2 \\ \vdots \\ \vdots \\ h_N \end{bmatrix}^T \begin{bmatrix} R & S_1 & S_2 & \cdots & S_{N-1} \\ * & R & S_N & \cdots & S_{2N-3} \\ * & * & \ddots & \ddots & \vdots \\ * & * & * & R & S_{\frac{N(N-1)}{2}} \\ & * & * & * & R \end{bmatrix} \begin{bmatrix} h_1 \\ h_2 \\ \vdots \\ \vdots \\ h_N \end{bmatrix}.$$

Proof We denote $[h_1, \ldots, h_N]$ as δ^T, $[h_i, h_j]$ as δ_{ij}^T, then we have

$$
\begin{aligned}
&\sum_{i=1}^{N} \frac{1}{\alpha_i} h_i^T R h_i - \delta^T \begin{bmatrix} R & S_1 & S_2 & \cdots & S_{N-1} \\ * & R & S_N & \cdots & S_{2N-3} \\ * & * & \ddots & \ddots & \vdots \\ * & * & * & R & S_{\frac{N(N-1)}{2}} \\ * & * & * & * & R \end{bmatrix} \delta \\
&= \delta^T \begin{bmatrix} c_1 R & -S_1 & -S_2 & \cdots & -S_{N-1} \\ * & c_2 R & -S_N & \cdots & -S_{2N-3} \\ * & * & \ddots & \ddots & \vdots \\ * & * & * & c_{N-1} R & -S_{\frac{N(N-1)}{2}} \\ * & * & * & * & c_N R \end{bmatrix} \delta,
\end{aligned} \tag{5.4}
$$

where $c_i = \frac{1-\alpha_i}{\alpha_i}, i = 1, \ldots, N$.

Since $\frac{1-\alpha_i}{\alpha_i} = \frac{\sum_{j=1}^{N} \alpha_j - \alpha_i}{\alpha_i}$, then from (5.4) we have,

$$
\begin{aligned}
&\sum_{\{i=1,\ldots,N-1, j=i+1,\ldots,N\}} \delta_{ij}^T \begin{bmatrix} \frac{\alpha_j}{\alpha_i} R & -S_{(i-1)N - \frac{i(i+1)}{2} + j} \\ * & \frac{\alpha_i}{\alpha_j} R \end{bmatrix} \delta_{ij} \\
&= \sum_{\{i=1,\ldots,N-1,\ j=i+1,\ldots,N\}} \delta_{ij}^T \begin{bmatrix} \sqrt{\frac{\alpha_j}{\alpha_i}} & 0 \\ 0 & -\sqrt{\frac{\alpha_i}{\alpha_j}} \end{bmatrix} \\
&\quad \times \begin{bmatrix} R & S_{(i-1)N - \frac{i(i+1)}{2} + j} \\ * & R \end{bmatrix} \begin{bmatrix} \sqrt{\frac{\alpha_j}{\alpha_i}} & 0 \\ 0 & -\sqrt{\frac{\alpha_i}{\alpha_j}} \end{bmatrix} \delta_{ij} \\
&= \sum_{\{i=1,\ldots,N-1,\ j=i+1,\ldots,N\}} \begin{bmatrix} \sqrt{\frac{\alpha_j}{\alpha_i}} h_i \\ -\sqrt{\frac{\alpha_i}{\alpha_j}} h_j \end{bmatrix}^T \\
&\quad \times \begin{bmatrix} R & S_{(i-1)N - \frac{i(i+1)}{2} + j} \\ * & R \end{bmatrix} \begin{bmatrix} \sqrt{\frac{\alpha_j}{\alpha_i}} h_i \\ -\sqrt{\frac{\alpha_i}{\alpha_j}} h_j \end{bmatrix} \\
&\geq 0.
\end{aligned}
$$

This ends the proof.

The following lemma is about the time derivative of a double integral term with variable upper and lower limits, which will be used in the proof of our main result.

Lemma 5.4 *For continuous functions $\bar{f}(t)$, $a(t)$ and $b(t)$, the double integration $w(t) = \int_{a(t)}^{b(t)} \int_{t-\theta}^{t} \bar{f}(s)\mathrm{d}s\mathrm{d}\theta$ is well-defined. Then the following equality holds:*

$$\frac{\mathrm{d}}{\mathrm{d}t}w(t) = (b(t) - a(t))\bar{f}(t) - (1 - \dot{b}(t)) \int_{t-b(t)}^{t-a(t)} \bar{f}(s)\mathrm{d}s + (\dot{b}(t) - \dot{a}(t)) \int_{t-a(t)}^{t} \bar{f}(s)\mathrm{d}s. \quad (5.5)$$

Proof According to the derivative formula of variable upper and lower limits of integral,

$$\frac{\mathrm{d}}{\mathrm{d}t} \int_{a(t)}^{b(t)} g(\theta, t)\mathrm{d}\theta = \int_{a(t)}^{b(t)} \frac{\partial g(\theta, t)}{\partial t}\mathrm{d}\theta + g(b(t), t)\frac{\mathrm{d}}{\mathrm{d}t}b(t) - g(a(t), t)\frac{\mathrm{d}}{\mathrm{d}t}a(t), \quad (5.6)$$

and let $g(\theta, t) = \int_{t-\theta}^{t} \bar{f}(s)\mathrm{d}s$, we can derive the Lemma 5.4.

5.3 Global Asymptotical Stability Result

Now we present a novel delay-dependent stability criterion for (5.3) in this section.

Theorem 5.5 *The origin of system (5.3) is globally asymptotically stable, if for given diagonal matrices Δ_1, Δ_2, and positive scalars τ_m, τ_M, ρ_m, ρ_M, β_k, γ_j, n_1 and n_2, there exist symmetric definite matrices $P > 0$, $W_s > 0$, $S_s > 0$, $Q_k > 0$, $R_j > 0$, positive definite diagonal matrices $V > 0$, $U > 0$, $\Lambda_1 > 0$ and $\Lambda_2 > 0$, and matrices G_r and J such that the following inequalities hold, $s = 1, 2, 3$:*

$$\max\{\beta_k, \gamma_j\}\rho_M < 1, \quad (5.7)$$

$$\Omega - \Xi_1 - \Xi_2 < 0, \quad (5.8)$$

$$\begin{bmatrix} S_3 & J \\ * & S_3 \end{bmatrix} \geq 0, \quad (5.9)$$

$$\begin{bmatrix} S_2 & G_r \\ * & S_2 \end{bmatrix} \geq 0, \quad (5.10)$$

where $0 \leq \beta_k < \beta_{k+1} \leq 1$, $0 \leq \gamma_{j+1} < \gamma_j \leq 1$, $k = 0, \ldots, n_1 - 1$, $j = 0, \ldots, n_2 - 1$, $r = 1, \ldots, \frac{(n_1+n_2)(n_1+n_2-1)}{2}$,

$$\Omega = \begin{bmatrix} \omega_1 & \chi_1 & 0 & 0 & 0 & 0 & \chi_2 & \chi_3 \\ * & \omega_2 & 0 & 0 & 0 & 0 & 0 & 0 \\ * & * & \omega_3 & 0 & 0 & 0 & 0 & 0 \\ * & * & * & \omega_4 & 0 & 0 & 0 & \chi_4 \\ * & * & * & * & \omega_5 & 0 & 0 & 0 \\ * & * & * & * & * & \omega_6 & 0 & 0 \\ * & * & * & * & * & * & \omega_7 & \chi_9 \\ * & * & * & * & * & * & * & \omega_8 \end{bmatrix}, \tag{5.11}$$

$\Xi_1 = \{\Xi_{1_{pq}}\} = \{\Xi'_{pq} - \Xi'_{(p-n)q} - \Xi'_{p(q-n)} + \Xi'_{(p-n)(q-n)}\}$, $p, q = 1, 2, \ldots, (n_1 + n_2 + 4)n$ *and* $\Xi_{pq} = 0$ *when* p *or* $q \leq 0$, ($\Xi_1 = \{\Xi_{1_{pq}}\}$ *means the element of matrix* Ξ_1 *in row* p, *column* q *is* $\Xi_{1_{pq}}$), *and*

$$\begin{aligned}
\omega_1 &= PA + A^T P^T + W_1 + W_2 - S + A^T S' A - \Delta_1 V \\
&\quad - \Sigma \Lambda_1 A - (\Sigma \Lambda_1 A)^T + \Gamma \Lambda_2 A + (\Gamma \Lambda_2 A)^T, \\
\omega_2 &= -W_1 + Q_0 - S_1, \omega_6 = -W_2 - R_{n_2-1}, \\
\omega_4 &= -(1 - \rho_M) Q_{n_1-1} + (1 - \rho_m) R_0 - \Delta_1 U, \\
\omega_7 &= B^T S' B + W_3 - V - \Sigma \Lambda_1 B - (\Sigma \Lambda_1 B)^T \\
&\quad + \Gamma \Lambda_2 B + (\Gamma \Lambda_2 B)^T, \\
\omega_8 &= C^T S' C - (1 - \rho_M) W_3 - U, \quad \chi_1 = S_1, \\
\chi_2 &= A^T S' B + PB + \Delta_2 V + A^T \Lambda_1^T - A^T \Lambda_2^T \\
&\quad - \Sigma \Lambda_1 B + \Gamma \Lambda_2 B, \\
\chi_3 &= A^T S' C + PC - \Sigma \Lambda_1 C + \Gamma \Lambda_2 C, \\
\chi_4 &= \Delta_2 U, \chi_9 = B^T S' C + C^T \Lambda_1^T - C^T \Lambda_2^T, \\
S' &= \tau_m^2 S_1 + (\tau_M - \tau_m)^2 S_2 + \frac{(1 - \gamma_1)^2 (\tau_M - \tau_m)^2}{1 - \gamma_1 \rho_M} S_3, \\
\omega_3 &= \mathrm{diag}\Big(- (1 - \beta_1 \rho_M) Q_0 + (1 - \beta_1 \rho_m) Q_1, \ldots, \\
&\quad - (1 - \beta_{n_1-1} \rho_M) Q_{n_1-2} + (1 - \beta_{n_1-1} \rho_m) Q_{n_1-1} \Big), \\
\omega_5 &= \mathrm{diag}\Big(- (1 - \gamma_1 \rho_M) R_0 + (1 - \gamma_1 \rho_m) R_1, \ldots, \\
&\quad - (1 - \gamma_{n_2-1} \rho_M) R_{n_2-2} + (1 - \gamma_{n_2-1} \rho_m) R_{n_2-1} \Big),
\end{aligned}$$

$$\Xi_2 = \begin{bmatrix} 0_{n_1 n \times n_1 n} & 0_{n_1 n \times n} & 0_{n_1 n \times n} & 0_{n_1 n \times n} & \Theta_1 \\ * & S_3 & J - S_3 & -J & \Theta_2 \\ * & * & -2S_3 - J & J - S_3 & \Theta_3 \\ * & * & * & -S_3 & \Theta_4 \\ * & * & * & * & \Theta_5 \end{bmatrix},$$

$$\Theta_1 = 0_{n_1 n\times(n_2+1)n},\quad \Theta_2 = 0_{n\times(n_2+1)n}, \Theta_3 = 0_{n\times(n_2+1)n},$$
$$\Theta_4 = 0_{n\times(n_2+1)n},\quad \Theta_5 = 0_{(n_2+1)n\times(n_2+1)n},$$

$$\{\Xi'_{pq}\} = \Xi' = \begin{bmatrix} 0_{2n\times 2n} & 0_{2n\times(n_1+n_2)} & 0_{2n\times 2n} \\ * & \Xi'' & 0_{(n_1+n_2)\times 2n} \\ * & * & 0_{2n\times 2n} \end{bmatrix},$$

$$\Xi'' = \begin{bmatrix} S_2 & G_1 & G_2 & \cdots & G_{(n_1+n_2-1)} \\ * & S_2 & G_{(n_1+n_2)} & \cdots & G_{(2n_1+2n_2-3)} \\ * & * & \ddots & \ddots & \vdots \\ * & * & * & S_2 & G_{\frac{(n_1+n_2-1)(n_1+n_2)}{2}} \\ * & * & * & * & S_2 \end{bmatrix}.$$

Proof Define a *Lyapunov function* as

$$V_0(t) = V_1(t) + V_2(t) + V_3(t), \tag{5.12}$$

where

$$\begin{aligned} V_1(t) &= x^T(t)Px(t) + \int_{t-\tau_m}^{t} x^T(s)W_1x(s)\mathrm{d}s \\ &\quad + \int_{t-\tau_M}^{t} x^T(s)W_2x(s)\mathrm{d}s \\ &\quad + 2\sum_{i=1}^{n}\lambda_{1i}\int_0^{x_i(t)}(f_i(s) - \delta_i^- s)\mathrm{d}s \\ &\quad + 2\sum_{i=1}^{n}\lambda_{2i}\int_0^{x_i(t)}(\delta_i^+ s - f_i(s))\mathrm{d}s, \\ V_2(t) &= \int_{t-\tau(t)}^{t} f^T(x(s))W_3f(x(s))\mathrm{d}s \\ &\quad + \tau_m\int_{-\tau_m}^{0}\int_{t+s}^{t}\dot{x}^T(\theta)S_1\dot{x}(\theta)\mathrm{d}\theta\mathrm{d}s \\ &\quad + (\tau_M - \tau_m)\int_{-\tau_M}^{-\tau_m}\int_{t+s}^{t}\dot{x}^T(\theta)S_2\dot{x}(\theta)\mathrm{d}\theta\mathrm{d}s, \end{aligned}$$

$$V_3(t) = \sum_{k=0}^{n_1-1}\left[\int_{t-\beta_{k+1}\tau(t)-(1-\beta_{k+1})\tau_m}^{t-\beta_k\tau(t)-(1-\beta_k)\tau_m} x^T(s)Q_kx(s)\mathrm{d}s\right]$$

$$+\sum_{j=0}^{n_2-1}\left[\int_{t-\gamma_{j+1}\tau(t)-(1-\gamma_{j+1})\tau_M}^{t-\gamma_j\tau(t)-(1-\gamma_j)\tau_M} x^T(s)R_jx(s)\mathrm{d}s\right]$$
$$+\frac{(1-\gamma_1)(\tau_M-\tau_m)}{1-\gamma_1\rho_M}\int_{-\gamma_1\tau(t)-(1-\gamma_1)\tau_M}^{-\gamma_1\tau(t)-(1-\gamma_1)\tau_m}\int_{t+s}^{t}\dot{x}^T(\theta)S_3\dot{x}(\theta)\mathrm{d}\theta\mathrm{d}s,$$

where $\beta_0=0,\ \beta_{n_1}=1, \gamma_0=1,\ \gamma_{n_2}=0, \beta_{n_1-1}=\gamma_1, n_1$ and n_2 are the numbers of partitioning the intervals $[t-\tau(t), t-\tau_m]$ and $[t-\tau_M, t-\tau(t)]$, respectively.

Time derivative of the Lyapunov function (5.12) along the solutions to (5.3) is computed as follows:

$$\begin{aligned}\dot{V}_1+\dot{V}_2 \le{}& 2x^T(t)P\Big[Ax(t)+Bf(x(t))+Cf(x(t-\tau(t)))\Big]\\
&+\sum_{l=1}^{2}x^T(t)W_lx(t)-x^T(t-\tau_m)W_1x(t-\tau_m)\\
&-x^T(t-\tau_M)W_2x(t-\tau_M)\\
&+2[f(x(t))-\Sigma x(t)]^T\Lambda_1\dot{x}(t)\\
&+2[\Gamma x(t)-f(x(t))]^T\Lambda_2\dot{x}(t)\\
&+\dot{x}^T(t)\Big[\tau_m^2S_1+(\tau_M-\tau_m)^2S_2\Big]\dot{x}(t)\\
&-\tau_m\int_{t-\tau_m}^{t}\dot{x}^T(s)S_1\dot{x}(s)\mathrm{d}s+f^T(x(t))W_3f(x(t))\\
&-(\tau_M-\tau_m)\int_{t-\tau_M}^{t-\tau_m}\dot{x}^T(s)S_2\dot{x}(s)\mathrm{d}s\\
&-(1-\rho_M)f^T(x(t-\tau(t)))W_3f(x(t-\tau_i(t))),\end{aligned}\tag{5.13}$$

$$\begin{aligned}\dot{V}_3={}&\sum_{k=0}^{n_1-1}\Big[(1-\beta_k\dot{\tau}(t))x^T\Big(t-\beta_k\tau(t)-(1-\beta_k)\tau_m\Big)\\
&\times Q_kx\Big(t-\beta_k\tau(t)-(1-\beta_k)\tau_m\Big)\\
&-(1-\beta_{k+1}\dot{\tau}(t))x^T\Big(t-\beta_{k+1}\tau(t)-(1-\beta_{k+1})\tau_m\Big)\\
&\times Q_kx\Big(t-\beta_{k+1}\tau(t)-(1-\beta_{k+1})\tau_m\Big)\Big]\\
&+\sum_{j=0}^{n_2-1}\Big[(1-\gamma_j\dot{\tau}(t))x^T\Big(t-\gamma_j\tau(t)-(1-\gamma_j)\tau_m\Big)\\
&\times R_jx\Big(t-\gamma_j\tau(t)-(1-\gamma_j)\tau_m\Big)\end{aligned}$$

$$
\begin{aligned}
&- (1-\gamma_{j+1}\dot{\tau}(t))x^T\Big(t-\gamma_{j+1}\tau(t)-(1-\gamma_{j+1})\tau_m\Big)\\
&\times R_j x\Big(t-\gamma_{j+1}\tau(t)-(1-\gamma_{j+1})\tau_m\Big)\Big]\\
&+\frac{(1-\gamma_1)^2(\tau_M-\tau_m)^2}{1-\gamma_1\rho_M}\dot{x}^T(t)S_3\dot{x}(t)\\
&-\frac{(1-\gamma_1)(\tau_M-\tau_m)}{1-\gamma_1\rho_M}\int_{t-\gamma_1\tau(t)-(1-\gamma_1)\tau_M}^{t-\gamma_1\tau(t)-(1-\gamma_1)\tau_m}\dot{x}^T(s)S_3\dot{x}(s)\mathrm{d}s\\
&+\frac{(1-\gamma_1)(\tau_M-\tau_m)}{1-\gamma_1\rho_M}\gamma_1\dot{\tau}(t)\int_{t-\gamma_1\tau(t)-(1-\gamma_1)\tau_M}^{t-\gamma_1\tau(t)-(1-\gamma_1)\tau_m}\dot{x}^T(s)S_3\dot{x}(s)\mathrm{d}s\\
&\le\sum_{k=0}^{n_1-1}\Big[(1-\beta_k\rho_m)x^T\Big(t-\beta_k\tau(t)-(1-\beta_k)\tau_m\Big)\\
&\times Q_k x\Big(t-\beta_k\tau(t)-(1-\beta_k)\tau_m\Big)\\
&-(1-\beta_{k+1}\rho_M)x^T\Big(t-\beta_{k+1}\tau(t)-(1-\beta_{k+1})\tau_m\Big)\\
&\times Q_k x\Big(t-\beta_{k+1}\tau(t)-(1-\beta_{k+1})\tau_m\Big)\Big]
\end{aligned}
$$

$$
\begin{aligned}
&+\sum_{j=0}^{n_2-1}\Big[(1-\gamma_j\rho_m)x^T\Big(t-\gamma_j\tau(t)-(1-\gamma_j)\tau_m\Big)\\
&\times R_j x\Big(t-\gamma_j\tau(t)-(1-\gamma_j)\tau_m\Big)\\
&-(1-\gamma_{j+1}\rho_M)x^T\Big(t-\gamma_{j+1}\tau(t)-(1-\gamma_{j+1})\tau_m\Big)\\
&\times R_j x\Big(t-\gamma_{j+1}\tau(t)-(1-\gamma_{j+1})\tau_m\Big)\Big]\\
&+\frac{(1-\gamma_1)^2(\tau_M-\tau_m)^2}{1-\gamma_1\rho_M}\dot{x}^T(t)S_3\dot{x}(t)\\
&-(1-\gamma_1)(\tau_M-\tau_m)\int_{t-\gamma_1\tau(t)-(1-\gamma_1)\tau_M}^{t-\gamma_1\tau(t)-(1-\gamma_1)\tau_m}\dot{x}^T(s)S_3\dot{x}(s)\mathrm{d}s, \qquad (5.14)
\end{aligned}
$$

where we have used Lemma 5.4 to compute the derivative of double integral term in $V_3(t)$.

$$f_k = \Big[x\Big(t - \beta_k\tau(t) - (1-\beta_k)\tau_m\Big) - x\Big(t - \beta_{k+1}\tau(t) - (1-\beta_{k+1})\tau_m\Big)\Big]^T,$$

$$g_j = \Big[x\Big(t - \gamma_j\tau(t) - (1-\gamma_j)\tau_m\Big) - x\Big(t - \gamma_{j+1}\tau(t) - (1-\gamma_{j+1})\tau_m\Big)\Big]^T,$$

$$\bar{\Omega} = \begin{bmatrix} S_2 & G_1 & G_2 & G_3 & \cdots & \cdots & G_{n_1+n_2-2} & G_{n_1+n_2-1} \\ * & S_2 & G_{n_1+n_2} & G_{n_1+n_2+1} & G_{n_1+n_2+2} & \cdots & G_{2(n_1+n_2)-4} & G_{2(n_1+n_2)-3} \\ * & * & \ddots & \vdots & \vdots & \vdots & \vdots & \vdots \\ * & * & * & \ddots & \vdots & \vdots & \vdots & \vdots \\ * & * & * & * & \ddots & \vdots & \vdots & \vdots \\ * & * & * & * & * & \ddots & \vdots & \vdots \\ * & * & * & * & * & * & S_2 & G_{\frac{(n_1+n_2)(n_1+n_2-1)}{2}} \\ * & * & * & * & * & * & * & S_2 \end{bmatrix}, \tag{5.15}$$

Furthermore, according to Assumption 5.1, the following conditions hold [7, 10]

$$\bar{\omega}_1^T(t)\begin{bmatrix} \Delta_1 V & -\Delta_2 V \\ -\Delta_2 V & V \end{bmatrix}\bar{\omega}_1(t) \le 0, \tag{5.16}$$

$$\bar{\omega}_2^T(t)\begin{bmatrix} \Delta_1 U & -\Delta_2 U \\ -\Delta_2 U & U \end{bmatrix}\bar{\omega}_2(t) \le 0, \tag{5.17}$$

where $\Delta_1 = \text{diag}(\delta_i^-\delta_i^+)$ and $\Delta_2 = \text{diag}(\frac{\delta_i^-+\delta_i^+}{2})$, $\bar{\omega}_1(t) = [x^T(t), f^T(x(t))]^T$, $\bar{\omega}_2(t) = [x^T(t-\tau(t)), f^T(x(t-\tau(t)))]^T$, V and U are two positive diagonal matrices, respectively.

On the other hand, from Lemma 5.2 we know that

$$\begin{aligned} &-(\tau_M - \tau_m)\int_{t-\tau_M}^{t-\tau_m} \dot{x}^T(s)S_2\dot{x}(s)\mathrm{d}s \\ &= -\sum_{k=0}^{n_1-1}\Bigg\{\frac{\tau_M - \tau_m}{(\beta_{k+1}-\beta_k)\tau(t) + (\beta_k - \beta_{k+1})\tau_m} \\ &\times [(\beta_{k+1}-\beta_k)\tau(t) + (\beta_k - \beta_{k+1})\tau_m] \\ &\times \int_{t-\beta_{k+1}\tau(t)-(1-\beta_{k+1})\tau_m}^{t-\beta_k\tau(t)-(1-\beta_k)\tau_m} \dot{x}^T(s)S_2\dot{x}(s)\mathrm{d}s\Bigg\} \\ &-\sum_{j=0}^{n_2-1}\Bigg\{\frac{\tau_M - \tau_m}{(\gamma_{j+1}-\gamma_j)\tau(t) + (\gamma_j - \gamma_{j+1})\tau_M} \\ &\times [(\gamma_{j+1}-\gamma_j)\tau(t) + (\gamma_j - \gamma_{j+1})\tau_M] \end{aligned}$$

$$
\begin{aligned}
&\times \int_{t-\gamma_{j+1}\tau(t)-(1-\gamma_{j+1})\tau_M}^{t-\gamma_j\tau(t)-(1-\gamma_j)\tau_M} \dot{x}^T(s)S_2\dot{x}(s)\mathrm{d}s\Bigg\} \\
\leq &-\sum_{k=0}^{n_1-1}\Bigg\{\frac{\tau_M-\tau_m}{(\beta_{k+1}-\beta_k)\tau(t)+(\beta_k-\beta_{k+1})\tau_m} \\
&\times\Big[x(t-\beta_k\tau(t)-(1-\beta_k)\tau_m) \\
&-x(t-\beta_{k+1}\tau(t)-(1-\beta_{k+1})\tau_m)\Big]^T \\
&\times S_2\Big[x(t-\beta_k\tau(t)-(1-\beta_k)\tau_m) \\
&-x(t-\beta_{k+1}\tau(t)-(1-\beta_{k+1})\tau_m)\Big]\Bigg\} \\
&-\sum_{j=0}^{n_2-1}\Bigg\{\frac{\tau_M-\tau_m}{(\gamma_{j+1}-\gamma_j)\tau(t)+(\gamma_j-\gamma_{j+1})\tau_M} \\
&\times\Big[x(t-\gamma_j\tau(t)-(1-\gamma_j)\tau_M) \\
&-x(t-\gamma_{j+1}\tau(t)-(1-\gamma_{j+1})\tau_M)\Big]^T \\
&\times S_2\Big[x(t-\gamma_j\tau(t)-(1-\gamma_j)\tau_M) \\
&-x(t-\gamma_{j+1}\tau(t)-(1-\gamma_{j+1})\tau_M)\Big]\Bigg\},
\end{aligned} \tag{5.18}
$$

$$
\begin{aligned}
&-\tau_m\int_{t-\tau_m}^{t}\dot{x}^T(s)S_1\dot{x}(s)\mathrm{d}s \\
\leq &-[x(t)-x(t-\tau_m)]^T S_1[x(t)-x(t-\tau_m)],
\end{aligned} \tag{5.19}
$$

and

$$
\begin{aligned}
&-(1-\gamma_1)(\tau_M-\tau_m)\int_{t-\gamma_1\tau(t)-(1-\gamma_1)\tau_M}^{t-\gamma_1\tau(t)-(1-\gamma_1)\tau_m}\dot{x}^T(s)S_3\dot{x}(s)\mathrm{d}s \\
=&-(1-\beta_{n_1-1})(\tau_M-\tau_m)\int_{t-\tau(t)}^{t-\beta_{n_1-1}\tau(t)-(1-\beta_{n_1-1})\tau_m}\dot{x}^T(s)S_3\dot{x}(s)\mathrm{d}s \\
&-(1-\gamma_1)(\tau_M-\tau_m)\int_{t-\gamma_1\tau(t)-(1-\gamma_1)\tau_M}^{t-\tau(t)}\dot{x}^T(s)S_3\dot{x}(s)\mathrm{d}s \\
\leq &-\frac{(1-\gamma_1)(\tau_M-\tau_m)}{(1-\beta_{n_1-1})(\tau(t)-\tau_m)}\Big[x\Big(t-\beta_{n_1-1}\tau(t)
\end{aligned}
$$

$$
\begin{aligned}
&- (1-\beta_{n_1-1})\tau_m\Big) - x(t-\tau(t))\Big]^T S_3 \\
&\times \Big[x\Big(t-\beta_{n_1-1}\tau(t) - (1-\beta_{n_1-1})\tau_m\Big) - x(t-\tau(t))\Big] \\
&- \frac{(1-\gamma_1)(\tau_M-\tau_m)}{(1-\gamma_1)(\tau_M-\tau(t))}\Big[x\Big(t-\gamma_1\tau(t) \\
&- (1-\gamma_1)\tau_M\Big) - x(t-\tau(t))\Big]^T S_3 \\
&\times \Big[x\Big(t-\gamma_1\tau(t)-(1-\gamma_1)\tau_M\Big) - x(t-\tau(t))\Big].
\end{aligned}
\tag{5.20}
$$

Because

$$
\begin{aligned}
1 = &\sum_{k=0}^{n_1-1}\left[\frac{(\beta_{k+1}-\beta_k)\tau(t) + (\beta_k-\beta_{k+1})\tau_m}{\tau_M-\tau_m}\right] \\
&+ \sum_{j=0}^{n_2-1}\left[\frac{(\gamma_{j+1}-\gamma_j)\tau(t) + (\gamma_j-\gamma_{j+1})\tau_M}{\tau_M-\tau_m}\right],
\end{aligned}
\tag{5.21}
$$

$$
\beta_{n_1-1} = \gamma_1, \tag{5.22}
$$

$$
1 = \frac{\tau(t)-\tau_m}{\tau_M-\tau_m} + \frac{\tau_M-\tau(t)}{\tau_M-\tau_m}, \tag{5.23}
$$

and let $F_x = (f_0^T, \ldots, f_{n_1-1}^T, g_0^T, \ldots, g_{n_2-1}^T)^T$,

$$
\begin{aligned}
h_1 &= \Big[x\Big(t-\beta_{n_1-1}\tau(t) - (1-\beta_{n_1-1})\tau_M\Big) - x(t-\tau(t))\Big]^T, \\
h_2 &= \Big[x\Big(t-\gamma_1\tau(t) - (1-\gamma_1)\tau_M\Big) - x(t-\tau(t))\Big]^T,
\end{aligned}
$$

applying Lemma 5.3 to inequalities (5.18) and (5.20), we get the following inequalities, respectively:

$$
\text{Inequality}|_{(5.18)} \le F_x^T \bar{\Omega} F_x \tag{5.24}
$$

$$
\text{Inequality}|_{(5.20)} \le (h_1^T \;\; h_2^T)\begin{bmatrix} S_3 & J \\ J^T & S_3 \end{bmatrix}(h_1^T \;\; h_2^T)^T, \tag{5.25}
$$

where f_k, g_j and $\bar{\Omega}$ are defined in (5.15), respectively, $k = 1, \ldots, n_1 - 1$, $j = 1, \ldots, n_2 - 1$.

Combining (5.13), (5.14), (5.16), (5.19), (5.24), with (5.25), and by some operations, we have $\dot{V}_0(t) < \zeta(t)^T(\Omega + \Xi_1 + \Xi_2)\zeta(t) < 0$ for $\forall \zeta(t) \neq 0$, if the inequalities (5.7)–(5.10) hold, where

$$
\begin{aligned}
\zeta(t)^T = &\Big[x^T(t), x^T\Big(t-\beta_0\tau(t) - (1-\beta_0)\tau_m\Big), \ldots, \\
&x^T\Big(t-\beta_k\tau(t) - (1-\beta_k)\tau_m\Big), \ldots,
\end{aligned}
$$

$$
\begin{aligned}
& x^T\Big(t-\beta_{n_1}\tau(t)-(1-\beta_{n_1})\tau_m\Big), \\
& x^T\Big(t-\gamma_1\tau(t)-(1-\gamma_1)\tau_M\Big),\ldots, \\
& x^T\Big(t-\gamma_j\tau(t)-(1-\gamma_j)\tau_M\Big),\ldots, \\
& x^T\Big(t-\gamma_{n_2}\tau(t)-(1-\gamma_{n_2})\tau_M\Big), \\
& f^T(x(t)),\ f^T(x(t-\tau(t)))\Big].
\end{aligned} \tag{5.26}
$$

This completes the proof.

Remark 5.6 In [2, 18], it is assumed that $\tau(t)\in[\tau_{j-1},\tau_j]$ for some $j\in[1,2,\ldots,n_0]$, where n_0 is the number of splitting delay interval $[0,\tau_M]$. By deciding which interval $\tau(t)$ belongs to, the corresponding stability criterion is then obtained. In general, the n_0 sets of LMI-based conditions need to be checked for different j. In contrast, we use the secondary partitioning approach to dynamically divide the delay interval, and $\tau(t)$ is naturally incorporated into the state variables with different subinterval, which leads to a compact LMI-based stability criterion.

Remark 5.7 In this chapter, different from the existing construction of Lyapunov function, a double integral term with variable upper and lower limits of integral is used to be the Lyapunov function (see the last term in $V_3(t)$). In order to estimate the derivative of the Lyapunov function, Lemma 5.4 is presented. This kind of Lyapunov function can be effectively combined with the proposed secondary *delay partitioning method*. Since the partitioning coefficients satisfy the equalities (5.21)–(5.23), an extended RCC (see Lemma 5.3) is presented to establish the relations among the states. Combining the secondary delay partitioning method and extended RCC, our approach is effective for the RNN with fast time-varying delay.

Remark 5.8 Note that, among the ways of reducing the conservativeness of the estimation of the upper bound of time delay, delay-slope-dependent method is an effective one [27, 34]. This method includes more information on the slope of neuron activation functions, and can establish the relationship between the time delay upper bound and the slope of neuron activation functions. In contrast, this chapter establishes the relation between the time delay upper bound and the partitioning parameters of subinterval of time-varying delay. Obviously, the methods in [27, 34] and those in this chapter are different, and they reduce the conservativeness of upper bound of time delay in different ways.

Remark 5.9 When the delay interval $[\tau_m,\tau_M]$ is divided into two delay subintervals $[\tau_m,\tau(t)]$ and $[\tau(t),\tau_M]$, how to further divide these two intervals into many subintervals is the key problem to be dealt with. In this chapter, we apply the convex combination method to the intervals $[\tau_m,\tau(t)]$ and $[\tau(t),\tau_M]$ by involving some dynamic weighting parameters, and by constructing a novel functional (see $V_3(t)$), then the secondary delay partitioning is realized. In the proposed method, the interval $[\tau_m, \tau(t)]$ is partitioned into n_1 segments by dynamic parameters β_k $(k = 0, \ldots, n_1 - 1)$, while the interval $[\tau(t), \tau_M]$ is partitioned into n_2 segments by dynamic parameters $\gamma_j (j = 0, \ldots, n_2 - 1)$. Here, n_1 may not be equal to n_2, and β_k and γ_j may be different. In order to obtain some compact expressions when computing the derivative of $V_3(t)$, some restriction conditions are required, for example, $\beta_{n_1-1} = \gamma_1$. How to choose these scalars β_k and γ_i to get the better results is a constrained optimal problem, a challenging problem, and out of the scope of the purpose of the present chapter. In this chapter, we choose these parameters by trial and error.

Remark 5.10 In order to realize the secondary delay partitioning method, some Lyapunov functions should be constructed, for example, $V_2(t)$ and $V_3(t)$. The last terms in $V_2(t)$ and $V_3(t)$ are useful for reducing the conservatism of the results. When one of them is removed, the desired results cannot be obtained. In principle, the integration intervals in $V_3(t)$ are variant or changeable, which implies that this kind of Lyapunov function can be used to tackle the problem of partitioning the fixed delay interval by flexible terminal. This partitioning method in fact is a contraction method by changing the terminals. How to use this method systematically to study the stability problems for RNNs with delay is a future direction.

Remark 5.11 By using different inequality techniques, some different stability analysis methods were proposed [31–34]. For example, a matrix-based quadratic convex approach was introduced in [32]. A double integral term of activation function was utilized as an element of augmented vector in estimating the time-derivative of Lyapunov–Krasovskii functional [33]. A novel approach which divided the bounding of activation function into two subinterval was proposed in [34]. For the case of two additive time-varying delays, the domain of the integral terms in Lyapunov–Krasovskii functional was partitioned into three parts in [31]. Although the methods in [31–34] can reduce the conservativeness of the upper bound of time delay, the terminal of interval time delay is fixed and no further partition is conducted. In contrast, we use the secondary partitioning method to divide the interval time delay by involving some weights, which can lead to the changeable terminal of some subinterval of time delay. The method used in this chapter is different from those in [31–34].

Remark 5.12 In [4, 5], the relationship between the time-varying delay and its upper bounds was taken into account when estimating the upper bound of the derivative of Lyapunov function, in which the free weight matrix method played an important role in establishing the improved delay-dependent stability criteria for neural networks with time-varying interval delay. In [6], by dividing the bounding of activation functions into two parts, an improved stability criterion was proposed. In [7], the uniform delay partitioning method was used on both discrete and distributed delays and a triple integral term was used in the construction of Lyapunov function, in which more number of fractions were considered and some improved stability results were established for neural networks with constant delay. In general, there are many ways to be explored to further reduce the conservativeness of the stability results for neural networks with time-varying delay, for example, the methods in [4–7]. As far as the delay partitioning method is concerned, the method in [3-6] is the first delay partitioning method, while in this chapter we further partition the first delay partitioning interval (called secondary delay partitioning method). This is the main difference between the present chapter and [4–7].

Remark 5.13 Analogous to the analysis method of this chapter, some higher order partitions seem possible for the interval delay, for example, third-order delay partitioning, fourth-order delay partitioning. In this case, how to establish the corresponding stability criterion needs further to be investigated.

5.4 Illustrative Example

In this section, a numerical example will be used to check the effectiveness of the proposed stability criterion.

Example 5.14 Consider the neural networks (5.3), where

$$A = -\text{diag}(1.2769, 0.62310, 0.9230, 0.448),$$

$$B = \begin{bmatrix} -0.0373 & 0.4852 & -0.3351 & 0.2336 \\ -1.6033 & 0.5988 & -0.3224 & 1.2352 \\ 0.3394 & -0.086 & -0.3824 & -0.5785 \\ -0.1311 & 0.3253 & -0.9534 & -0.5015 \end{bmatrix},$$

$$C = \begin{bmatrix} 0.8674 & -1.2405 & -0.5325 & 0.022 \\ 0.0474 & -0.9164 & 0.036 & 0.9816 \\ 1.8495 & 2.6117 & -0.3788 & 0.8428 \\ -2.0413 & 0.5179 & 1.1734 & -0.2775 \end{bmatrix},$$

$$\Sigma = \text{diag}(0, 0, 0, 0),$$

$$\Gamma = \text{diag}(0.1137, 0.1279, 0.7994, 0.2368),$$

This example has often been used to compare the conservativeness of the stability results in the literature, in which the larger the maximum bound of time delay is, the less conservative the derived stability result is. Pertaining to this example, our obtained delay bounds and the detailed comparisons with other results are listed in Table 5.1 ($n_1 = 2, n_2 = 2, \beta_1 = \gamma_1 = 10^{-4}, \tau_m = 0, \rho_m = 0$) and Table 5.2 ($n_1 = 2, n_2 = 2, \beta_1 = \gamma_1 = 0.9999, \tau_m = 1, \rho_m = -\rho_M$), respectively. From these two tables we can see that when the change rate of time-varying delay is small, for example, $\rho_M \le 0.5$, our result will be slightly conservative comparing with some existing results. This point can be shown by the terms ω_3 and ω_5 in Theorem 5.5. When ρ_M is very small, the term $-(1 - \beta_{n_1-1}\rho_M)Q_{n_1-2}$ or $-(1 - \gamma_{n_2-1}\rho_M)R_{n_2-2}$ is nearly equal to $-Q_{n_1-2}$ or $-R_{n_2-2}$, respectively, which means that the parameters β_{n_1-1} and γ_{n_2-1} have not much influence on the change rate of time-varying delay. On the contrary, for the case of fast change rate of time-varying delay, for example,

Table 5.1 Comparisons of delay bound τ_M with the results in some references for different ρ_M

Method	$\dot{\tau}(t)$	0.1	0.5	0.9
Theorem 1 [4]	$\lvert\dot{\tau}(t)\rvert \le \rho_M$	3.2793	2.2245	1.5847
Theorem 2 [5]	$\lvert\dot{\tau}(t)\rvert \le \rho_M$	3.3039	2.5376	2.0853
Theorem 1 [8]	$\lvert\dot{\tau}(t)\rvert \le \rho_M$	3.2819	2.2261	1.6035
Theorem 2 [24]	$\lvert\dot{\tau}(t)\rvert \le \rho_M$	3.4183	2.5943	2.1306
Theorem 2 [20]	$\lvert\dot{\tau}(t)\rvert \le \rho_M$	3.3574	2.5915	2.0779
Theorem 2 [23]	$\lvert\dot{\tau}(t)\rvert \le \rho_M$	3.3981	2.6711	2.1783
Proposition 2 [18]	$\lvert\dot{\tau}(t)\rvert \le \rho_M$	3.5204	2.7167	2.2141
Theorem 1 [2]	$\lvert\dot{\tau}(t)\rvert \le \rho_M$	3.7665	2.6814	2.2274
Theorem 1 [16]	$\lvert\dot{\tau}(t)\rvert \le \rho_M$	3.8428	2.7081	2.2485
Theorem 5.5 in this chapter	$\lvert\dot{\tau}(t)\rvert \le \rho_M$	3.4886	2.6056	2.2522

Table 5.2 Comparisons of delay bound τ_M with the results in some references for different ρ_M

Method	$\dot{\tau}(t)$	0.1	0.9	$\ge$1
Theorem 1 [5]	$\lvert\dot{\tau}(t)\rvert \le \rho_M$	3.3068	2.2736	2.2393
Proposition 3 [18]	$\lvert\dot{\tau}(t)\rvert \le \rho_M$	3.5235	2.3510	2.2740
Proposition 2 [18]	$\lvert\dot{\tau}(t)\rvert \le \rho_M$	3.8025	2.3811	2.3114
Theorem 5.5 in this chapter	$\lvert\dot{\tau}(t)\rvert \le \rho_M$	3.7515	2.4628	2.4554

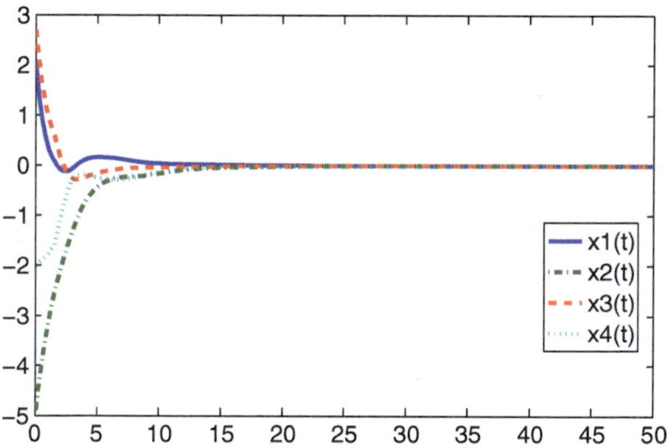

Fig. 5.2 State trajectories of $x_1(t)$ $x_2(t)$ $x_3(t)$ and $x_4(t)$

$\rho_M \ge 0.9$, our result will be more effective than the existing results. In this case, the parameters β_{n_1-1} and γ_{n_2-1} in Theorem 5.5 will have much influence on the change rate of time-varying delay, which also verifies the statements in Remark 5.7. In general, it is impossible to use one method to obtain the best upper bound of time delay for both slow and fast time-varying case simultaneously. Each method is generally suitable for a kind of time-varying delay. Therefore, in this example, we further confirm that our proposed method is suitable for the case of fast time-varying delay.

When $\tau(t) = 1.1261 + 0.3536\sin(0.4620t) + 0.7725\cos(0.5\sqrt{3}\,t)$ (i.e., $\tau_M = 2.2477$, $\rho_M = 0.8324$) and the initial values are randomly chosen, the simulation result is given in Fig. 5.2. Obviously, the concerned neural network is globally asymptotically stable.

5.5 Summary

In this chapter, we have studied the global asymptotic stability for a class of RNNs with time-varying delay. By using the secondary delay partitioning method, extended RCC approach, and a double integral term with variable upper and lower limits as a Lyapunov functional, an LMI-based stability criterion has been established. The proposed secondary delay partitioning method is mainly used to reduce the conservativeness of the stability criterion for RNN with fast change rate of time-varying delay. A numerical example is used to demonstrate the effectiveness of the derived criteria. However, there are many adjustable parameters to be determined in Theorem 5.5, how to optimize these parameters is not an easy work. One way is to design optimal algorithms to solve these parameters, which will be a further research direction.

References

1. Y. Chen, W. Zheng, Stability analysis of time-delay neural networks subject to stochastic perturbations. IEEE Trans. Cybern. **43**(6), 2122–2134 (2013)
2. C. Ge, C. Hua, X. Guan, New delay-dependent stability criteria for neural networks with time-varying delay using delay-decomposition approach. IEEE Trans. Neural Netw. **25**(7), 1378–1383 (2014)
3. C. Ge, C. Hua, X. Guan, New delay-dependent stability criteria for neutral systems with time-varying delay using delay-decomposition approach. Int. J. Control, Autom. Syst. **12**(4), 786–793 (2014)
4. Y. He, G. Liu, D. Rees, M. Wu, Stability analysis for neural networks with time-varying interval delay. IEEE Trans. Neural Netw. **18**(6), 1850–1854 (2007)
5. Y. He, G. Liu, D. Rees, New delay-dependent stability criteria for neural networks with time-varying delay. IEEE Trans. Neural Netw. **18**(1), 310–314 (2007)
6. O. Kwon, M. Park, S. Lee, J. Park, E. Cha, Stability for neural networks with time-varying delays via some new approaches. IEEE Trans. Neural Netw. Learn. Syst. **24**(2), 181–193 (2013)
7. S. Lakshmanan, J. Park, H. Jung, O. Kwon, R. Rakkiyappan, A delay partitioning approach to delay-dependent stability analysis for neutral type neural networks with discrete and distributed delays. Neurocomputing **111**, 81–89 (2013)
8. T. Li, L. Guo, C. Sun, C. Lin, Further result on delay-dependent stability criterion of neural networks with time-varying delays. IEEE Trans. Neural Netw. **19**(4), 726–730 (2008)

9. T. Li, T. Wang, A. Song, S. Fei, Combined convex technique on delay-dependent stability for delayed neural networks. IEEE Trans. Neural Netw. Learn. Syst. **24**(9), 1459–1466 (2013)
10. Y. Liu, Z. Wang, X. Liu, Global exponential stability of generalized recurrent neural networks with discrete and distributed delays. Neural Netw. **19**(5), 667–675 (2006)
11. S. Mou, H. Gao, J. Lam, W. Qiang, A new criterion of delay dependent asymptotic stability for Hopfield neural networks with time delay. IEEE Trans. Neural Netw. **19**(3), 532–535 (2008)
12. S. Mou, H. Gao, W. Qiang, K. Chen, New delay-dependent exponential stability for neural networks with time delay. IEEE Trans. Syst., Man, Cybern. Part B: Cybern. **38**(2), 571–576 (2008)
13. H. Shao, Delay-dependent stability for recurrent neural networks with time-varying delays. IEEE Trans. Neural Netw. **19**(9), 1647–1651 (2008)
14. Z. Wu, J. Lam, H. Su, J. Chu, Stability and dissipativity analysis of static neural networks with time delay. IEEE Trans. Neural Netw. Learn. Syst. **23**(2), 199–210 (2012)
15. J. Jian, B. Wang, Stability analysis in Lagrange sense for a class of BAM neural networks of neutral type with multiple time-varying delays. Neurocomputing **149**, 930–939 (2015)
16. H. Zeng, Y. He, M. Wu, C. Zhang, Complete delay-decomposing approach to asymptotic stability for neural networks with time-varying delays. IEEE Trans. Neural Netw. **22**(5), 806–812 (2011)
17. C. Zhang, Y. He, L. Jiang, Q. Wu, M. Wu, Delay-dependent stability criteria for generalized neural networks with two delay components. IEEE Trans. Neural Netw. Learn. Syst. **25**(7), 1263–1276 (2014)
18. X. Zhang, Q. Han, New Lyapunov-Krasovskii functionals for global asymptotic stability of delayed neural networks. IEEE Trans. Neural Netw. **20**(3), 533–539 (2009)
19. X. Zhang, Q. Han, Global asymptotic stability for a class of generalized neural networks with interval time-varying delays. IEEE Trans. Neural Netw. **22**(8), 1180–1192 (2011)
20. H. Zhang, Z. Liu, G. Huang, Z. Wang, Novel weighting delay- based stability criteria for recurrent neural networks with time varying delay. IEEE Trans. Neural Netw. **21**(1), 91–106 (2010)
21. H. Zhang, F. Yang, X. Liu, Q. Zhang, Stability analysis for neural networks with time-varying delay based on quadratic convex combination. IEEE Trans. Neural Netw. Learn. Syst. **24**(4), 513–521 (2013)
22. B. Wang, Y. Zeng, J. Cheng, Further improvement in delay-dependent stability criteria for continuous-time systems with time-varying delays. Neurocomput. **147**, 324–329 (2015)
23. C. Zheng, H. Zhang, Z. Wang, New delay-dependent global exponential stability criterion for cellular-type neural networks with time-varying delays. IEEE Trans. Circuits Syst. II **56**(3), 250–254 (2009)
24. X. Zhu, G. Yang, New delay-dependent stability results for neural networks with time-varying delay. IEEE Trans. Neural Netw. **19**(10), 1783–1791 (2008)
25. H. Chen, S. Zhong, J. Shao, Exponential stability criterion for interval neural networks with discrete and distributed delays. Appl. Math. Comput. **250**, 121–130 (2015)
26. Z. Wang, L. Liu, Q. Shan, H. Zhang, Stability criteria for recurrent neural networks with time-varying delay based on secondary delay partitioning method. IEEE Transactions on Neural Networks and Learning Systems (2015). doi:10.1109/TNNLS.2014.2387434, (to appear in)
27. T. Li, W. Xing Zheng, C. Lin, Delay-slope-dependent stability results of recurrent neural networks. IEEE Trans. Neural Netw. **22**(12), 2138–2143 (2011)
28. P. Park, J. Ko, C. Jeong, Reciprocally convex approach to stability of systems with time-varying delays. Automatica **47**(1), 235–238 (2011)
29. L. Wu, Z. Feng, J. Lam, Stability and synchronization of discrete-time neural networks with switching parameters and time-varying delays. IEEE Trans. Neural Netw. Learn. Syst. **24**(12), 1957–1972 (2013)
30. T. Wang, M. Xue, S. Fei, T. Li, Triple Lyapunov functional technique on delay-dependent stability for discrete-time dynamical networks. Neurocomputing **122**, 221–228 (2013)
31. J. Ko, P. Park, Reciprocally convex approach for the stability of networked control systems. Lecture Notes in Electrical Engineering **110**, 1–9 (2012)

32. X. Zhang, Q. Han, Global asymptotic stability analysis for delayed neural networks using a matrix-based quadratic convex approach. Neural Netw. **54**, 57–69 (2014)
33. O. Kwon, M. Park, J. Park, S. Lee, E. Cha, On stability analysis for neural networks with interval time-varying delays via some new augmented Lyapunov-Krasovskii functional. Commun. Nonlinear Sci. Numer Simul. **19**, 3184–3201 (2014)
34. O. Kwon, S. Lee, J. Park, E. Cha, New approaches on stability criteria for neural networks with interval time-varying delays. Appl. Math. Comput. **218**, 9953–9964 (2012)

Chapter 6
LMI-Based Stability Criteria for Static Neural Networks

In Chaps. 1 and 5, stability problems based on LMI have been investigated for a kind of local field RNNs with delay. This kind of local field RNNs is usually used in pattern recognition and associative memory. In fact, there is another kind of RNNs, which is called static neural networks. Static neural networks have widely been applied in optimization problems. Therefore, in this chapter, based on the result in [25], we will establish some LMI-based stability results for static neural networks.

6.1 Introduction

In the recent years, neural networks have been found a lot of successful applications in many fields, such as pattern recognition, associative memories, signal processing, fixed-point computations, and so on. When designing a neural network or implementing it by VLSI (very large scale integrated) electronic circuit in practice, stability is frequently one of the preconditions which have to be concerned with. Since time delay is unavoidably in real-world systems and is often an important source of oscillation and instability, the stability analysis of neural networks with time delays has emerged as a research topic of primary significance in the past years [1–25].

As reported in [6, 12], recurrent neural networks can be classified as static neural networks and local field networks. In the past few decades, the stability analysis problem has been studied thoroughly for the latter. However, it has received little attention for the former. As reviewed in [3, 4, 7], many neural networks exhibiting short-term memory are modeled by non-invertible networks, such as the oculomotor integrator or the head direction system [5]. That is, local field neural network models and static neural network models are not always equivalent. For the static neural networks without delays, exponential stability conditions were obtained in [2] when the connection weigh matrix is symmetric, while the robustly exponential stability of the unique equilibrium was studied in [11], where an LMI method was employed. As for the static neural networks with delays, reference [4] obtains a

Z. Wang et al., *Qualitative Analysis and Control of Complex Neural Networks with Delays*, Studies in Systems, Decision and Control 34,
DOI 10.1007/978-3-662-47484-6_6

delay-independent stability criterion which ensures the globally asymptotic stability of the unique equilibrium of the network with a constant time delay. In [3], by use of nonlinear measure approach, a delay-independent exponential stability criterion is developed for recurrent neural networks with constant time delay. Reference [7] researches the uniqueness and *delay-dependent* exponential stability condition for neural networks with a constant time delay. Reference [5] discusses the globally asymptotic stability of the network with a time-varying delay which is also delay-dependent. However, there is still some conservatism in these analysis results.

In this chapter, in order to reduce the conservativeness of the previous results, we construct a Lyapunov–Krasovskii functional and derive some new sufficient conditions for the *globally exponential stability* of the unique equilibrium of the static neural network, which are delay-dependent and computationally efficient [25]. Since the free weight matrix approach [1, 9] and the Jensen integral inequality [26] are involved, the results are less conservative than some existing ones. An illustrative example is given to demonstrate the effectiveness of the proposed results.

6.2 Problem Formulation

Considering the following *static neural networks* with time-varying delays:

$$\dot{x}(t) = -Ax(t) + f(Wx(t - \tau(t)) + J), \tag{6.1}$$

where $x(t) = (x_1(t), x_2(t), ..., x_n(t))^T \in \mathbb{R}^n$ is the neural state vector, $A = \text{diag}\{a_1, a_2, ..., a_n\}$ is a positive diagonal matrix, $W = (w_{ij})_{n\times n}$ is a known constant matrix, $0 \leq \tau(t) \leq h$ is the time-varying delay, where h is constant. $J = (J_1, J_2, ..., J_n)^T$ is the constant external input vector, and $f(Wx(t - \tau(t))) = (f_1(W_1x(t - \tau(t))), f_2(W_2x(t - \tau(t))), ..., f_n(W_nx(t - \tau(t))))^T \in \mathbb{R}^n$ denotes the neural activation function, where W_j denotes the jth row of matrix W. It is assumed that $f_j(s)$ is continuous, and there exist constants σ_j such that

$$0 \leq \frac{f_j(s_1) - f_j(s_2)}{s_1 - s_2} \leq \sigma_j, \ j = 1, 2, ..., n, \tag{6.2}$$

for any $s_1, s_2 \in \mathbb{R}, s_1 \neq s_2$.

Moreover, we assume that the initial condition of system (6.1) has the form

$$x_i(t) = \phi_i(t), \quad t \in [-h, 0]$$

where $\phi_i(t)(i = 1, 2, ..., n)$ is continuous function.

In this chapter, we assume that system (6.1) always has an equilibrium point x^*. In order to prove the *exponential stability* of the equilibrium point x^* of system (6.1), we will first simplify system (6.1) as follows. Let $u(\cdot) = x(\cdot) - x^*$, then we have the *error system*,

$$\dot{u}(t) = -Au(t) + g(Wu(t - \tau(t))), \tag{6.3}$$

where $g(Wu(t-\tau(t))) = (g_1(W_1u(t-\tau(t))), g_2(W_2u(t-\tau(t))), ..., g_n(W_nu(t-\tau(t))))^T$, $g_j(W_ju(t)) = f_j(W_ju(t)+W_jx^*+J_j) - f_j(W_jx^*+J_j)$. By assumption (6.2), we can see that

$$0 \leq \frac{g_j(u_j(t))}{u_j(t)} \leq \sigma_j. \tag{6.4}$$

The definition of exponential stability is now given.

Definition 6.1 The system (6.1) is said to be globally exponentially stable if there exist constants $k > 0$ and $K > 1$ such that

$$||x(t)|| \leq K \sup_{-h \leq \theta \leq 0} ||x(\theta)||\, e^{-kt},$$

where k is called the exponential convergence rate.

Clearly, the equilibrium point of system (6.1) is exponentially stable if and only if the zero solution of system (6.3) is exponentially stable.

In order to obtain the results, we need the following lemma:

Lemma 6.2 (see [21, 22]) *Assuming that function $g_j(s)$ satisfies inequality (6.4), then the following inequality holds*

$$\int_v^u (g_j(s) - g_j(v))\mathrm{d}s \leq (u - v)(g_j(u) - g_j(v)).$$

6.3 Main Results

First, we present a new exponential stability result for static neural network (6.3) for $\dot{\tau}(t) \leq \eta < 1$.

Theorem 6.3 *Under the assumption (6.2) and $0 \leq \tau(t) \leq h,\ 0 \leq \dot{\tau}(t) \leq \eta < 1$, given a constant $k > 0$, suppose that there exist positive definite symmetric matrices P_1, nonnegative definite symmetric matrices P_i $(i = 2, 3, 4)$, positive diagonal matrices $T_1, T_2, D = \mathrm{diag}\{d_1, d_2, ..., d_n\}$, real symmetric matrices $Q = [Q_{ij}]_{2\times 2} \geq 0$, $S = [S_{ij}]_{3\times 3} \geq 0$ and real matrices $U^T = \begin{bmatrix} U_1^T & U_2^T & U_3^T \end{bmatrix}$, $Z^T = \begin{bmatrix} Z_1^T & Z_2^T & Z_3^T \end{bmatrix}$ such that the following LMIs hold:*

$$\Xi_1 = \begin{bmatrix} S & U \\ * & e^{-2kh}P_4 \end{bmatrix} \geq 0, \tag{6.5}$$

$$\Xi_2 = \begin{bmatrix} S & Z \\ * & e^{-2kh} P_4 \end{bmatrix} \geq 0, \tag{6.6}$$

$$\Omega = \begin{bmatrix} \Omega_{11} & \Omega_{12} & \Omega_{13} & \Omega_{14} & \Omega_{15} \\ * & \Omega_{22} & 0 & \Omega_{24} & \Omega_{25} \\ * & 0 & Q_{22} - 2T_1 & DW & 0 \\ * & * & * & \Omega_{44} & 0 \\ * & * & 0 & 0 & \Omega_{55} \end{bmatrix} < 0, \tag{6.7}$$

where "" are entries readily inferred by symmetry, and*

$$\begin{aligned}
\Omega_{11} &= 2k(P_1 + 2W^T D\Sigma W) - P_1 A - AP_1 + P_2 + hS_{11} \\
&\quad + hA(P_3 + P_4)A - \frac{1}{h} e^{-2kh} P_3 + U_1 + U_1^T + Q_{11}, \\
\Omega_{12} &= \frac{1}{h} e^{-2kh} P_3 - U_1 + U_2^T + Z_1 + hS_{12}, \\
\Omega_{13} &= -AW^T D + Q_{12} + W^T \Sigma T_1, \\
\Omega_{14} &= P_1 - hA(P_3 + P_4), \\
\Omega_{15} &= U_3^T - Z_1 + hS_{13}, \\
\Omega_{22} &= -\frac{2}{h} e^{-2kh} P_3 - (1-\eta) e^{-2kh} Q_{11} - U_2 - U_2^T + Z_2 + Z_2^T + hS_{22}, \\
\Omega_{24} &= -(1-\eta) e^{-2kh} Q_{12} + W^T \Sigma T_2, \\
\Omega_{25} &= \frac{1}{h} e^{-2kh} P_3 - U_3^T - Z_2 + Z_3^T + hS_{23}, \\
\Omega_{44} &= h(P_3 + P_4) - (1-\eta) e^{-2kh} Q_{22} - 2T_2, \\
\Omega_{55} &= -e^{-2kh} P_2 - \frac{1}{h} e^{-2kh} P_3 - Z_3 - Z_3^T + hS_{33}, \\
\Sigma &= \mathrm{diag}\{\sigma_1, \sigma_2, \ldots, \sigma_n\},
\end{aligned}$$

then the origin of neural network (6.3) is the unique equilibrium point, and it is exponentially stable with convergence rate k.

Proof First, we show the uniqueness of the equilibrium point by contradiction. To end this, let $\hat{u}$ be the equilibrium point of the delayed recurrent neural network (6.3), then we have

$$-A\hat{u} + g(W\hat{u}) = 0.$$

Now suppose $\hat{u} \neq 0$. It is easy to see that

$$2\Big(\hat{u}^T P_1 + g^T(W\hat{u}) DW\Big)\Big(-A\hat{u} + g(W\hat{u})\Big) = 0. \tag{6.8}$$

By inequality (6.4), the following inequality holds

$$2\hat{u}^T W^T \Sigma(T_1 + T_2)g(W\hat{u}) \geq 2g^T(W\hat{u})(T_1 + T_2)g(W\hat{u}).$$

This together with Eq. (6.8) gives

$$-\hat{u}^T(P_1 A + AP_1)\hat{u} + 2\hat{u}^T E_1 g(\hat{u}) + g^T(\hat{u})E_2 g(\hat{u}) \geq 0,$$

i.e.,

$$\begin{bmatrix} \hat{u}^T & g^T(\hat{u}) \end{bmatrix} \begin{bmatrix} -P_1 A - AP_1 & E_1 \\ * & E_2 \end{bmatrix} \begin{bmatrix} \hat{u} \\ g(\hat{u}) \end{bmatrix} \geq 0, \tag{6.9}$$

where $E_1 = P_1 + W^T \Sigma(T_1 + T_2) - AW^T D$, $E_2 = DW + W^T D - 2(T_1 + T_2)$.

On the other hand, let

$$\Pi = \begin{bmatrix} I & I & 0 & 0 & I \\ 0 & 0 & I & I & 0 \end{bmatrix},$$

multiplying (6.7) by Π and Π^T on its left and right side respectively, we obtain

$$\begin{bmatrix} E_0 & E_1 - hA(P_3 + P_4) \\ * & E_2 + h(P_3 + P_4) \end{bmatrix} + \left(1 - (1 - \eta)e^{-2kh}\right)Q < 0,$$

where

$$\begin{aligned} E_0 = & -P_1 A - AP_1 + 2k(P_1 + 2W^T D\Sigma W) \\ & + \left(1 - e^{-2kh}\right)P_2 + hA(P_3 + P_4)A \\ & + h\{S_{11} + S_{12} + S_{12}^T + S_{13} + S_{13}^T + S_{22} + S_{23} + S_{23}^T + S_{33}\}. \end{aligned}$$

That is

$$\begin{aligned} & \begin{bmatrix} -P_1 A - AP_1 & E_1 \\ * & E_2 \end{bmatrix} + \left(1 - (1 - \eta)e^{-2kh}\right)Q \\ & + \begin{bmatrix} 2k(P_1 + 2W^T D\Sigma W) + \left(1 - e^{-2kh}\right)P_2 & 0 \\ 0 & 0 \end{bmatrix} \\ & + \begin{bmatrix} -A \\ I \end{bmatrix} h(P_3 + P_4) \begin{bmatrix} -A & I \end{bmatrix} \\ & + \begin{bmatrix} I & I & I \\ 0 & 0 & 0 \\ 0 & 0 & 0 \end{bmatrix} hS \begin{bmatrix} I & 0 & 0 \\ I & 0 & 0 \\ I & 0 & 0 \end{bmatrix} < 0. \end{aligned}$$

Note that $P_1 > 0$, $P_i \geq 0$ $(i = 2, 3, 4)$, $Q \geq 0$, $S \geq 0$, $D > 0$, therefore

$$\begin{bmatrix} -P_1 A - AP_1 & E_1 \\ * & E_2 \end{bmatrix} < 0.$$

Obviously, this contradicts with (6.9). The contradiction implies that $\hat{u} = 0$. That is, the origin of the delayed recurrent neural networks (6.3) is the unique *equilibrium* point.

Next, we show the unique equilibrium point of networks (6.3) is exponentially stable. Consider the following Lyapunov–Krasovskii functional:

$$\begin{aligned} V(u(t)) = {} & e^{2kt} u^T(t) P_1 u(t) + \int_{t-h}^{t} e^{2ks} u^T(s) P_2 u(s) \mathrm{d}s \\ & + \int_{t-h}^{t} \int_{\theta}^{t} e^{2ks} \dot{u}^T(s)(P_3 + P_4)\dot{u}(s) \mathrm{d}s \mathrm{d}\theta \\ & + \int_{t-\tau(t)}^{t} e^{2ks} \xi^T(u(s)) Q \xi(u(s)) \mathrm{d}s \\ & + 2e^{2kt} \sum_{i=1}^{n} d_i \int_0^{W_i u(t)} g_i(s) \mathrm{d}s, \end{aligned} \tag{6.10}$$

where $\xi^T(u(s)) = (u^T(s), g^T(Wu(s)))$.

For convenience, we denote $u_\tau = u(t - \tau(t))$. The time derivative of functional (6.10) along the trajectories of system (6.3) is obtained as follows:

$$\begin{aligned} \dot{V}(u(t)) = {} & e^{2kt} \Big\{ 2ku^T(t) P_1 u(t) + 2u^T(t) P_1 \dot{u}(t) \\ & + u^T(t) P_2 u(t) - e^{-2kh} u^T(t-h) P_2 u(t-h) \\ & - \int_{t-h}^{t} e^{2k(s-t)} \dot{u}^T(s)(P_3 + P_4)\dot{u}(s) \mathrm{d}s \\ & + \xi^T(u(t)) Q \xi^T(u(t)) + h\dot{u}^T(t)(P_3 + P_4)\dot{u}(t) \\ & - (1 - \dot{\tau}(t)) e^{-2k\tau(t)} \xi^T(u_\tau) Q \xi^T(u_\tau) \\ & + 4k \sum_{i=1}^{n} d_i \int_0^{W_i u(t)} g_i(s) \mathrm{d}s + 2g^T(Wu(t)) D W \dot{u}(t) \Big\}. \end{aligned} \tag{6.11}$$

By Lemma 6.2 and inequality (6.4), we have

$$\begin{aligned} 4k \sum_{i=1}^{n} d_i \int_0^{W_i u(t)} g_i(s) \mathrm{d}s & \leq 4ku^T(t) W^T D g(Wu(t)) \\ & \leq 4ku^T(t) W^T D \Sigma W u(t). \end{aligned} \tag{6.12}$$

It is clear that the following equation is true:

$$\int_{t-h}^{t} \dot{u}^T(s)(P_3+P_4)\dot{u}(s)\mathrm{d}s = \int_{t-\tau(t)}^{t} \dot{u}^T(s)(P_3+P_4)\dot{u}(s)\mathrm{d}s + \int_{t-h}^{t-\tau(t)} \dot{u}^T(s)(P_3+P_4)\dot{u}(s)\mathrm{d}s. \quad (6.13)$$

By using Lemma 5.2 (i.e., the *Jensen integral inequality*), we obtain

$$\begin{aligned} &-\int_{t-\tau(t)}^{t} \dot{u}^T(s)P_3\dot{u}(s)\mathrm{d}s \\ &\le -\frac{1}{\tau(t)}\Big(\int_{t-\tau(t)}^{t} \dot{u}(s)\mathrm{d}s\Big)^T P_3 \int_{t-\tau(t)}^{t} \dot{u}(s)\mathrm{d}s \\ &\le -\frac{1}{h}[u(t)-u_\tau]^T P_3[u(t)-u_\tau], \end{aligned} \quad (6.14)$$

$$\begin{aligned} &-\int_{t-h}^{t-\tau(t)} \dot{u}^T(s)P_3\dot{u}(s)\mathrm{d}s \\ &\le -\frac{1}{h-\tau(t)}\Big(\int_{t-h}^{t-\tau(t)} \dot{u}(s)\mathrm{d}s\Big)^T P_3 \int_{t-h}^{t-\tau(t)} \dot{u}(s)\mathrm{d}s \\ &\le -\frac{1}{h}[u_\tau-u(t-h)]^T P_3[u_\tau-u(t-h)]. \end{aligned} \quad (6.15)$$

On the other hand, based on the *Leibniz-Newton formula*, for any real matrix $U_i, Z_i (i = 1, 2, 3)$ with compatible dimensions, we get

$$\begin{aligned} 0 = {} & 2e^{2kt}\Big[u^T(t)U_1 + u_\tau^T U_2 + u^T(t-h)U_3\Big] \\ & \times\Big[u(t) - u_\tau - \int_{t-\tau(t)}^{t} \dot{u}(s)\mathrm{d}s\Big], \end{aligned} \quad (6.16)$$

$$\begin{aligned} 0 = {} & 2e^{2kt}\Big[u^T(t)Z_1 + u_\tau^T Z_2 + u^T(t-h)Z_3\Big] \\ & \times\Big[u_\tau - u(t-h) - \int_{t-h}^{t-\tau(t)} \dot{u}(s)\mathrm{d}s\Big]. \end{aligned} \quad (6.17)$$

From inequality (6.4), the following matrix inequalities hold for any positive diagonal matrices T_1, T_2 with compatible dimensions

$$0 \le 2e^{2kt}\{u^T(t)W^T\Sigma T_1 g(Wu(t)) - g^T(Wu(t))T_1 g(Wu(t))\}, \quad (6.18)$$

$$0 \le 2e^{2kt}\{u_\tau^T W^T\Sigma T_2 g(Wu_\tau) - g^T(Wu_\tau)T_2 g(Wu_\tau)\}. \quad (6.19)$$

Moreover, for any real positive definite symmetric matrix S with compatible dimension, we have

$$0 = e^{2kt}\Big(h\kappa^T(t)S\kappa(t) - \int_{t-\tau(t)}^{t} \kappa^T(t)S\kappa(t)\mathrm{d}s - \int_{t-h}^{t-\tau(t)} \kappa^T(t)S\kappa(t)\mathrm{d}s\Big), \quad (6.20)$$

where $\kappa^T(t) = (u^T(t), u_\tau^T, u^T(t-h))$.

Adding the terms on the right-hand side of Eqs. (6.11), (6.13), (6.16), (6.17), (6.20), inequalities (6.12), (6.14), (6.15), (6.18) and (6.19), we obtain

$$\begin{aligned}\dot{V}(u(t)) \le e^{2kt}\Big(&\zeta^T(t)\Omega\zeta(t) - \int_{t-\tau(t)}^{t} \varsigma^T(t,s)\Xi_1\varsigma(t,s)\mathrm{d}s \\ &- \int_{t-h}^{t-\tau(t)} \varsigma^T(t,s)\Xi_2\varsigma(t,s)\mathrm{d}s\Big), \end{aligned} \quad (6.21)$$

where Ξ_1, Ξ_2, Ω are defined in (6.5), (6.6), (6.7) respectively, $\varsigma^T(t,s) = (\kappa^T(t), \dot{u}^T(s))$ and $\zeta^T(t) = \big[u^T(t),\ u_\tau^T,\ g^T(Wu(t)),\ g^T(Wu_\tau),\ u^T(t-h)\big]$. Therefore $V(u(t)) \le V(u(0))$.

Furthermore, similar to [22, 23], from Lemma 5.2, (6.3) and inequality (6.4), we have

$$V(u(0)) \le \Lambda||\phi(t) - x^*||^2,$$

where

$$\begin{aligned}\Lambda = {} & \lambda_M(P_1) + \lambda_M(W^T D\Sigma W) + h\Big[\lambda_M(Q_{12}Q_{22}^{-1}Q_{12}^T) \\ & + 2\lambda_M(W^T\Sigma Q_{22}\Sigma W) + \lambda_M(P_2) + \lambda_M(Q_{11})\Big] \\ & + h^2\lambda_M(P_3 + P_4)\Big[\lambda_M(A^T A) + \lambda_M(W^T\Sigma^2 W)\Big].\end{aligned}$$

Meanwhile $V(u(t)) \ge e^{2kt}||u(t)||^2\lambda_m(P_1)$, by Lyapunov stability theory, the proof of Theorem 6.3 is completed.

Remark 6.4 It is easy to see that the stability results given in [3, 4] are delay-independent. As well known, delay-dependent criteria make use of information on the length of delays, and are usually less conservative than delay-independent ones especially when the time delay is small. An example will show that the condition in Theorem 6.3 is more effective than those in [3, 4].

Remark 6.5 For $\eta \ge 1$, Q will no longer be helpful to improve the stability condition since $-(1-\eta)Q$ is nonnegative definite. Therefore, by setting $Q = 0$ in Theorem 6.3, an easy delay-dependent and rate-independent criterion can be derived for unknown η.

Next, we estimate the upper bound of $\dot{V}(u(t))$ by following the idea of *convex combination*, and another stability criterion can be developed as follows:

Theorem 6.6 *Under the assumption (6.2) and $0 \le \tau(t) \le h,\ 0 \le \dot{\tau}(t) \le \eta < 1$, given a constant $k > 0$, suppose that there exist positive definite symmetric matrices P_1, nonnegative definite symmetric matrices $P_i (i = 2, 3, 4)$, positive diagonal matrices T_1, T_2, D, real symmetric matrices $Q \ge 0, S \ge 0$ and real matrices $U, Z, M^T = \begin{bmatrix} M_1^T & M_2^T & 0 & 0 & M_3^T \end{bmatrix}, N^T = \begin{bmatrix} N_1^T & N_2^T & 0 & 0 & N_3^T \end{bmatrix}$, such that (6.5), (6.6) and the following LMIs hold:*

$$\Upsilon_1 = \begin{bmatrix} \Upsilon & hN \\ * & -hP_3 \end{bmatrix} < 0, \tag{6.22}$$

$$\Upsilon_2 = \begin{bmatrix} \Upsilon & hM \\ * & -hP_3 \end{bmatrix} < 0, \tag{6.23}$$

where

$$\Upsilon = \begin{bmatrix} \Upsilon_{11} & \Upsilon_{12} & \Omega_{13} & \Omega_{14} & \Upsilon_{15} \\ * & \Upsilon_{22} & 0 & \Omega_{24} & \Upsilon_{25} \\ * & 0 & Q_{22} - 2T_1 & DW & 0 \\ * & * & * & \Omega_{44} & 0 \\ * & * & 0 & 0 & \Upsilon_{55} \end{bmatrix},$$

$$\begin{aligned}
\Upsilon_{11} &= 2k(P_1 + 2W^T D \Sigma W) - P_1 A - A P_1 + P_2 + hS_{11} \\
&\quad + hA(P_3 + P_4)A + U_1 + U_1^T + Q_{11} + M_1 + M_1^T, \\
\Upsilon_{12} &= -U_1 + U_2^T + Z_1 + hS_{12} - M_1 + M_2^T + N_1, \\
\Upsilon_{15} &= U_3^T - Z_1 + hS_{13} + M_3^T - N_1, \\
\Upsilon_{22} &= -(1-\eta)e^{-2kh} Q_{11} - U_2 - U_2^T + Z_2 + Z_2^T \\
&\quad + hS_{22} - M_2 - M_2^T + N_2 + N_2^T, \\
\Upsilon_{25} &= -U_3^T - Z_2 + Z_3^T + hS_{23} - M_3^T - N_2 + N_3^T, \\
\Upsilon_{55} &= -e^{-2kh} P_2 - Z_3 - Z_3^T + hS_{33} - N_3 - N_3^T,
\end{aligned}$$

other parameters are the same defined in Theorem 6.3, then the origin of neural network (6.3) is the unique equilibrium point, and is exponentially stable with convergence rate k.

Proof The proof of the uniqueness of the equilibrium point is similar to Theorem 6.3. Now we prove the exponential stability of the unique equilibrium point. From [27], the following inequalities hold

$$-\int_{t-\tau(t)}^{t} \dot{u}^T(s) P_3 \dot{u}(s)\mathrm{d}s \le \tau(t)\zeta^T(t) M P_3^{-1} M^T \zeta(t) + 2\zeta^T(t) M[u(t) - u_\tau], \tag{6.24}$$

$$-\int_{t-h}^{t-\tau(t)} \dot{u}^T(s) P_3 \dot{u}(s)\mathrm{d}s \le (h - \tau(t))\zeta^T(t) N P_3^{-1} N^T \zeta(t) + 2\zeta^T(t) N[u_\tau - u(t-h)]. \tag{6.25}$$

Thus, combining Eqs. (6.11), (6.13), (6.16), (6.17), (6.20) and inequalities (6.12), (6.18), (6.19), (6.24) and (6.25), it yields

$$\begin{aligned}\dot{V}(u(t)) \le e^{2kt}\Big\{&\zeta^T(t)\big[\Upsilon + \tau(t) M P_3^{-1} M^T \\ &+ (h-\tau(t)) N P_3^{-1} N^T\big]\zeta(t) \\ &- \int_{t-\tau(t)}^{t} \varsigma^T(t,s)\Xi_1\varsigma(t,s)\mathrm{d}s \\ &- \int_{t-h}^{t-\tau(t)} \varsigma^T(t,s)\Xi_2\varsigma(t,s)\mathrm{d}s\Big\}.\end{aligned}$$

Note that $0 \le \tau(t) \le h$, thus,

$$\Upsilon + \tau(t) M P_5^{-1} M^T + (h-\tau(t)) N P_3^{-1} N^T < 0,$$

holds if and only if

$$\Upsilon + h N P_3^{-1} N^T < 0, \tag{6.26}$$

and

$$\Upsilon + h M P_3^{-1} M^T < 0 \tag{6.27}$$

hold. From the Schur complement (see [22, 27]), inequalities (6.26) and (6.27) are equivalent to (6.21) and (6.22) respectively, thus $\dot{V}(u(t)) < 0$ holds if (6.5), (6.6), (6.22) and (6.23) hold. This completes the proof.

Remark 6.7 Similar to Remark 6.5, by setting $Q = 0$, we can employ the criterion of Theorem 6.6 to analyze the stability of neural network when $\tau(t)$ is not differentiable or $\dot{\tau}(t)$ is unknown.

6.4 Illustrative Example

In this section, we provide a numerical example to demonstrate the effectiveness and less conservativeness of our delay-dependent stability criteria.

Consider system (6.3) with

$$A = \text{diag}\{4.1989, 0.7160, 1.9985\},$$
$$W = \begin{pmatrix} -0.1052 & -0.5069 & -0.1121 \\ -0.0257 & -0.2808 & 0.0212 \\ 0.1205 & -0.2153 & 0.1315 \end{pmatrix},$$
$$\sigma_1 = 0.4219, \sigma_2 = 3.8993, \sigma_3 = 1.0160.$$

For this neural network, it is verified that all the asymptotic stability results in [4, 5] and the exponential stability result in [3] fail to ascertain the stability of the equilibrium point. Further, the criteria in [2, 11] can only be applied to confirm the stability without time delay, thus none of the criteria in [2, 11] is applicable to ascertain the exponential stability for $\tau(t) \neq 0$. Moreover, the criterion in [7] can only be applied to ensure the stability with constant time delay, so the result of [7] cannot be applicable to verify the stability for $\eta \neq 0$. Therefore, all of the criteria given in [2–5, 7, 11] fail to conclude whether this given neural network is asymptotically stable or not for $\eta \neq 0$. When the time delay is constant, the exponential stability criterion of [7] gives the upper bound of time delay being 1.9518. However, by use of the MATLAB LMI Control Toolbox, from Theorems 6.3 and 6.6, the calculated maximal upper bounds of time delay being all 4.5872.

If we set $h = 1$, the calculated maximal convergence rates of k for various η with Theorems 6.3, 6.6, Remarks 6.5, and 6.7 in this chapter and that of [7] are listed in Table 6.1.

Therefore, we can say that for this system the results in this chapter are much effective and less conservative than those in [2–5, 7, 11].

Table 6.1 Calculated maximal convergence rate of k for $h = 1$ and various η (where "–" denotes that the condition cannot give the exponential convergence rate

η	0	0.1	0.2	Unknown
Ref. [7]	–	Fail	Fail	Fail
Theorems 6.3 and 6.6	0.2505	0.2378	0.2273	–
Remarks 6.5 and 6.7	–	–	–	0.4424

6.5 Summary

Based on free weight matrix method and Jensen integral inequality, two sufficient conditions have been derived for the globally exponential stability of static neural networks with time-varying delays. The obtained results can be expressed in the form of LMI and are easy to be verified. The effectiveness is demonstrated by an illustrative example.

References

1. Y. He, G.P. Liu, D. Rees, M. Wu, Stability analysis for neural networks with time-varying interval delay. IEEE Trans. Neural Netw. **18**(6), 1850–1854 (2007)
2. S. Hu, J. Wang, Global stability of a class of continuous-time recurrent neural networks. IEEE Trans. Circuits Syst. I Fundam. Theory Appl. **49**(9), 1334–1347 (2002)
3. P. Li, J. Cao, Stability in static delayed neural networks: a nonlinear measure approach. Neurcomputing **69**, 1776–1781 (2006)
4. J. Liang, J. Cao, A based-on LMI stability criterion for delayed recurrent neural networks. Chaos Solitons Fractals **28**, 154–160 (2006)
5. C.Y. Lu, T.J. Su, S.C. Huang, Delay-dependent stability analysis for recurrent neural networks with time-varying delay. IET Control Theory Appl. **2**(8), 736–742 (2008)
6. H. Qiao, J. Peng, Z.B. Zhang, A reference model approach to stability analysis of neural network. IEEE Trans. Syst. Man Cybern. B Cybern. **33**, 925–936 (2003)
7. H. Shao, Delay-dependent approaches to globally exponential stability for recurrent neural networks. IEEE Trans. Circuits Syst. II Express Br. **55**(6), 591–595 (2008)
8. Z. Wang, H. Zhang, W. Yu, Robust exponential stability analysis of neural networks with multiple time delays. Neurocomputing **70**, 2534–2543 (2007)
9. M. Wu, F. Liu, P. Shi, Y. He, R. Yokoyama, Exponential stability analysis for neural networks with time-varying delay. IEEE Trans. Syst. Man Cybern. B Cybern. **38**, 1152–1156 (2008)
10. K. Shi, H. Zhu, S. Zhong, Less conservative stability criteria for neural networks with discrete and distributed delays using a delay-partitioning approach. Neurocomputing **140**, 273–282 (2014)
11. S. Xu, J. Lam, D.W.C. Ho, Y. Zou, Global robust exponential stability analysis for interval recurrent neural networks. Phys. Lett. A **325**, 124–133 (2004)
12. Z.B. Xu, H. Qiao, J. Peng, A comparative study on two modeling approaches in neural networks. Neural Netw. **17**, 73–85 (2004)
13. Z. Zeng, J. Wang, Global exponential stability of recurrent neural networks with time-varying delays in the presence of strong external stimuli. Neural Netw. **19**, 1528–1537 (2006)
14. Z. Zeng, J. Wang, Improved conditions for global exponential stability of recurrent neural networks with time-varying delays. IEEE Trans. Neural Netw. **17**(3), 623–635 (2006)
15. H. Zhang, Z. Wang, Global asymptotic stability of delayed cellular neural networks. IEEE Trans. Neural Netw. **18**(3), 947–950 (2007)
16. H. Zhang, Z. Wang, D. Liu, Robust stability analysis for interval Cohen-Grossberg neural networks with unknown time-varying delays. IEEE Trans. Neural Netw. **19**(11), 1942–1955 (2008)
17. R. Rakkiyappan, G. Velrnurugan, J. Cao, Multiple μ-stability analysis of complex-valued neural networks with unbounded time-varying delays. Neurocomputing **149**, 594–607 (2015)
18. H. Zhang, Z. Wang, D. Liu, Global asymptotic stability of recurrent neural networks with multiple time-varying delays. IEEE Trans. Neural Netw. **19**(5), 855–873 (2008)
19. H. Zhang, Z. Wang, D. Liu, Robust exponential stability of cellular neural networks with multiple time varying delays. IEEE Trans. Circuits Syst. II Express Br. **54**(8), 730–734 (2007)

20. H. Zhang, Z. Wang, D. Liu, Global asymptotic stability and robust stability of a class of Cohen-Grossberg neural networks with mixed delays. IEEE Trans. Circuits Syst. I Fundam. Theory Appl. **56**(3), 616–629 (2009)
21. Q. Zhang, X. Wei, J. Xu, Delay-dependent exponential stability of cellular neural networks with time-varying delays. Chaos Solitons Fractals **23**, 1363–1369 (2005)
22. C.D. Zheng, H. Zhang, Z. Wang, Novel delay-dependent criteria for global robust exponential stability of delayed cellular neural networks with norm-bounded uncertainties. Neurocomputing **72**(9), 1744–1754 (2009)
23. C.D. Zheng, H. Zhang, Z. Wang, New delay-dependent global exponential stability criterion for cellular type neural networks with time-varying delays. IEEE Trans. Circuits Syst. II Express Br. **56**(3), 250–254 (2009)
24. Y. Du, S. Zhong, N. Zhou, Global asymptotic stability of Markovian jumping stochastic Cohen-Grossberg BAM neural networks with discrete and distributed time-varying delays. Appl. Math. Comput. **243**, 624–636 (2014)
25. C.D. Zheng, H. Zhang, Z. Wang, Delay-dependent globally exponential stability criteria for static neural networks: an LMI approach. IEEE Trans. Circuits Syst. II: Express Br. **56**(7), 605–609 (2009)
26. K. Gu, V.L. Kharitonov, J. Chen, *Stability of Time-Delay Systems* (Birkhauser, Boston, 2003)
27. S. Boyd, L.E.I. Ghaoui, E. Feron, V. Balakrishnan, *Linear Matrix Inequality in System and Control Theory* (SIAM, Philadelphia, 1994)

Chapter 7
Multiple Stability for Discontinuous RNNs

In Chaps. 4–6, global stability problems are considered for some kinds of RNNs with delays. In those network models, the considered activation functions in RNNs are continuous. However, in some cases, discontinuous activation can be used in some associative memory and pattern storage problems. Such kind of applications require more equilibrium to restore so many patterns. In this case, local stability or multiple stability is meaningful and important. In this chapter, we will discuss the local stability of multiple equilibrium points for time-varying delayed recurrent neural networks with discontinuous activation functions. The contents in this chapter are from the result in [1].

7.1 Introduction

In the past few decades, many efforts have been made on the applications of neural networks, such as, signal processing, image processing, pattern recognition, associative memories, optimization problems, and so on. Such applications rely heavily on the dynamical properties of neural network systems. Therefore, the theoretical study on these dynamical systems is of great importance [2–8]. The notion of "multistability" of a neural network describes coexistence of multiple stable patterns such as equilibrium points or periodic orbits. In an associative memory neural network, the addressable memories or patterns are stored as stable equilibrium points or stable periodic orbits. Thus, it is required that neural networks exhibit more than one stable equilibrium point or more than one exponentially attractive periodic orbit instead of a single globally stable equilibrium point. In recent years, some sufficient conditions for multistability and multiperiodicity of recurrent neural networks have been studied in the literature. References [9–11] investigated multiple stable equilibrium points or multiple stable periodic orbits of recurrent neural networks by geometrical observation. References [12–16] discussed multiple stable equilibrium points or multiple stable periodic orbits of *recurrent neural networks* based on decomposition

Z. Wang et al., *Qualitative Analysis and Control of Complex Neural Networks with Delays*, Studies in Systems, Decision and Control 34,
DOI 10.1007/978-3-662-47484-6_7

of state space. A general n-dimensional neural networks with piecewise linear *activation functions* with two corner points have 3^n equilibrium points or periodic orbits, of which 2^n are stable equilibrium points or stable periodic orbits. References [17–23] discussed multistability and multiperiodicity of recurrent neural networks with unsaturating piecewise linear transfer functions. For more references, see [24–43] and references therein.

It is well known that the activation functions play an important role in the dynamics analysis of recurrent neural networks. The storage capacity of patterns and associative memories relies heavily on the structures of activation functions [44]. It has been shown that n-neuron recurrent neural networks with one step piecewise linear activation functions could have 2^n locally exponentially stable equilibrium points located in saturation regions (see [9, 10, 12, 42]). In order to increase storage capacity, a stair-style activation function can be redefined with k steps. In [16], multistability for n-neuron neural networks with k-stair activation functions was discussed. It was shown that this system could have $(4k - 1)^n$ equilibrium points and $(2k)^n$ of them were locally exponentially stable. In [15], the authors investigated the neural networks with a class of nondecreasing piecewise linear activation functions with $2r$ corner points, and n-neuron dynamical system could have and only have $(2r + 1)^n$ equilibrium points under some conditions, of which $(r + 1)^n$ were locally exponentially stable and the others were unstable. Most of the above results were based on the assumption that the activation functions are continuous. Whereas, as mentioned in [28], a brief overview on some common neural network models reveals that neural networks described by differential equations with a discontinuous right-hand side are of importance and do frequently arise in practice. When dealing with dynamical systems possessing high-slope nonlinear elements, it is often advantageous to model them with a system of differential equations with discontinuous right-hand side. Global convergence criteria of neural networks with discontinuous activation functions were first introduced by Forti and Nistri [28] and Lu and Chen [22]. Much efforts have been devoted to analyzing the dynamical properties of neural networks with discontinuous activation functions. See [22, 29–36] and references therein. The dynamics of Cohen–Grossberg neural networks with discontinuous activations functions was discussed in [22]. In [29–31], some sufficient conditions were obtained for the global convergence of neural networks with discontinuous neuron activations. The global asymptotic stability of delayed neural networks with discontinuous activation functions was derived in [32]. In [24], the authors discussed multistability of neural networks without delay and with discontinuous activation functions. The value of activation functions was located in $[-1, 1]$.

Motivated by the above discussions, the purpose of this chapter is to explore the *multistability* of n-neuron recurrent neural networks with time-varying delays and k-level discontinuous activation functions. Based on the decomposition of state space, we will establish some new sufficient conditions for the existence of multiple equilibrium points in this chapter. Under the configuration of *discontinuous activation functions*, recurrent neural networks have k^n locally exponentially stable equilibrium points. The activation functional value is located in $[c_1, c_k]$, where c_1 and c_k can be any constants. The result is more general than that in the existing literature (for example [24]). Moreover, conditions for the existence of sets of stable equilibrium points and unstable equilibrium points are derived for recurrent neural networks without delay. Our result is a more comprehensive discussion of multistability for neural networks with discontinuous activation functions [1]. Finally, three examples are given to illustrate the effectiveness of the obtained results.

7.2 Problem Formulations and Preliminaries

In this chapter, we consider the following n-neuron recurrent neural networks with time-varying delays:

$$\frac{\mathrm{d}x_i(t)}{\mathrm{d}t} = -x_i(t) + \sum_{j=1}^{n} a_{ij} f_j(x_j(t)) + \sum_{j=1}^{n} b_{ij} f_j(x_j(t-\tau_{ij}(t))) + u_i, \qquad (7.1)$$

where $\mathbf{x}(t) = (x_1(t), x_2(t), \cdots, x_n(t))^T \in \mathbb{R}^n$, $x_i(t)$ represents the state of the ith neuron at time t, $i = 1, 2, \cdots, n$; a_{ij}, b_{ij} correspond to the connection weights of the jth unit on the ith unit at time t and time $t - \tau_{ij}(t)$, respectively; $\tau_{ij}(t)$ denotes time-varying delay that satisfies $0 \leq \tau_{ij}(t) \leq \tau := \max_{1 \leq i,j \leq n}\{\sup\{\tau_{ij}(t)\}\}$; $f_j(\cdot)$ is the activation function; $\mathbf{u} = (u_1, u_2, \cdots, u_n)^T \in \mathbb{R}^n$ is an input vector.

We consider a general class of *activation function* as follows:

$$f_j(\xi) = \begin{cases} c_1, & \xi \in (-\infty, p_1), \\ c_2, & \xi \in (p_1, p_2), \\ \cdots & \\ c_{k-1}, & \xi \in (p_{k-2}, p_{k-1}), \\ c_k, & \xi \in (p_{k-1}, +\infty), \\ \text{undefined}, & \xi \in \{p_1, p_2, \cdots, p_{k-1}\}, \end{cases} \qquad (7.2)$$

where $j = 1, 2, \cdots, n$; k is an integer that satisfies $k \geq 1$; $c_1, c_2, \cdots, c_k$, $p_1, p_2, \cdots, p_{k-1}$ are constants with $c_1 < c_2 < \cdots < c_k$, $p_1 < p_2 < \cdots < p_{k-1}$. Denote $p_0 = -\infty$, $p_k = +\infty$. Typical configuration of the activation function is depicted in Fig. 7.1.

Since $f_j(\cdot)$ is allowed to have points of discontinuity, we need to specify what mean by solution of the system having a discontinuous right-side. For this purpose, we consider the solution of delayed neural networks (7.1) in the sense of Filippov [37].

Definition 7.1 ([29]) A function $\mathbf{x}(t) = (x_1(t), x_2(t), \cdots, x_n(t))^T : [t_0 - \tau, T) \to \mathbb{R}^n$, $T \in (t_0, +\infty]$ is a solution of system (7.1) on $[t_0 - \tau, T)$ if, (1). $\mathbf{x}(t)$ is continuous on $[t_0, T)$; (2). there exists a measurable function $\gamma(t) = (\gamma_1(t), \gamma_2(t), \cdots, \gamma_n(t))^T$: $[t_0 - \tau, T) \to \mathbb{R}^n$, such that $\gamma_i(t) \in K[f_i(x_i(t))]$ for almost all (a.a.) $t \in [t_0 - \tau, T)$, and

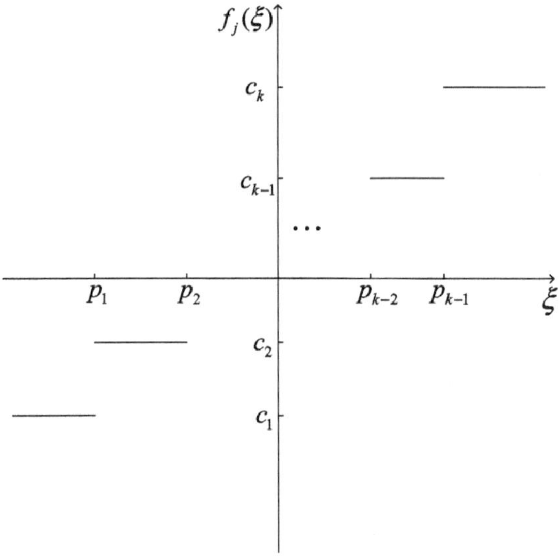

Fig. 7.1 The graph for activation function (7.2)

$$\frac{\mathrm{d}x_i(t)}{\mathrm{d}t} = -x_i(t) + \sum_{j=1}^{n} a_{ij}\gamma_j(t) + \sum_{j=1}^{n} b_{ij}\gamma_j(t - \tau_{ij}(t)) + u_i,$$

for a.a. $t \in [t_0, T)$, where $K[E]$ represents closure of the convex hull of E.

Definition 7.2 ([29]) For any continuous function $\varphi(\theta) = (\varphi_1(\theta), \varphi_2(\theta), \cdots, \varphi_n(\theta))^T : [t_0 - \tau, t_0] \to \mathbb{R}^n$ and any measurable selection $\lambda(\theta) = (\lambda_1(\theta), \lambda_2(\theta), \cdots, \lambda_n(\theta))^T : [t_0 - \tau, t_0] \to \mathbb{R}^n$, such that $\lambda_i(\theta) \in K[f_i(\varphi_i(\theta))]$ for a.a. $\theta \in [t_0 - \tau, t_0]$ by an initial value problem associated with system (7.1) having initial condition $\varphi(\theta)$, $\lambda(\theta)$, we mean the following problem: find a couple of functions $[\mathbf{x}(t), \gamma(t)] : [t_0 - \tau, T] \to \mathbb{R}^n \times \mathbb{R}^n$, such that $\mathbf{x}(t)$ is a solution of system (7.1) on $[t_0 - \tau, T)$ for some $T > t_0$, $\gamma(t)$ is an output associated with $\mathbf{x}(t)$ and

$$\begin{cases} \frac{\mathrm{d}x_i(t)}{\mathrm{d}t} = -x_i(t) + \sum\limits_{j=1}^{n} a_{ij}\gamma_j(t) + \sum\limits_{j=1}^{n} b_{ij}\gamma_j(t - \tau_{ij}(t)) + u_i, & \text{for a.a. } t \in [t_0, T), \\ x_i(\theta) = \varphi_i(\theta), & \forall\, \theta \in [t_0 - \tau, t_0], \\ \gamma_i(\theta) = \lambda_i(\theta), & \text{for a.a. } \theta \in [t_0 - \tau, t_0]. \end{cases}$$

Definition 7.3 ([37]) A solution $x = \phi(t)$ ($t_0 \le t < \infty$) of a differential inclusion $(\mathrm{d}x)/(\mathrm{d}t) \in F(t, x)$ is called *stable* (respectively, weakly stable) if for any $\varepsilon > 0$, there exists $\delta > 0$, which possesses the following property. For each $\bar{x}_0$ such that $|\bar{x}_0 - \phi(t_0)| < \delta$, each solution (respectively, some solution) $\bar{x}(t)$ with the initial data $\bar{x}(t) = \bar{x}_0$ for $t_0 \le t < \infty$ exists and satisfies the inequality $|\bar{x}(t) - \phi(t)| < \varepsilon$ ($t_0 \le t < \infty$).

Definition 7.4 ([41]) The equilibrium $\mathbf{x}^*$ of system (7.1) is said to be locally *exponentially stable* in region $\mathcal{D}$, if there exist constants $\alpha > 0$ and $M > 0$ such that for a.a. $t \geq t_0$

$$\|\mathbf{x}(t, t_0, \phi) - \mathbf{x}^*\| \leq M\|\phi - \mathbf{x}^*\| \exp\{-\alpha(t - t_0)\},$$

where $\mathbf{x}(t, t_0, \phi)$ is the state of system (7.1) with any initial condition $\phi(\theta) \in C([t_0 - \tau, t_0], \mathcal{D})$.

Lemma 7.5 ([38]) *Let $\mathcal{D}$ be a bounded and closed set in $\mathbb{R}^n$, and H be a mapping on complete matric space $(\mathcal{D}, \|\cdot\|)$, where for any $\boldsymbol{x}, \boldsymbol{y} \in \mathcal{D}$, $\|\boldsymbol{x} - \boldsymbol{y}\| = \max_{1 \leq i \leq n}\{|x_i - y_i|\}$ is measurement in $\mathcal{D}$. If $H(\mathcal{D}) \subset \mathcal{D}$ and there exists a constant $\alpha < 1$ such that for any $\boldsymbol{x}, \boldsymbol{y} \in \mathcal{D}$, $\|H(\boldsymbol{x}) - H(\boldsymbol{y})\| \leq \alpha\|\boldsymbol{x} - \boldsymbol{y}\|$, then there exists a unique $\boldsymbol{x}^* \in \mathcal{D}$ such that $H(\boldsymbol{x}^*) = \boldsymbol{x}^*$.*

Some denotations are required as follows:

$$\begin{aligned}
(-\infty, p_1) =&(-\infty, p_1)^1 \times (p_1, p_2)^0 \times \cdots \times (p_{k-1}, +\infty)^0, \\
(p_1, p_2) =&(-\infty, p_1)^0 \times (p_1, p_2)^1 \times \cdots \times (p_{k-1}, +\infty)^0, \\
&\vdots \\
(p_{k-1}, +\infty) =&(-\infty, p_1)^0 \times (p_1, p_2)^0 \times \cdots \times (p_{k-1}, +\infty)^1.
\end{aligned}$$

and denote

$$\Omega = \left\{ \prod_{i=1}^{n} (-\infty, p_1)^{\delta_1^i} \times (p_1, p_2)^{\delta_2^i} \times \cdots \times (p_{k-1}, +\infty)^{\delta_k^i} : \right\},$$

where $(\delta_1^i, \delta_2^i, \cdots, \delta_k^i) = (1, 0, \cdots, 0)$ or $(0, 1, \cdots, 0)$ or $\cdots$ or $(0, 0, \cdots, 1)$. It is easy to see that Ω is composed of k^n parts. For example, when $n = 2$, $k = 4$, Ω is made up of 4^2 parts, and all parts of Ω are depicted in Fig. 7.2.

7.3 Main Results

In this section, we will discuss the existence and stability for recurrent neural networks with discontinuous activation functions. Some sufficient conditions are established to ensure system (7.1) with activation function (7.2) can have k^n equilibrium points in Ω and they are locally exponentially stable. Moreover, conditions for the existence of sets of stable equilibrium points and unstable equilibrium points are derived for system (7.1) without delay.

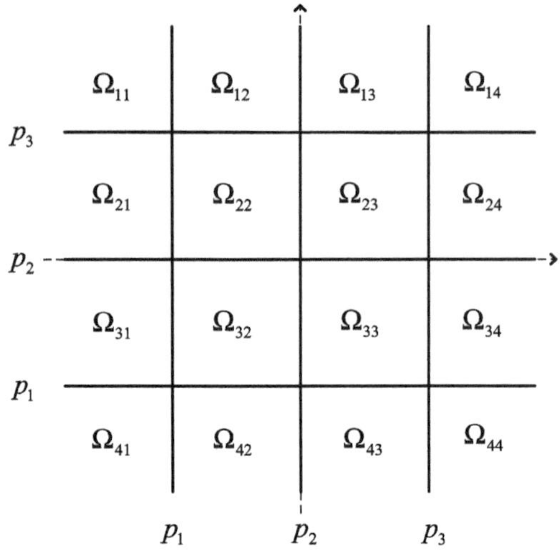

Fig. 7.2 Ω is made up of 4^2 parts

For any $\Omega_\kappa \subset \Omega$, we have the following results.

Theorem 7.6 *There exists one equilibrium point in Ω_κ for system (7.1) with activation function (7.2), if the following conditions hold:*

$$p_{m-1} < (a_{ii} + b_{ii})c_m + \sum_{j=1, j\neq i}^{n} (a_{ij} + b_{ij})c_{q_j} + u_i < p_m, \tag{7.3}$$

where $i = 1, 2, \cdots, n$, $m = 1, 2, \cdots, k$, $q_j = 1, 2, \cdots, k$, $j = 1, 2, \cdots, n$.

Proof From condition (7.3), there exists a small constant ε $(0 < \varepsilon \ll \min\{|1/p_1|, |1/p_{k-1}|\})$ such that

$$p_{m-1} + \varepsilon \leq (a_{ii} + b_{ii})c_m + \sum_{j=1, j\neq i}^{n} (a_{ij} + b_{ij})c_{q_j} + u_i \leq p_m - \varepsilon. \tag{7.4}$$

Denote

$$[-1/\varepsilon, p_1 - \varepsilon] = [-1/\varepsilon, p_1 - \varepsilon]^1 \times [p_1 + \varepsilon, p_2 - \varepsilon]^0 \times \cdots \times [p_{k-1} + \varepsilon, 1/\varepsilon]^0,$$
$$[p_1 + \varepsilon, p_2 - \varepsilon] = [-1/\varepsilon, p_1 - \varepsilon]^0 \times [p_1 + \varepsilon, p_2 - \varepsilon]^1 \times \cdots \times [p_{k-1} + \varepsilon, 1/\varepsilon]^0,$$
$$\cdots$$
$$[p_{k-1} + \varepsilon, 1/\varepsilon] = [-1/\varepsilon, p_1 - \varepsilon]^0 \times [p_1 + \varepsilon, p_2 - \varepsilon]^0 \times \cdots \times [p_{k-1} + \varepsilon, 1/\varepsilon]^1,$$

$$\Phi = \left\{ \prod_{i=1}^{n} [-1/\varepsilon, p_1 - \varepsilon]^{\delta_1^i} \times [p_1 + \varepsilon, p_2 - \varepsilon]^{\delta_2^i} \times \cdots \times [p_{k-1} + \varepsilon, 1/\varepsilon]^{\delta_k^i} \right\},$$

where $(\delta_1^i, \delta_2^i, \cdots, \delta_k^i) = (1, 0, \cdots, 0)$ or $(0, 1, \cdots, 0)$ or $\cdots$ or $(0, 0, \cdots, 1)$.

It is obvious that Φ is made up of k^n parts and Φ is bounded in Ω. There must exist $\Phi_\iota \subset \Phi$ such that $\Phi_\iota \subset \Omega_\kappa$. For any $\mathbf{x} = (x_1, x_2, \cdots, x_n)^T \in \Phi_\iota$, consider the following algebraic equations

$$x_i = (a_{ii} + b_{ii}) f_i(x_i) + \sum_{j=1, j\neq i}^{n} (a_{ij} + b_{ij}) f_j(x_j) + u_i, \; i = 1, 2, \cdots, n.$$

Denote

$$H_i(\mathbf{x}) = (a_{ii} + b_{ii}) f_i(x_i) + \sum_{j=1, j\neq i}^{n} (a_{ij} + b_{ij}) f_j(x_j) + u_i. \tag{7.5}$$

Let $H(\mathbf{x}) = (H_1(\mathbf{x}), H_2(\mathbf{x}), \cdots, H_n(\mathbf{x}))^T$. We will show that $H(\Phi_\iota) \subset \Phi_\iota$. We can get from (7.2) that

$$f_j(x_j) = \begin{cases} c_m, & j = i, \\ c_{q_j}, & j \neq i, \end{cases} \tag{7.6}$$

where $i = 1, 2, \cdots, n, m = 1, 2, \cdots, k, q_j = 1, 2, \cdots, k, j = 1, 2, \cdots, n$. Hence, it follows from (7.5) and (7.6) that

$$\begin{aligned} H_i(\mathbf{x}) &= (a_{ii} + b_{ii}) f_i(x_i) + \sum_{j=1, j\neq i}^{n} (a_{ij} + b_{ij}) f_j(x_j) + u_i \\ &= (a_{ii} + b_{ii}) c_m + \sum_{j=1, j\neq i}^{n} (a_{ij} + b_{ij}) c_{q_j} + u_i. \end{aligned} \tag{7.7}$$

From (7.4) and (7.7), we can get that

$$p_{m-1} + \varepsilon \leq H_i(\mathbf{x}) \leq p_m - \varepsilon.$$

Hence,

$$H(\Phi_\iota) \subset \Phi_\iota.$$

For any $\mathbf{x}, \mathbf{y} \in \Phi_\iota$,

$$\begin{aligned} |H_i(\mathbf{x}) - H_i(\mathbf{y})| &= |(a_{ii} + b_{ii})(f_i(x_i) - f_i(y_i)) \\ &\quad + \sum_{j=1, j\neq i}^{n} (a_{ij} + b_{ij})(f_j(x_j) - f_j(y_j))| \\ &= 0. \end{aligned}$$

Therefore, there exists a constant $\alpha < 1$ such that

$$\|H(\mathbf{x}) - H(\mathbf{y})\| \leq \alpha\|\mathbf{x} - \mathbf{y}\|.$$

According to Lemma 7.5, there exists a unique solution $\mathbf{x}^*$ located in Φ_ι for system (7.1) with activation function (7.2) such that $H(\mathbf{x}^*) = \mathbf{x}^*$; i.e., $\mathbf{x}^*$ is an isolated equilibrium point located in Ω_ι for system (7.1). By arbitrariness of ε, $\mathbf{x}^*$ is an isolated equilibrium point located in Ω_κ for system (7.1) with activation function (7.2). This completes the proof.

According to Theorem 7.6, we can obtain the following corollary.

Corollary 7.7 *There exists one equilibrium point in Ω_κ for system (7.1) with activation function (7.2), if*

$$\begin{aligned}
&(a_{ii} + b_{ii})c_m + \sum_{j=1, j\neq i}^{n} |a_{ij} + b_{ij}| \max\{|c_1|, |c_k|\} + |u_i| < p_m, \\
&(a_{ii} + b_{ii})c_m - \sum_{j=1, j\neq i}^{n} |a_{ij} + b_{ij}| \max\{|c_1|, |c_k|\} - |u_i| > p_{m-1},
\end{aligned} \tag{7.8}$$

where $i = 1, 2, \ldots, n$; $m = 1, 2, \ldots, k$.

Proof It is noted that

$$\begin{aligned}
&(a_{ii} + b_{ii})c_m + \sum_{j=1, j\neq i}^{n} (a_{ij} + b_{ij})c_{q_j} + u_i \\
&\leq (a_{ii} + b_{ii})c_m + \sum_{j=1, j\neq i}^{n} |a_{ij} + b_{ij}| \max\{|c_1|, |c_k|\} + |u_i|,
\end{aligned} \tag{7.9}$$

$$\begin{aligned}
&(a_{ii} + b_{ii})c_m + \sum_{j=1, j\neq i}^{n} (a_{ij} + b_{ij})c_{q_j} + u_i \\
&\geq (a_{ii} + b_{ii})c_m - \sum_{j=1, j\neq i}^{n} |a_{ij} + b_{ij}| \max\{|c_1|, |c_k|\} - |u_i|.
\end{aligned} \tag{7.10}$$

where $i = 1, 2, \ldots, n$, $m = 1, 2, \ldots, k$, $q_j = 1, 2, \ldots, k$, $j = 1, 2, \cdots, n$. From (7.8)–(7.10), we can get that

$$p_{m-1} < (a_{ii} + b_{ii})c_m + \sum_{j=1, j\neq i}^{n} (a_{ij} + b_{ij})c_{q_j} + u_i < p_m.$$

According to Theorem 7.6, there exists one equilibrium point in Ω_κ for system (7.1) with activation function (7.2).

Remark 7.8 It is noted that the number of such region Ω_κ is k^n in Ω. Hence there exist k^n equilibrium points for system (7.1) with activation function (7.2) in Ω.

Next, we will investigate the stability of multiple equilibrium points for system (7.1) with activation function (7.2).

Theorem 7.9 *If one of conditions (7.3) and (7.8) holds, then system (7.1) with activation function (7.2) has k^n locally exponentially stable equilibrium points in Ω.*

Proof For any $\Omega_\nu \subset \Omega$, according to Theorem 7.6, there exists one equilibrium point $\mathbf{x}^* = (x_1^*, x_2^*, \cdots, x_3^*)^T \in \Omega_\nu$. Let $\mathbf{x}(t)$ be the solution of system (7.1) with the initial condition $\mathbf{x}(t_0) \in \Omega_\nu$. Denote $z_i(t) = x_i(t) - x_i^*$, $i = 1, 2, \cdots, n$. Then

$$\frac{\mathrm{d}z_i(t)}{\mathrm{d}t} = -z_i(t). \tag{7.11}$$

Thus, we obtain $|x_i(t) - x_i^*| = |x_i(t_0) - x_i^*| \exp\{-(t - t_0)\}$. $\mathbf{x}^*$ is a locally exponentially stable equilibrium point for system (7.1) with activation function (7.2) in Ω_ν. Due to the arbitrariness of Ω_ν, we conclude that system (7.1) has k^n locally exponentially stable equilibrium points in Ω.

Remark 7.10 Activation function (7.2) is composed of k sections. Neural networks with this class of *activation functions* can store more patterns than those with a stair-style activation function (such as [10, 12]) in practical applications. Based on the configuration of activation function (7.2), the space $\mathbb{R}^n$ can contain k^n parts. Each of these parts is an attractive basin of one equilibrium point. If the length of any one section (for example (p_r, p_{r-1})) in activation function (7.2) increases, then the attractive basin of the corresponding equilibrium point will increase. One large attractive basin can contain sufficient information about the stable pattern. We can conclude that the attractive basins of equilibrium points rely heavily on the configuration of activation functions.

It is worth noting that the above results can be applied to the following system without delay:

$$\frac{\mathrm{d}x_i(t)}{\mathrm{d}t} = -x_i(t) + \sum_{j=1}^{n} a_{ij} f_j(x_j(t)) + u_i, \; i = 1, 2, \cdots, n. \tag{7.12}$$

Corollary 7.11 *There exists one equilibrium point in Ω_κ for system (7.12) with activation function (7.2), if the following conditions hold:*

$$p_{m-1} < a_{ii} c_m + \sum_{j=1, j \neq i}^{n} a_{ij} c_{q_j} + u_i < p_m, \tag{7.13}$$

where $i = 1, 2, \cdots, n$, $m = 1, 2, \cdots, k$, $q_j = 1, 2, \cdots, k$, $j = 1, 2, \cdots, n$.

Corollary 7.12 *There exists one equilibrium point in* Ω_κ *for system (7.12) with activation function (7.2), if*

$$
\begin{aligned}
a_{ii}c_m + \sum_{j=1, j\neq i}^{n} |a_{ij}| \max\{|c_1|, |c_k|\} + |u_i| &< p_m, \\
a_{ii}c_m - \sum_{j=1, j\neq i}^{n} |a_{ij}| \max\{|c_1|, |c_k|\} - |u_i| &> p_{m-1},
\end{aligned} \tag{7.14}
$$

where $i = 1, 2, \cdots, n$; $m = 1, 2, \cdots, k$.

Corollary 7.13 *If one of conditions (7.13) and (7.14) holds, then system (7.12) with activation function (7.2) has* k^n *locally exponentially stable equilibrium points in* Ω.

Remark 7.14 Reference [24] discussed multistability of neural networks without delay and with discontinuous activation functions. When we denote $c_1 = -2$ and $c_k = 2$ in this chapter, it is easy to find that the parameters go beyond the definition of activation functions in [24]. Hence, it cannot be applied in [24]. But it can be applied in our chapter. So, our results are superior to that in [24].

Remark 7.15 In this chapter, we can regard all subregions in Ω as a whole region. Based on sufficient conditions in Theorems 7.6 and 7.9 and Corollaries 7.12–7.13, there exist k^n equilibrium points which are exponentially stable in the whole region. For any initial condition ϕ in the whole region, the system will be exponentially convergence toward an equilibrium point corresponding to initial condition ϕ. An important application of *multistability* of recurrent neural networks is to implement pattern memory [39]. A recalling probe, which is sufficiently similar to the pattern to be retrieved, is set as an initial state and the state variables converge to one of the locally stable equilibrium points, which corresponds to the pattern to be retrieved. However, in this case, it is difficult to avoid spurious equilibrium points, and accurate pattern recalling cannot be guaranteed using initial conditions. Therefore, it is necessary to reduce the number of spurious equilibrium points. In [40], analysis and design of associative memories based on recurrent neural networks with linear saturation activation functions was presented, and a formula for the number of spurious equilibrium points was derived. In further research work, reducing number of spurious equilibrium points in attractive regions for recurrent neural networks with discontinuous activation functions will be an issue to be further studied.

Now, we will discuss the existence of sets of stable equilibrium points and unstable equilibrium points for system (7.12) with activation function (7.2). Denote

$$
\begin{aligned}
N_1 &= \{j | x_j = p_{q_j}, q_j = 1, 2, \cdots, k-1, j \in \{1, 2, \cdots, n\}\}, \\
N_2 &= \{j | x_j \in (p_{q_j}, p_{q_j+1}), q_j = 0, 1, \cdots, k-1, j \in \{1, 2, \cdots, n\}\}.
\end{aligned}
$$

where x_j represents the state of the jth neuron for system (7.1).

Theorem 7.16 *There exists a set of equilibrium points for system (7.12) with activation function (7.2) in $\mathbb{R}^n$ if*

$$\left\{\sum_{j\in N_1} a_{ij}[c_{q_j}, c_{q_j+1}] + \sum_{j\in N_2} a_{ij}c_{q_j} + u_i\right\} \cap \left\{p_1, p_2, \cdots, p_{k-1}\right\} \neq \emptyset; \quad (7.15)$$

moreover, an equilibrium point $\mathbf{x}^*$ in the set of equilibrium points is stable if

$$|x_i^* - \sum_{j=1}^{n} a_{ij}\zeta_j - u_i| < |x_i^* - x_i(t_0)|; \quad (7.16)$$

$\mathbf{x}^*$ is unstable if

$$|x_i^* - \sum_{j=1}^{n} a_{ij}\zeta_j - u_i| > |x_i^* - x_i(t_0)|, \quad (7.17)$$

and one of the following conditions is true

$$\sum_{j=1}^{n} a_{ij}\zeta_j + u_i - x_i^* < 0, \quad (7.18)$$

$$\sum_{j=1}^{n} a_{ij}\zeta_j + u_i - x_i^* > 0, \quad (7.19)$$

where $i, j = 1, 2, \cdots, n$, $\zeta_j \in [c_1, c_k]$, $\boldsymbol{x}(t_0) = (x_1(t_0), x_2(t_0), \ldots, x_n(t_0))^T \in \mathbb{R}^n$ is any initial state.

Proof Let $x_i(t)$ be a solution of system (7.12) in the sense of Filippov with initial condition $x_i(t_0)$. It satisfies the differential inclusion

$$\frac{\mathrm{d}x_i(t)}{\mathrm{d}t} \in -x_i(t) + \sum_{j=1}^{n} a_{ij}K[f_j(x_j(t))] + u_i \quad (7.20)$$

The stationary equation for (7.20) is

$$0 \in -x_i^* + \sum_{j=1}^{n} a_{ij}K[f_j(x_j^*)] + u_i, \quad (7.21)$$

where

$$K[f_j(x_j^*)] = \begin{cases} c_1, & x_j^* \in (-\infty, p_1), \\ [c_1, c_2], & x_j^* = p_1, \\ c_2, & x_j^* \in (p_1, p_2), \\ \cdots & \\ c_{k-1}, & x_j^* \in (p_{k-2}, p_{k-1}), \\ \left[c_{k-1}, c_k\right], & x_j^* = p_{k-1}, \\ c_k, & x_j^* \in (p_{k-1}, +\infty). \end{cases}$$

We claim that $N_1 \neq \emptyset$. If it is not true, then $x_i^* = \sum_{j=1}^{n} a_{ij} c_{q_j} + u_i$. From (7.15), we have $x_i^* \in \{p_1, p_2, \cdots, p_{k-1}\}$. Therefore, $f_i(x_i^*) \in [c_{q_i}, c_{q_i+1}]$ ($q_i \in \{1, 2, \cdots, k-1\}$). This yields a contradiction to $N_1 = \emptyset$. So, $N_1 \neq \emptyset$ and $x_i^* \in \sum_{j \in N_1} a_{ij}[c_{q_j}, c_{q_j+1}] + \sum_{j \in N_2} a_{ij} c_{q_j} + u_i$. There exists a set of equilibrium points for system (7.12) with activation function (7.2). Now, we discuss the stability of equilibrium points. There exist $\zeta_j \in [c_1, c_k]$ ($j = 1, 2, \cdots, n$) such that

$$\frac{\mathrm{d}x_i(t)}{\mathrm{d}t} = -x_i(t) + \sum_{j=1}^{n} a_{ij}\zeta_j + u_i. \tag{7.22}$$

The solution of Eq. (7.22) with initial condition $x_i(t_0)$ is given by

$$x_i(t) = \left(x_i(t_0) - \sum_{j=1}^{n} a_{ij}\zeta_j - u_i \right) e^{-(t-t_0)} + \sum_{j=1}^{n} a_{ij}\zeta_j + u_i. \tag{7.23}$$

Case 1. Based on condition (7.16), we have

$$\begin{aligned} |x_i(t) - x_i^*| &= | \left(x_i(t_0) - \sum_{j=1}^{n} a_{ij}\zeta_j - u_i \right) e^{-(t-t_0)} + \sum_{j=1}^{n} a_{ij}\zeta_j + u_i - x_i^* | \\ &< |x_i(t_0) - \sum_{j=1}^{n} a_{ij}\zeta_j - u_i| + |x_i^* - \sum_{j=1}^{n} a_{ij}\zeta_j - u_i| \\ &< |x_i^* - x_i(t_0)| + 2|x_i^* - \sum_{j=1}^{n} a_{ij}\zeta_j - u_i| \\ &< 3|x_i^* - x_i(t_0)|. \end{aligned} \tag{7.24}$$

Therefore, $\mathbf{x}^*$ is a stable equilibrium point.

Case 2. Based on condition (7.17), we can derive the following result.

$$|x_i^* - \sum_{j=1}^{n} a_{ij}\zeta_j - u_i| > |x_i^* - x_i(t_0)|$$
$$\geq |x_i(t_0) - \sum_{j=1}^{n} a_{ij}\zeta_j - u_i| - |x_i^* - \sum_{j=1}^{n} a_{ij}\zeta_j - u_i|.$$

Therefore,

$$2|x_i^* - \sum_{j=1}^{n} a_{ij}\zeta_j - u_i| > |x_i(t_0) - \sum_{j=1}^{n} a_{ij}\zeta_j - u_i|. \tag{7.25}$$

From (7.25), we have

$$\left(x_i(t_0) - \sum_{j=1}^{n} a_{ij}\zeta_j - u_i\right) e^{-(t-t_0)} + \sum_{j=1}^{n} a_{ij}\zeta_j + u_i - x_i^*$$
$$< 2e^{-(t-t_0)}|x_i^* - \sum_{j=1}^{n} a_{ij}\zeta_j - u_i| + \sum_{j=1}^{n} a_{ij}\zeta_j + u_i - x_i^*, \tag{7.26}$$
$$\left(x_i(t_0) - \sum_{j=1}^{n} a_{ij}\zeta_j - u_i\right) e^{-(t-t_0)} + \sum_{j=1}^{n} a_{ij}\zeta_j + u_i - x_i^*$$
$$> -2e^{-(t-t_0)}|x_i^* - \sum_{j=1}^{n} a_{ij}\zeta_j - u_i| + \sum_{j=1}^{n} a_{ij}\zeta_j + u_i - x_i^*. \tag{7.27}$$

That is

$$x_i(t) - x_i^* < 2e^{-(t-t_0)}|x_i^* - \sum_{j=1}^{n} a_{ij}\zeta_j - u_i| + \sum_{j=1}^{n} a_{ij}\zeta_j + u_i - x_i^*, \tag{7.28}$$

$$x_i(t) - x_i^* > -2e^{-(t-t_0)}|x_i^* - \sum_{j=1}^{n} a_{ij}\zeta_j - u_i| + \sum_{j=1}^{n} a_{ij}\zeta_j + u_i - x_i^*. \tag{7.29}$$

From (7.18), there exists $T_1 > t_0$ such that, for $t > T_1$,

$$2e^{-(t-t_0)}|x_i^* - \sum_{j=1}^{n} a_{ij}\zeta_j - u_i| + \sum_{j=1}^{n} a_{ij}\zeta_j + u_i - x_i^*$$

$$< 2e^{-(T_1-t_0)}|x_i^* - \sum_{j=1}^{n} a_{ij}\zeta_j - u_i| + \sum_{j=1}^{n} a_{ij}\zeta_j + u_i - x_i^* < 0.$$

And from (7.28), we have

$$|x_i(t) - x_i^*| > |2e^{-(T_1-t_0)}|x_i^* - \sum_{j=1}^{n} a_{ij}\zeta_j - u_i| + \sum_{j=1}^{n} a_{ij}\zeta_j + u_i - x_i^*|.$$

Similarly, from (7.19) and (7.29), there exists $T_2 > t_0$ such that, for $t > T_2$,

$$|x_i(t) - x_i^*| > | - 2e^{-(T_2-t_0)}|x_i^* - \sum_{j=1}^{n} a_{ij}\zeta_j - u_i| + \sum_{j=1}^{n} a_{ij}\zeta_j + u_i - x_i^*|.$$

Therefore, $\mathbf{x}^*$ is an *unstable* equilibrium point. This completes the proof.

Remark 7.17 Throughout this book, only this chapter does not use the well-known *Lyapunov stability theory* in the stability analysis of RNNs with discontinuous *activation function*. In the discontinuous case, the concept of *solution in the sense of Filippov* is used instead of the *stability in the sense of Lyapunov*. Therefore, for the continuous case, Lyapunov theory can play an important role in the qualitative analysis of RNNs, while for the discontinuous case, some other analysis method or theory should be adopted.

7.4 Illustrative Examples

In this section, three examples are presented to illustrate our results.

Example 7.18 Consider the following 2-neuron delayed recurrent neural networks with 3-level discontinuous activation function

$$\begin{aligned}\dot{x}_1(t) =& - x_1(t) + 0.7f(x_1(t)) + 0.5f(x_2(t)) \\ &+ 0.5f(x_1(t - \sin(t))) - 0.51f(x_2(t - \sin(t))) + 0.1, \\ \dot{x}_2(t) =& - x_2(t) - 0.62f(x_1(t)) + 0.6f(x_2(t)) \\ &+ 0.6f(x_1(t - \sin(t))) + 0.7f(x_2(t - \sin(t))) - 0.2. \end{aligned} \tag{7.30}$$

The activation function is described by

$$f(\xi) = \begin{cases} -1, & \xi \in (-\infty, -1), \\ 0, & \xi \in (-1, 1), \\ 1, & \xi \in (1, +\infty). \end{cases} \tag{7.31}$$

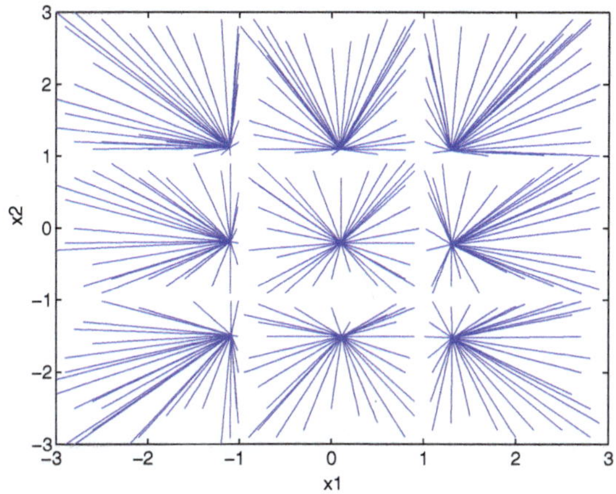

Fig. 7.3 Transient behaviors of $x_1(t)$ and $x_2(t)$ of system (7.30)

This system satisfies the conditions of Theorem 7.6. According to Theorem 7.6, system (7.30) has $3^2 = 9$ locally exponentially stable equilibrium points. The dynamics of this system are depicted in Fig. 7.3, where evolutions of 270 initial conditions have been tracked.

Example 7.19 Consider the following 2-neuron recurrent neural networks without delay and with 3-level discontinuous activation function

$$\begin{aligned} \dot{x}_1(t) &= -x_1(t) + 0.7f(x_1(t)) + 0.1f(x_2(t)) - 0.1, \\ \dot{x}_2(t) &= -x_2(t) - 0.15f(x_1(t)) + 0.8f(x_2(t)) - 0.2. \end{aligned} \tag{7.32}$$

The activation function is described by

$$f(\xi) = \begin{cases} -2, & \xi \in (-\infty, -1), \\ 1, & \xi \in (-1, 1), \\ 2, & \xi \in (1, +\infty). \end{cases} \tag{7.33}$$

The value of activation function (7.33) is located in $[-2, 2]$, which goes beyond the definition of activation functions in [24]. Hence, it cannot be applied in [24]. Whereas, the activation function (7.33) is useful in this chapter. System (7.32) with activation function (7.33) satisfies the conditions of Corollary 7.11. According to Corollarys 7.11 and 7.13, system (7.32) has $3^2 = 9$ locally exponentially stable equilibrium points. The dynamics of this system are depicted in Fig. 7.4, where evolutions of 270 initial conditions have been tracked.

Example 7.20 Design a recurrent neural network with 25 neurons to store three patterns shown in Fig. 7.5 as stable memories (black $= -1$ and white $= 1$).

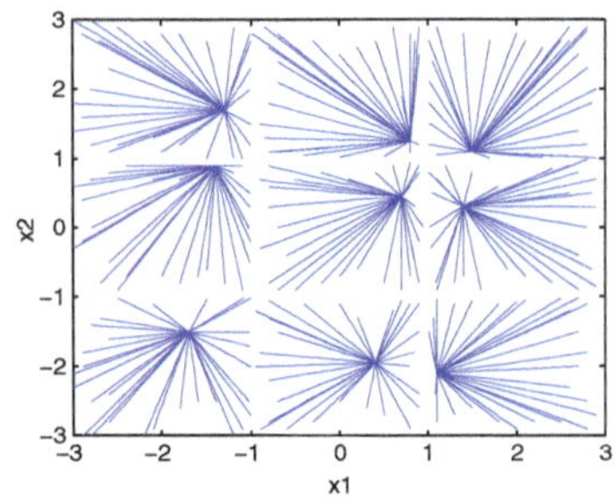

Fig. 7.4 Transient behaviors of $x_1(t)$ and $x_2(t)$ of system (7.32)

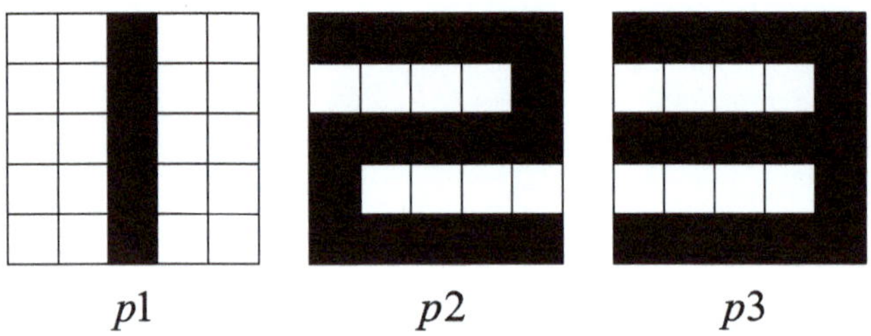

Fig. 7.5 Three desired memory patterns for Example 7.20

Step 1. Choose activation function as follows:

$$f(\xi) = \begin{cases} -1, & \xi \in (-\infty, 0), \\ 1, & \xi \in (0, +\infty). \end{cases}$$

Step 2. Choose connection weight matrix $A = (a_{ij})_{25\times 25}$,

$$a_{ij} = \begin{cases} 5, & i = j, \\ 0.2, & i \neq j, \end{cases}$$

and input vector $\mathbf{u} = 0$.

According to Corollary 7.11, the following recurrent neural network

$$\frac{\mathrm{d}\mathbf{x}(t)}{\mathrm{d}t} = -\mathbf{x}(t) + Af(\mathbf{x}(t)) \tag{7.34}$$

can store 2^{25} stable memory patterns. When one initial condition is sufficiently near to the equilibrium point corresponding to the pattern which is retrieved, the desired output pattern can be derived. Specifically, simulation results with six random initial values, which are sufficiently near to the equilibrium point corresponding to the pattern $p3$, are depicted in Fig. 7.6. We can see that Fig. 7.6d is the desired output pattern, and Fig. 7.6a–c,e,f are spurious memory patterns. Removing spurious memory patterns is a very difficult problem at present. In further research work, reducing number of spurious memory patterns is a challenging topic.

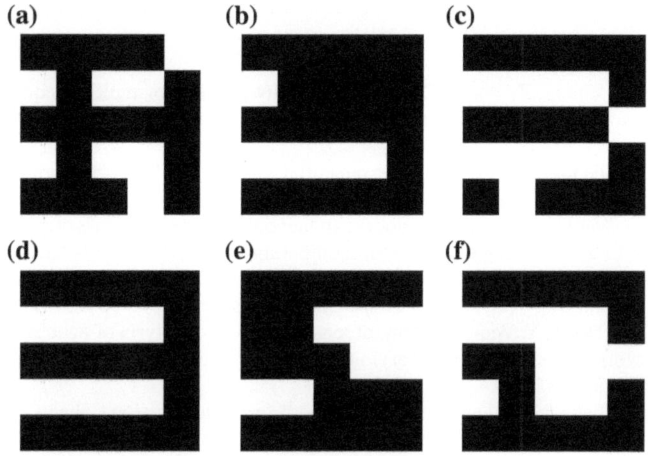

Fig. 7.6 Transient behaviors of output variables with six initial conditions

7.5 Summary

In this chapter, we have discussed multistability for n-dimensional recurrent neural networks with time-varying delays and k-level discontinuous activation functions. Some sufficient conditions have been established to ensure that system could have k^n locally exponentially stable equilibrium points. Moreover, we have also derived conditions for the existence of sets of stable equilibrium points and unstable equilibrium points for recurrent neural networks without delay and with discontinuous activation functions. The activation functions are k-level discontinuous. Neural network models with this class of activation functions can store more patterns than those with a stair-style activation function (such as [10, 12]) in practical applications. Compared with [24], our activation function value is located in larger range. Hence, our results are less conservative. In multi-equilibrium associative memories, it is necessary that each dynamic trajectory converges to one of locally asymptotically stable equilibrium points. However, when the initial state of a neural network is located at arbitrary given attractive region, it is difficult to avoid spurious equilibrium points. So, our further research work might continue to reduce the number of spurious equilibrium points in attractive regions for recurrent neural networks with discontinuous activation functions. It might be possible to extend the current results to synchronization and state estimation of neural networks with discontinuous activation functions.

References

1. Y. Huang, H. Zhang, Z. Wang, Dynamical stability analysis of multiple equilibrium points in time-varying delayed recurrent neural networks with discontinuous activation functions. Neurocomputing **91**, 21–28 (2012)
2. Z. Wang, H. Zhang, W. Yu, Robust exponential stability analysis of neural networks with multiple time delays. Neurocomputing **70**(15), 2534–2543 (2007)
3. H. Zhang, G. Wang, New criteria of global exponential stability for a class of generalized neural networks with time-varying delays. Neurocomputing **70**(15), 2486–2494 (2007)
4. H. Zhang, Z. Wang, D. Liu, Robust exponential stability of recurrent neural networks with multiple time-varying delays. IEEE Trans. Circuits Syst. II **54**(8), 730–734 (2007)
5. H. Zhang, M. Dong, Y. Wang, N. Sun, Stochastic stability analysis of neutral-type impulsive neural networks with mixed time-varying delays and Markovian jumping. Neurocomputing **73**(15), 2689–2695 (2010)
6. Z. Wang, H. Zhang, Global asymptotic stability of reaction-diffusion Cohen-Grossberg neural networks with continuously distributed delays. IEEE Trans. Neural Netw. **21**(1), 39–49 (2010)
7. H. Zhang, Z. Liu, G. Huang, Z. Wang, Novel weighting-delay-based stability criteria for recurrent neural networks with time-varying delay. IEEE Trans. Neural Netw. **21**(1), 91–106 (2010)
8. H. Schmidt, R. Petkov, M. Richardson, Dynamics on networks: the role of local dynamics and global networks on the emergence of hypersynchronous neural activity. Plos Comput. Biol. **10**(11), e1003947 (2014)
9. C. Cheng, K. Lin, C. Shih, Multistability in recurrent neural networks. SIAM J. Appl. Math. **66**(4), 1301–1320 (2006)
10. C. Cheng, K. Lin, C. Shih, Multistability and convergence in delayed neural networks. Phys. D: Nonlinear Phenom. **225**(1), 61–74 (2007)
11. G. Huang, J. Cao, Multistability in bidirectional associative memory neural networks. Phys. Lett. A **372**(16), 2842–2854 (2008)
12. J. Cao, G. Feng, Y. Wang, Multistability and multiperiodicity of delayed Cohen-Grossberg neural networks with a general class of activation functions. Phys. D: Nonlinear Phenom. **237**(13), 1734–1749 (2008)
13. Y. Huang, Z. Wang, H. Zhang, T. Zhang, Multistability and enlarged basins of attraction in bidirectional associative memory neural networks, in *2010 IEEE International Conference on Networking, Sensing and Control* (2010), pp. 376–381
14. L. Wang, W. Lu, T. Chen, Multistability and new attraction basins of almost periodic solutions of delayed neural networks. IEEE Trans. Neural Netw. **20**(10), 1581–1593 (2009)
15. L. Wang, W. Lu, T. Chen, Coexistence and local stability of multiple equilibria in neural networks with piecewise linear nondecreasing activation functions. Neural Netw. **23**(2), 189–200 (2010)
16. Z. Zeng, T. Huang, W. Zheng, Multistability of recurrent neural networks with time-varying delays and the piecewise linear activation function. IEEE Trans. Neural Netw. **21**(8), 1371–1377 (2010)
17. Y. Huang, X. Zhang, Multistability properties of linear threshold discrete-time recurrent neural networks. Int. J. Inf. Syst. Sci. **7**(1), 1–10 (2011)
18. Z. Yi, K. Tan, Multistability of discrete-time recurrent neural networks with unsaturating piecewise linear activation functions. IEEE Trans. Neural Netw. **15**(2), 329–336 (2004)
19. L. Zhang, Z. Yi, J. Yu, Multiperiodicity and attractivity of delayed recurrent neural networks with unsaturating piecewise linear transfer functions. IEEE Trans. Neural Netw. **19**(1), 158–167 (2008)
20. L. Zhang, Z. Yi, J. Yu, P. Heng, Some multistability properties of bidirectional associative memory recurrent neural networks with unsaturating piecewise linear transfer functions. Neurocomputing **72**(18), 3809–3817 (2009)
21. T. Chen, S. Amari, New theorems on global convergence of some dynamical systems. Neural Netw. **14**(3), 251–255 (2001)

22. W. Lu, T. Chen, Dynamical behaviors of Cohen-Grossberg neural networks with discontinuous activation functions. Neural Netw. **18**(3), 231–242 (2005)
23. L. Zhang, Z. Yi, S. Zhang, P.A. Heng, Activity invariant sets and exponentially stable attractors of linear threshold discrete-time recurrent neural networks. IEEE Trans. Autom. Control **54**(6), 1341–1347 (2009)
24. G. Huang, J. Cao, Multistability of neural networks with discontinuous activation function. Commun. Nonlinear Sci. Numer. Simulation **13**(10), 2279–2289 (2008)
25. Z. Zeng, D. Huang, Z. Wang, Memory pattern analysis of cellular neural networks. Phys. Lett. A **342**(2), 114–128 (2005)
26. Z. Zeng, J. Wang, Multiperiodicity of discrete-time delayed neural networks evoked by periodic external inputs. IEEE Trans. Neural Netw. **17**(5), 1141–1151 (2006)
27. M. Morita, Associative memory with nonmonotone dynamics. Neural Netw. **6**(1), 115–126 (1993)
28. M. Forti, P. Nistri, Global convergence of neural networks with discontinuous neuron activation. IEEE Trans. Circuits Syst. I **50**(11), 1421–1435 (2003)
29. M. Forti, P. Nistri, D. Papini, Global exponential stability and global convergence in finite time of delayed neural networks with infinite gain. IEEE Trans. Neural Netw. **16**(6), 1449–1463 (2005)
30. M. Forti, M. Grazzini, P. Nistri, L. Pancioni, Generalized Lyapunov approach for convergence of neural networks with discontinuous or non-Lipschitz activations. Phys. D: Nonlinear Phenom. **214**(1), 88–99 (2006)
31. M. Forti, M-matrices and global convergence of discontinuous neural networks. Int. J. Circuit Theory Appl. **35**(2), 105–130 (2007)
32. L. Li, L. Huang, Global asymptotic stability of delayed neural networks with discontinuous neuron activations. Neurocomputing **72**(16–18), 3726–3733 (2009)
33. X. Liu, J. Cao, On periodic solutions of neural networks via differential inclusions. Neural Netw. **22**(4), 329–334 (2009)
34. X. Liu, J. Cao, Robust state estimation for neural networks with discontinuous activations. IEEE Trans. Syst. Man Cybern.: Part B **40**(6), 1425–1437 (2010)
35. X. Liu, J. Cao, Local synchronization of one-to-one coupled neural networks with discontinuous activations. Cogn. Neurodynamics **5**(1), 13–20 (2011)
36. X. Liu, T. Chen, J. Cao, W. Lu, Dissipativity and quasi-synchronization for neural networks with discontinuous activations and parameter mismatches. Neural Netw. **24**(10), 1013–1021 (2011)
37. A. Filippov, *Differential Equations with Discontinuous Righthand Sides* (Kluwer Academic, Boston, 1988)
38. K. Yosida, *Functional Analysis* (Springer, Berlin, 1978)
39. Z. Zeng, J. Wang, Associative memories based on continuous-time cellular neural networks designed using space-invariant cloning templates. Neural Netw. **22**(6), 651–657 (2009)
40. Z. Zeng, J. Wang, Analysis and design of associative memories based on recurrent neural networks with linear saturation activation functions and time-varying delays. Neural Comput. **19**(8), 2149–2182 (2007)
41. Y. Huang, H. Zhang, Z. Wang, Dynamical stability analysis of multiple equilibrium points in time-varying delayed recurrent neural networks with discontinuous activation functions. Neurocomputing **91**, 21–28 (2012)
42. Z. Zeng, J. Wang, Multiperiodicity and exponential attractivity evoked by periodic external inputs in delayed cellular neural networks. Neural comput. **18**(4), 848–870 (2006)
43. B. Wang, Y. Zeng, J. Cheng, Further improvement in delay-dependent stability criteria for continuous-time systems with time-varying delays. Neurocomputing **147**, 324–329 (2015)
44. W. Feng, S. Yang, H. Wu, Further results on robust stability of bidirectional associative memory neural networks with norm-bounded uncertainties. Neurocomputing **148**, 535–543 (2015)

Chapter 8
LMI-based Passivity Criteria for RNNs with Delays

As one of most widely used qualitative characteristics, stability property are studied for the equilibrium point of some kinds of RNNs with delays in Chaps. 4–7. For a dynamical system, there are many qualitative characteristics to be studied. In this chapter, we will study the passivity problem for neural networks with discrete and unbounded distributed time-varying delays. The contents in this chapter is mainly from the authors' previous paper [1].

8.1 Introduction

In the past two decades, neural networks have been extensively studied in many aspects and successfully applied to many fields such as pattern identifying, voice recognizing, system controlling, signal processing systems, static image treatment, and solving nonlinear algebraic equations, etc. In hardware implementation of neural networks, time delays are inevitably present due to the finite switching speeds of the amplifiers. It is well known that time delays not only deteriorate dynamical performance such as the boundary of the basin of attraction of the stable equilibria but also affect the stability of a network creating oscillatory and unstable characteristics. Hence, it is of primary importance to investigate the stability of delayed neural networks. There exist some results of stability for delayed neural networks, see, for example, [2–12]. Among them, *delay-dependent* criteria (see, e.g. [2, 3, 5]) make use of information on the length of delays, therefore they are usually less conservative than delay-independent ones (see, e.g. [9, 11, 13]) especially when the time delay is small. Much attention has been paid to the delay-dependent type recently.

Passivity can be used to demonstrate that passive circuits will be stable under specific criteria. In addition, passive circuits will not necessarily be stable under all stability criteria. For instance, a resonant series LC circuit will have unbounded voltage output for a bounded voltage input, but will be *stable* in the sense of Lyapunov, and

Z. Wang et al., *Qualitative Analysis and Control of Complex Neural Networks with Delays*, Studies in Systems, Decision and Control 34,
DOI 10.1007/978-3-662-47484-6_8

given bounded energy input will have bounded energy output. Passivity is frequently used in control systems to design stable control systems or to show stability in control systems. This is especially important in the design of large, complex control systems. Since neural networks are implemented in hardware based on large-scale integrated circuits, the concept of passivity has played an important role in the analysis of the stability of dynamical systems, nonlinear control, and other research areas [14–16]. The essence of the passivity theory is that the passive properties of a system can keep the system internal *stability*. Recently, the problem of passivity analysis for delayed neural networks has been addressed in [17–22], where sufficient conditions for passivity were established. Note that the passivity conditions in both [17] and [18] are delay-independent, which are usually more conservative than delay-dependent ones, particularly in the case when the delay size is small [5, 12]. Considering this, several delay-dependent passivity conditions for delayed neural networks were proposed in [13, 15, 23–26], which are based on linear matrix inequalities techniques (LMIs) and Jensen integral inequality or free-weighting matrix method. Especially, delay-dependent passivity results were obtained in [27] for neural networks with discrete and bounded distributed time-varying delays, which are also based on LMIs techniques and free-weighting matrix method. On the other hand, neural networks usually have a spatial extent due to the presence of a multitude of parallel pathways with a variety of axon sizes and lengths. Therefore, there will be a distribution of conduction velocities along these pathways. In this case, it is not suitable to model a distribution of propagation by discrete delays, and a more appropriate way is to incorporate continuously distributed delays. To the best of our knowledge, delay-dependent sufficient conditions for passivity of uncertain neural networks with discrete and unbounded distributed time-varying delays have not yet been established in the literature, mainly due to the mathematical difficulties in dealing with discrete and unbounded distributed delays simultaneously. Hence, it is our intention in this chapter to tackle such an important yet challenging problem.

Motivated by the preceding discussions, our objective in this chapter is to study the passivity of a class of uncertain neural networks with discrete and unbounded distributed time-varying delays. By combining the Gu's discretized procedure [2, 3, 28–30] on complete LKF (Lyapunov–Krasovskii Functional) with the free-weighting matrix technique, we develop a new discretized LKF method for analyzing the passivity of delayed neural networks. Moreover, due to the presence of modeling error, external disturbance or parameter fluctuation during the physical implementation, uncertainty is unavoidable and may affect the passivity of the whole system. Thus, it is necessary to further extend our main result to the uncertain case. In this regard, we investigate the robust passivity of neural networks by considering two kinds of uncertainties: time-varying structured uncertainty and the interval uncertainty. In the form of LMIs, delay-dependent passivity conditions are obtained for two kinds of uncertain neural networks, respectively. These conditions can be easily checked by using recently developed algorithms in solving LMIs. The less conservatism of the proposed conditions is demonstrated via a numerical example.

In the following, we let the shorthand $\mathrm{col}\{M_1, M_2, \ldots, M_k\}$ denote a column matrix with the matrices $M_1, M_2, \ldots, M_k$.

8.2 Problem Formulation

Considering the following cellular neural networks with time-varying delay:

$$\begin{aligned}\dot{x}(t) &= -Ax(t) + W_0 f(x(t)) + W_1 f(x(t-\tau(t))) \\ &\quad + W_2 \int_{-\infty}^{t} K(t-s) f(x(s))\mathrm{d}s + u(t), \\ y(t) &= f(x(t)), \\ x(t) &= \phi(t), \quad t \in [-\bar{\tau}, 0]\end{aligned} \tag{8.1}$$

where $x(t) = (x_1(t), x_2(t), \ldots, x_n(t))^T \in \mathbb{R}^n$ is the neural state vector, $A = \mathrm{diag}(a_1, a_2, \ldots, a_n)$ is a positive diagonal matrix, $W_\iota = (w_{ij}^{\iota})_{n\times n}$ $(\iota = 0, 1, 2)$ are known constant matrices, $0 \le \tau(t) \le \bar{\tau}$ is the time-varying delay, and $\bar{\tau}$ is a constant. $u(t)$ is the external input vector, and $f(x(t)) = (f_1(x(t)), f_2(x_2(t)), \ldots, f_n(x_n(t)))^T \in \mathbb{R}^n$ denotes the neural activation function, continuous function $\phi(t) = (\phi_1(t), \phi_2(t), \ldots, \phi_n(t))^T \in \mathbb{R}^n$ is the initial condition. $K(t-s) = \mathrm{diag}(k_1(t-s), k_2(t-s), \ldots, k_n(t-s))$ denotes the delay kernel. It is assumed that k_i is a real value nonnegative continuous function defined in $[0, \infty)$ satisfying

$$\int_0^{\infty} k_i(s)\mathrm{d}s = 1, \; i = 1, 2, \ldots, n.$$

In this chapter, it is assumed that the neural activation function satisfies the following condition.

Assumption 8.1 Neural activation function is bounded and satisfies the following condition:

$$|f_j(s_1) - f_j(s_2)| \le \sigma_j |s_1 - s_2|, \tag{8.2}$$

where $f_j(0) = 0, \; j = 1, 2, \ldots, n, \; s_1, s_2 \in \mathbb{R}, s_1 \ne s_2$.

For notational simplicity, we denote $\Sigma = \mathrm{diag}\{\sigma_1, \sigma_2, \ldots, \sigma_n\}$.

We now introduce the following definition of passivity.

Definition 8.2 ([17]) The system in (8.1) is said to be *passive* if there exists a scalar $\gamma > 0$ such that for all $t_f \ge 0$

$$2\int_0^{t_f} y^T(s)u(s)\mathrm{d}s \ge -\gamma \int_0^{t_f} u^T(s)u(s)\mathrm{d}s. \tag{8.3}$$

under the zero initial condition.

In order to obtain the results, we need the following lemma.

Lemma 8.3 (see [9, 31]) *Let $\mathbb{H}$, $F(t)$ be real matrices of appropriate dimensions with $\mathbb{H} > 0$ and $F^T(t)F(t) \leq I$. Then for any matrices X and Y with appropriate dimensions, the following matrix inequality holds:*

$$X^T F(t)Y + Y^T F^T(t)X \leq X^T \mathbb{H}^{-1} X + Y^T \mathbb{H} Y.$$

8.3 Passivity for RNNs Without Uncertainty

Now, we present the passivity result for system (8.1) with $\dot{\tau}(t) \leq \eta < 1$.

Theorem 8.4 *Under Assumption 8.1, system (8.1) is passive for $0 \leq \tau(t) \leq \bar{\tau} = \nu h$ (where ν is an positive integer), $\dot{\tau}(t) \leq \eta < 1$, if there exist constant scalar $\gamma > 0$, positive definite symmetric matrices B, $P = [P_{\alpha\kappa}]_{2\times 2}$, nonnegative definite symmetric matrices $Q = [Q_{pq}]_{3\times 3}$, R, S, E_j, positive diagonal matrices Ω, Ξ, Λ, T_1, T_2, real matrices $D_{lj} = D_{jl}^T$, $C_j (j, l = 0, 1, \ldots, \nu)$, $X_\iota (\iota = 1, \ldots, 4)$, Y with compatible dimensions such that the following LMIs hold ($\kappa = 1, 2$):*

$$\begin{bmatrix} B & \mathbb{C} \\ \mathbb{C}^T & \mathbb{D} + \mathbb{E} \end{bmatrix} > 0, \tag{8.4}$$

$$\begin{bmatrix} \Gamma + \overline{\Gamma} + \overline{\Gamma}^T & \mathcal{C} & \mathfrak{C} & \bar{\tau}\Phi_\kappa \\ \mathcal{C}^T & -\mathcal{D} - \mathcal{E} & 0 & 0 \\ \mathfrak{C}^T & 0 & -3\mathcal{E} & 0 \\ \bar{\tau}\Phi_\kappa^T & 0 & 0 & -S \end{bmatrix} < 0, \tag{8.5}$$

where

$$\Gamma = \begin{bmatrix} \Gamma_{11} & 0 & -C_\nu & 0 & P_{12} & Q_{12} & 0 & 0 & 0 \\ * & \Gamma_{22} & 0 & 0 & P_{22} - Q_{13} & 0 & -(1-\eta)Q_{12} & 0 & 0 \\ * & * & -R - E_\nu & 0 & 0 & 0 & 0 & 0 & 0 \\ * & * & * & \Gamma_{44} & 0 & 0 & 0 & 0 & 0 \\ * & * & * & * & -Q_{33} & 0 & -Q_{23}^T & 0 & 0 \\ * & * & * & * & * & \Gamma_Q & 0 & 0 & -I \\ * & * & * & * & * & * & -(1-\eta)Q_{22} - T_2 & 0 & 0 \\ * & * & * & * & * & * & * & -\Omega & 0 \\ * & * & * & * & * & * & * & * & -\gamma I \end{bmatrix}$$

with $\Gamma_Q = Q_{22} + \Omega - T_1$,

$$\mathbb{C} = [\, C_0 \; C_1 \; \ldots \; C_\nu \,], \; \mathbb{D} = [\, D_{lj} \,]_{0 \leq l, j \leq \nu}, \;\; \mathbb{E} = \frac{1}{h}\mathrm{diag}\{\, E_0 \; E_1 \; \ldots \; E_\nu \,\},$$

$$\Gamma_{11} = Q_{11} + R + C_0 + C_0^T + E_0 + \Sigma^2 T_1,$$

$$\Gamma_{22} = -(1-\eta)Q_{11} + \Sigma^2 T_2, \;\; \Gamma_{44} = Q_{33} + \bar{\tau}^2 S - Y - Y^T,$$

$$\overline{\Gamma} = \mathcal{P}[-A\ 0\ 0\ 0\ 0\ W_0\ W_1\ W_2\ I\,] + \bar{\tau}[\,\Phi_1\ -\Phi_1 + \Phi_2\ -\Phi_2\ 0\ 0\ 0\ 0\ 0\ 0\,],$$
$$\mathcal{P}^T = [\,B + P_{11} + Q_{13}^T + (\Xi + \Lambda)\Sigma\ \ P_{12}\ \ 0\ \ Y^T\ \ 0\ \ Q_{23}^T + \Xi - \Lambda\ \ 0\ 0\ 0\,],$$
$$\Phi_1^T = [\,X_1^T\ X_2^T\ 0\ 0\ 0\ 0\ 0\ 0\ 0\,],\quad \Phi_2^T = [\,0\ X_3^T\ X_4^T\ 0\ 0\ 0\ 0\ 0\ 0\,],$$
$$\mathcal{C} = [\,\mathcal{C}_{ij}\,]_{9\times\nu},\quad \mathfrak{C} = [\,\mathfrak{C}_{ij}\,]_{9\times\nu},\quad \mathcal{D} = [\,\mathcal{D}_{lj}\,]_{\nu\times\nu},$$
$$\mathcal{C}_{1j} = \frac{h}{2}(D_{0,j-1} + D_{0j}) - (C_{j-1} - C_j),$$
$$\mathcal{C}_{3j} = -\frac{h}{2}(D_{\nu,j-1} + D_{\nu j}),\quad \mathcal{C}_{4j} = \frac{h}{2}(C_{j-1} + C_j),$$
$$\mathfrak{C}_{1j} = -\frac{h}{2}(D_{0,j-1} - D_{0j}),$$
$$\mathfrak{C}_{3j} = \frac{h}{2}(D_{\nu,j-1} - D_{\nu j}),\quad \mathfrak{C}_{4j} = -\frac{h}{2}(C_{j-1} - C_j),$$
$$\mathcal{D}_{lj} = h(D_{l-1,j-1} - D_{lj}),\quad \mathcal{E} = \mathrm{diag}\{\,E_0 - E_1\ E_1 - E_2\ \ldots\ E_{\nu-1} - E_\nu\,\},$$

other parameters $\mathcal{C}_{ij}, \mathfrak{C}_{ij}\,(i = 1, \ldots, 9;\ j = 1, \ldots, \nu)$ *are all equal to zeros.*

Proof Employ the following LKF:

$$\begin{aligned}
V(t) =& V_1(t) + V_2(t), \qquad (8.6)\\
V_1(t) =& x^T(t)Bx(t) + 2x^T(t)\int_{-\bar{\tau}}^{0} C(s)x(t+s)\mathrm{d}s\\
&+ \int_{-\bar{\tau}}^{0}\int_{-\bar{\tau}}^{0} x^T(t+s)D(s,v)x(t+v)\mathrm{d}s\mathrm{d}v\\
&+ \int_{-\bar{\tau}}^{0} x^T(t+s)E(s)x(t+s)\mathrm{d}s,\\
V_2(t) =& \Theta^T(t)P\Theta(t) + \int_{-\tau(t)}^{0} \Upsilon^T(t+s)Q\Upsilon(t+s)\mathrm{d}s\\
&+ \int_{-\bar{\tau}}^{0} x^T(t+s)Rx(t+s)\mathrm{d}s\\
&+ \bar{\tau}\int_{-\bar{\tau}}^{0} (s+\bar{\tau})\dot{x}^T(t+s)S\dot{x}(t+s)\mathrm{d}s\\
&+ \sum_{i=1}^{n}\omega_i\int_0^{\infty} k_i(v)\int_{t-v}^{t} f_i^2(u_i(s))\mathrm{d}s\mathrm{d}v\\
&+ 2\sum_{i=1}^{n}\left\{\xi_i\int_0^{x_i(t)}(\sigma_i s + f_i(s))\mathrm{d}s + \lambda_i\int_0^{x_i(t)}(\sigma_i s - f_i(s))\mathrm{d}s\right\},
\end{aligned}$$

where $\Omega = \mathrm{diag}\{\omega_1, \omega_2, \ldots, \omega_n\}$, $\Xi = \mathrm{diag}\{\xi_1, \xi_2, \ldots, \xi_n\}$, $\Lambda = \mathrm{diag}\{\lambda_1, \lambda_2, \ldots, \lambda_n\}$, and $C(s)$, $D(s, v) = D^T(v, s)$, $E(s) = E^T(s)$ are continuous matrix functions,

$$\Theta^T(t) = \left[x^T(t)\ x^T(t-\tau(t)) \right],\ \Upsilon^T(s) = \left[x^T(s)\ f^T(x(s))\ \dot{x}^T(s) \right].$$

The time derivative of functional (8.6) along the trajectories of system (8.1) is obtained as follows:

$$\begin{aligned}
\dot{V}_1(t) =& 2x^T(t)B\dot{x}(t) + 2\dot{x}^T(t)\int_{-\bar{\tau}}^{0} C(s)x(t+s)\mathrm{d}s \\
&+ 2x^T(t)\int_{-\bar{\tau}}^{0} C(s)\dot{x}(t+s)\mathrm{d}s \\
&+ 2\int_{-\bar{\tau}}^{0}\int_{-\bar{\tau}}^{0} \dot{x}^T(t+s)D(s,v)x(t+v)\mathrm{d}s\mathrm{d}v \\
&+ 2\int_{-\bar{\tau}}^{0} \dot{x}^T(t+s)E(s)x(t+s)\mathrm{d}s, \\
\dot{V}_2(t) =& 2\Theta^T(t)P\dot{\Theta}(t) + \Upsilon^T(t)Q\Upsilon(t) - (1-\dot{\tau}(t))\Upsilon^T(t-\tau(t))Q\Upsilon(t-\tau(t)) \\
&+ x^T(t)Rx(t) - x^T(t-\bar{\tau})Rx(t-\bar{\tau}) + \bar{\tau}^2\dot{x}^T(t)S\dot{x}(t) \\
&- \bar{\tau}\int_{t-\bar{\tau}}^{t} \dot{x}^T(s)S\dot{x}(s)\mathrm{d}s \\
&+ \sum_{i=1}^{n}\omega_i\int_0^{\infty} k_i(s)\left\{f_i^2(u_i(t)) - f_i^2(u_i(t-s))\right\}\mathrm{d}s \\
&+ 2\{x^T(t)\Sigma + f^T(x(t))\}\Xi\dot{x}(t) + 2\{x^T(t)\Sigma - f^T(x(t))\}\Lambda\dot{x}(t).
\end{aligned} \tag{8.7}$$

From Cauchy–Schwarz inequality we have

$$\begin{aligned}
&\sum_{i=1}^{n}\omega_i\int_0^{\infty} k_i(s)\left(f_i^2(x_i(t)) - f_i^2(x_i(t-s))\right)\mathrm{d}s \\
&= f^T(x(t))\Omega f(x(t)) - \sum_{i=1}^{n}\omega_i\int_0^{\infty} k_i(s)\mathrm{d}s\int_0^{\infty} k_i(s)f_i^2(x_i(t-s))\mathrm{d}s \\
&\le f^T(x(t))\Omega f(x(t)) - \sum_{i=1}^{n}\omega_i\left(\int_0^{\infty} k_i(s)f_i(x_i(t-s))\mathrm{d}s\right)^2 \\
&= f^T(x(t))\Omega f(x(t)) \\
&\quad - \left(\int_{-\infty}^{t} K(t-s)f(x(s))\mathrm{d}s\right)^T \Omega\left(\int_{-\infty}^{t} K(t-s)f(x(s))\mathrm{d}s\right).
\end{aligned} \tag{8.8}$$

Now we use the Gu's discretized LKF method in [29] to choose continuous matrix functions $C(s)$, $D(s,v)$, $E(s)$. This technique consists in dividing the delay interval $[-\bar{\tau}, 0]$ into ν segments $[\rho_j, \rho_{j-1}]$ $(j = 1, 2, \ldots, \nu)$ of equal length h, then $\rho_j = -jh$. This method also divides the square $[-\bar{\tau}, 0] \times [-\bar{\tau}, 0]$ into ν^2 small squares

$[\rho_j, \rho_{j-1}] \times [\rho_l, \rho_{l-1}]$ $(j, l = 1, 2, \ldots, \nu)$. Each square is further divided into two triangles, which will be used in the expression of $D(s, v)$.

The continuous matrix functions $C(s)$ and $E(s)$ are chosen to be linear within each segment and the continuous matrix function $D(s, v)$ is chosen to linear with each triangular. Then, they can be expressed in terms of their values at the dividing points using a linear interpolation formula as

$$
\begin{aligned}
&C(\rho_j + \varsigma h) = \varsigma C_{j-1} + (1-\varsigma)C_j,\\
&E(\rho_j + \varsigma h) = \varsigma E_{j-1} + (1-\varsigma)E_j,\\
&D(\rho_j + \varsigma h, \rho_l + \epsilon h) = \begin{cases} \epsilon D_{j-1,l-1} + (\varsigma - \epsilon)D_{j-1,l} + (1-\varsigma)D_{jl}, \ \varsigma \geq \epsilon,\\ \varsigma D_{j-1,l-1} + (\epsilon - \varsigma)D_{j,l-1} + (1-\epsilon)D_{jl}, \ \varsigma < \epsilon, \end{cases}
\end{aligned}
$$

for $0 \leq \varsigma, \epsilon \leq 1,\ j, l = 1, \ldots, \nu$. Thus, $V_1(t)$ is completely determined by matrices B, C_j, D_{jl}, E_j $(j, l = 0, 1, \ldots, \nu)$.

Dividing the integral interval $[-\bar{\tau}, 0]$ into ν intervals $[\rho_j, \rho_{j-1}]$ $(j = 1, 2, \ldots, \nu)$ and integrating by parts, yields

$$
\begin{aligned}
\dot{V}_1(t) =& 2x^T(t)B\dot{x}(t) + 2\dot{x}^T(t)\sum_{j=1}^{\nu}\int_{\rho_j}^{\rho_{j-1}} C(s)x(t+s)\mathrm{d}s\\
&+ 2x^T(t)\sum_{j=1}^{\nu}\Big(C(s)x(t+s)\Big|_{\rho_j}^{\rho_{j-1}} - \int_{\rho_j}^{\rho_{j-1}} \dot{C}(s)x(t+s)\mathrm{d}s\Big)\\
&+ 2\sum_{j=1}^{\nu}\sum_{l=1}^{\nu}\Big(\int_{\rho_j}^{\rho_{j-1}} x^T(t+s)D(s,v)x(t+v)\Big|_{s=\rho_l}^{s=\rho_{l-1}}\mathrm{d}v\\
&- \int_{\rho_j}^{\rho_{j-1}}\int_{\rho_l}^{\rho_{l-1}} x^T(t+s)\Big(\frac{\partial D(s,v)}{\partial s} + \frac{\partial D(s,v)}{\partial v}\Big)x(t+v)\mathrm{d}s\mathrm{d}v\Big)\\
&+ \sum_{j=1}^{\nu}\Big(x^T(t+s)E(s)x(t+s)\Big|_{\rho_j}^{\rho_{j-1}} - \int_{\rho_j}^{\rho_{j-1}} x^T(t+s)\dot{E}(s)x(t+s)\mathrm{d}s\Big)\\
=& 2x^T(t)B\dot{x}(t) + x^T(t)\big(C(0) + C^T(0) + E(0)\big)x(t) - 2x^T(t)C(-\bar{\tau})x(t-\bar{\tau})\\
&- x^T(t-\bar{\tau})E(-\bar{\tau})x(t-\bar{\tau}) + 2\dot{x}^T(t)\sum_{j=1}^{\nu}\int_{\rho_j}^{\rho_{j-1}} C(s)x(t+s)\mathrm{d}s\\
&+ 2x^T(t)\sum_{j=1}^{\nu}\int_{\rho_j}^{\rho_{j-1}} \big(-\dot{C}(s) + D(0,s)\big)x(t+s)\mathrm{d}s\\
&- \sum_{j=1}^{\nu}\int_{\rho_j}^{\rho_{j-1}} \big[2x^T(t-\bar{\tau})D(-\bar{\tau},s) + x^T(t+s)\dot{E}(s)\big]x(t+s)\mathrm{d}s\\
&- \sum_{j=1}^{\nu}\sum_{l=1}^{\nu}\int_{\rho_j}^{\rho_{j-1}}\int_{\rho_l}^{\rho_{l-1}} x^T(t+s)\Big(\frac{\partial D(s,v)}{\partial s} + \frac{\partial D(s,v)}{\partial v}\Big)x(t+v)\mathrm{d}s\mathrm{d}v. \quad (8.9)
\end{aligned}
$$

It is noted that the following equations hold for any $s \in [\rho_j, \rho_{j-1}], v \in [\rho_l, \rho_{l-1}]$ $(j, l = 1, 2, \ldots, \nu)$,

$$\begin{aligned}\dot{C}(s) =& \frac{1}{h}(C_{j-1} - C_j),\\ \dot{E}(s) =& \frac{1}{h}(E_{j-1} - E_j),\\ \frac{\partial D(s, v)}{\partial s} + \frac{\partial D(s, v)}{\partial v} =& \frac{1}{h}(D_{j-1,l-1} - D_{jl}).\end{aligned}$$

Through some calculations, we can obtain

$$\begin{aligned}&2\int_{\rho_j}^{\rho_{j-1}} C(s)x(t+s)\mathrm{d}s\\ =&h\int_0^1 \big[(C_j + C_{j-1}) + (1-2\varsigma)(C_j - C_{j-1})\big]x(t+\rho_j+\varsigma h)\mathrm{d}\varsigma, \qquad (8.10)\\ &2\int_{\rho_j}^{\rho_{j-1}} \big(-\dot{C}(s) + D(0, s)\big)x(t+s)\mathrm{d}s\\ =&\int_0^1 \big[2(C_j - C_{j-1}) + h(D_{0j} + D_{0,j-1})\\ &+ (1-2\varsigma)h(D_{0j} - D_{0,j-1})\big]x(t+\rho_j+\varsigma h)\mathrm{d}\varsigma, \qquad (8.11)\end{aligned}$$

$$\begin{aligned}&2\int_{\rho_j}^{\rho_{j-1}} D(-\bar{\tau}, s)x(t+s)\mathrm{d}s\\ &= h\int_0^1 \big[(D_{\nu j} + D_{\nu,j-1}) + (1-2\varsigma)(D_{\nu j} - D_{\nu,j-1})\big]x(t+\rho_j+\varsigma h)\mathrm{d}\varsigma,\\ &\qquad (8.12)\\ &\int_{\rho_j}^{\rho_{j-1}}\int_{\rho_l}^{\rho_{l-1}} x^T(t+s)\Big(\frac{\partial D(s, v)}{\partial s} + \frac{\partial D(s, v)}{\partial v}\Big)x(t+v)\mathrm{d}s\mathrm{d}v\\ &= h\int_0^1\int_0^1 x^T(t+\rho_j+\epsilon h)(D_{j-1,l-1} - D_{jl})x(t+\rho_l+\varsigma h)\mathrm{d}\varsigma\mathrm{d}\epsilon, \qquad (8.13)\\ &\int_{\rho_j}^{\rho_{j-1}} x^T(t+s)\dot{E}(s)x(t+s)\mathrm{d}s\\ &= \int_0^1 x^T(t+\rho_j+\varsigma h)(E_{j-1} - E_j)x(t+\rho_j+\varsigma h)\mathrm{d}\varsigma. \qquad (8.14)\end{aligned}$$

On the other hand, from Assumption 8.1 that the following matrix inequalities hold for any positive diagonal matrices T_1, T_2 with compatible dimensions

$$x^T(t)T_1\Sigma^2x(t) - f^T(x(t))T_1f(x(t)) \geq 0, \tag{8.15}$$

$$x^T(t-\tau(t))T_2\Sigma^2x(t-\tau(t)) - f^T(x(t-\tau(t)))T_2f(x(t-\tau(t))) \geq 0. \tag{8.16}$$

Furthermore, it is easy to see that the following equation holds for any real matrix Y with compatible dimension

$$\begin{aligned} 0 = {} & 2\dot{x}^T(t)Y\Big\{-\dot{x}(t) - Ax(t) + W_0f(x(t)) + W_1f(x(t-\tau(t))) \\ & + W_2\int_{-\infty}^{t} K(t-s)f(x(s))\mathrm{d}s + u(t)\Big\}. \end{aligned} \tag{8.17}$$

Noting that $0 \leq \tau(t) \leq \bar{\tau}$, therefore the following inequality holds

$$\int_{t-\bar{\tau}}^{t} \dot{x}^T(s)S\dot{x}(s)\mathrm{d}s = \int_{t-\tau(t)}^{t} \dot{x}^T(s)S\dot{x}(s)\mathrm{d}s + \int_{t-\bar{\tau}}^{t-\tau(t)} \dot{x}^T(s)S\dot{x}(s)\mathrm{d}s. \tag{8.18}$$

From Lemma 8.3 and the *Leibniz–Newton formula*, for any real matrices $X_i (i = 1, \ldots, 4)$ with compatible dimensions, we get

$$\begin{aligned} & -\int_{t-\tau(t)}^{t} \dot{x}^T(s)S\dot{x}(s)\mathrm{d}s \\ & \quad \leq \int_{t-\tau(t)}^{t} \{\zeta^T(t)\Phi_1S^{-1}\Phi_1^T\zeta(t) + 2\zeta^T(t)\Phi_1\dot{x}(s)\}\mathrm{d}s \\ & \quad = \tau(t)\zeta^T(t)\Phi_1S^{-1}\Phi_1^T\zeta(t) + 2\zeta^T(t)\Phi_1(x(t) - x(t-\tau(t))), \end{aligned} \tag{8.19}$$

$$\begin{aligned} & -\int_{t-\bar{\tau}}^{t-\tau(t)} \dot{x}^T(s)S\dot{x}(s)\mathrm{d}s \\ & \quad \leq \int_{t-\bar{\tau}}^{t-\tau(t)} \{\zeta^T(t)\Phi_2S^{-1}\Phi_2^T\zeta(t) + 2\zeta^T(t)\Phi_2\dot{x}(s)\}\mathrm{d}s \\ & \quad = (\bar{\tau} - \tau(t))\zeta^T(t)\Phi_2S^{-1}\Phi_2^T\zeta(t) \\ & \quad + 2\zeta^T(t)\Phi_2(x(t-\tau(t)) - x(t-\bar{\tau})), \end{aligned} \tag{8.20}$$

where

$$\begin{aligned} \zeta^T(t) = {} & [x^T(t)\ x^T(t-\tau(t))\ x^T(t-\bar{\tau})\ \dot{x}^T(t)\ (1-\dot{\tau}(t))\dot{x}^T(t-\tau(t)) \\ & f^T(x(t))\ f^T(x(t-\tau(t)))\ \Big(\textstyle\int_{-\infty}^{t} K(t-s)f(x(s))\mathrm{d}s\Big)^T\ u^T(t)\Big]. \end{aligned}$$

Now, to show the passivity of the delayed neural network in (8.1), we set

$$J(t_f) = \int_0^{t_f} [-\gamma u^T(t)u(t) - 2y^T(t)u(t)]\mathrm{d}t.$$

where $t_f \geq 0$. Noting the zero initial condition, we can deduce that

$$\begin{aligned} J(t_f) &= \int_0^{t_f} \left[\dot{V}(t) - \gamma u^T(t)u(t) - 2y^T(t)u(t)\right]\mathrm{d}t - V(x_{t_f}) \\ &\leq \int_0^{t_f} \left[\dot{V}(t) - \gamma u^T(t)u(t) - 2y^T(t)u(t)\right]\mathrm{d}t. \end{aligned}$$

From (8.6)–(8.20), we obtain

$$\begin{aligned} &\dot{V}(t) - \gamma u^T(t)u(t) - 2y^T(t)u(t) \\ &\leq \zeta^T(t)\big(\Gamma + \bar{\tau}\tau(t)\Phi_1 S^{-1}\Phi_1^T + \bar{\tau}(\bar{\tau} - \tau(t))\Phi_2 S^{-1}\Phi_2^T\big)\zeta(t) \\ &\quad + 2\zeta^T(t)\int_0^1 \big(\mathcal{C} + (1 - 2\varsigma)\mathfrak{C}\big)\vartheta(t, \varsigma)\mathrm{d}\varsigma \\ &\quad - \int_0^1\int_0^1 \vartheta^T(t, \varsigma)\mathcal{D}\vartheta(t, \epsilon)\mathrm{d}\varsigma - \int_0^1 \vartheta^T(t, \varsigma)\mathcal{E}\vartheta(t, \varsigma)\mathrm{d}\varsigma, \end{aligned} \tag{8.21}$$

where

$$\vartheta^T(t, \varsigma) = [\,x^T(t - h + \varsigma h)\ x^T(t - 2h + \varsigma h)\ \ldots\ x^T(t - \nu h + \varsigma h)\,].$$

Applying Prop. 5.21 of [29] to (8.21), we conclude that $\dot{V}(t) - \gamma u^T(t)u(t) - 2y^T(t)u(t) < 0$ if (8.4) and the following matrix inequality hold:

$$\begin{bmatrix} \Gamma + \overline{\Gamma} + \overline{\Gamma}^T + \bar{\tau}\tau(t)\Phi_1 S^{-1}\Phi_1^T + \bar{\tau}(\bar{\tau} - \tau(t))\Phi_2 S^{-1}\Phi_2^T & \mathcal{C} & \mathfrak{C} \\ \mathcal{C}^T & -\mathcal{D} - \mathcal{E} & 0 \\ \mathfrak{C}^T & 0 & -3\mathcal{E} \end{bmatrix} < 0. \tag{8.22}$$

Note that $0 \leq \tau(t) \leq \bar{\tau}$, so inequality (8.22) holds if and only if the following inequalities are true, $(\kappa = 1, 2)$

$$\begin{bmatrix} \Gamma + \overline{\Gamma} + \overline{\Gamma}^T + \bar{\tau}^2\Phi_\kappa S^{-1}\Phi_\kappa^T & \mathcal{C} & \mathfrak{C} \\ \mathcal{C}^T & -\mathcal{D} - \mathcal{E} & 0 \\ \mathfrak{C}^T & 0 & -3\mathcal{E} \end{bmatrix} < 0. \tag{8.23}$$

From the well known *Schur complement*, inequalities (8.23) are equivalent to inequalities (8.5), thus $J(t_f) \leq 0$ holds for any $t_f \geq 0$ if inequalities (8.4) and (8.5) are true. From the Definition 8.2 of passivity, the proof of Theorem 8.4 is completed.

Remark 8.5 Compared with the existing augmented LKFs [13, 15, 17, 26, 27], the proposed one contains the following terms: $\int_{t-\tau(t)}^{t} \Upsilon^T(s)Q\Upsilon(s)\mathrm{d}s\ \ \Theta^T(t)P\Theta(t)$, other than $\int_{t-\tau(t)}^{t} \left[x^T(s)Q_{11}x(s) + f^T(x(s))Q_{22}f(x(s))\right]\mathrm{d}s$ and $x^T(t)P_{11}x(t)$.

Through the following numerical examples, it will be found that these terms play important roles in the reduction of conservativeness.

Remark 8.6 It is easy to see that the derivatives of $\int_{t-\tau(t)}^{t} \Upsilon^T(s)Q\Upsilon(s)\mathrm{d}s$ and $\Theta^T(t)P\Theta(t)$ have some terms containing $1-\dot{\tau}(t)$. In order to absorb some $1-\dot{\tau}(t)$, we introduce $(1-\dot{\tau}(t))\dot{x}^T(t-\tau(t))$ in $\zeta(t)$ but not $\dot{x}^T(t-\tau(t))$, so Γ contains fewer $1-\dot{\tau}(t)$, which leads to a more effective result than that in [5].

Remark 8.7 It is easy to see that the passivity result in [17] is *delay-independent*. As is well known, delay-dependent criteria make use of information on the length of delay, and are usually less conservative than delay-independent ones especially when the size of the time delay is small. Numerical examples will show the effectiveness of the conditions in Theorem 8.4.

Remark 8.8 In Theorem 8.4, by setting $P_{12} = P_{22} = 0$, $Q = 0$, we can employ this criterion to analyze the passivity of neural network when $\tau(t)$ is not differentiable or $\dot{\tau}(t)$ is unknown.

8.4 Passivity for RNNs with Uncertainty

In this section, we will study the following uncertain RNNs,

$$\begin{aligned}\dot{x}(t) = &-\tilde{A}x(t) + \widetilde{W}_0 f(x(t)) + \widetilde{W}_1 f(x(t-\tau(t)))\\ &+ \widetilde{W}_2 \int_{-\infty}^{t} K(t-s)f(x(s))\mathrm{d}s + u(t),\end{aligned} \tag{8.24}$$

in which two kinds of uncertainties will be considered respectively, i.e., time-varying structured uncertainty and interval uncertainty.

Case I: Time-varying structured uncertainty

First, we consider the neural networks (8.24) with time-varying structured uncertainties as follows:

$$\begin{aligned}&\tilde{A} = A + \Delta A(t),\ \widetilde{W}_\iota = W_\iota + \Delta W_\iota(t),\\ &\Delta A(t) = HF(t)G,\ \Delta W_\iota(t) = H_\iota F_\iota(t)G_\iota,\quad (\iota = 0, 1, 2),\end{aligned} \tag{8.25}$$

where H, H_ι and G, $G_\iota(\iota = 0, 1, 2)$ are known constant matrices, and $F(t)$, $F_\iota(t)$ $(\iota = 0, 1, 2)$ are unknown time-varying matrices satisfying

$$F^T(t)F(t) \le I,\ F_\iota^T(t)F_\iota(t) \le I\ (\iota = 0, 1, 2).$$

Then, for passivity of system (8.24) with (8.25), we have the following theorem:

Theorem 8.9 *Under Assumption 8.1, system* (8.24) *with* (8.25) *is passive for* $0 \le \tau(t) \le \bar{\tau} = \nu h$, $\dot{\tau}(t) \le \eta < 1$, *if there exist constant scalar* $\gamma > 0$, *matrices*

$B > 0, P > 0, Q > 0, R > 0, S > 0, E_j > 0$, *positive diagonal matrices* $\Omega, \Xi, \Lambda, T_1, T_2$, *real matrices* $D_{lj} = D_{jl}^T$, $Y, C_j(j, l = 0, 1, \ldots, \nu)$, $X_\iota(\iota = 1, \ldots, 4)$, *and positive scalars* $\varepsilon, \varepsilon_\iota(\iota = 0, 1, 2)$ *such that* (8.4) *and the following LMIs hold* ($\kappa = 1, 2$):

$$\begin{bmatrix} \Gamma + \overline{\Gamma} + \overline{\Gamma}^T + \mathfrak{G}\mathfrak{G}^T & \mathcal{C} & \mathfrak{C} & \bar{\tau}\Phi_\kappa & \mathcal{P}\mathcal{H}^T \\ \mathcal{C}^T & -\mathcal{D} - \mathcal{E} & 0 & 0 & 0 \\ \mathfrak{C}^T & 0 & -3\mathcal{E} & 0 & 0 \\ \bar{\tau}\Phi_\kappa^T & 0 & 0 & -S & 0 \\ \mathcal{H}\mathcal{P}^T & 0 & 0 & 0 & -\Psi \end{bmatrix} < 0, \quad (8.26)$$

where

$$\begin{aligned} \mathfrak{G}^T &= [\sqrt{\varepsilon}G\ 0\ 0\ 0\ 0\ \sqrt{\varepsilon_0}G_0\ \sqrt{\varepsilon_1}G_1\ \sqrt{\varepsilon_2}G_2\ 0], \\ \mathcal{H}^T &= [H\ H_0\ H_1\ H_2], \quad \Psi = \text{diag}\{\varepsilon I,\ \varepsilon_0 I,\ \varepsilon_1 I,\ \varepsilon_2 I\}, \end{aligned}$$

and other parameters are all defined in Theorem 8.4.

Proof Replacing $A, W_\iota(\iota = 0, 1, 2)$ in (8.1) and (8.17) with $\tilde{A}, \widetilde{W}_\iota(\iota = 0, 1, 2)$ respectively, we have that system (8.24) is passive if (8.4) and the following LMIs hold ($\kappa = 1, 2$) :

$$\begin{bmatrix} \Omega_a & \mathcal{C} & \mathfrak{C} \\ \mathcal{C}^T & -\mathcal{D} - \mathcal{E} & 0 \\ \mathfrak{C}^T & 0 & -3\mathcal{E} \end{bmatrix} < 0, \quad (8.27)$$

where

$$\begin{aligned} \Omega_a &= \Gamma + \overline{\Gamma} + \overline{\Gamma}^T + \bar{\tau}^2\Phi_\kappa S^{-1}\Phi_\kappa^T + \mathcal{P}\mathcal{H}^T\mathcal{F}(t)\mathcal{G}^T + \mathcal{G}\mathcal{F}^T(t)\mathcal{H}\mathcal{P}^T, \\ \mathcal{F}(t) &= \text{diag}\{\, F(t)\ F_0(t)\ F_1(t)\ F_2(t)\,\}, \\ \mathcal{G}^T &= [\,\mathcal{G}\ \mathcal{G}_0\ \mathcal{G}_1\ \mathcal{G}_2\,], \\ \mathcal{G}^T &= [-G^T\ 0\ 0\ 0\ 0\ 0\ 0\ 0\ 0], \quad \mathcal{G}_0^T = [0\ 0\ 0\ 0\ 0\ G_0^T\ 0\ 0\ 0], \\ \mathcal{G}_1^T &= [0\ 0\ 0\ 0\ 0\ 0\ G_1^T\ 0\ 0], \quad \mathcal{G}_2^T = [0\ 0\ 0\ 0\ 0\ 0\ 0\ G_2^T\ 0]. \end{aligned}$$

By Lemma 8.3 and Theorem 8.4, from (8.27) the neural networks in (8.24) is passive for all time-varying uncertainties satisfying (8.25) if (8.4) and the following LMIs hold ($\kappa = 1, 2$):

$$\begin{bmatrix} \Gamma + \overline{\Gamma} + \overline{\Gamma}^T + \bar{\tau}^2\Phi_\kappa S^{-1}\Phi_\kappa^T + \mathcal{P}\mathcal{H}^T\Psi^{-1}\mathcal{H}\mathcal{P}^T + \mathcal{G}\Psi\mathcal{G}^T & \mathcal{C} & \mathfrak{C} \\ \mathcal{C}^T & -\mathcal{D} - \mathcal{E} & 0 \\ \mathfrak{C}^T & 0 & -3\mathcal{E} \end{bmatrix} < 0,$$

Applying the *Schur complement* to above inequalities, we obtain (8.26) and the proof is completed.

Case II: ***Interval uncertainty***

Now we consider systems with an uncertainty due to bounded parameter variations, that is, $\tilde{A}$, $\widetilde{W}_{\iota}(\iota = 0, 1, 2)$ in (8.24) may take any constant values within two bounding matrices and their elements a_i, w^{ι}_{ij} can be described as follows:

$$\begin{aligned} A_I &= \{\tilde{A} = \mathrm{diag}\{\tilde{a}_1, \tilde{a}_2, \ldots, \tilde{a}_n\} : 0 < \underline{a}_i \le \tilde{a}_i \le \bar{a}_i\}, \\ W^{\iota}_I &= \{\widetilde{W}_{\iota} = (\widetilde{w}^{\iota}_{ij})_{n\times n} : \underline{w}^{\iota}_{ij} \le \widetilde{w}^{\iota}_{ij} \le \bar{w}^{\iota}_{ij}\}, \ (\iota = 0, 1, 2). \end{aligned} \tag{8.28}$$

For convenience, we define

$$\begin{aligned} &\underline{A} = \mathrm{diag}\{\underline{a}_1, \underline{a}_2, \ldots, \underline{a}_n\}, \ \bar{A} = \mathrm{diag}\{\bar{a}_1, \bar{a}_2, \ldots, \bar{a}_n\}, \ \underline{W}_{\iota} = (\underline{w}^{\iota}_{ij})_{n\times n}, \\ &\overline{W}_{\iota} = (\bar{w}^{\iota}_{ij})_{n\times n}, A = \frac{1}{2}(\underline{A} + \bar{A}), \ L_A = \frac{1}{2}(\bar{A} - \underline{A}) = \mathrm{diag}\{\alpha_1, \alpha_2, \ldots, \alpha_n\}, \\ &W_{\iota} = \frac{1}{2}(\underline{W}_{\iota} + \overline{W}_{\iota}), \ L_{W_{\iota}} = \frac{1}{2}(\overline{W}_{\iota} - \underline{W}_{\iota}) = (\beta^{\iota}_{ij})_{n\times n} \ (\iota = 0, 1, 2). \end{aligned}$$

Since each element of $L_A, L_{W_{\iota}} (\iota = 0, 1, 2)$ is nonnegative, then we can define

$$\begin{aligned} &H_A = G_A = \mathrm{diag}\big\{\sqrt{\alpha_1}, \sqrt{\alpha_2}, \ldots, \sqrt{\alpha_n}\big\}_{n\times n}, \\ &H_{W_{\iota}} = \Big[\sqrt{\beta^{\iota}_{11}}e_1, \ldots, \sqrt{\beta^{\iota}_{1n}}e_1, \ldots\ldots, \sqrt{\beta^{\iota}_{n1}}e_n, \ldots, \sqrt{\beta^{\iota}_{nn}}e_n\Big]_{n\times n^2}, \\ &G_{W_{\iota}} = \Big[\sqrt{\beta^{\iota}_{11}}e_1, \ldots, \sqrt{\beta^{\iota}_{1n}}e_n, \ldots, \sqrt{\beta^{\iota}_{n1}}e_1, \ldots, \sqrt{\beta^{\iota}_{nn}}e_n\Big]^T_{n^2\times n}, \ (\iota = 0, 1, 2), \end{aligned}$$

where $e_i (i = 1, \ldots, n)$ denotes the ith standard basis of $\mathbb{R}^n$. Furthermore, we denote

$$\begin{aligned} \Theta &= \{Z \in \mathbb{R}^{n\times n} : Z = \mathrm{diag}\{\theta_1, \ldots, \theta_n\}, \ |\theta_i| \le 1, \ i = 1, \ldots, n\}, \\ \Delta &= \{Z \in \mathbb{R}^{n^2\times n^2} : Z = \mathrm{diag}\{\delta_{11}, \ldots, \delta_{1n}, \ldots\ldots, \delta_{n1}, \ldots, \delta_{nn}\}, \\ &\qquad |\delta_{ij}| \le 1, \ i, j = 1, \ldots, n\}. \end{aligned}$$

Then it is easy to check that

$$\begin{aligned} A_I &= \{\tilde{A} = A + H_A\Theta_A G_A : \Theta_A \in \Theta\}, \\ W^{\iota}_I &= \{\widetilde{W}_{\iota} = W_{\iota} + H_{W_{\iota}}\Delta_{W_{\iota}}G_{W_{\iota}} : \Delta_{W_{\iota}} \in \Delta\}, \end{aligned}$$

with $\Theta^T_A\Theta_A \le I$, $\Delta^T_{W_{\iota}}\Delta_{W_{\iota}} \le I$, $(\iota = 0, 1, 2)$. Then by Theorem 8.9, we obtain the following result for passivity of neural networks (8.24) with interval uncertainty (8.28).

Theorem 8.10 *Under Assumption 8.1, system* (8.24) *with* (8.28) *is passive for* $0 \le \tau(t) \le \bar{\tau} = \nu h$, $\dot{\tau}(t) \le \eta < 1$, *if there exist constant scalar* $\gamma > 0$, *matrices*

$B > 0, P > 0, Q > 0, R > 0, S > 0, E_j > 0$, *positive diagonal matrices* $\Omega, \Xi, \Lambda, T_1, T_2$, *real matrices* $D_{lj} = D_{jl}^T$, $Y, C_j (j, l = 0, 1, \ldots, \nu)$, $X_\iota(\iota = 1, \ldots, 4)$, *and positive scalars* $\varepsilon, \varepsilon_\iota(\iota = 0, 1, 2)$ *such that* (8.4) *and the following LMIs hold* ($\kappa = 1, 2$)*:*

$$\begin{bmatrix} \Gamma + \overline{\Gamma} + \overline{\Gamma}^T + \mathbb{G}\mathbb{G}^T & \mathcal{C} & \mathfrak{C} & \bar{\tau}\Phi_\kappa & \mathcal{P}\mathfrak{H}^T \\ \mathcal{C}^T & -\mathcal{D} - \mathcal{E} & 0 & 0 & 0 \\ \mathfrak{C}^T & 0 & -3\mathcal{E} & 0 & 0 \\ \bar{\tau}\Phi_\kappa^T & 0 & 0 & -S & 0 \\ \mathfrak{H}\mathcal{P}^T & 0 & 0 & 0 & -\Psi \end{bmatrix} < 0, \tag{8.29}$$

where

$$\mathbb{G}^T = [\, \sqrt{\varepsilon} G_A \; 0\; 0\; 0\; 0 \; \sqrt{\varepsilon_0} G_{W_0} \; \sqrt{\varepsilon_1} G_{W_1} \; \sqrt{\varepsilon_2} G_{W_2} \; 0 \,],$$
$$\mathfrak{H}^T = [\, H_A \; H_{W_0} \; H_{W_1} \; H_{W_2} \,],$$

and other parameters are all defined in Theorem 8.4.

Remark 8.11 Similar to Remark 8.8, by setting $P_{12} = P_{22} = 0$, $Q = 0$ in Theorems 8.9 and 8.10, we can employ these criteria to analyze the passivity of neural network with uncertainty when $\tau(t)$ is not differentiable or $\dot{\tau}(t)$ is unknown.

8.5 Illustrative Examples

In this section, we provide three numerical examples to demonstrate the effectiveness and less conservativeness of our delay-dependent passivity criteria over some recent results in the literature.

Example 8.12 Consider the system (8.24) with

$$\bar{A} = \text{diag}\{2.4225, 2.3225\}, \; \bar{A} = \text{diag}\{2.1775, 2.0775\}, \; \Sigma = I,$$
$$\overline{W}_0 = \begin{bmatrix} 0.4024 & 0.3024 \\ -0.2976 & 0.2024 \end{bmatrix}, \; \underline{W}_0 = \begin{bmatrix} 0.1976 & 0.0976 \\ -0.5024 & -0.0024 \end{bmatrix},$$
$$\overline{W}_1 = \begin{bmatrix} 0.5784 & 0.7784 \\ 0.7784 & -0.3216 \end{bmatrix}, \; \underline{W}_1 = \begin{bmatrix} 0.4216 & 0.6216 \\ 0.6216 & -0.4784 \end{bmatrix},$$
$$\overline{W}_2 = \begin{bmatrix} 0.5676 & 0.2676 \\ -0.2324 & 1.2676 \end{bmatrix}, \; \underline{W}_2 = \begin{bmatrix} 0.4324 & 0.1324 \\ -0.3676 & 1.1324 \end{bmatrix}.$$

For this model with $\tau(t) = 0.5 + 0.5\sin(t)$, that is $\bar{\tau} = 1, \eta = 0.5$, from Remark 8.11 of Theorem 8.10 with $\nu = 3$ we can conclude that this system is passive. For different η's, Table 8.1 gives the results on the maximum $\bar{\tau}$ allowed by the method in

Table 8.1 Calculated maximal upper bounds of time delays for various η in Example 8.12

Methods	$\eta = 0$	$\eta = 0.1$	$\eta = 0.5$	$\eta = 0.9$	Unknown η
This chapter ($\nu = 1$)	1.2190	0.9429	–	–	–
This chapter ($\nu = 2$)	1.5152	1.3560	0.9202	0.6504	0.6386
This chapter ($\nu = 3$)	1.6070	1.4466	1.0130	0.7292	0.7178
This chapter ($\nu = 4$)	1.6651	1.5032	1.0654	0.7724	0.7610

Theorem 8.10 or Remark 8.11 in this chapter, where "–"means that the result is not applicable to the corresponding case, and "unknown η" means that η can be arbitrary value or $\tau(t)$ can be not differentiable.

Moreover, it is easy to see that the larger ν is, the larger $\bar{\tau}$ becomes. We also compute the number of the decision variables involved in Theorem 8.10 for $\nu = 3$ and $\nu = 4$, the results are 142 and 184 respectively. When $\eta = 0$, the difference between the values of maximal upper bounds for $\nu = 4$ and $\nu = 3$ is just 3.6 %, but the number of the decision variables involved by the former is 29.6 % larger than that by the latter. As a compromise, taking $\nu = 3$ is a good choice for the obtained maximal upper bounds and for the low cost of the CPU time.

Example 8.13 Consider the uncertain neural networks (8.24), where the parameters are as follows: [17],

$$A = \mathrm{diag}\{1, 2\},\ W_2 = 0,\ G = G_0 = G_1 = [0.1\ 0.2],\ \Sigma = I,$$
$$W_0 = \begin{bmatrix} 1 & -1 \\ -0.5 & 2 \end{bmatrix},\ W_1 = \begin{bmatrix} 0.1 & 0.5 \\ 0.2 & 0.4 \end{bmatrix},$$
$$H_0 = \begin{bmatrix} 0.2 \\ 0.3 \end{bmatrix},\ H_1 = \begin{bmatrix} 0.3 \\ 0.2 \end{bmatrix},\ H_2 = \begin{bmatrix} 0.2 \\ 0.1 \end{bmatrix}.$$

It is verified that none of the conditions given in [13, 15, 26, 27, 32] can conclude whether this model is passive or not for any time delays. On the other hand, the author of [13] pointed out that the results of Examples 1 and 2 given in [17] were wrong. In fact, the criterion of [17] fails to assure the passivity of this model. However, for this model with $\tau(t) = 0.18 + 0.55\sin(2t)$, that is $\bar{\tau} = 0.73$, $\eta = 1.1$, from Remark 8.11 of Theorem 8.9 with $\nu = 4$ we can conclude that this system is passive. For different η's, Table 8.2 gives the results on the maximum $\bar{\tau}$ allowed via the method in Theorem 8.9 or Remark 8.11 in this chapter.

Example 8.14 Consider the uncertain neural networks (8.24), where the parameters are as follows [27],

$$A = \begin{bmatrix} 2.2 & 0 \\ 0 & 1.5 \end{bmatrix},\ W_0 = \begin{bmatrix} 1 & 0.6 \\ 0.1 & 0.3 \end{bmatrix},\ W_1 = \begin{bmatrix} 1 & -0.1 \\ 0.1 & 0.2 \end{bmatrix},\ W_2 = 0,$$
$$\Sigma = I,\ H = H_0 = H_1 = 0.1I,\ G = 0.1I, G_0 = 0.2I, G_1 = 0.3I, G_2 = 0.$$

Table 8.2 Calculated maximal upper bounds of time delays for various η in Example 8.13

Methods	$\eta = 0$	$\eta = 0.1$	$\eta = 0.5$	$\eta = 0.9$	Unknown η
This chapter ($\nu = 1$)	–	–	–	–	–
This chapter ($\nu = 2$)	0.8025	0.7678	0.7212	0.6411	0.6000
This chapter ($\nu = 3$)	0.8934	0.8604	0.8107	0.7258	0.6806
This chapter ($\nu = 4$)	0.9422	0.9100	0.8606	0.7736	0.7312

Table 8.3 Calculated maximal upper bounds of time delays for various η in Example 8.14

Methods	$\eta = 0.1$	$\eta = 0.5$	$\eta = 0.9$	Unknown η
Ref. [13] (corrected)	0.0847	0.0785	0.0698	–
Ref. [26]	0.2338	0.2242	0.2213	0.2152
Ref. [15]	0.4148	0.4004	0.3954	0.3846
Ref. [27]	0.7841	0.4145	0.4082	0.3994
This chapter ($\nu = 1$)	0.6806	0.5506	0.4464	0.4038
This chapter ($\nu = 2$)	3.9066	1.7338	1.4358	1.3228
This chapter ($\nu = 3$)	4.4928	1.8674	1.5199	1.3961
This chapter ($\nu = 4$)	4.8502	1.9434	1.5668	1.4350

It is proved that the conditions given in [17, 32] fail to conclude whether this model is passive or not for $\eta \geq 0.1$. For different η's, Table 8.3 gives the comparison results on the maximum $\bar{\tau}$ allowed via the methods in [13, 15, 26, 27] and Theorem 8.9 or Remark 8.11 in this chapter, where the results of [13] (corrected) are the corrected ones of [13] based on [26].

From this table, one can see that Theorem 8.9 provides the larger upper bounds than those criteria in [13, 15, 26, 27]. In particular, when $\eta = 0.1$, the achieved maximal upper bound $\bar{\tau}$ by Theorem 8.9 with $\nu = 3$ is 5204.4, 1821.6, 983.1 and 472.8 % larger than those in [13, 15, 26] and [27], respectively.

Therefore, we can say that for these three systems the results in this chapter are much effective and less conservative than those in [13, 15, 17, 26, 27, 32].

8.6 Summary

In this chapter we have investigated the passivity problem of recurrent neural networks with discrete and unbounded distributed time-varying delays. By integrating the Gu's discretization technique with the free-weighting matrix approaches, we proposed novel passivity criteria for the considered systems. The obtained results are expressed in the form of LMI, which can be easily optimized. Finally, numerical examples have showed the superiority of our proposed passivity conditions to some existing ones. One of the future research topics would be an extension of the present results to more general cases, for example, the case with impulsive effects, the case with stochastic terms or the case with reaction–diffusion terms.

References

1. C.-D. Zheng, Z. Wang, Novel passivity criteria for neural networks with discrete and unbounded distributed time-varying delays. Int. J. Comput. Math. **90**(9), 1782–1795 (2013)
2. W.H. Chen, W.X. Zheng, Improved delay-dependent asymptotic stability criteria for delayed neural networks. IEEE Trans. Neural Netw. **19**(12), 2154–2161 (2008)
3. F.O. Souza, R.M. Palhares, P.Y. Ekel, Improved asymptotic stability analysis for uncertain delayed state neural networks. Chaos Solitons Fractals **39**, 240–247 (2009)
4. Y. Shen, J. Wang, An improved algebraic criterion for global exponential stability of recurrent neural networks with time-varying delays. IEEE Trans. Neural Netw. **19**(3), 528–531 (2008)
5. J. Sun, G.P. Liu, J. Chen, D. Rees, Improved stability criteria for neural networks with time-varying delay. Phys. Lett. A **373**, 342–348 (2009)
6. Z. Zeng, J. Wang, Global exponential stability of recurrent neural networks with time-varying delays in the presence of strong external stimuli. Neural Netw. **19**, 1528–1537 (2006)
7. Z. Zeng, J. Wang, Improved conditions for global exponential stability of recurrent neural networks with time-varying delays. IEEE Trans. Neural Netw. **17**(3), 623–635 (2006)
8. H. Zhang, Z. Wang, Global asymptotic stability of delayed cellular neural networks. IEEE Trans. Neural Netw. **18**(3), 947–950 (2007)
9. H. Zhang, Z. Wang, D. Liu, Robust stability analysis for interval Cohen-Grossberg neural networks with unknown time-varying delays. IEEE Trans. Neural Netw. **19**(11), 1942–1955 (2008)
10. H. Zhang, Z. Wang, D. Liu, Global asymptotic stability of recurrent neural networks with multiple time-varying delays. IEEE Trans. Neural Netw. **19**(5), 855–873 (2008)
11. H. Zhang, Z. Wang, D. Liu, Robust exponential stability of cellular neural networks with multiple time varying delays. IEEE Trans. Circuits Syst. II **54**(8), 730–734 (2007)
12. X.L. Zhu, G.H. Yang, New delay-dependent stability results for neural networks with time-varying delay. IEEE Trans. Neural Netw. **19**(10), 1783–1791 (2008)
13. J.H. Park, Further results on passivity analysis of delayed cellular neural networks. Chaos Solitons Fractals **34**, 1546–1551 (2007)
14. D.H. Ji, J.H. Koo, S.C. Won, S.M. Lee, J.H. Park, Passivity-based control for Hopfield neural networks using convex representation. Appl. Math. Comput. **217**(13), 6168–6175 (2011)
15. Z. Zhang, S. Mou, J. Lam, H. Gao, New passivity criteria for neural networks with time-varying delay. Neural Netw. **22**, 864–868 (2009)
16. L.O. Chua, Passivity and complexity. IEEE Trans. Circuits Syst.-I: Fundam. Theory Appl. **46**(1), 71–82 (1999)
17. C. Li, X. Liao, Passivity analysis of neural networks with time delay. IEEE Trans. Circuits Syst. II **52**(8), 471–475 (2005)
18. X. Lou, B. Cui, Passivity analysis of integro-differential neural networks with time-varying delays. Neurocomputing **70**(6), 1071–1078 (2007)
19. B. Zhang, S. Xu, J. Lam, Relaxed passivity conditions for neural networks with time-varying delays. Neurocomputing **142**, 299–306 (2014)
20. R. Raja, U. Raja, R. Samidurai, Passivity analysis for uncertain discrete-time stochastic BAM neural networks with time-varying delays. Neural Comput. Appl. **25**(4), 751–766 (2014)
21. H. Zeng, Y. He, M. Wu, Improved conditions for passivity of neural networks with a time-varying delay. IEEE Trans. Cybern. **44**(6), 785–792 (2014)
22. H. Zeng, J. Park, H. Shen, Robust passivity analysis of neural networks with discrete and distributed delays. Neurocomputing **149**, 1092–1097 (2015)

23. O.M. Kwon, S.M. Lee, J.H. Park, On improved passivity criteria of uncertain neural networks with time-varying delays. Nonlinear Dyn. **67**(2), 1261–1271 (2012)
24. Z.-G. Wu, J.H. Park, H. Su, J. Chu, Passivity analysis of Markov jump neural networks with mixed time-delays and piecewise-constant transition rates. Nonlinear Anal. Real World Appl. **13**(5), 2423–2431 (2012)
25. Z.-G. Wu, J.H. Park, H. Su, J. Chu, New results on exponential passivity of neural networks with time-varying delays. Nonlinear Anal. Real World Appl. **13**(4), 1593–1599 (2012)
26. S. Xu, W.X. Zheng, Y. Zou, Passivity analysis of neural networks with time-varying delays. IEEE Trans. Circuits Syst. II **56**(4), 325–329 (2009)
27. B. Chen, H. Li, C. Lin, Q. Zhou, Passivity analysis for uncertain neural networks with discrete and distributed time-varying delays. Phys. Lett. A. **373**, 1242–1248 (2009)
28. K. Gu, Discretized LMI set in the stability problem of linear time delay systems. Int. J. Control **68**, 923–976 (1997)
29. K. Gu, An integral inequality in the stability problem of time-delay systems, in: Proc. 39th IEEE Conf. Decis. Control, Sydney, Australia, pp. 2805–2810, (2000)
30. E. Fridman, Descriptor discretized Lyapunov functional method: analysis and design. IEEE Trans. Automat. Control **51**(5), 890–897 (2006)
31. E.N. Sanchez, J.P. Perez, Input-to-state stability (ISS) analysis for dynamic NN. IEEE Trans. Circ. Syst. I **46**, 1395–1398 (1999)
32. C.Y. Lu, H.H. Tsai, T.J. Su, J.S.H. Tsai, C.W. Liao, A delay-dependent approach to passivity analysis for uncertain neural networks with time-varying delay. Neural Proc. Lett. **27**(3), 237–246 (2008)

Chapter 9
Dissipativity and Invariant Sets for Neural Networks with Delay

Energy dissipation is a fundamental concept in *dynamical systems*. Passivity and dissipativity characterize the "energy" consumption of a dynamical system and form a powerful tool in many real applications. Passivity is closely related to stability and exhibits a compositional property for parallel and feedback interconnections. Passivity-based control is especially useful in the analysis of complex coupled systems. In Chap. 8, passivity problem is studied for a kind of neural networks with delays. Dissipativity and invariant sets are also the qualitative characteristics of a dynamical system. Such qualitative characteristics as passivity, dissipativity, and invariant sets are extensions and upgrades of the stability property, which can characterize the dynamics of dynamical systems more general. Based on such analysis, in this chapter some sufficient conditions for dissipativity and invariant sets have been established for a kind of RNNs with delay. The contents in this chapter are some extensions of the stability results of previous chapters.

9.1 Delay-Dependent Dissipativity Conditions for Delayed RNNs

9.1.1 Introduction

Since the study of *dissipative systems* was initiated by Willems [1], and further addressed by Hill and Moylan [2], there has been a steady increase in the interest of dissipative systems in the past several decades. The reasons are as follows: (1) The dissipative theory gives a framework for the design and analysis of control systems using an input–output description based on energy-related considerations [3, 4]. (2) The dissipative theory serves as a powerful or even indispensable tool in characterizing important system behaviors, such as stability and passivity, and has close connections with passivity theorem, bounded real lemma, Kalman–Yakubovich

Z. Wang et al., *Qualitative Analysis and Control of Complex Neural Networks with Delays*, Studies in Systems, Decision and Control 34,
DOI 10.1007/978-3-662-47484-6_9

lemma, and the circle criterion [5]. Many significant advances on this issue have been reported in the literature. To name a few, by using a linear matrix inequality (LMI) approach, the problem of quadratic dissipative control for linear systems with or without uncertainty was studied in [6], where some necessary and sufficient conditions were presented for synthesis of feedback controllers to ensure the *dissipativity* of the resulting closed-loop system. In [7], by proposing multiple storage functions and multiple supply rates, a framework of dissipativity theory for switched systems was established. More recently, some dissipativity conditions were presented in [8] for singular systems. When stochastic noise was taken into consideration in studying dissipative systems, the problems of sliding mode control were tackled in [9]. In [10], the robust reliable dissipative filtering problem has been investigated for uncertain discrete-time singular system with interval time-varying delays and sensor failures. The problem of static output-feedback dissipative control has been studied for linear continuous-time system based on an augmented system approach in [11]. A necessary and sufficient condition for the existence of a desired controller has been given, and a corresponding iterative algorithm has also been developed to solve the condition.

On the other hand, as a special class of nonlinear dynamical systems, neural networks have achieved much attention in the last years due to their extensive applications in various fields, including combinatorial optimization, associated memory, pattern recognition, and so on [12–17]. In practice, time delays are unavoidably encountered in the electronic implementation of neural works, which usually bring negative and undesirable effects, such as performance degradation or even instability of neural networks. Therefore, it is not surprising that the past decades have witnessed fruitful literature about performance analysis of delayed neural networks (DNNs). In the existing literature, compared with stability and passivity problems, the problem of dissipativity analysis for DNNs has also been investigated, which is also an aspect of qualitative analysis for nonlinear system. For instance, by introducing an integral partitioning technique, a sufficient delay-dependent dissipativity condition was given in [18] for DNNs, where the time delay under consideration was a constant distributed delay. However, neither the time-varying delay nor parameters uncertainty was taken into account in [18]. Robust dissipativity analysis for DNNs with time-varying delay was addressed in [19–21], where some effective delay-dependent dissipativity conditions were established by introducing some free-weighting matrices. Reference [22] addresses the global dissipativity of a general class of continuous-time recurrent neural networks, and the set of global dissipativity is characterized using the parameters of recurrent neural network models. In particular, it is shown that the *Hopfield neural networks* and cellular neural networks with or without time delays are dissipative systems. It should be pointed out that these dissipativity conditions based on the *free-weighting matrix* approach and the integral partitioning technique, still have some space to be improved. As an extension of stability and passivity analysis, it is necessary to investigate the dissipativity problem for RNNs. These motivate the present study.

In this section, attention is focused on the derivation of improved dissipativity conditions for a class of DNNs. To this end, a reciprocally convex approach combined with an extended Wirtinger inequality is employed. Some delay-dependent sufficient conditions that guarantee the dissipativity of the considered DNNs are established.

Note that, in this section, $|\cdot|$ refers to the Euclidean vector norm; $\|\cdot\|$ stands for the usual $L_2[0,\infty)$ norm. The signal space under consideration is L_2 space or the extended L_2 space. Denote the truncation of $u(t)$ up to time $F(0 \le F < \infty)$ by $u_F(t)$. The inner product of truncated signals $u_F(t)$, $y_F(t)$ is denoted by $\langle u, y\rangle_F$, where $\langle u, y\rangle_F = \int_0^F y^T(t)u(t)\mathrm{d}t$.

9.1.2 Problem Formulation

Consider the following uncertain delayed neural networks (DNNs),

$$\dot{x}_i(t) = -a_i x_i(t) + \sum_{j=1}^{n} b_{ij} f_j(x_j(t)) + \sum_{j=1}^{n} c_{ij} f_j(x_j(t-h(t))) + u_i(t), \quad (9.1)$$

$$y_i(t) = f_i(x_i(t)), \quad (9.2)$$

or equivalently

$$\dot{x}(t) = -Ax(t) + Bf(x(t)) + Cf(x(t-h(t))) + u(t), \quad (9.3)$$

$$y(t) = f(x(t)), \quad (9.4)$$

where $x(t) = (x_1(t), x_2(t), \ldots, x_n(t))^T \in \mathbb{R}^n$, $x_i(t)$ is the state vector of the ith network at time t; $f(x(t)) = (f_1(x_1(t)), f_2(x_2(t)), \ldots, f_n(x_n(t)))^T \in \mathbb{R}^n$ denotes the neuron activation function; $u(t) = (u_1(t), u_2(t), \ldots, u_n(t))^T$ stands for the external inputs at time t and $y(t)$ is the output; $A = \mathrm{diag}\{a_1, a_2, \ldots, a_n\}$ is a diagonal matrix with positive entries; $B = (b_{ij})_{n\times n}$ and $C = (c_{ij})_{n\times n}$ are the interconnection matrices representing the weight coefficients of the neurons. $h(t)$ is a bounded function denoting the time-varying delay, $0 < h(t) \le h$, $\dot{h} \le \mu$. $x(t) = \phi(t)$, $t \in [-h, 0]$, where $\phi(t)$ is the initial function. Throughout this chapter, we shall use the following assumptions.

Assumption 9.1 (*see* [23]) Every activation function $f_i(\cdot)$ in (9.3) is continuous and bounded, and satisfies

$$\underline{F}_i \le \frac{f_i(x) - f_i(y)}{x - y} \le \overline{F}_i, \quad i = 1, 2, \ldots, n, \quad (9.5)$$

where $f_i(0) = 0$, $x, y \in \mathbb{R}$, $x \ne y$, and $\underline{F}_i$ and $\overline{F}_i$ are known real scalars and they may be positive, negative, or zero.

Assumption 9.1 means that the resulting *activation functions* may be non-monotonic and more general than the usual sigmoid functions and Lipschitz-type condition [19]. For the sake of simplicity, in what follows we denote

$$\underline{F} = \text{diag}\{\underline{F}_1, \underline{F}_2, \ldots, \underline{F}_n\}, \quad \overline{F} = \text{diag}\{\overline{F}_1, \overline{F}_2, \ldots, \overline{F}_n\}.$$

Definition 9.2 The *energy supply function* of DNNs (9.3) and (9.4) is defined as

$$\mathcal{G}(u, y, \digamma) = \langle y, Qy\rangle_\digamma + 2\langle y, Su\rangle_\digamma + \langle u, Ru\rangle_\digamma, \quad \forall \digamma > 0, \tag{9.6}$$

where Q, R, and S are real matrices of appropriate dimensions, with Q and R being symmetric matrices. The notation $\langle y, Su\rangle_\digamma$ represents $\int_0^\digamma y^T(t)Su(t)\mathrm{d}t$, and the other symbols are similarly defined.

Without loss of generality, as noted in [24], in this section we assume that $Q \leq 0$ and denoted that $-Q = \hat{Q}^T\hat{Q}$ for some $\hat{Q}$.

Definition 9.3 ([25]) Consider the DNNs (9.3) and (9.4) with input $u(t)$ and output $y(t)$, where $u(t), y(t) \in \mathbb{R}^n$. It is called

(1) *passive*, if there is a constant $\beta < 0$ such that

$$\langle y, u\rangle_\digamma \geq \beta. \tag{9.7}$$

(2) *input strictly passive* (ISP), if there exist $\nu > 0$ and a constant $\beta \leq 0$ such that

$$\langle y, u\rangle_\digamma \geq \beta + \nu\langle u, u\rangle_\digamma. \tag{9.8}$$

(3) *output strictly passive* (OSP), if there exist $\rho > 0$ and a constant $\beta \leq 0$ such that

$$\langle y, u\rangle_\digamma \geq \beta + \rho\langle y, y\rangle_\digamma. \tag{9.9}$$

(4) *very strictly passive* (VSP), if there exist $\rho > 0$ and $\nu > 0$ and a constant $\beta \leq 0$ such that

$$\langle y, u\rangle_\digamma \geq \beta + \rho\langle y, y\rangle_\digamma + \nu\langle u, u\rangle_\digamma. \tag{9.10}$$

In all cases, the inequality should hold for $\forall u(t), \forall \digamma > 0$ and the corresponding $y(t)$.

The constant β is related to the initial condition of the DNNs (9.3) and (9.4) and plays an important role in the stability analysis. The inner product $\langle y, u\rangle_\digamma$ may be interpreted as the externally supplied energy to DNNs during the interval $[0, \digamma]$. The above Definition 9.3 can be viewed as special cases of (Q, R, S)−dissipative systems defined in Definition 9.2, where $\mathcal{G}(u, y, \digamma)$ is called the supply rate function for DNNs (9.3) and (9.4) [25].

Note that, if a system is ISP for $\nu > 0$, it is also ISP for $\nu - \epsilon$, where $0 \leq \epsilon < \nu$. Analogously, if the system is OSP for $\rho > 0$, it is also OSP for $\rho - \epsilon$, where $0 \leq \epsilon < \rho$. If the system is VSP for (ρ, ν), it is also VSP for $(\rho - \epsilon, \nu - \epsilon)$, where $0 \leq \epsilon < \min(\rho, \nu)$. A positive value of ρ or ν can thus be interpreted as an excess of *passivity* and these two values (called passivity levels) characterize "how passive" the system is. If ρ or ν is negative, we say the system has a shortage of passivity. More details can also be referred to [26].

Definition 9.4 ([20]) Given a scalar $\theta > 0$, real matrices $Q = Q^T$ and $R = R^T$ and matrix S, neural network (9.3) and (9.4) is said to be strictly $(Q, R, S)-\theta$*-dissipative*, if for any $\digamma \geq 0$, under zero initial state, the following condition is satisfied:

$$\mathcal{G}(u, y, \digamma) \geq \theta \langle u, u \rangle_{\digamma}, \quad \forall \digamma \geq 0, \tag{9.11}$$

Lemma 9.5 (Extended Wirtinger inequality [27]) *Let ζ be any continuously differentiable function on interval $[c, a]$ and $\zeta(a) = \zeta(c) = 0$. Then, for any matrix $M \in \mathbb{R}^{n \times n}$, $M = M^T > 0$, the following inequality holds*

$$\int_c^a \dot{\zeta}^T(s) M \dot{\zeta}(s) ds \geq \frac{\pi^2}{(a-c)^2} \int_c^a \zeta^T(s) M \zeta(s) ds$$

The problem concerned in this section is formulated as follows: for given matrices Q, S, R and scalars $\theta > 0, h > 0$, consider the DNNs (9.3) and (9.4), determine under what condition the considered neural network is strictly $(Q, R, S) - \theta$-dissipative by using a reciprocally convex approach combined with an extended Wirtinger inequality.

9.1.3 θ-dissipativity Result

In this subsection, along the similar routine in [20], we will present a delay-dependent sufficient condition, which ensures that delayed neural network (9.3) and (9.4) is strictly $(Q, R, S) - \theta$-dissipative.

Theorem 9.6 *Given a scalar $h > 0$, $\mu > 0$, and matrices $Q = Q^T$, S, $R = R^T$, DNNs (9.3) and (9.4) is strictly $(Q, R, S) - \theta$-dissipative, if there exist matrices $P > 0$, $X = \begin{bmatrix} X_1 & X_2 \\ X_2^T & X_3 \end{bmatrix} > 0$, $Z = \begin{bmatrix} Z_1 & Z_2 \\ Z_2^T & Z_3 \end{bmatrix} > 0$, $W > 0$, Y_1, Y_2, diagonal matrices $\Phi_l > 0$, $l = 1, 2, 3$, $\Omega_1 = \text{diag}\{\lambda_{11}, \lambda_{12}, \ldots, \lambda_{1n}\} > 0$, $\Omega_2 = \text{diag}\{\lambda_{21}, \lambda_{22}, \ldots, \lambda_{2n}\} > 0$, and a scalar $\theta > 0$ such that the following LMIs hold:*

$$\Xi = \begin{bmatrix} J_1 & J_2 & J_4 & J_7 & J_{11} & J_{13} \\ * & J_3 & J_5 & J_8 & 0 & J_{14} \\ * & * & J_6 & J_9 & 0 & 0 \\ * & * & * & J_{10} & 0 & 0 \\ * & * & * & * & J_{12} & hW \\ * & * & * & * & * & -W \end{bmatrix} < 0, \tag{9.12}$$

$$\begin{bmatrix} \mathrm{diag}\{W, \pi^2 W\} & Y \\ * & \mathrm{diag}\{W, \pi^2 W\} \end{bmatrix} \geq 0 \tag{9.13}$$

where

$$J_1 = \begin{bmatrix} -\mathcal{P}A - A^T\mathcal{P}^T - (0.25\pi^2 + 1)W & \mathcal{P}B - A^T\Omega_{12} \\ * & \Omega_{12}B + B^T\Omega_{12} - \Phi_1 - Q \end{bmatrix}$$
$$+ X + hZ + \Lambda_1,$$

$$J_2 = \begin{bmatrix} (1 - 0.25\pi^2)W - Y_1 - 1.25Y_2 & \mathcal{P}C \\ 0 & \Omega_{12}C \end{bmatrix},$$

$$J_3 = -(1-\mu)Z + \Lambda_2 + \mathrm{diag}\{-(2 + 0.5\pi^2)W + Y_1 + Y_1^T - 0.25(Y_2 + Y_2^T), 0\},$$

$$J_3 = -X + \Lambda_2,$$

$$J_4 = \mathrm{diag}\{Y_1 - 0.25Y_2, 0\},$$

$$J_5 = \mathrm{diag}\{(1 - 0.25\pi^2)W - Y_1 + 0.75Y_2, 0\},$$

$$J_6 = -X + \Lambda_3 + \mathrm{diag}\{(0.25\pi^2 + 1)W, 0\},$$

$$J_7 = \begin{bmatrix} 0.5\pi^2 W & 1.5Y_2 \\ 0 & 0 \end{bmatrix}, \quad J_8 = \begin{bmatrix} 0.5\pi^2 W + 1.5Y_2 & 0.5\pi^2 W - 0.5Y_2 \\ 0 & 0 \end{bmatrix}$$

$$J_9 = \begin{bmatrix} -0.5Y_2 & 0.5\pi^2 W \\ 0 & 0 \end{bmatrix}, \quad J_{10} = \begin{bmatrix} -\pi^2 W & -Y_2 \\ 0 & -\pi^2 W \end{bmatrix}$$

$$J_{11} = \begin{bmatrix} \mathcal{P} \\ \Omega_{12} - S \end{bmatrix}, \quad J_{12} = -(R - \theta I)$$

$$J_{13} = \begin{bmatrix} -hA^T W \\ hB^T W \end{bmatrix}, \quad J_{14} = \begin{bmatrix} 0 \\ hC^T W \end{bmatrix}$$

$$\mathcal{P} = P + \overline{F}\Omega_2 - \underline{F}\Omega_1, \quad \Omega_{12} = \Omega_1 - \Omega_2,$$

$$Y = \begin{bmatrix} Y_1 & Y_2 \\ Y_2 & Y_2 \end{bmatrix}, \quad \Lambda_l = \begin{bmatrix} -\underline{F}\overline{F}\Phi_l & \frac{\underline{F}+\overline{F}}{2}\Phi_l \\ * & -\Phi_l \end{bmatrix}, \quad l = 1, 2, 3.$$

Proof First, we define some new vectors

$$\begin{aligned}\varsigma(t) &= \left[\xi^T(t),\ \xi^T(t-h(t)),\ \xi^T(t-h),\ \varsigma_1^T(t),\ u^T(t)\right]^T,\\ \xi(t) &= \left[x^T(t),\ f^T(x(t))\right]^T,\\ \varsigma_1 &= \left[\tfrac{1}{h(t)}\int_{t-h(t}^{t} x^T(\vartheta)\mathrm{d}\vartheta,\ \tfrac{1}{h-h(t)}\int_{t-h}^{t-h(t)} x^T(\vartheta)\mathrm{d}\vartheta\right]^T,\end{aligned}$$

and choose the following *Lyapunov function* candidate for DNNs (9.3) and (9.4):

$$V(t) = V_1(t) + V_2(t) + V_3(t), \tag{9.14}$$

where

$$\begin{aligned}V_1(t) &= x^T(t)Px(t) + 2\sum_{i=1}^{n}\int_0^{x_i(t)}\left\{\lambda_{1i}[f_i(\vartheta) - \underline{F}_i\vartheta] + \lambda_{2i}[\overline{F}_i\vartheta - f_i(\vartheta)]\right\}\mathrm{d}\vartheta,\\ V_2(t) &= \int_{t-h}^{t}\xi^T(\vartheta)X\xi(\vartheta)\mathrm{d}\vartheta + \int_{t-h(t)}^{t}\xi^T(\vartheta)Z\xi(\vartheta)\mathrm{d}\vartheta,\\ V_3(t) &= h\int_{-h}^{0}\int_{t+s}^{t}\dot{x}^T(\vartheta)W\dot{x}(\vartheta)\mathrm{d}\vartheta\mathrm{d}s.\end{aligned}$$

Then, it can be deduced that,

$$\dot{V}_1(t) = 2x^T(t)P\dot{x}(t) + 2[(f(x(t)) - \underline{F}x(t))^T\Omega_1 + (\overline{F}x(t) - f(x(t)))^T\Omega_2]\dot{x}(t), \tag{9.15}$$

Similarly, we have

$$\begin{aligned}\dot{V}_2(t) &= \xi^T(t)X\xi(t) - \xi^T(t-h)X\xi(t-h) + \xi^T(t)Z\xi(t)\\ &\quad - (1-\mu)x^T(t-h(t))Z\xi(t-h(t)),\end{aligned} \tag{9.16}$$

and

$$\dot{V}_3(t) = h^2\dot{x}^T(t)W\dot{x}(t) - h\int_{t-h}^{t}\dot{x}^T(\vartheta)W\dot{x}(\vartheta)\mathrm{d}\vartheta, \tag{9.17}$$

Inspired by Seuret and Gouaisbaut [27], for the vector function $x : [t-h(t), t] \to \mathbb{R}^n$, we set a new vector function $\eta : [t-h(t), t] \to \mathbb{R}^n$

$$\eta(\vartheta) = x(\vartheta) - \frac{\vartheta - t + h(t)}{h(t)}x(t) - \frac{t-\vartheta}{h(t)}x(t-h(t)).$$

It is easy to see that $\eta(t) = 0$ and $\eta(t - h(t)) = 0$, thus we use the Lemma 9.5, and have

$$\begin{aligned}
& -h\int_{t-h(t)}^{t} \dot{x}^T(\vartheta)W\dot{x}(\vartheta)\mathrm{d}\vartheta \\
= & -h\int_{t-h(t)}^{t} \dot{\eta}^T(\vartheta)W\dot{\eta}(\vartheta)\mathrm{d}\vartheta - \frac{h}{h(t)}[x(t) - x(t-h(t))]^T W[x(t) - x(t-h(t))] \\
\leq & -\frac{h\pi^2}{h^3(t)}\int_{t-h(t)}^{t} \eta^T(\vartheta)\mathrm{d}\vartheta W \int_{t-h(t)}^{t} \eta(\vartheta)\mathrm{d}\vartheta \\
& -\frac{h}{h(t)}[x(t) - x(t-h(t))]^T W[x(t) - x(t-h(t))].
\end{aligned} \tag{9.18}$$

Noting that

$$\begin{aligned}
& \int_{t-h(t)}^{t} \eta(\vartheta)\mathrm{d}\vartheta \\
& \quad = \int_{t-h(t)}^{t} \left[x(\vartheta) - \frac{\vartheta - t + h(t)}{h(t)}x(t) - \frac{t-\vartheta}{h(t)}x(t-h(t))\right]\mathrm{d}\vartheta \\
& \quad = \int_{t-h(t)}^{t} x(\vartheta)\mathrm{d}\vartheta - 0.5h(t)x(t) - 0.5h(t)x(t-h(t)),
\end{aligned} \tag{9.19}$$

it follows from (9.18) that

$$\begin{aligned}
& -h\int_{t-h(t)}^{t} \dot{x}^T(\vartheta)W\dot{x}(\vartheta)\mathrm{d}\vartheta \\
\leq & -\frac{h}{h(t)}[x(t) - x(t-h(t))]^T W[x(t) - x(t-h(t))] \\
& -\frac{\pi^2 h}{h(t)}\left[\frac{1}{h(t)}\int_{t-h(t)}^{t} x(\vartheta)\mathrm{d}\vartheta - 0.5x(t) - 0.5x(t-h(t))\right]^T \\
& \times W\left[\frac{1}{h(t)}\int_{t-h(t)}^{t} x(\vartheta)\mathrm{d}\vartheta - 0.5x(t) - 0.5x(t-h(t))\right] \\
= & -\frac{h}{h(t)}\varsigma^T(t)\mathcal{H}_1^T W\mathcal{H}_1\varsigma(t) - \frac{\pi^2 h}{h(t)}\varsigma^T(t)\mathcal{H}_2^T W\mathcal{H}_2\varsigma(t) \\
= & -\frac{h}{h(t)}\varsigma^T(t)\mathcal{H}_{12}^T\mathrm{diag}\{W, \pi^2 W\}\mathcal{H}_{12}\varsigma(t) \\
= & -\left(1 + \frac{h - h(t)}{h(t)}\right)\varsigma^T(t)\mathcal{H}_{12}^T\mathrm{diag}\{W, \pi^2 W\}\mathcal{H}_{12}\varsigma(t).
\end{aligned} \tag{9.20}$$

Similarly, we have

$$
\begin{aligned}
&-h\int_{t-h}^{t-h(t)} \dot{x}^T(\vartheta)W\dot{x}(\vartheta)\mathrm{d}\vartheta \\
&\le -\frac{h}{h-h(t)}[x(t-h(t))-x(t-h)]^T W[x(t-h(t))-x(t-h)] \\
&\quad -\frac{\pi^2 h}{h-h(t)}\left[\frac{1}{h-h(t)}\int_{t-h}^{t-h(t)} x(\vartheta)\mathrm{d}\vartheta - 0.5x(t-h(t)) - 0.5x(t-h)\right]^T \\
&\quad \times W\left[\frac{1}{h-h(t)}\int_{t-h}^{t-h(t)} x(\vartheta)\mathrm{d}\vartheta - 0.5x(t-h(t)) - 0.5x(t-h)\right] \\
&= -\frac{h}{h-h(t)}\varsigma^T(t)\mathcal{H}_3^T W\mathcal{H}_3\varsigma(t) - \frac{\pi^2 h}{h-h(t)}\varsigma^T(t)\mathcal{H}_4^T W\mathcal{H}_4\varsigma(t) \\
&= -\frac{h}{h-h(t)}\varsigma^T(t)\mathcal{H}_{34}^T \mathrm{diag}\{W,\pi^2 W\}\mathcal{H}_{34}\varsigma(t) \\
&= -\left(1+\frac{h(t)}{h-h(t)}\right)\varsigma^T(t)\mathcal{H}_{34}^T \mathrm{diag}\{W,\pi^2 W\}\mathcal{H}_{34}\varsigma(t),
\end{aligned} \tag{9.21}
$$

where

$$
\begin{aligned}
&\mathcal{H}_{12} = \begin{bmatrix}\mathcal{H}_1\\ \mathcal{H}_2\end{bmatrix}, \quad \mathcal{H}_{34} = \begin{bmatrix}\mathcal{H}_3\\ \mathcal{H}_4\end{bmatrix}, \\
&\mathcal{H}_1 = \begin{bmatrix} I & 0 & -I & 0 & 0 & 0 & 0 & 0\end{bmatrix}, \quad \mathcal{H}_2 = \begin{bmatrix} 0.5I & 0 & 0.5I & 0 & 0 & 0 & -I & 0\end{bmatrix}, \\
&\mathcal{H}_3 = \begin{bmatrix} 0 & 0 & I & 0 & -I & 0 & 0 & 0\end{bmatrix}, \quad \mathcal{H}_4 = \begin{bmatrix} 0 & 0 & 0.5I & 0 & 0.5I & 0 & 0 & -I\end{bmatrix}.
\end{aligned}
$$

Now, in view of reciprocally convex approach introduced in [28], it is easy to see that from (9.13),

$$
\begin{aligned}
&\varsigma^T(t)\begin{bmatrix}\sqrt{\frac{h-h(t)}{h(t)}}\mathcal{H}_{12}\\ -\sqrt{\frac{h-h(t)}{h(t)}}\mathcal{H}_{34}\end{bmatrix}^T \begin{bmatrix}\mathrm{diag}\{W,\pi^2 W\} & Y\\ * & \mathrm{diag}\{W,\pi^2 W\}\end{bmatrix} \\
&\quad \times \begin{bmatrix}\sqrt{\frac{h-h(t)}{h(t)}}\mathcal{H}_{12}\\ -\sqrt{\frac{h-h(t)}{h(t)}}\mathcal{H}_{34}\end{bmatrix}\varsigma(t) \ge 0,
\end{aligned} \tag{9.22}
$$

which is equivalent to the following inequality as

$$
\frac{h-h(t)}{h(t)}\varsigma^T(t)\mathcal{H}_{12}^T \mathrm{diag}\{W,\pi^2 W\}\mathcal{H}_{12}\varsigma(t) \tag{9.23}
$$

$$
+\frac{h(t)}{h-h(t)}\varsigma^T(t)\mathcal{H}_{34}^T \mathrm{diag}\{W,\pi^2 W\}\mathcal{H}_{34}\varsigma(t) \tag{9.24}
$$

$$
\ge \varsigma^T(t)\mathcal{H}_{12}^T Y\mathcal{H}_{34}\varsigma(t) + \varsigma^T(t)\mathcal{H}_{34}^T Y^T\mathcal{H}_{12}\varsigma(t), \tag{9.25}
$$

Together with (9.18)–(9.21), it yields

$$\begin{aligned}
&-h\int_{t-h}^{t}\dot{x}^T(\vartheta)W\dot{x}(\vartheta)\mathrm{d}\vartheta\\
&\le -\varsigma^T(t)\mathcal{H}_{12}^T\mathrm{diag}\{W,\pi^2W\}\mathcal{H}_{12}\varsigma(t)-\varsigma^T(t)\mathcal{H}_{12}^TY\mathcal{H}_{34}\varsigma(t)\\
&\quad -\varsigma^T(t)\mathcal{H}_{12}^T\mathrm{diag}\{W,\pi^2W\}\mathcal{H}_{12}\varsigma(t)-\varsigma^T(t)\mathcal{H}_{34}^TY^T\mathcal{H}_{12}\varsigma(t)\\
&= -\varsigma^T(t)\begin{bmatrix}\mathcal{H}_{12}\\ \mathcal{H}_{34}\end{bmatrix}^T\begin{bmatrix}\mathrm{diag}\{W,\pi^2W\} & Y\\ * & \mathrm{diag}\{W,\pi^2W\}\end{bmatrix}\begin{bmatrix}\mathcal{H}_{12}\\ \mathcal{H}_{34}\end{bmatrix}\varsigma(t).
\end{aligned} \tag{9.26}$$

Therefore, it is not difficult to obtain that

$$\begin{aligned}
\dot{V}_3(t) &\le h^2\dot{x}^TM\dot{x}(t)\\
&\quad -\varsigma^T(t)\begin{bmatrix}\mathcal{H}_{12}\\ \mathcal{H}_{34}\end{bmatrix}^T\begin{bmatrix}\mathrm{diag}\{W,\pi^2W\} & Y\\ * & \mathrm{diag}\{W,\pi^2W\}\end{bmatrix}\begin{bmatrix}\mathcal{H}_{12}\\ \mathcal{H}_{34}\end{bmatrix}\varsigma(t).
\end{aligned} \tag{9.27}$$

On the other hand, it follows from (9.5) that

$$(f_i(x_i(t))-\underline{F}_ix_i(t))(\overline{F}_ix_i(t)-f_i(x_i(t)))\le 0, \tag{9.28}$$

which means that for any appropriately dimensioned diagonal matrices $\Phi_l > 0$, $l = 1, 2, 3$, the following inequalities hold:

$$\xi^T(t)\Lambda_1\xi(t)\ge 0, \tag{9.29}$$

$$\xi^T(t-h)\Lambda_2\xi(t-h)\ge 0, \tag{9.30}$$

$$\int_{t-h}^{t}\xi^T(\vartheta)\Lambda_1\xi(\vartheta)\mathrm{d}\vartheta\ge 0, \tag{9.31}$$

where Λ_l, $l = 1, 2, 3$, have been denoted in (9.12).

From (9.15)–(9.27), it can be concluded that

$$\dot{V}(t)-y^T(t)Qy(t)-2y^T(t)Su(t)-u^T(t)Ru(t)+\theta u^T(t)u(t)\le\varsigma^T(t)\Xi_1\varsigma(t).$$

where

$$\Xi_1=\begin{bmatrix} J_1 & J_2 & J_4 & J_7 & J_{11}\\ * & J_3 & J_5 & J_8 & 0\\ * & * & J_6 & J_9 & 0\\ * & * & * & J_{10} & 0\\ * & * & * & * & J_{12}\end{bmatrix}+h^2\mathcal{A}^T(t)W\mathcal{A}(t)$$

and

$$\mathcal{A}(t) = \begin{bmatrix} -A & B & 0 & C & 0 & 0 & 0 & 0 & I \end{bmatrix}$$

On the other hand, by using *Schur complement* and Lemma 2 in [19], it follows from (9.12) that $\varsigma^T(t)\Xi_1\varsigma(t) < 0$, which implies that

$$\dot{V}(t) - y^T(t)Qy(t) - 2y^T(t)Su(t) - u^T(t)Ru(t) + \theta u^T(t)u(t) < 0. \quad (9.32)$$

Noting that $-Q > 0$, we also have $\dot{V}(t) < 0$ for $u(t) = 0$. Then, in light of (9.14), it can be shown that $\dot{V}(t) < \delta\|x(t)\|^2$ for a sufficiently small $\delta > 0$ and $x(t) \neq 0$, that is, neural network (9.3) and (9.4) is stable.

In what follows, we will prove that neural network (9.3) and (9.4) is strictly $(Q, R, S)-\theta$-dissipative. To this end, we introduce the following performance index

$$J_F = -\langle y, Qy\rangle_F - \langle y, Sy\rangle_F - \langle u, Ru\rangle_F + \theta\langle u, u\rangle_F, \quad (9.33)$$

where $F > 0$. Under the zero initial condition, it follows from (9.32) and (9.33) that

$$\begin{aligned} J_F = \int_0^F & [\dot{V}(t) - y^T(t)Qy(t) - 2y^T(t)Su(t) - u^T(t)Ru(t) \\ & + \theta u^T(t)u(t)]\mathrm{d}t < 0. \end{aligned}$$

for any nonzero $u(t) \in \mathcal{L}_2[0, \infty)$. This implies

$$\int_0^F [y^T(t)Qy(t) + 2y^T(t)Su(t) + u^T(t)Ru(t)]\mathrm{d}t > \theta \int_0^F u^T(t)u(t)\mathrm{d}t,$$

then the condition in (9.11) is satisfied. Hence, DNNs (9.3) and (9.4) is strictly $(Q, R, S) - \theta$-dissipative according to Definition 9.4. This completes the proof.

Remark 9.7 Many existing works (for example, [19, 29]) employed $V_3(t) = h\int_{-h}^{0}\int_{t+s}^{t}\dot{x}^T(\vartheta)W\dot{x}(\vartheta)\mathrm{d}\vartheta\mathrm{d}s$ as a part of Lyapunov functional to derive delay-dependent conditions, where the derivation of $V_3(t)$ was enlarged as $\dot{V}_3(t) = h^2\dot{x}^T(t)W\dot{x}(t) - [x(t) - x(t-h)]^T W[x(t) - x(t-h)]$ by using the noted *Jensen's inequality*. It should be pointed out that, however, some useful information was ignored in those papers. In this work, an extended Wirtinger inequality is introduced and then the derivation of $V_3(t)$ is enlarged as shown in (9.27), it is easy to see that the conservatism is reduced due to $W > 0$.

Remark 9.8 Theorem 9.31 provides a sufficient condition of strictly $(Q, R, S) - \theta$-dissipative for DNNs (9.3) and (9.4), it is worth mentioning that the proposed condition can be directly used to analyze the passivity of DNNs (9.3) and (9.4). By following a similar line in the proof of Theorem 9.6, the passivity condition of DNNs (9.3) and (9.4) can be obtained readily, which is shown as the following corollary.

Corollary 9.9 *Given a scalar $h > 0$, $\mu > 0$, and matrices $Q = Q^T = 0$, $S = I$, $R = R^T = 2\theta I$, DNNs (9.3) and (9.4) is passive, if there exist matrices $P > 0$, $X = \begin{bmatrix} X_1 & X_2 \\ X_2^T & X_3 \end{bmatrix} > 0$, $Z = \begin{bmatrix} Z_1 & Z_2 \\ Z_2^T & Z_3 \end{bmatrix} > 0$, $W > 0$, Y_1, Y_2, diagonal matrices $\Phi_l > 0$, $l = 1, 2, 3$, $\Omega_1 = \mathrm{diag}\{\lambda_{11}, \lambda_{12}, \ldots, \lambda_{1n}\} > 0$, $\Omega_2 = \mathrm{diag}\{\lambda_{21}, \lambda_{22}, \ldots, \lambda_{2n}\} > 0$, and a scalar $\theta > 0$ such that the LMIs in (9.12) and (9.13) hold.*

Remark 9.10 In fact, the conditions (i.e., stability criterion, passivity condition, and strictly $(Q, R, S) - \theta$-dissipative condition) for DNNs (9.3) and (9.4)) given in [19, 29, 30] were based on the idea of free-weighting matrix. However, our method outperforms those in [19, 29, 30]. The reduced conservatism of Corollary 9.9 benefits from reciprocally convex approach combined with an extended Wirtinger inequality. In addition, the proposed conditions require less computational complexity than the methods in [19, 29]. More specifically, there are $9.5n^2 + 5.5n + 1$ variables and $8n^2 + 8n + 1$ variables needed to solve in Theorem 1 in [29] and in Corollary 3 in [19], respectively, compared with $6n^2 + 8n + 1$ variables needed in our Corollary 9.9.

9.2 Positive Invariant Sets and Attractive Sets of DNN

9.2.1 Introduction

From a systems-theoretic point of view, the *global stability* of recurrent neural networks (RNNs) is a very interesting issue for research because of the special nonlinear structure of RNNs. From a practical point of view, the global stability of RNNs is also very important because it is a prerequisite in many neural network applications. In the past two decades, there has been arousing widespread concern on neural networks, due to their successful applications in many areas, such as pattern recognition, paralleled computation, associative memory, optimization, control, and signal processing, etc. It is well known that the integration and communication delays are unavoidably encountered both in biological and artificial neural systems. On the other hand, it has also been shown that the process of moving images requires the introduction of delay in the signal transmitted through the networks. Therefore, *Lyapunov stability* is one of the important properties of dynamic systems and there are a large number of results on the *stability* in Lyapunov sense for neural networks with bounded or unbounded time delays [31–39]. For example, Ref. [33] gets some sufficient conditions for globally exponential stability of unique equilibrium of the neural work with unbounded time delay by assuming that the activation functions are bounded and satisfy the global Lipschitz condition.

The studies of periodicity and almost periodicity are also important to *dynamical systems* as well as neural networks. A necessary condition of existence on global attractive periodicity and almost periodicity is that the system is ultimately bounded.

Global attractive periodic states and almost periodic states should be within global attractive sets, so that it would provide specific bounds on the existence of them.

It is worth mentioning that *Lyapunov stability* refers to the stability of equilibrium points which requires the existence and uniqueness of equilibrium points, while Lagrange stability refers to the stability of the total system which does not require the information of equilibrium points. Moreover, the global stability in Lyapunov sense can be viewed as a special case of stability in Lagrange sense by regarding an equilibrium point as an attractive set [40]. It is generally considered that a nonlinear dynamic systems may have chaos when the system is dissipative (or ultimately bounded) in large scale and has positive Lyapunov index in small scale. The notion of global stability in Lagrange sense is to extend the global stability in Lyapunov sense by including many dynamic behaviors such as stability, periodicity, and chaos. Basically, the study of global stability in Lagrange sense is to determine global attractive sets. Once a global attractive set is found, a rough bound of equilibria, periodic states, and chaotic attractors can be estimated. Thus global stability in Lagrange sense for RNNs can provide more prior knowledge. In addition, the global *stability in Lyapunov sense* on unique equilibrium point and the *stability in Hopfield sense* (i.e., complete stability) can be viewed as a special case of *stability in Lagrange sense* by regarding an equilibrium point as an attractive set. For this reason, the Lagrange stability has been extensively studied [40–47]. Reference [40] gives the detailed estimation of global exponential attractive sets and positive invariant sets of RNNs without any hypothesis on the existence, which is only based on the parameters of the systems. Meanwhile, it is also verified that outside the global exponential *attractive set* there is no equilibrium state, periodic state, almost periodic state, and chaos attractor of the neural network. Though delays occur frequently in practical applications, it is difficult to measure them precisely. In most situations, delays are time-varying, and in fact unbounded. That is, the entire history of the DNNs affects the present states [34]. For example, when the time delay is $\tau(t) = \frac{t}{3}$, the time delay $\tau(t)$ is unbounded and time-varying. Therefore, the studies of neural network with variable and unbounded time delays are more important and actual to practical neural networks than those with bounded delays. Therefore, it is necessary to study the Lagrange stability and invariant set of neural network with unbounded time delays.

Motivated by the above analysis, the aim of this section is to study the positive invariant sets and global exponential attractive sets of a class of neural networks with unbounded time delays.

9.2.2 Problem Formulation and Preliminaries

In this section, denote $\mathbb{R}^+ = (0, +\infty)$ and $\Gamma = \{1, 2, \ldots, n\}$. $C[X, Y]$ is a class of continuous mapping set from the topological space X to the topological space Y. Especially, $C \triangleq C((-\infty, t_0], \mathbb{R}^n)$.

Consider the following neural networks with unbounded time delays

$$\dot{x}_i(t) = -c_i x_i(t) + \sum_{j=1}^{n} a_{ij} g_j(x_j(t)) + \sum_{j=1}^{n} b_{ij} f_j(x_j(t - \tau(t))) + u_i, \tag{9.34}$$

where $i, j \in \Gamma = \{1, 2, \ldots, n\}$; $c_i \in \mathbb{R}^+$ is the self-feedback connection weight; $a_{ij}, b_{ij} \in \mathbb{R}$ are connection weights related to the neurons without and with delays, respectively; $u_i \in \mathbb{R}$ is an external input; $\tau(t) \geq 0$ is a continuous time delay; $g_j(\cdot)$ and $f_j(\cdot)$ which meet the conditions $g_j(0) = 0$ and $f_j(0) = 0$ are activation functions related to the neurons without and with delays, respectively. In this section, our results hold under the following assumption:

Assumption 9.11 $g_j(\cdot)$ and $f_j(\cdot)$ are Lipschitz continuous, $g_j(0) = 0$, $f_j(0) = 0$, $j \in \Gamma$. That is, there exist constants h_j, L_j such that for any $x_1, x_2, y_1, y_2 \in \mathbb{R}$ the following inequalities hold,

$$\mid g_j(x_1) - g_j(x_2) \mid \leq h_j \mid x_1 - x_2 \mid, \quad \mid f_j(y_1) - f_j(y_2) \mid \leq L_j \mid y_1 - y_2 \mid.$$

For any initial function $\varphi(s) \in C$, $s \in (-\infty, t_0]$, the solution of (9.34) that starts from the initial condition φ will be denoted by $x(t; t_0, \varphi)$ or simply $x(t)$ if no confusion occurs.

In the following, some definitions and lemmas are given so that our main results can be expediently explained.

Definition 9.12 (*see* [22]) The neural network model (9.34) is said to be a *dissipative system*, if there exists a compact set $S \subset \mathbb{R}^n$, such that $\forall x_0 \in \mathbb{R}^n$, $\exists F > 0$, when $t \geq t_0 + F$, $x(t; t_0, x_0) \subset S$, where $x(t; t_0, x_0)$ denotes the solution of Eq. (9.34) from initial state x_0 and initial time t_0, $x(t) = (x_1(t), \ldots, x_n(t))^T$. In this case, S is called a globally attractive set. A set S is called positive invariant, if $\forall x_0 \in S$ implies $x(t; t_0, x_0) \in S$ for $t > t_0$.

Definition 9.13 (*see* [22]) Let S is a globally attractive set of neural network model (9.34). The neural network model (9.34) is said to be globally exponentially dissipative system, if there exists a compact set $S^* \supset S$ in $\mathbb{R}^n$ such that $\forall x_0 \in \mathbb{R}^n \backslash S^*$, there exists a constant $M(x_0) > 0$ and $\alpha > 0$ such that

$$\inf_{x \in \mathbb{R}^n \backslash S^*} \{\|x(t; t_0, x_0) - \tilde{x}\| | \tilde{x} \in S^*\} \leq M(x_0) e^{-\alpha(t - t_0)}, \tag{9.35}$$

where the set S^* is called globally exponentially *attractive set*, $x \in \mathbb{R}^n \backslash S^*$ means $x \in \mathbb{R}^n$ but $x \notin S^*$.

Definition 9.14 (*see* [46, 48]) A set $S \subseteq \mathbb{R}^n$ is said to be a *positive invariant set* of (9.34), if for $\forall s \in [t_0 - \tau(t_0), t_0]$, $x(s) \in S$ implies $x(t; t_0, \varphi) \subseteq S$, $t \geq t_0$.

Definition 9.15 (*see* [46, 48]) A set $S \subseteq \mathbb{R}^n$ is said to be a *attractive set* of (9.34), if for $\forall s \in [t_0 - \tau(t_0), t_0]$, $x(s) \in \mathbb{R}^n \backslash S$, $\lim_{t \to +\infty} \rho(x(t), S) = 0$ holds, where $\mathbb{R}^n \setminus S$ is the complement set of S, $\rho(x, S) = \inf_{y \in S} \parallel x - y \parallel$ is the distance between x and S.

Definition 9.16 (*see* [45]) The compact set $\Omega \triangleq \{x \in \mathbb{R}^n | V(x) \leq l\}$ is said to be a global exponential attractive set of (9.34), where $V(x)$ is a radially unbounded and positive definite function, if there exists a nonnegative continuous function $K(\cdot)$, and two positive constants l and α such that for any solution $x(t) = x(t; t_0, \varphi)$ of (9.34), $V(x(t)) > l$, implies

$$V(x(t)) - l \leq K(\varphi) \exp\{-\alpha t\}, \ t \geq t_0.$$

Definition 9.17 (*see* [49]) The trajectory of network (9.34) is said to be *uniformly stable* in Lagrange sense (or uniformly bounded), if for any $H > 0$, there exists a constant $K = K(H) > 0$ such that $|x(t; t_0, \phi)| < K$ for all $\phi \in C_H$ and $t \geq 0$, where C_H is defined as the subset $\{\sigma \in C : \|\sigma\| \leq H\}$.

Definition 9.18 (*see* [49]) The trajectory of network (9.34) is called *globally exponentially stable* in Lagrange sense, if it is both uniformly stable in Lagrange sense and globally exponentially attractive.

Lemma 9.19 (see [47, 49]) *Let* $a > 0, b > 0, p > 1, q > 1$, *and* $\frac{1}{p} + \frac{1}{q} = 1$. *Then we have the inequality* $ab \leq \frac{1}{p}a^p + \frac{1}{q}b^q$, *and the equality holds if and only if* $a^p = b^q$.

9.2.3 Invariant Set Results

In the following, we will establish some invariant set results for neural network (9.34).

Theorem 9.20 *Under Assumption 9.11, and assuming that time delay satisfies* $0 < \dot{\tau}(t) \leq \tau_M < 1$ *as well as* $c_i > h_i \sum_{j=1}^n |a_{ji}| + \frac{L_i}{1-\tau_M} \sum_{j=1}^n |b_{ji}|, i \in \Gamma$. *Then the set*

$$S_1 = \left\{ x \in \mathbb{R}^n | \sum_{i=1}^n |x_i(t)| \leq \frac{\sum_{i=1}^n |u_i|}{\min\limits_{1 \leq i \leq n} \{c_i - h_i \sum_{j=1}^n |a_{ji}| - \frac{L_i}{1-\tau_M} \sum_{j=1}^n |b_{ji}|\}} \right\}$$

is a positive invariant set and global attractive set of the system (9.34).

Proof We consider the following radially unbounded and positive definite Lyapunov function

$$V(t)=\sum_{i=1}^{n}\left(|x_i(t)|+\frac{1}{1-\tau_M}\sum_{j=1}^{n}\left(|b_{ij}|\int_{t-\tau(t)}^{t}|f_j(x_j(s)\right)|\mathrm{d}s\right).$$

Calculating the Dini derivative of $V(t)$ along the positive semi trajectory of (9.34), it follows that

$$\begin{aligned}
&D^{+}V(t)|_{(9.34)}\\
\le&\sum_{i=1}^{n}\left(-c_i|x_i(t)|+\sum_{j=1}^{n}|a_{ij}||g_j(x_j(t))|+\sum_{j=1}^{n}|b_{ij}||f_j(x_j(t-\tau(t)))|+|u_i|\right)\\
&+\sum_{i=1}^{n}\left(\frac{1}{1-\tau_M}\sum_{j=1}^{n}|b_{ij}||f_j(x_j(t))|\right.\\
&\left.-\frac{1}{1-\tau_M}(1-\dot{\tau}(t))\sum_{j=1}^{n}|b_{ij}||f_j(x_j(t-\tau(t)))|\right)\\
\le&\sum_{i=1}^{n}\left(-c_i|x_i(t)|+\sum_{j=1}^{n}|a_{ij}||g_j(x_j(t))|\right.\\
&\left.+\frac{1}{1-\tau_M}\sum_{j=1}^{n}|b_{ij}||f_j(x_j(t))|+|u_i|\right)\\
\le&\sum_{i=1}^{n}\left(-c_i|x_i(t)|+\sum_{j=1}^{n}h_j|a_{ij}||x_j(t)|+\frac{1}{1-\tau_M}\sum_{j=1}^{n}L_j|b_{ij}||x_j(t)|+|u_i|\right)\\
=&\sum_{i=1}^{n}\left(-(c_i-h_i\sum_{j=1}^{n}|a_{ji}|-\frac{L_i}{1-\tau_M}\sum_{j=1}^{n}|b_{ji}|)|x_i(t)|+|u_i|\right)\\
\le&-\min_{1\le i\le n}\left\{c_i-h_i\sum_{j=1}^{n}|a_{ji}|-\frac{L_i}{1-\tau_M}\sum_{j=1}^{n}|b_{ji}|\right\}\sum_{i=1}^{n}|x_i(t)|+\sum_{i=1}^{n}|u_i|. \quad (9.36)
\end{aligned}$$

When $x\in\mathbb{R}_n\setminus S_1$, that is $x\notin S_1$, then $D^{+}V(t)|_{(9.34)}<0$, which implies that for $\forall\varphi\in S_1, t\ge t_0, x(t;t_0,\varphi)\subseteq S_1$ holds. And for $x_0\notin S_1$, there exists $F0$ such that $x(t;t_0,\varphi)\subseteq S_1$ holds for all $t\ge t_0+F$. Following the Definitions 9.14 and 9.15, we can know S_1 is a positive invariant and global attractive set of (9.34).

From Theorem 9.20, we can derive the following corollary:

Corollary 9.21 *Under Assumption 9.11 and assuming that time delay satisfies* $0 < \dot{\tau}(t) \leq \tau_M < 1$ *as well as* $c_i > h_i \sum_{j=1}^n |a_{ji}| + \frac{L_i}{1-\tau_M} \sum_{j=1}^n |b_{ji}|, i \in \Gamma$. *Then the set*

$$\tilde{S}_1 = \left\{ x \in \mathbb{R}^n | |x_i(t)| \leq \frac{\sum_{i=1}^n |u_i|}{c_i - h_i \sum_{j=1}^n |a_{ji}| - \frac{L_i}{1-\tau_M} \sum_{j=1}^n |b_{ji}|}, \ i \in \Gamma \right\}$$

is a positive invariant and global attractive set of the system (9.34).

Choosing $\xi_1 + \xi_2 = 1, 0 < \xi_1 < 1$, and let $A = (a_{ij})_{n \times n}$, $B = (b_{ij})_{n \times n}$,
$Q_{11} = \text{diag}(-\xi_1 c_1, \ldots, -\xi_1 c_n)$, $Q = \begin{pmatrix} Q_{11} & \frac{A}{2} & \frac{B}{2} \\ \frac{A^T}{2} & 0 & 0 \\ \frac{B^T}{2} & 0 & 0) \end{pmatrix}$, $\tilde{Q} = \begin{pmatrix} Q_{11} + \varepsilon I_n & \frac{A}{2} & \frac{B}{2} \\ \frac{A^T}{2} & 0 & 0 \\ \frac{B^T}{2} & 0 & 0) \end{pmatrix}$,
where $0 \leq \varepsilon \ll 1$. Then we have the following theorem.

Theorem 9.22 *Under Assumption 9.11 and assume that* Q *is negative semidefinite. Then the set*

$$S_2 = \left\{ x \in \mathbb{R}^n | \sum_{i=1}^n c_i x_i^2(t) \leq \frac{\sum_{i=1}^n u_i^2}{4\xi_2 \varepsilon} \right\}$$

is a positive invariant and global attractive set of the system (9.34).

Proof Introducing the radially unbounded and positive definite Lyapunov function

$$V(t) = \sum_{i=1}^n \frac{x_i^2(t)}{2},$$

it leads to

$$\begin{aligned}
&\frac{dV(t)}{dt}|_{(9.34)} \\
&= \sum_{i=1}^n x_i(t) \left(-c_i x_i(t) + \sum_{j=1}^n a_{ij} g_j(x_j(t)) + \sum_{j=1}^n b_{ij} f_j(x_j(t - \tau(t))) + u_i \right) \\
&= -\sum_{i=1}^n \xi_1 c_i x_i^2(t) + \sum_{i=1}^n x_i(t) \sum_{j=1}^n a_{ij} g_j(x_j(t)) \\
&\quad + \sum_{i=1}^n x_i(t) \sum_{j=1}^n b_{ij} f_j(x_j(t - \tau(t))) + \sum_{i=1}^n \left(-\xi_2 c_i x_i^2(t) + 2\sqrt{\varepsilon} x_i(t) \frac{u_i}{2\sqrt{\varepsilon}} \right) \\
&\leq (x^T(t), g^T(x(t)), f^T(x(t - \tau(t))))^T Q (x^T(t), g^T(x(t)), f^T(x(t - \tau(t))))
\end{aligned}$$

$$
\begin{aligned}
&+\sum_{i=1}^{n}\left(-\xi_2 c_i x_i^2(t)+\varepsilon x_i^2(t)+\frac{u_i^2}{4\varepsilon}\right)\\
&=(x^T(t),g^T(x(t)),f^T(x(t-\tau(t))))^T\tilde{Q}(x^T(t),g^T(x(t)),f^T(x(t-\tau(t))))\\
&+\sum_{i=1}^{n}\left(-\xi_2 c_i x_i^2(t)+\frac{u_i^2}{4\varepsilon}\right). \qquad (9.37)
\end{aligned}
$$

Since Q is negative semidefinite, by the continuation of Q, $\tilde{Q}$ is negative semidefinite too. So

$$
D^+V(t)|_{(9.34)} \le -\xi_2\sum_{i=1}^{n} c_i x_i^2(t)+\sum_{i=1}^{n}\frac{u_i^2}{4\varepsilon}.
$$

When $x \in \mathbb{R}^n \setminus S_2$, i.e., $x \notin S_2$, then $D^+V(t)|_{(9.34)} \le 0$, which implies that for $\forall\varphi \in S_2$, $t \ge t_0$, $x(t;t_0,\varphi) \subseteq S_2$ holds. And for $x_0 \notin S_2$, there exists $F > 0$ such that $x(t;t_0,\varphi) \subseteq S_2$ holds for all $t \ge t_0 + F$. Following the Definitions 9.14 and 9.15, S_2 is a positive invariant and globally attractive set of (9.34).

From Theorem 9.22, we have the following corollary:

Corollary 9.23 *Under Assumption 9.11 and assume that Q is negative semidefinite. Then the set*

$$
\tilde{S}_2=\left\{x \in \mathbb{R}^n | x_i^2(t) \le \frac{u_i^2}{4\xi_2 c_i \varepsilon},\ i \in \Gamma\right\}
$$

is a positive invariant and global attractive set of the system (9.34).

Now, similar to [46], we will establish a sufficient condition of global exponential attractive set for the system (9.34).

Theorem 9.24 *Under Assumption 9.11, and assume that:*

1. *there exist $p > 1$, $\varepsilon_i > 0$, $i = 1, 2, 3$, such that $I_{11} > I_{12} > 0$, where*

$$
\begin{aligned}
I_{11}=\min_{1\le i\le n}&\left\{pc_i-\sum_{j=1}^{n}\left((p-1)\varepsilon_1 h_j|a_{ij}|\right.\right.\\
&\left.\left.+\frac{1}{\varepsilon_1^{p-1}}h_i|a_{ji}|+(p-1)\varepsilon_2 L_j|b_{ij}|\right)-(p-1)\varepsilon_3\right\}, \qquad (9.38)
\end{aligned}
$$

$$
I_{12}=\max_{1\le i\le n}\left\{\sum_{j=1}^{n}\frac{1}{\varepsilon_2^{p-1}}L_i|b_{ji}|\right\}. \qquad (9.39)
$$

2. *there exist $M_1 > 1$ and $\beta_1 > 0$ such that*

$$
e^{\int_{t_0}^{t}(-I_{11}+I_{12}e^{I_{11}\tau(s)})ds} \le M_1 e^{-\beta_1(t-t_0)}.
$$

Then the set $S = \{x \in \mathbb{R}^n | \sum_{i=1}^n |x_i(t)| \le \frac{\eta}{I_{11}}\}$ *is a global exponential attractive set of the system (9.34), where* $\eta \ge \frac{I_{11} \sum_{i=1}^n \frac{|u_i|^p}{p\varepsilon_3^{p-1}}}{I_{11} - I_{12}}$.

Proof Consider the radially unbounded and positive definite Lyapunov function

$$V(t) = \sum_{i=1}^n \frac{|x_i(t)|^p}{p}.$$

Calculating the *Dini derivative* of $V(t)$ along the positive semi trajectory of (9.34) and employing Lemma 9.19, we obtain

$$\begin{aligned}
&D^+ V(t)|_{(9.34)} \\
&\le \sum_{i=1}^n |x_i(t)|^{p-1} \left(-c_i |x_i(t)| + \sum_{j=1}^n |a_{ij}||g_j(x_j(t))| \right. \\
&\quad \left. + \sum_{j=1}^n |b_{ij}||f_j(x_j(t - \tau(t)))| + |u_i| \right) \\
&\le \sum_{i=1}^n |x_i(t)|^{p-1} \left(-c_i |x_i(t)| + \sum_{j=1}^n h_j |a_{ij}||x_j(t)| \right. \\
&\quad \left. + \sum_{j=1}^n L_j |b_{ij}||x_j(t - \tau(t))| + |u_i| \right) \\
&= \sum_{i=1}^n \left(-c_i |x_i(t)|^p + \sum_{j=1}^n h_j |a_{ij}||x_i(t)|^{p-1} |x_j(t)| \right) \\
&\quad + \sum_{i=1}^n \sum_{j=1}^n \left(L_j |b_{ij}||x_i(t)|^{p-1} |x_j(t - \tau(t))| + |x_i(t)|^{p-1} |u_i| \right) \\
&\le \sum_{i=1}^n \left[-c_i |x_i(t)|^p + \sum_{j=1}^n h_j |a_{ij}| \left(\frac{p-1}{p} \varepsilon_1 |x_i(t)|^p + \frac{1}{p\varepsilon_1^{p-1}} |x_j(t)|^p \right) \right] \\
&\quad + \sum_{i=1}^n \sum_{j=1}^n L_j |b_{ij}| \left(\frac{p-1}{p} \varepsilon_2 |x_i(t)|^p + \frac{1}{p\varepsilon_2^{p-1}} |x_j(t - \tau(t))|^p \right) \\
&\quad + \sum_{i=1}^n \left(\frac{p-1}{p} \varepsilon_3 |x_i(t)|^p + \frac{|u_i|^p}{p\varepsilon_3^{p-1}} \right) \\
&= \sum_{i=1}^n \left[-pc_i + \sum_{j=1}^n \left((p-1)\varepsilon_1 h_j |a_{ij}| + \frac{1}{\varepsilon_1^{p-1}} h_i |a_{ji}| + (p-1)\varepsilon_2 L_j |b_{ij}| \right) \right.
\end{aligned}$$

$$
\begin{aligned}
&\left.+(p-1)\varepsilon_3\right]\frac{|x_i(t)|^p}{p}\\
&+\sum_{i=1}^{n}\sum_{j=1}^{n}\frac{1}{p\varepsilon_2^{p-1}}L_j|b_{ij}||x_j(t-\tau(t))|^p+\sum_{i=1}^{n}\frac{|u_i|^p}{p\varepsilon_3^{p-1}}\\
\le&-I_{11}V(t)+I_{12}V(t-\tau(t))+\sum_{i=1}^{n}\frac{|u_i|^p}{p\varepsilon_3^{p-1}}.
\end{aligned}
\tag{9.40}
$$

Let

$$
y(t)=\left(V(t)-\frac{\eta}{I_{11}}\right)e^{I_{11}(t-t_0)},\quad y(t_0)=\sup_{-\infty<s\le t_0}\left(V(s)-\frac{\eta}{I_{11}}\right),
$$

then

$$
\begin{aligned}
&D^+y(t)|_{(9.34)}\\
\le&\left(-I_{11}V(t)+I_{12}V(t-\tau(t))+\sum_{i=1}^{n}\frac{|u_i|^p}{p\varepsilon_3^{p-1}}+I_{11}V(t)-\eta\right)e^{I_{11}(t-t_0)}\\
=&I_{12}\left(V(t-\tau(t))-\frac{\eta-\sum_{i=1}^{n}\frac{|u_i|^p}{p\varepsilon_3^{p-1}}}{I_{12}}\right)e^{I_{11}(t-t_0)}.
\end{aligned}
\tag{9.41}
$$

Since $\eta\ge\frac{I_{11}\sum_{i=1}^{n}\frac{|u_i|^p}{p\varepsilon_3^{p-1}}}{I_{11}-I_{12}}$, one has

$$
V(t-\tau(t))-\frac{\eta-\sum_{i=1}^{n}\frac{|u_i|^p}{p\varepsilon^{p-1}}}{I_{12}}\le V(t-\tau(t))-\frac{\eta}{I_{11}}.
\tag{9.42}
$$

From (9.41) and (9.42), one gets

$$
D^+y(t)|_{(9.34)}\le I_{12}\left(V(t-\tau(t))-\frac{\eta}{I_{11}}\right)e^{I_{11}(t-t_0)}=I_{12}y(t-\tau(t))e^{I_{11}\tau(t)}.
$$

Then

$$
y(t)\le y(t_0)+\int_{t_0}^{t}I_{12}y(s-\tau(s))e^{I_{11}\tau(s)}\mathrm{d}s,\ t\ge t_0.
\tag{9.43}
$$

Let

$$
U(t)=\begin{cases}\bar{y}(t_0)+\int_{t_0}^{t}I_{12}y(s-\tau(s))e^{I_{11}\tau(s)}\mathrm{d}s,\ t\ge t_0,\\ \bar{y}(t_0),\ t\le t_0.\end{cases}
\tag{9.44}
$$

Since $I_{12} > 0$, when $V(t) > \frac{\eta}{I_{11}}$, then $y(t) \geq 0$, $U(t)$ is a monotone increasing function and we have for $\forall t \geq t_0$,

$$\frac{dU(t)}{dt} = I_{12} y(t - \tau(t)) e^{I_{11}\tau(t)}. \tag{9.45}$$

It follows from $I_{11} > 0$, (9.44) and (9.45), one gets $y(t) \leq U(t)$ for $\forall t \geq t_0$. Hence, $y(t - \tau(t)) \leq U(t - \tau(t)) \leq U(t)$. So

$$I_{12} y(t - \tau(t)) e^{I_{11}\tau(t)} \leq I_{12} U(t) e^{I_{11}\tau(t)}, \ t \geq t_0. \tag{9.46}$$

Combining (9.45) and (9.46), one gets

$$\frac{dU(t)}{dt} \leq I_{12} U(t) e^{I_{11}\tau(t)}.$$

Therefore

$$V(t) - \frac{\eta}{I_{11}} = y(t) e^{-I_{11}(t-t_0)} \leq U(t) e^{-I_{11}(t-t_0)} \leq \bar{y}(t_0) e^{\int_{t_0}^{t}(-I_{11} + I_{12} e^{I_{11}\tau(s)}) ds}.$$

From (2) in Theorem 9.24, we can obtain

$$V(t) - \frac{\eta}{I_{11}} \leq M_1 \bar{y}(t_0) e^{-\beta_1 (t-t_0)}.$$

It follows from Definition 9.16, the set S_3 is a global exponential attractive set of the system (9.34).

Note that Theorem 9.24 establishes a global exponential attractive set of the system (9.34) by Young Inequality (see Lemma 9.19). In Theorem 9.24, $p \neq 1$. If $p = 1$, how about the case? In other words, if we do not use the Young Inequality, can we derive a global exponential attractive set of the system (9.34)? In the following, we will present such a kind of result without using Lemma 9.19).

Theorem 9.25 *Under Assumption 9.11, and assume that:*
(a) there exist $W_i > 0$, $i \in \Gamma$, such that

$$I_{21} = \min_{1 \leq j \leq n} \left\{ c_j - \sum_{i=1}^{n} \frac{W_i}{W_j} h_j |a_{ij}| \right\} > I_{22} = \max_{1 \leq j \leq n} \left\{ \sum_{i=1}^{n} \frac{W_i}{W_j} L_j |b_{ij}| \right\}.$$

And let
(b) there exist $M_2 > 1$ and $\beta_2 > 0$ such that

$$e^{\int_{t_0}^{t}(-I_{21} + I_{22} e^{I_{21}\tau(s)}) ds} \leq M_2 e^{-\beta_2 (t-t_0)}.$$

Then the set $S_4 = \{x \in \mathbb{R}^n | \sum_{i=1}^n W_i |x_i(t)| \leq \frac{\eta}{l_{21}}\}$ *is a global exponential attractive set of the system (9.34), where* $\eta \geq \frac{l_{21} \sum_{i=1}^n W_i |u_i|}{l_{21} - l_{22}}$.

Proof Consider the radially unbounded and positive definite Lyapunov function

$$V(t) = \sum_{i=1}^n W_i |x_i(t)|,$$

then

$$\begin{aligned} & D^+ V(t)|_{(9.34)} \\ & \leq \sum_{i=1}^n \Bigg(-W_i c_i |x_i(t)| + \sum_{j=1}^n W_i |a_{ij}| |g_j(x_j(t))| \\ & \quad + \sum_{j=1}^n W_i |b_{ij}| |f_j(x_j(t-\tau(t)))| + W_i |u_i| \Bigg) \\ & \leq \sum_{i=1}^n \Bigg(-W_i c_i |x_i(t)| + \sum_{j=1}^n W_i h_j |a_{ij}| |x_j(t)| \\ & \quad + \sum_{j=1}^n W_i L_j |b_{ij}| |x_j(t-\tau(t))| + W_i |u_i| \Bigg) \\ & \leq -l_{21} V(t) + l_{22} V(t-\tau(t)) + \sum_{i=1}^n W_i |u_i|, \ t \geq t_0. \end{aligned} \tag{9.47}$$

Similar to the last part of the proof of Theorem 9.24, S_4 is a global exponential attractive set of the system (9.34).

If both $f_i(\cdot)$ and $g_i(\cdot)$ are bounded, it is well known that there exists an equilibrium of the system (9.34) by Schauder fixed point theorem. Thus we have the following corollary which is the Theorem 9.20 in [33].

Corollary 9.26 *Under Assumption 9.11, if both* $f_i(\cdot)$ *and* $g_i(\cdot)$ *are bounded,* $i \in \Gamma$, *and the (a) and (b) in Theorem 9.25 hold, then the equilibrium point of the system (9.34) is globally exponentially stable.*

Choosing $A' = (a_{ij} h_j)_{n \times n}$, $B' = (b_{ij} L_j)_{n \times n}$, we denote that $\lambda_{A'}$ and $\lambda_{B'}$ are the maximal eigenvalues of following matrices

$$A^* = \begin{pmatrix} 0 & A' \\ A'^T & 0 \end{pmatrix}_{2n \times 2n} \text{ and } B^* = \begin{pmatrix} 0 & B' \\ B'^T & 0 \end{pmatrix}_{2n \times 2n},$$

respectively. It is easy to prove that if λ is an eigenvalue of the matrix A^*, then $-\lambda$ is also an eigenvalue of the matrix A^*, thus $\lambda_{A'} > 0$, $\lambda_{B'} > 0$ [46].

Theorem 9.27 *Under Assumption 9.11, and assume that:*
(A) there exists $\tilde{\varepsilon} > 0$*, such that*

$$I_{31} = 2 \min_{1 \le i \le n} \{c_i\} - \tilde{\varepsilon} - 2\lambda_{A'} - \lambda_{B'} > I_{32} = \lambda_{B'}.$$

(B) there exist $M_3 > 1$ *and* $\beta_3 > 0$ *such that*

$$e^{\int_{t_0}^{t}(-I_{31}+I_{32}e^{I_{31}\tau(s)})\mathrm{d}s} \le M_3 e^{-\beta_3(t-t_0)}.$$

Then the set $S_5 = \{x \in \mathbb{R}^n | \sum_{i=1}^{n} |x_i(t)| \le \frac{\eta}{I_{31}}\}$ *is a global exponential attractive set of the system (9.34), where* $\eta \ge \frac{I_{31}\sum_{i=1}^{n}\frac{u_i^2}{\tilde{\varepsilon}}}{I_{31}-I_{32}}$.

Proof Consider the following function

$$V(t) = \sum_{i=1}^{n} x_i^2(t),$$

and let

$$\tilde{f}(x(t)) = \left(\frac{1}{h_1} f_1(x_1(t)), \ldots, \frac{1}{h_n} f_n(x_n(t))\right)^T,$$

$$\tilde{g}(x(t)) = \left(\frac{1}{L_1} g_1(x_1(t)), \ldots, \frac{1}{L_n} g_n(x_n(t))\right)^T,$$

one has

$$\begin{aligned}
&\frac{\mathrm{d}V(t)}{\mathrm{d}t}|_{(9.34)} \\
&= \sum_{i=1}^{n} 2x_i(t)\left(-c_i x_i(t) + \sum_{j=1}^{n} a_{ij} g_j(x_j(t)) + \sum_{j=1}^{n} b_{ij} f_j(x_j(t-\tau(t))) + u_i\right) \\
&= -\sum_{i=1}^{n} 2c_i x_i^2(t) + \sum_{i=1}^{n} 2x_i(t) \sum_{j=1}^{n} a_{ij} g_j(x_j(t)) \\
&\quad + \sum_{i=1}^{n} 2x_i(t) \sum_{j=1}^{n} b_{ij} f_j(x_j(t-\tau(t))) + \sum_{i=1}^{n} 2x_i(t)\sqrt{\tilde{\varepsilon}}\frac{u_i}{\sqrt{\tilde{\varepsilon}}} \\
&\le \sum_{i=1}^{n}(-2c_i + \tilde{\varepsilon})x_i^2(t) + (x^T(t), \tilde{g}^T(x(t))) \begin{pmatrix} 0 & A' \\ A'^T & 0 \end{pmatrix} (x^T(t), \tilde{g}^T(x(t)))^T \\
&\quad + (x^T(t), \tilde{f}^T(x(t-\tau(t)))) \begin{pmatrix} 0 & B' \\ B'^T & 0 \end{pmatrix} (x^T(t), \tilde{f}^T(x(t-\tau(t))))^T + \sum_{i=1}^{n} \frac{u_i^2}{\tilde{\varepsilon}}
\end{aligned}$$

$$
\begin{aligned}
\leq &(-2 \min_{1\leq i\leq n}\{c_i\} + \tilde{\varepsilon} + 2\lambda_{A'} + \lambda_{B'})V(t) + \lambda_{B'}V(t-\tau(t)) + \sum_{i=1}^{n}\frac{u_i^2}{\tilde{\varepsilon}} \\
= &- I_{31}V(t) + I_{32}V(t-\tau(t)) + \sum_{i=1}^{n}\frac{u_i^2}{\tilde{\varepsilon}}.
\end{aligned} \tag{9.48}
$$

Similar to the last part of the proof of Theorem 9.24, S_5 is a globally exponentially attractive set of the system (9.34).

9.3 Attracting and Invariant Sets of CGNN with Delays

9.3.1 Introduction

Since *Cohen–Grossberg neural networks* (CGNN) were first proposed by Cohen and Grossberg [50] in 1983, many researchers have done extensive works on this subject due to their extensive applications in many fields such as pattern recognition, parallel computing, associative memory, signal and image processing, and combinatorial optimization. In such applications, it is of prime importance to ensure that the designed neural networks is stable. In reality, time delays often occur due to finite switching speeds of the amplifiers and communication time. Moreover, it was observed both experimentally and numerically that time delay could destroy a stable network and cause sustained oscillations, bifurcation or chaos, and thus could be harmful. In recent years, the dynamical behaviors of Cohen–Grossberg neural networks with delays have been studied by many researchers [51–58].

Similar to the study of invariant set of recurrent neural networks with delay in Sect. 9.2, it is necessary to study the global *attracting set* and invariant set of CGNNs with time-varying delays. This motivates us to write this section. Different from [59–61], we will introduce a new nonlinear differential inequality, which is more effective than the linear differential inequalities for studying the asymptotic behavior of some nonlinear differential equations. Applying this new nonlinear delay differential inequality, the attracting set and *invariant set* of CGNNs are obtained. Meanwhile, using the properties of M-cone and a generalization of *Barbalat's lemma*, the boundedness and asymptotic behavior for the solution of the inequality are obtained. Furthermore, without using Lyapunov functional, the proposed method is shown to be simple yet effective for analyzing the asymptotic behavior of CGNNs with time-varying delays.

For convenience, in the following, the meanings of notations Γ, $\mathbb{R}_+$ and $C[X, Y]$ are the same defined in Sect. 9.2.2. For $A, B \in \mathbb{R}^{m\times n}$ or $A, B \in \mathbb{R}^n$, $A \geq B (A > B)$ means that each pair of corresponding elements of A and B satisfies the inequality "$\geq$ ($>$)." In particular, let $C \triangleq C[[-\tau, 0], \mathbb{R}^n]$ denote the family of all bounded continuous $\mathbb{R}^n$-valued functions ϕ defined on $[-\tau, 0]$ with the norm $\| \phi \| = \sup_{-\tau\leq\theta\leq 0} |$

$\phi(\theta) \mid$, where $\mid \cdot \mid$ is Euclidean norm of $\mathbb{R}^n$. For any $x \in \mathbb{R}^n$, $A \in \mathbb{R}^{n\times n}$, $\phi \in C$, we define

$$
\begin{aligned}
[x]^+ &= (\mid x_1 \mid, \mid x_2 \mid, \ldots, \mid x_n \mid)^T \triangleq col\{\mid x_i \mid\},\\
[A]^+ &= (\mid a_{ij} \mid)_{n\times n},\\
[\phi(t)]_\tau &= col\{[\phi_i(t)]_\tau\},\\
[\phi(t)]_\tau^+ &= col\{[\phi(t)]^+\}_\tau,\\
[\phi_i(t)]_\tau &= \sup_{-\tau\le\theta\le 0}\{\phi_i(t+\theta)\}.
\end{aligned}
$$

For an M-matrix D ([62, 63], p. 114), we denote $D \in M$ and $\Omega_M(D) \triangleq \{z \in \mathbb{R}^n | Dz > 0, z > 0\}$.

9.3.2 Problem Formulation and Preliminaries

Now, we will consider the following neural networks,

$$
\begin{aligned}
\dot{x}_i(t) = \alpha_i(x_i(t))\Bigg[&-\beta_i(x_i(t)) + E_i + \sum_{j=1}^{n} a_{ij} g_j(x_j(t))\\
&+\sum_{j=1}^{n} b_{ij} g_j(x_j(t-\tau_{ij}(t)))\Bigg],\\
x_i(t_0+s) = \phi_i(s),\ &-\tau \le s \le 0, i \in N,
\end{aligned}
\tag{9.49}
$$

where $x_i(t)$ is the ith neuron state, $\alpha_i(\cdot)$ represents an amplification function, $\beta_i(\cdot)$ is an appropriately behaved function, $g_j(\cdot)$ denotes the activation function, $\tau_{ij}(t)$ is a time-varying delay with $0 \le \tau_{ij}(t) \le \tau$, τ is a constant. E_i denotes the input and bias of the ith neuron, $i, j \in \Gamma$.

Throughout the subsection, we always assume that for any $\phi \in C$, system (9.49) has least one solution through (t_0, ϕ), denoted by $x(t; t_0, \phi)$ or $x_t(t_0, \phi)$ (simply $x(t)$ or x_t if no confusion occurs), where $x_t(t_0, \phi) = x(t+s, t_0, \phi) \in C$, $s \in [-\tau, 0]$.

To prove our results, the following generalization of *Barbalat's lemma* are necessary.

Lemma 9.28 (See [64]) *Let $f(t)$ be defined, continuous, and piecewise continuously differentiable for $t \ge 0$, and let $f(t)$ and $\dot{f}(t)$ be bounded. Let $G(x)$ be defined, continuous, and positive definite for all x. Further, let*

$$\int_0^\infty G(f(t))\,\mathrm{d}t < \infty,$$

then $f(t) \to 0$ as $t \to \infty$.

In the following, referring to [48], a useful nonlinear differential inequality will be given.

Lemma 9.29 *Assume* $u(t) \in C[[t_0, \infty), \mathbb{R}^n]$ *satisfies that*

$$\begin{cases} D^+[u(t)]^+ \leq R(u(t))\left(P[u(t)]^+ + Q[u(t)]_\tau^+ + E\right), \ t \geq t_0, \\ u(t_0+\theta) = \qquad \phi(\theta) \in C, \theta \in [-\tau, 0], \end{cases} \tag{9.50}$$

where D^+ *denotes the* Dini derivative *operator,* $P = (p_{ij})_{n\times n}$ *and* $p_{ij} \geq 0$ *for* $i \neq j$, $Q = (q_{ij})_{n\times n} \geq 0$, $E = col\{E_i\} \geq 0$, $R(u) = \text{diag}\{R_i(u)\}$, $R_i(\cdot) \in C[\mathbb{R}^n, \mathbb{R}_+]$. *If* $D = -(P+Q) \in M$ *and* $L = D^{-1}E$ *, then:*

(1) For any constant $\bar{d} \geq 1$, *the solution* $u(t)$ *of (9.50) satisfies*

$$[u(t)]^+ \leq L, \ t \geq t_0, \tag{9.51}$$

provided that $[\phi]_\tau^+ \leq \bar{d}L$.

(2)

$$[u(t)]^+ \leq ze^{-\lambda \int_{t_0}^t \hat{R}(u(s))\mathrm{d}s} + L, \quad t \geq t_0, \tag{9.52}$$

provided that the initial conditions satisfy

$$[u(t)]^+ \leq ze^{-\lambda \int_{t_0}^t \hat{R}(u(s))\mathrm{d}s} + L, \quad t \in [t_0 - \tau, t_0], \tag{9.53}$$

where $\hat{R}(u) \leq \min_{1\leq i\leq n}\{R_i(u)\}$ *with* $\hat{R}(u) \in C[\mathbb{R}^n, \mathbb{R}_+]$, $z = col\{z_i\} \in \Omega_M(D)$ *and the positive constant* λ *is determined by the following inequality*

$$[\lambda I + P + Qe^{\lambda H\tau}]z < 0, \tag{9.54}$$

where $H = \max_{[u]^+\leq \bar{d}L} \hat{R}(u) < \infty$.

Proof Since $D = -(P+Q) \in M$, we have $D^{-1} \geq 0$. Let $F = D^{-1}col\{1\}\epsilon$ ($\epsilon > 0$ small enough), then $F > 0$. In order to prove (9.51), we will first prove that

$$[u(t)]^+ \leq \bar{d}L + F \triangleq col\{\bar{x}_i\} = \bar{x}, \quad \forall t \geq t_0, \tag{9.55}$$

for any given initial function $\phi \in C$ with $[\phi]_\tau^+ \leq \bar{d}L$.

If (9.55) does not hold, then there exist $i \in \Gamma$ and $t_1 > t_0$ such that

$$|\, u_i(t_1)\, |= \bar{x}_i, \quad [u(t)]^+ \leq \bar{x}, \qquad \text{for } t \leq t_1, \tag{9.56}$$

and

$$D^+ \,|\, u_i(t_1)\, | \geq 0. \tag{9.57}$$

It follows from (9.50) and (9.56) that

$$\begin{aligned} D^+ \mid u_i(t_1) \mid^+ &\le R(u(t_1)) \big[P[u(t_1)]^+ + Q[u(t_1)]^+_\tau + E\big] \\ &\le R(u(t_1))[(P+Q)\bar{x} + E] \\ &= -R(u(t_1))[\bar{d}E + col\{1\}\epsilon - E] \\ &\le -R(u(t_1))col\{1\}\epsilon < 0, \end{aligned}$$

which contradicts with the inequality (9.57). So (9.55) holds for all $t \ge t_0$. Letting $\epsilon \to 0$ in (9.55), we have

$$[u(t)]^+ \le \bar{d}L, t \ge t_0.$$

The proof of part (1) is completed.

Since $L = D^{-1}E$, we have $(P+Q)L + E = 0$. Then

$$\sum_{j=1}^{n}(p_{ij} + q_{ij})L_j + E_i = 0, \quad i \in \Gamma. \tag{9.58}$$

From (9.54), we can get

$$\sum_{j=1}^{n}(p_{ij} + q_{ij}e^{\lambda H\tau})z_j < -\lambda z_i, \quad i \in \Gamma. \tag{9.59}$$

In the following, we at fist shall prove that for any positive constant ϵ,

$$\mid u_i(t) \mid \le (1+\epsilon)\left(z_i e^{-\lambda\int_{t_0}^{t}\hat{R}(u(s))ds} + L_i\right) \triangleq w_i(t), t \ge t_0, \ i \in N. \tag{9.60}$$

We let

$$\begin{aligned} \mathfrak{F} &= \{i \in \Gamma \mid \ |u_i(t)| > w_i(t) \text{ for some } \ t \in [t_0, \infty]\}, \\ \theta_i &= \inf\{t \in [t_0, \infty] \mid \ |u_i(t)| > w_i(t), i \in \mathfrak{F}\}. \end{aligned}$$

If inequality (9.60) is not true, then $\mathfrak{F}$ is a nonempty set and there must exist some integer $m \in \mathfrak{F}$ such that $\theta_m = \min_{i\in\mathfrak{F}}\{\theta_i\} \in [t_0, \infty)$.

By $u_m(t) \in C[[t_0, \infty), \mathbb{R}]$ and the inequality (9.60), we can get

$$\theta_m > t_0, |u_m(\theta_m)| = w_m(\theta_m), D^+|u_m(\theta_m)| = \dot{w}_m(\theta_m) \tag{9.61}$$

$$|u_i(t)| \le w_i(t), t \in [t_0 - \tau, \theta_m], i \in \Gamma. \tag{9.62}$$

By using (9.50) and (9.58)–(9.62), we obtain

$$\begin{aligned}
&D^+|u_m(\theta_m)| \\
&\le R_m(u(\theta_m))\left[\sum_{j=1}^{n}(1+\varepsilon)z_j e^{-\lambda\int_{t_0}^{\theta_m}\hat{R}(u(s))\mathrm{d}s}\left[p_{mj}+q_{mj}e^{\lambda\int_{\theta_m-\tau}^{\theta_m}\hat{R}(u(s))\mathrm{d}s}\right]-\varepsilon E_m\right] \\
&\le R_m(u(\theta_m))\left[\sum_{j=1}^{n}\left(p_{mj}+q_{mj}e^{\lambda H\tau}\right)(1+\varepsilon)z_j e^{-\lambda\int_{t_0}^{\theta_m}\hat{R}(u(s))\mathrm{d}s}\right] \\
&< -\lambda R(u(\theta_m))(1+\varepsilon)z_m e^{-\lambda\int_{t_0}^{\theta_m}\hat{R}(u(s))\mathrm{d}s} = \dot{w}_m(\theta_m),
\end{aligned}$$

which contradicts with the inequality in (9.61). Thus the inequality (9.60) holds. Therefore, letting $\varepsilon \to 0$, we have (9.52). The proof is completed.

Remark 9.30 When the initial conditions $\phi \in PC$ (PC is an abbreviation of piecewise continuous function) defined in [60], Lemma 9.29 still holds. Therefore, many known results are easily obtained. For example, Lemma 2.1 in [61], Theorem 3.1 in [60], and Lemma 1 in [65] can be derived by Lemma 9.29 if we choose $R(u) \ge \mathrm{diag}(s_1, \ldots, s_n) > 0$, $E = 0$; $R(u) = F$, $E = 0$; and $R(u) = F$ in (9.50), respectively.

According to Lemma 9.29, we can establish the following result:

Proposition 9.31 *Under the conditions of Lemma 9.29, if $\hat{R}(u)$ is positive definite, then*

$$\lim_{t\to+\infty}[u(t)]^+ \le L. \tag{9.63}$$

Proof We only need to consider the following two possible cases:

(i) If $\int_{t_0}^{+\infty}\hat{R}(u(s))\mathrm{d}s = +\infty$. Then from (9.52), we have $\lim_{t\to+\infty}[u(t)]^+ \le L$.

(ii) If $\int_{t_0}^{+\infty}\hat{R}(u(s))\mathrm{d}s < +\infty$. Since $R(\cdot)$ is continuous, we can get $\int_{t_0}^{+\infty}\hat{R}(u(s))$ $\mathrm{d}s < +\infty$.

On the other hand, from (9.51), for any given initial function $\phi \in C$, we have that $u(t)$ is bounded. Furthermore, $\dot{u}(t)$ is bounded by (9.50). Thus by Lemma 9.28, we have $\lim_{t\to+\infty}u(t) = 0 \le L$. Therefore, the conclusion holds and the proof is completed.

9.3.3 Invariant Set Result

In this subsection, we always suppose the following assumptions to be true.

(A1) *Amplification functions* $\alpha_i(\cdot)$ are positive and continuous. Furthermore, there is a continuous and positive definite function $\hat{\alpha}(s)$ such that $\min_{1\le i\le n}\{\alpha_i\} \ge \hat{\alpha}(s)$.

(A2) There exists a positive diagonal matrix $\beta = \text{diag}\{\beta_i\}$ such that

$$\frac{\beta_i(s_1) - \beta_i(s_2)}{s_1 - s_2} \geq \beta_i > 0,$$

for all $i \in \Gamma, \quad s_1 \neq s_2, \quad s_1, s_2 \in \mathbb{R}$.

(A3) The *activation function* $g_j(x(t))$ with $g_j(0) = 0$ satisfies the Lipschitz condition, that is, for any $u_j \in \mathbb{R}, \ j \in \Gamma$, there exist nonnegative constants G_j such that $| g_j(u_j) | \leq G_j | u_j |$.

(A4) Let $\hat{D} = -(\hat{P} + \hat{Q}) \in M$, where $\hat{P} = (\hat{p}_{ij})_{n\times n}$, $\hat{p}_{ij} = -\beta_i + |a_{ii}|G_i$, $p_{ij} = |a_{ij}|G_j$, for $i \neq j$; $\hat{Q} = (\hat{q}_{ij})_{n\times n}$, $\hat{q}_{ij} = |b_{ij}|G_j$; $\hat{E} = \text{col}\{\beta_i(0) + |E_i|\}$.

Theorem 9.32 *Assume that (A1)–(A4) hold, then* $S = \{\phi \in C | [\phi]_\tau^+ \leq \hat{D}^{-1}\hat{E}\}$ *is a positive invariant and global attracting set of (9.49).*

Proof Calculating the upper right *Dini derivative* $D^+[x(t)]^+$ along system (9.49), from conditions (A1)–(A3), we obtain

$$\begin{aligned}
D^+[x_i(t)]^+ &= \text{sgn}(x_i(t))\dot{x}_i(t) \\
&\leq \alpha_i(x_i(t)) \left[-\beta_i|x_i(t)| + \beta_i(0) + \sum_{j=1}^{n} |a_{ij}|G_j|x_j(t)| \right. \\
&\quad \left. + \sum_{j=1}^{n} |b_{ij}|G_j|x_j(t - \tau_{ij}(t))| + |E_i| \right] \\
&\leq \alpha_i(x_i(t)) \left[(-\beta_i + |a_{ii}|G_i|)x_i(t)| + \sum_{j=1, j\neq i}^{n} |a_{ij}|G_j|x_j(t)| \right. \\
&\quad \left. + \sum_{j=1, j\neq i}^{n} |b_{ij}|G_j|x_j(t - \tau_{ij}(t))| + \beta_i(0) + |E_i| \right].
\end{aligned} \tag{9.64}$$

From (A4), we get that

$$D^+[x(t)]^+ \leq \alpha(x(t)) \left[\hat{P}[x(t)]^+ + \hat{Q}[x(t)]_\tau^+ + \hat{E} \right], t \geq t_0, \tag{9.65}$$

where $\alpha(x(t)) = \text{diag}(\alpha_1(x_1(t)), \ldots, \alpha_n(x_n(t)))$. Then from the conclusion 1) of Lemma 9.29, we can obtain

$$[x(t)]^+ \leq \hat{L}, \quad t \geq t_0, \tag{9.66}$$

provided $[\phi]_\tau^+ \leq \hat{L}$, where $\hat{L} = \hat{D}^{-1}\hat{E}$. So S is a positive invariant set of (9.49).

On the other hand, since $\hat{D} \in M$, there exists a positive vector $z = (z_1, \ldots, z_n)^T$ such that

$$\hat{D}z > 0, \quad \text{or} \quad [\hat{P} + \hat{Q}]z < 0.$$

According to continuity property, we know that there must exist a positive scalar λ such that

$$\left[\lambda I + \hat{P} + \hat{Q}e^{\lambda \hat{H}\tau}\right]z < 0, \tag{9.67}$$

where $\hat{H} = \max_{[u]^+ \leq \hat{d}\hat{L}} \hat{\alpha}(u) < \infty$ and $\hat{d} \geq 1$ is a constant such that $[\phi]_\tau^+ \leq \hat{d}\hat{L}$.

Then by (9.65), (9.67) and (A4), all the conditions of Theorem 9.31 are satisfied, we have

$$\lim_{t \to \infty} [x(t)]^+ \leq \hat{L}. \tag{9.68}$$

According to the Definitions 9.14 and 9.15, S is also a global attracting set of (9.49). The proof is completed.

Remark 9.33 We can easily find that our result improves the earlier criteria on the amplification functions. In [61, 66, 67], amplification functions in the systems are continuous and satisfy $\alpha_i(u) > \alpha_i > 0$ (α_i is a constant) for all $u \in \mathbb{R}^n$. Moreover, conditions ensuring the asymptotic behavior of systems in [51, 52, 68, 69] require that the amplification functions $\alpha_i(u)$ is bounded, positive and continuous, i.e., there exist constants $\bar{\alpha}_i$, $\underline{\alpha}_i$ such that $0 < \underline{\alpha}_i \leq \alpha_i(u) \leq \bar{\alpha}_i < \infty$. However, in this subsection, we only require that amplification functions $\alpha_i(u)$ are positive, continuous, and there is a continuous and positive definite function $\hat{\alpha}(u)$ such that $\min_{1 \leq i \leq n}\{\alpha_i(u)\} \geq \hat{\alpha}(u)$.

Remark 9.34 Dissipativity and invariant set are discussed for a kind of neural networks with delays in this chapter. In the assumptions on the *activation function* of the neural networks, Lipschitz conditions are required and the activation function equals zero at origin. This is a fundamental assumption in the application of Lyapunov stability. Similarly, passivity problem for RNNs discussed in Chap. 8 has the same assumption on the activation. Therefore, we can also conclude that besides the origin or the fixed equilibrium point of the isolated systems, the concepts such as passivity, dissipativity, and invariant sets are some generalizations of the fixed equilibrium point, in which fixed equilibrium point can be the center of a circle or a focus of an ellipse. The underlying *error system* formed by the coordinate transformation required in the stability analysis of RNNs still exists in the analysis of passivity and invariant sets of RNNs.

9.4 Summary

In this chapter, we have discussed the dissipativity and invariant set of neural networks with delay, which is an extension of stability and passivity. In this chapter, the adopted method is not in the LMI form, while the algebraic inequality method is used. Indeed, different methods can be chosen according to the requirement of the concerned problems. Each method has its own features and advantages.

References

1. J. Willems, Dissipative dynamical systems part I: general theory. Arch. Ration. Mech. Anal. **45**, 321–351 (1972)
2. D. Hill, P. Moylan, Dissipative dynamical systems: basic input output and state properties. J. Frankl. Inst. **309**, 327–357 (1980)
3. B. Zhang, W. Zheng, S. Xu, Filtering of Markovian jump delay systems based on a new performance index. IEEE Trans. Circuits Syst. I, Regul. Pap. **60**(5), 1250–1263 (2013)
4. W.M. Haddad, V. Chellaboina, *Nonlinear Dynamical Systems and Control: A Lyapunov-Based Approach* (Princeton University Press, Princeton, 2008)
5. D.J. Hill, P.J. Moylan, The stability of nonlinear dissipative systems. IEEE Trans. Autom. Control **21**(5), 708–711 (1976)
6. S. Xie, L. Xie, C.D. Souza, Robust dissipative control for linear systems with dissipative uncertainty. Int. J. Control **70**, 169–191 (1998)
7. J. Zhao, D. Hill, Dissipativity theory for switched systems. IEEE Trans. Autom. Control **53**, 941–953 (2008)
8. Z. Feng, J. Lam, H. Gao, α-dissipativity analysis of singular time delay systems. Automatica **47**, 2548–2552 (2011)
9. L. Wu, W. Zheng, H. Gao, Dissipativity-based sliding mode control of switched stochastic systems. IEEE Trans. Autom. Control **58**, 785–793 (2013)
10. Z. Feng, J. Lam, Robust reliable dissipative filtering for discrete delay singular systems. Signal Process. **92**(12), 3010–3025 (2012)
11. Z. Feng, J. Lam, Z. Shu, Dissipative control for linear systems by static output feedback. Int. J. Syst. Sci. **44**(8), 1566–1576 (2013)
12. M. Mahmoud, G. Khan, Dissipativity analysis for discrete stochastic neural networks with Markovian delays and partially known transition matrix. Appl. Math. Comput. **228**, 292–310 (2014)
13. J. Cao, J. Wang, Global asymptotic and robust stability of recurrent neural networks with time delays. IEEE Trans. Circuits Syst. I, Regul. Pap. **52**(2), 417–426 (2005)
14. D. Zhou, J. Cao, Globally exponential stability conditions for cellular neural networks with time-varying delays. Appl. Math. Comput. **131**, 487–496 (2002)
15. L. Duan, L. Huang, Global dissipativity of mixed time-varying delayed neural networks with discontinuous activations. Commun. Nonlinear Sci. Numer. Simul. **19**(12), 4122–4134 (2014)
16. Z. Wu, P. Shi, H. Su, J. Chu, Passivity analysis for discrete-time stochastic Markovian jump neural networks with mixed time delays. IEEE Trans. Neural Netw. **22**, 1566–1575 (2011)
17. X. Li, R. Rakkiyappan, G. Velmurugan, Dissipativity analysis of memristor-based complex-valued neural networks with time-varying delays. Inf. Sci. **294**, 645–665 (2015)
18. Z. Feng, J. Lam, Stability and dissipativity analysis for distributed delay cellular neural networks. IEEE Trans. Neural Netw. **22**, 976–981 (2011)
19. Z. Wu, J. Park, H. Su, J. Chu, Robust dissipativity analysis of neural networks with time-varying delay and randomly occurring uncertainties. Nonlinear Dyn. **74**, 730–734 (2012)

20. J. Wang, J.H. Park, H. Shen, J. Wang, Delay-dependent robust dissipativity conditions for delayed neural networks with random uncertainties. Appl. Math. Comput. **221**, 710–719 (2013)
21. J. Hu, Z. Wang, H. Gao, L. Stergioulas, Robust sliding mode control for discrete stochastic systems with mixed time-delays, randomly occurring uncertainties and nonlinearities. IEEE Trans. Ind. Electron. **59**, 3008–3015 (2012)
22. X.X. Liao, J. Wang, Global dissipativity of continuous-time recurrent neural networks with time delay. Phys. Rev. E **68**, 16118 (2003)
23. Y. Liu, Z. Wang, X. Liu, Global exponential stability of generalized recurrent neural networks with discrete and distributed delays. Neural Netw. **19**, 667–675 (2006)
24. Z. Wu, H. Su, P. Shi, *Analysis and Synthesis of Singular Systems with Time-Delays* (Springer, Berlin, 2013)
25. R. Lozano, B. Brogliato, O. Egeland, B. Maschke, *Dissipative Systems Analysis and Control: Theory and Applications*, 1st edn. (Springer, London, 2000)
26. J. Bao, P.L. Lee, *Process Control: The Passive Systems Approach*, 1st edn. (Advances in Industrial Control (Springer, London, 2007)
27. A. Seuret, F. Gouaisbaut, Jensen's and Wirtinger's inequalities for time-delay systems, in *11th IFAC Workshop on Time-Delay Systems*, France, Grenoble, pp. 343–348 (2013)
28. P. Park, J. Ko, C. Jeong, Reciprocally convex approach to stability of systems with time-varying delays. Automatica **47**, 235–238 (2011)
29. H. Zeng, Y. He, M. Wu, S. Xiao, Passivity analysis for neural networks with a time-varying delay. Neurocomputing **74**, 730–734 (2011)
30. S. Xu, W. Zheng, Y. Zou, Passivity analysis of neural networks with time-varying delays. IEEE Trans. Circuits Syst. II, Express Briefs **56**, 325–329 (2009)
31. X. Liao, J. Wang, Z. Zeng, Global asymptotic stability and global exponential stability of delayed cellular neural networks. IEEE Trans. Circuits. Syst II: Express Briefs **52**, 403–409 (2005)
32. B. Wang, J. Jian, C. Guo, Global exponential stability of a class of BAM networks with time-varying delays and continuously distributed delays. Neurocomputing **71**, 495–501 (2008)
33. Z. Zeng, C. Fu, X. Liao, Stability analysis of neural networks with infinite time-varying delay. J. Math. **22**(4), 391–396 (2002)
34. J. Zhang, Y. Suda, T. Iwas, Absolutely exponential stability of a class of neural networks with unbounded delay. Neural Netw. **17**(3), 391–397 (2004)
35. Z. Zeng, J. Wang, X. Liao, Global asymptotic stability and global exponential stability of neural networks with unbounded time-varying delays. IEEE Trans. Circuits Syst. II: Express. Briefs **52**(3), 168–173 (2005)
36. Z. Zeng, X. Liao, Global stability for neural networks with unbounded time varying delays. Acta Math. Sci. **25A**(5), 621–626 (2005)
37. Y. Zhang, P. Heng, K. Leung, Convergence analysis of cellular neural networks with unbounded delay. IEEE Trans. Circuits Syst. I: Fundam. Theory Appl. 48(6), 680–687 (2001)
38. Q. Zhang, X. Wei, J. Xu, Global exponential stability for nonautonomous cellular neural networks with unbounded delays. Chaos Solitons Fractals **39**(3), 1144–1151 (2009)
39. W. Zhao, A. Yan, Stability analysis of neural networks with both variable and unbounded delays. Chaos Solitons Fractals **40**(2), 697–707 (2009)
40. X. Liao, Q. Luo, Z. Zeng, Positive invariant and global exponential attractive sets of neural networks with time-varying delays. Neurocomputing **71**, 513–520 (2008)
41. H. Zhao, W. Cui, Global attracting set of a class of neutral Hopfield neural networks. J. Biomath. **18**(2), 197–203 (2003)
42. J. Pei, B. Li, Z. Yang, Invariant set and attracting property of second order Hopfield neural networks with time delays. J. Sichuan Univ. (Natural Science Edition) **42**(4), 688–691 (2005)
43. D. Xu, H. Zhao, Invariant set and attractivity of nonlinear differential equations with delays. J. Appl. Math. **15**, 321–325 (2002)
44. J. Jian, Z. Tu, H. Yu, Estimating the globally attractive set and positively invariant set of a new Lorenz-like chaotic system and its applications, in *2009 International Workshop on Chaos-fractals Theories and Applications*, pp. 241–245 (2009)

45. X. Liao, Q. Luo, Z. Zeng, Y. Guo, Global exponential stability in Lagrange sense for recurrent neural networks with time delays. Nonlinear Anal.: Real World Appl. 9, 1535–1557 (2008)
46. Z. Tu, J. Jian, B. Wang, Positive invariant sets and global exponential attractive sets of a class of neural networks with unbounded time-delays. Commun. Nonlinear Sci. Numer. Simul. **16**, 3738–3745 (2011)
47. B. Wang, J. Jian, M. Jiang, Stability in Lagrange sense for Cohen-Grossberg neural networks with time-varying delays and finite distributed delays. Nonlinear Anal.: Hybrid Syst. **4**, 65–78 (2010)
48. D. He, D. Xu, Attracting and invariant sets of fuzzy Cohen-Grossberg neural networks with time-varying delays. Phys. Lett. A **372**, 7057–7062 (2008)
49. A. Wu, Z. Zeng, Lagrange stability of memristive neural networks with discrete and distributed delays. IEEE Trans. Neural Netw. Learn. Syst. **25**(4), 690–703 (2014)
50. M. Cohen, S. Grossberg, Absolute stability of global pattern formation and parallel memory storage by competitive neural networks. IEEE Trans. Syst. Man Cybern. **13**(5), 815–816 (1983)
51. T. Li, S. Fei, Stability analysis of Cohen-Grossberg neural networks with time-varying and distributed delays. Neurocomputing **71**(6), 1069–1081 (2008)
52. C. Huang, L. Huang, Dynamics of a class of Cohen-Grossberg neural networks with time-varying delays. Nonlinear Anal. Real World Appl. **8**(1), 40–52 (2007)
53. J. Cao, G. Feng, Y. Wang, Multistability and multiperiodicity of delayed Cohen-Grossberg neural networks with a general class of activation functions. Phys. D **237**(13), 1734–1749 (2008)
54. J. Cao, Q. Song, Stability in Cohen-Grossberg-type bidirectional associative memory neural networks with time-varying delays. Nonlinearity **19**(7), 1601–1617 (2006)
55. J. Cao, X. Li, Stability in delayed Cohen-Grossberg neural networks: LMI optimization approach. Phys. D **212**(2), 54–65 (2005)
56. Z. Zhao, J. Jian, Positive invariant sets and global exponential attractive sets of BAM neural networks with time-varying and infinite distributed delays. Neurocomputing **142**, 447–457 (2014)
57. Z. Zhao, J. Jian, Attracting and quasi-invariant sets for BAM neural networks of neutral-type with time-varying and infinite distributed delays. Neurocomputing **140**, 265–272 (2014)
58. P. Zhang, Fundamental domain of invariant sets and applications. Ergod. Theory Dyn. Syst. **34**, 341–352 (2014)
59. A. Halanay, *Differential Equations: Stability, Oscillations, Time Lages* (Academic Press, New York, 1966)
60. D. Xu, Z. Yang, Impulsive delay differential inequality and stability of neural networks. J. Math. Anal. Appl. **305**(1), 107–120 (2005)
61. Z. Yang, D. Xu, Impulsive effects on stability of Cohen-Grossberg neural networks with variable delays. Appl. Math. Comput. **177**(1), 63–78 (2006)
62. R.A. Horn, C.R. Johnson, *Topics in Matrix Analysis*, vol. 2 (Cambridge University Press, England, 1991)
63. Z. Wang, *Stability and Synchronization for Complex Dynamical Neural Networks* (in Chinese) (China Science Press, Beijing, 2014)
64. M.A. Aizerman, F.R. Gantmakher, Absolute Stability of Control Systems, (Akad. Nauk, Moscow, 1963). (Translation: Holden-Day, San Francisco-London, 1964)
65. D. Xu, Z. Yang, Attracting and invariant sets for a class of impulsive functional differential equations. J. Math. Anal. Appl. **329**(2), 1036–1044 (2007)
66. Q. Song, J. Cao, Impulsive effects on stability of fuzzy Cohen-Grossberg neural networks with time-varying delays. IEEE Trans. Syst. Man Cybern. Part B 37(3), 733–741 (2007)
67. J. Zhang, K. Zhang, D. Ren, Global exponential stability of fuzzy Cohen-Grossberg neural networks with variable delays. Lect. Notes Comput. Sci. **4491**(1), 890–896 (2007)
68. C. Bai, Stability analysis of Cohen-Grossberg BAM neural networks with delays and impulses. Chaos Solitons Fractals **35**(2), 263–267 (2008)
69. K. Li, Q. Song, Exponential stability of impulsive Cohen-Grossberg neural networks with time-varying delays and reaction-diffusion terms. Neurocomputing **72**(3), 231–240 (2008)

Chapter 10
Synchronization Stability in Complex Neural Networks

In Chaps. 4–9, all the considered RNNs models can be described as some state-vector differential equations, which can also be regarded as isolated systems or node systems. With the information communications and region interconnections being quickly developed, some isolated systems are coupled together closely, no matter they are coupled passively or actively. In this case, the dynamics of complex interconnected dynamical systems should be deeply studied. The interconnected complex dynamical systems are the upgraded version of isolated dynamical systems, therefore, some dynamics are different from those in isolated dynamical systems. Among the dynamics of complex dynamical systems, synchronization problem has been hotly investigated in recent years. In theory, synchronization conception is an upgrade of stability conception in vector space. Based on this discussion, this chapter will discuss the synchronization stability of complex interconnected neural networks with nonsymmetric coupling. The main contents in this chapter are from the research result in [1].

10.1 Introduction

There exist increasing interests in the study of dynamical properties of delayed recurrent neural networks due to its potential applications in various fields, including online optimization, pattern recognition, signal and image processing, and associative memories [1–5]. Most of the previous studies mainly concentrated on stability analysis, periodic or almost periodic attractors, and dissipativity of recurrent neural networks with or without delays [6–11]. Since it has been reported that there are *synchronization* phenomena in many real systems, such as in an array composing of identical delayed neural networks, it is important to study the synchronization problem in coupled networks and systems in engineering applications such as secure communication and signal generators design [12–18]. Therefore, the study of synchronization of coupled neural networks is an important step for both understanding brain science and designing coupled neural networks for practical use [1, 19–28].

Z. Wang et al., *Qualitative Analysis and Control of Complex Neural Networks with Delays*, Studies in Systems, Decision and Control 34,
DOI 10.1007/978-3-662-47484-6_10

Nowadays, there are mainly two kinds of coupling matrix structure to study the synchronization in an array composing of identical delayed neural networks,

(1) symmetric *coupling matrix* $G = (G_{ij}) \in \mathbb{R}^{N\times N}$, which means that for two connected nodes the influences to each other are the same. That is, $G_{ij} = G_{ji} \geq 0$ for $i \neq j$ and $G_{ii} = -\sum_{j=1,j\neq i}^{N} G_{ij}$;

(2) nonsymmetric coupling matrix, $G_{ij} \geq 0$ for $i \neq j$ and $G_{ii} = -\sum_{j=1,j\neq i}^{N} G_{ij}$, $i, j = 1, \ldots, N$, N is the number of couple term, $G = (G_{ij}) \in \mathbb{R}^{N\times N}$ is an irreducible coupling matrix.

For the case of symmetric coupling matrix, the synchronization problems have been investigated in [12, 18, 29–33], and some synchronization criteria have been derived based on LMI method or other methods. All the LMI-based synchronization results are on the basis of Kronecker product expression, in which the symmetric and irreducible feature of couple matrix plays an important role in the derivation.

For the case of nonsymmetric coupling matrix, the synchronization problems have been studied in [34–40], in which the coupling matrix G is irreducible. In [35], the coupling matrix elements $G_{ij} \geq 0$ for $i \neq j$, and $G_{ii} = -\sum_{j=1,j\neq i} G_{ij}$, and G is irreducible. The methods in [36–38] are to use the Jacobian matrix of nonlinear function at synchronization state, which can only ensure the local synchronization. Meanwhile, all the results in [36–39] are in the Kronecker form, which enhance the difficulty to check. The results in [40] are on the basis of eigenvalue approach, in which some relations between the coupling matrix of linear couple term and the parameters of isolated system are established.

In this chapter, we will extend the requirement condition of the coupling matrix $G = (G_{ij})$, and study an array of linearly delay coupled system consisting of N identical delayed neural networks with each network being an n-dimensional *dynamical system*. Synchronization stability problem of the coupled interconnected large-scale system is first studied on the basis of LMI, and some discussions on the coupling matrix G are compared with the synchronization problem of the existing complex networks.

The main contributions of this chapter are as follows:

(1) Without the requirement of symmetric and irreducible conditions on coupling matrix G, some global asymptotical synchronization stability criteria are established for an array of linearly coupled neural networks with delays, which make the synchronization criteria of complex networks more flexible.

(2) The relations between the stability of isolated neural networks and the synchronization of an array of linearly coupled neural networks with delays are discussed, which present a deep insight into the research on the dynamics of complex systems.

(3) Some discussions are made on the coordinate transformation method in dealing with synchronization problem of complex networks, which reveal the necessity of the uniqueness assumption on the equilibrium point as required in [29].

10.2 Problem Formulation and Preliminaries

We consider an array of linearly coupled complex neural networks consisting of N identical delayed neural networks with each isolated node network being an n-dimensional *dynamical system* as follows:

$$\frac{\mathrm{d}x_i(t)}{\mathrm{d}t} = - Dx_i(t) + Ag(x_i(t)) + Bg(x_i(t-\tau_1)) + a_1 \sum_{j=1}^{N} G_{ij} C x_j(t) + a_2 \sum_{j=1}^{N} G_{ij} \Gamma x_j(t-\tau_2) + U, \tag{10.1}$$

with isolated node networks

$$\frac{\mathrm{d}x_i(t)}{\mathrm{d}t} = - Dx_i(t) + Ag(x_i(t)) + Bg(x_i(t-\tau_1)) + U, \tag{10.2}$$

where $x_i(t) = (x_{i,1}(t), \ldots, x_{i,n}(t))^T \in \mathbb{R}^n$ denotes the state vector of the neurons in the ith neural networks, $g(x_i(t)) = (g_1(x_{i,1}(t)), \ldots, g_n(x_{i,n}(t)))^T, i = 1, \ldots, N$, $D = \text{diag}(d_1, \ldots, d_n) > 0$, $A = (a_{ij})_{n\times n}$, $B = (b_{ij})_{n\times n}$, τ_1 and τ_2 are the positive and constant delays, respectively, $G = (G_{ij})_{N\times N}$ is the coupling matrix representing the coupling strength and topological structure of the networks, $C = \text{diag}(c_1, \ldots, c_n)$ and $\Gamma = \text{diag}(\gamma_1, \ldots, \gamma_n)$ are the positive diagonal matrices representing the inner-linking strengths, a_1 and a_2 are the strengths of the constant coupling and delayed coupling, $U = (U_1, \ldots, U_n)^T$ denotes an external input.

Remark 10.1 In the synchronization analysis of coupled networks (10.1), if all the states are the same or synchronous, i.e., $x_1(t) = x_2(t) = \cdots = x_N(t) = s(t)$, where $s(t)$ is a common solution of node networks (10.2), a restriction condition must be required on the coupled matrix $G = (G_{ij})$. This is the fundamental reason why a zero-row-sum $\sum_{j=1}^{N} G_{ij} = 0$ is required. In this case, *synchronization state* of coupled system (10.1) is just the same as that isolated node networks (10.2). A usually used zero-row-sum condition is expressed as $G_{ii} = -\sum_{j=1,j\neq i}^{N} G_{ij}$ [31–40]. Obviously, the assumption $G_{ii} = -\sum_{j=1,j\neq i}^{N} G_{ij}$ is only an alternative of guaranteeing the zero-row-sum condition $\sum_{j=1}^{N} G_{ij} = 0$.

In the literature [12, 18, 29–33], symmetric coupling matrix G is required, which also means that the zero-column-sum condition of coupling matrix G holds. The underlying reason is the requirement of mathematical techniques, such as Kroneker product method, in dealing with the synchronization problem. Therefore, along the similar routine of symmetrically connected neural networks toward the asymmetrically connected neural networks, the development of symmetrically coupled complex systems is being toward the asymmetrically/nonsymmetrically coupled complex systems [34–40].

In reference [41], the synchronization problem is investigated in an array of linearly coupled identical networks, in which the coupling configuration matrix G is not required to be nonnegative condition on its off-diagonal elements, and the zero-row sum condition $\sum_{j=1}^{N} G_{ij} = 0$ is only required for ensuring the synchronization. Obviously, the restriction on coupling configuration matrix G is more relaxed. This naturally arises a more general question: if no any restriction is made on coupling matrix G, what would the complex networks be like? In general, this question may resort to the stability problem of complex large-scale systems. In the sense of above discussions, synchronization problem is only a special case of stability problem of complex large-scale systems.

The conception of synchronization in coupled systems or complex networks should be clarified clearly before discussing the following problems. For the synchronization studied in [42–44], an external input control action is executed or a *synchronization controller* is designed to implement the state synchronization between two same/different kinds of isolated systems. How to guarantee these two systems synchronous is to study the synchronization *error system* for the sake of designing the synchronization controller. For the synchronization of complex coupled networks, the synchronization conception should fall into the stability field of dynamical system [45–48], which is only concerned with the internal interactions of system parameters of the coupled systems. That is to say, *synchronization* is the internal self-organizing dynamical behavior of the coupled interconnected large-scale systems. In general, the final synchronization states of the coupled networks are unknown in advance.

Remark 10.2 In the case $\sum_{j=1}^{N} G_{ij} = 0$, synchronous state $s(t)$ in (10.1) should be a solution of isolated node network (10.2), which may be stable or unstable. One way to analyze the global synchronization is directly to consider the coupled systems, for example, Kroneker product method [12, 18, 29], no matter what the synchronous states are. Another commonly used method is to linearize the system at the interested point [23], which often results in local synchronization. Moreover, similar to stability analysis of recurrent neural networks [6–11], in which the existence condition of the solution of isolated system (10.2) must be guaranteed in advance, a coordinate transformation method is also used to analyze the synchronization problem of complex networks [34, 36]. The coordinate transformation method seems like the synchronization control of chaotic systems [42–44], but the underlying principle is different. As pointed out in [49], if $s(t)$ is a constant states, then $s(t)$ is an *equilibrium* point of node networks (10.2); if $s(t)$ is the nonconstant states, variational analysis and linearization can be done only near the trajectories $s(t)$ and this trajectory must contain an attracting set. For the case that $s(t)$ is a solution of the chaotic node networks (10.2), there does not seem to be any rigorous results in the literature on networks synchronization to a chaotic trajectory. For example, the approach of fixing $x_1(t) = s(t)$ in [50] is only heuristic but not completely rigorous [49]. Even for a node neural networks with multiple equilibrium point, no a universal stability analysis method is used [51–56], at least the linear coordinate transformation of equilibrium point is not used. This is the reason why there is an assumption condition required in [29].

In this chapter, we regard the coupling matrix G as an arbitrary matrix to study the global stability problem of the coupled neural networks. However, for the synchronization problem, the zero-row-sum condition on G is still required. Note that for the complex networks with nonsymmetrical coupling matrix G, synchronization problems are also studied in the literature. For example, in [41], an adaptive controller is designed to realize the controlled synchronization without the non-negative condition on its off-diagonal elements of G. In [34–40], Jacobian matrix method, Kronecker product method, and eigenvalue approach are used, respectively, to establish the synchronization criteria for the complex networks with nonsymmetrical coupling matrix G. In contrast, we will use LMI method to establish some novel synchronization criteria for the complex neural networks with nonsymmetrical G, in which a relation between the stability criteria of both isolated node networks and the coupled networks will be established. Therefore, in the aspects of analyzing methods and the synchronization criteria, the present results are different from the existing research.

The initial condition associated with (10.1) are given as follows,

$$x_i(s) = \phi_{i0}(s) \in \mathcal{C}([-\tau, 0], \mathbb{R}^n), i = 1, \ldots, n,$$

where $\tau = \max\{\tau_1, \tau_2\}$.

For the activation functions $g_k(\cdot)$, the following condition is required, $k = 1, \ldots, n$.

Assumption 10.3 The bounded activation function $g_k(\cdot)$ is Lipschitz continuous and monotonically nondecreasing, i.e., there exist constants $\delta_k > 0$ such that

$$0 \leq \frac{g_k(\zeta) - g_k(\xi)}{\zeta - \xi} \leq \delta_k,$$

for any $\zeta, \xi \in \mathbb{R}$ and $\zeta \neq \xi$, $k = 1, \ldots, n$. Let $\Delta = \operatorname{diag}(\delta_1, \ldots, \delta_n)$.

A *synchronized network* has the property that the state trajectories of the coupled agents converge to a common trajectory. More precisely, we have the following definition.

Definition 10.4 (*see* [25]) System (10.1) is said to be globally asymptotically synchronized if for any $x_i(t_0), x_j(t_0) \in \mathbb{R}^n$, when $t \to \infty$, $\|x_i(t) - x_j(t)\| \to 0, t \geq t_0$, $i, j = 1, \ldots, N$.

Thus, different from network stability, network synchronization requires that the differences of the states of the isolated neural networks converge to zero as time runs off to infinity.

Remark 10.5 According to the neural networks stability results in [6–11], the bounded activation function can always guarantee the existence of the solution, for example $s(t)$, of isolated node networks (10.2). In the zero-row-sum condition of G, $s(t)$ is also the synchronization state of coupled system (10.1).

Similar to the stability analysis methods used in [6–11], we suppose that $s(t)$ is an equilibrium point of the coupled system (10.1), that is,

$$\frac{\mathrm{d}s(t)}{\mathrm{d}t} = - Ds(t) + Ag(s(t)) + Bg(s(t-\tau_1)) + a_1 \sum_{j=1}^{N} G_{ij} Cs(t) + a_2 \sum_{j=1}^{N} G_{ij} \Gamma s(t-\tau_2) + U. \tag{10.3}$$

Define the linear coordinate transformation $e_i(t) = x_i(t) - s(t), i = 1, \ldots, N$, then the *error dynamical systems* can be described as follows:

$$\frac{\mathrm{d}e_i(t)}{\mathrm{d}t} = - De_i(t) + Af(e_i(t)) + Bf(e_i(t-\tau_1)) + a_1 \sum_{j=1}^{N} G_{ij} Ce_j(t) + a_2 \sum_{j=1}^{N} G_{ij} \Gamma e_j(t-\tau_2), \tag{10.4}$$

where $f(e_i(t)) = g(x_i(t)) - g(s(t))$, $f(e_i(t-\tau_1)) = g(x_i(t-\tau_1)) - g(s(t))$, $i = 1, \ldots, N$.

Remark 10.6 The main purpose of this study is to use the stability analysis method in neural network theory to study the dynamics of an array of linear coupled system consisting of N identical neural networks (10.1). In the stability analysis of complex networks (10.1), we do not require any assumption on coupling matrix G, which can be seen in the proof of Theorem 10.7 in sequel. However, if we discuss the synchronization problem of complex system (10.1), the zero-row-sum condition of coupling matrix G must be required. Comparing the stability problem and synchronization problem of complex system (10.1), we can conclude that synchronization stability is to some degree a special case of global stability of complex system (10.1). However, synchronization state $s(t)$ of isolated node networks (10.2) may be stable, chaotic or oscillated. In this sense, synchronization state in complex networks (10.1) is different from the equilibrium in node networks (10.2), which has its own features and blazes a new way in complex science. Just based on above argument we will establish a stability criterion for complex networks (10.1) in two separate forms (see Theorem 10.7). One condition is for the stability of node networks (10.2), and the other condition is for the constraints of coupling information. Without considering the information of coupling matrix G, using our proposed result we can judge whether the complex networks (10.1) is stable. If the zero-row-sum condition of coupling matrix G holds, we can judge whether the synchronization of complex networks (10.1) is achieved by our proposed results. Therefore, the proposed results in this chapter extend the application fields of the synchronization criteria in the literature, and bridge the stability theory of neural network and synchronization theory of complex neural networks.

10.3 Synchronization Results

Now we are in a position to state the main results.

Theorem 10.7 *Under Assumption 10.3, if there exist positive definite diagonal matrices* $P_i = diag(p_1^i, p_2^i, \ldots, p_n^i)$, $R_i = diag(r_1^i, r_2^i, \ldots, r_n^i)$, $\bar{P}_j = diag(p_j^1, p_j^2, \ldots, p_j^N)$, $\bar{Q}_j = diag(q_j^1, q_j^2, \ldots, q_j^N)$, *positive diagonal matrices* Q_{1i} *and* Q_{2i}, *such that the following linear matrix inequalities hold simultaneously,*

$$\Psi_{1i} = \begin{bmatrix} \Psi_i & P_i A + \Delta Q_{1i} & P_i B & 0 \\ * & -2Q_{1i} & 0 & 0 \\ * & * & -2Q_{2i} & \Delta Q_{2i} \\ * & * & * & -R_i \end{bmatrix} < 0, \tag{10.5}$$

$$\Psi_{2j} = \begin{bmatrix} \bar{Q}_j - d_j \bar{P}_j + 2a_1 c_j \bar{P}_j G & a_2 \gamma_j \bar{P}_j G \\ * & -\bar{Q}_j \end{bmatrix} < 0, \tag{10.6}$$

then the coupled system (10.4) is globally stable, where $\Psi_i = R_i - 0.5(P_i D + D^T P_i)$, $i = 1, \ldots, N$, $j = 1, \ldots, n$.

Proof Let us consider the following *Lyapunov–Krasovskii functional* candidate for system (10.4),

$$V(t) = V_1(t) + V_2(t) \tag{10.7}$$

where

$$V_1(t) = \sum_{i=1}^{N} e_i^T(t) P_i e_i(t), \tag{10.8}$$

$$V_2(t) = \sum_{i=1}^{N} \int_{t-\tau_1}^{t} e_i^T(s) R_i e_i(s) ds + \sum_{i=1}^{N} \int_{t-\tau_2}^{t} e_i^T(s) Q_i e_i(s) ds, \tag{10.9}$$

with $P_i = \text{diag}(p_1^i, p_2^i, \ldots, p_n^i)$, $R_i = \text{diag}(r_1^i, r_2^i, \ldots, r_n^i)$ and $Q_i = \text{diag}(q_1^i, q_2^i, \ldots, q_n^i)$, are positive definite diagonal matrix, $i = 1, 2, \ldots, N$.

Calculating the time derivatives of $V_1(t)$ and $V_2(t)$ along the trajectories of system (10.4), we have

$$
\begin{aligned}
\frac{\mathrm{d}V_1(t)}{\mathrm{d}t} &= 2\sum_{i=1}^{N} e_i^T(t)P_i\dot{e}_i(t) \\
&= 2\sum_{i=1}^{N} e_i^T(t)P_i\Big[-De_i(t)+Af(e_i(t))+Bf(e_i(t-\tau_1)) \\
&+a_1\sum_{j=1}^{N} G_{ij}Ce_j(t)+a_2\sum_{j=1}^{N} G_{ij}\Gamma e_j(t-\tau_2)\Big],
\end{aligned} \tag{10.10}
$$

$$
\begin{aligned}
\frac{\mathrm{d}V_2(t)}{\mathrm{d}t} &= \sum_{i=1}^{N}\Big[e_i^T(t)R_ie_i(t)-e_i^T(t-\tau_1)R_ie_i(t-\tau_1)\Big] \\
&+\sum_{i=1}^{N}\Big[e_i^T(t)Q_ie_i(t)-e_i^T(t-\tau_2)Q_ie_i(t-\tau_2)\Big].
\end{aligned} \tag{10.11}
$$

The following facts hold according to the requirement of activation function,

$$
2\Big[e_i^T(t)\Delta Q_{1i}f(e_i(t))-f^T(e_i(t))Q_{1i}f(e_i(t))\Big]\geq 0, \tag{10.12}
$$

$$
\begin{aligned}
&2\Big[e_i^T(t-\tau_1)\Delta Q_{2i}f(e_i(t-\tau_1)) \\
&\quad -f^T(e_i(t-\tau_1))Q_{2i}f(e_i(t-\tau_1))\Big]\geq 0.
\end{aligned} \tag{10.13}
$$

Meanwhile, the following equalities hold,

$$
2\sum_{i=1}^{N} e_i^T(t)P_ia_1\sum_{j=1}^{N} G_{ij}Ce_j(t)=2\sum_{j=1}^{n} a_1c_j\bar{e}_j^T(t)\bar{P}_jG\bar{e}_j(t), \tag{10.14}
$$

$$
2\sum_{i=1}^{N} e_i^T(t)P_ia_2\sum_{j=1}^{N} G_{ij}\Gamma e_j(t-\tau_2)=2\sum_{j=1}^{n} a_2\gamma_j\bar{e}_j^T(t)\bar{P}_jG\bar{e}_j(t-\tau_2), \tag{10.15}
$$

$$
\begin{aligned}
&\sum_{i=1}^{N}\Big[e_i^T(t)Q_ie_i(t)-e_i^T(t-\tau_2)Q_ie_i(t-\tau_2)\Big] \\
&\quad =\sum_{j=1}^{n}\Big[\bar{e}_j^T(t)\bar{Q}_j\bar{e}_j(t)-\bar{e}_j^T(t-\tau_2)\bar{Q}_j\bar{e}_j(t-\tau_2)\Big],
\end{aligned} \tag{10.16}
$$

$$
-\sum_{i=1}^{N} e_i^T(t)P_iDe_i(t)=-\sum_{j=1}^{n} d_j\bar{e}_j^T(t)\bar{P}_j\bar{e}_j(t), \tag{10.17}
$$

where $\bar{P}_j = \mathrm{diag}(p_j^1, p_j^2, \ldots, p_j^N)$, and the elements in $\bar{P}_j$ are all the same elements in P_i, $\bar{Q}_j = \mathrm{diag}(q_j^1, q_j^2, \ldots, q_j^N)$, and the elements in $\bar{Q}_j$ are all the same elements in Q_i, Q_{1i}, Q_{2i} are positive diagonal matrices, $e_i(t) = (e_i^1, e_i^2, \ldots, e_i^n)^T$, $\bar{e}_j(t) = (e_1^j, e_2^j, \ldots, e_N^j)^T$, $i = 1, \ldots, N$, $j = 1, \ldots, n$.

Substituting (10.12)–(10.17) into (10.10) and (10.11), it yields,

$$\begin{aligned}
\frac{\mathrm{d}V(t)}{\mathrm{d}t} \leq & \sum_{i=1}^{N}\Big[e_i^T(t)\Big(-\frac{P_iD+(P_iD)^T}{2}+R_i\Big)e_i^T(t) \\
& + 2e_i^T(t)(P_iA + \Delta Q_{1i})f(e_i(t)) \\
& + 2e_i^T(t)P_iBf(e_i(t-\tau_1)) - 2f^T(e_i(t))Q_{1i}f(e_i(t)) \\
& + 2e_i^T(t-\tau_1)\Delta Q_{2i}f(e_i(t-\tau_1)) - e_i^T(t-\tau_1)R_ie_i(t-\tau_1) \\
& - 2f^T(e_i(t-\tau_1))Q_{1i}f(e_i(t-\tau_1)) \\
& + \sum_{j=1}^{n}\Big[\bar{e}_j^T(t)\Big(\bar{Q}_j - d_j\bar{P}_j + 2a_1c_j\bar{P}_jG\Big)\bar{e}_j(t) \\
& + 2\bar{e}_j^T(t)a_2\gamma_j\bar{P}_jG\bar{e}_j(t-\tau_2) - \bar{e}_j(t-\tau_2)\bar{Q}_j\bar{e}_j(t-\tau_2)\Big] \\
= & \sum_{i=1}^{N}\eta_i^T(t)\Psi_{1i}\eta_i(t) + \sum_{j=1}^{n}(\bar{e}_j^T(t), \bar{e}_j^T(t-\tau_2))\Psi_{2j}(\bar{e}_j^T(t), \bar{e}_j^T(t-\tau_2))^T,
\end{aligned} \tag{10.18}$$

where Ψ_{1i} and Ψ_{2j} are the same as those defined in (10.5) and (10.6), $\eta_i^T(t) = (e_i^T(t), f^T(e_i(t)), f^T(e_i(t-\tau_1)), e_i^T(t-\tau_1))$. Therefore, if $\Psi_{1i} < 0$ and $\Psi_{2j} < 0$, one has $\mathrm{d}V(t)/\mathrm{d}t < 0$ for any $(e_i^T(t), f^T(e_i(t)), f^T(e_i(t-\tau_1)), e_i^T(t-\tau_1)) \neq 0$ and $(\bar{e}_j^T(t), \bar{e}_j^T(t-\tau_2)) \neq 0$. $\mathrm{d}V(t)/\mathrm{d}t = 0$ for $(e_i^T(t), f^T(e_i(t)), f^T(e_i(t-\tau_1)), e_i^T(t-\tau_1)) = 0$ and $(\bar{e}_j^T(t), \bar{e}_j^T(t-\tau_2))^T = 0$. Therefore, the origin of system (10.4) is globally stable. This completes the proof.

Noting that in the proof of Theorem 10.7, we have used the conditions of activation function by involving some adjustable parameters Q_{1i} and Q_{2i}, which lead to inequalities (10.12) and (10.13). If we take another way of activation function to derive some inequalities, for example,

$$\begin{aligned}
& 2e_i^T(t)P_iAf(e_i(t)) \\
\leq & e_i^T(t)P_iAQ_{1i}^{-1}A^TP_ie_i(t) + f^T(e_i(t))Q_{1i}f(e_i(t)) \\
\leq & e_i^T(t)P_iAQ_{1i}^{-1}A^TP_ie_i(t) + e_i^T(t)\Delta Q_{1i}\Delta e_i(t),
\end{aligned} \tag{10.19}$$

$$\begin{aligned}
&2e_i^T(t)P_iBf(e_i(t-\tau_1)) \\
&\leq e_i^T(t)P_iBQ_{2i}^{-1}B^TP_ie_i(t) + f^T(e_i(t-\tau_1))Q_{2i}f(e_i(t-\tau_1)) \\
&\leq e_i^T(t)P_iBQ_{2i}^{-1}B^TP_ie_i(t) + e_i^T(t-\tau_1)\Delta Q_{2i}\Delta e_i(t-\tau_1),
\end{aligned} \tag{10.20}$$

where Q_{1i} and Q_{2i} are positive diagonal matrices, $i = 1, \ldots, N$, then we can obtain the following result.

Theorem 10.8 *Under Assumption 10.3, if there exist positive definite diagonal matrices* $P_i = diag(p_1^i, p_2^i, \ldots, p_n^i)$, $R_i = diag(r_1^i, r_2^i, \ldots, r_n^i)$, $\bar{P}_j = diag(p_j^1, p_j^2, \ldots, p_j^N)$, $\bar{Q}_j = diag(q_j^1, q_j^2, \ldots, q_j^N)$, *positive diagonal matrices* Q_{1i} *and* Q_{2i}, *such that the following linear matrix inequalities hold simultaneously,*

$$\Psi_{1i}^1 = \begin{bmatrix} \Psi_i^1 & P_iA & P_iB & 0 \\ * & -Q_{1i} & 0 & 0 \\ * & * & -Q_{2i} & 0 \\ * & * & * & \Delta Q_{2i}\Delta - R_i \end{bmatrix} < 0, \tag{10.21}$$

$$\Psi_{2j}^1 = \begin{bmatrix} \bar{Q}_j - d_j\bar{P}_j + 2a_1c_j\bar{P}_jG & a_2\gamma_j\bar{P}_jG \\ * & -\bar{Q}_j \end{bmatrix} < 0, \tag{10.22}$$

then the coupled system (10.4) is globally stable, where $\Psi_i^1 = R_i + \Delta Q_{1i}\Delta - 0.5(P_iD + D^TP_i), i = 1, \ldots, N, j = 1, \ldots, n.$

Proof The proof of Theorem 10.8 is as the same as Theorem 10.7 except that the inequalities (10.12) and (10.13) are substituted by (10.19) and (10.20), respectively. The details are omitted.

Remark 10.9 The differences between Theorems 10.7 and 10.8 lie in the fact that the condition of activation function $f(\cdot)$ is used differently in the proof procedure (see (10.12) and (10.13), or (10.19) and (10.20)) although the selected Lyapunov functional is the same. That is, different inequality treatment leads to different expressions Ψ_{1i} and Ψ_{1i}^1, which are just the stability criteria of the node networks, while the coupled term expressions Ψ_{2i} and Ψ_{2i}^1 are the same. Because both Theorems 10.7 and 10.8 are sufficient conditions, in general, it is difficult to state which criterion would be better.

Remark 10.10 In both Theorems 10.7 and 10.8, the first condition Ψ_{1i} (or Ψ_{1i}^1) is independent of second condition Ψ_{2j} (or Ψ_{2j}^1), while the second condition Ψ_{2j} (or Ψ_{2j}^1) is dependent on the first condition Ψ_{1i} (or Ψ_{1i}^1). Therefore, when solving the conditions in Theorems 10.7 and 10.8, the first LMI (10.5) (or (10.21)) should be first computed, then computing the second LMI (10.6) (or (10.22)).

Remark 10.11 In both Theorems 10.7 and 10.8, the first condition Ψ_{1i} (or Ψ_{1i}^1) is only related to the dynamics of the isolated node networks, which is independent of the coupling connections. The second condition Ψ_{2j} (or Ψ_{2j}^1), on the contrary, depends on the coupling matrix G, coupling strengths a_1 and a_2. In general, the first condition Ψ_{1i} (or Ψ_{1i}^1) means a kind of stability condition of the node networks [8–11]. If the node networks (10.2) are stable, the coupled complex neural networks (10.1) may not be stable because of the connection condition Ψ_{2j} (or Ψ_{2j}^1). Therefore, Theorems 10.7 and 10.8 bridge the stability of isolated node recurrent neural networks and stability of the complex neural networks composing of the arrays of N identical recurrent neural networks. Different coupling configurations, e.g., G, a_1 and a_2, may influence the total dynamics of the coupled complex networks, despite the node networks are stable.

Remark 10.12 The proof of Theorems 10.7 and 10.8 is inspired by the stability analysis of recurrent neural networks and fuzzy systems, respectively. Instead of using Kroneker product method, which is often used to deal with the coupled complex networks (10.1) as a whole dynamical system, we deal with the isolated or node networks directly. Then in a similar way to weighting the fuzzy rule by different membership degree in a fuzzy system, the coupled terms in an isolated or node networks is weighted and the node networks is integrated as a total large-scale system. The advantage of this method is easy to use the stability analysis method of recurrent neural networks, which can usually derive two separate stability or synchronization conditions, i.e., one is for node dynamics and the other is for the coupled connections. The key difficulty of the analysis procedure is how to tackle the coupled connections G, a_1, and a_2. In this chapter, by using the relation (10.14)–(10.17), we have successfully established some stability criteria.

Remark 10.13 In the proof of Theorems 10.7 and 10.8, no any restriction is imposed on coupling matrix G, which means that the symmetry and irreducible condition on G can be canceled in this chapter. If we require the zero-row-sum condition of G, i.e., $\sum_{j=1}^{N} G_{ij} = 0$, then the conditions in Theorems 10.7 and 10.8 are converted into the global synchronization conditions. Meanwhile, we do not require the positivity of G_{ij} for $i \neq j$ and $G_{ii} = -\sum_{j=1}^{N} G_{ij}$, either. In this sense, the couple matrix G can be regarded as a weighted topology or an interconnected coefficient matrix.

In the proof of Theorems 10.7 and 10.8,

$$-2\sum_{i=1}^{N} e_i^T(t) P_i D e_i(t)$$

is divided into two parts, which leads to (10.17) and

$$-\sum_{i=1}^{N} e_i^T(t) \frac{(P_i D + D^T P_i)}{2} e_i(t)$$

while the former is used to compensate the effect of term $\sum_{j=1}^{n} \bar{e}_j^T(t)\bar{P}_j G\bar{e}_j(t)$. Keep this in mind and observe that Q_i (see Lyapunov functional (10.9)) does not appear in Theorem 10.7 (see (10.5)) and Theorem 10.8 (see (10.21)), which may lead to imbalance of unknown parameter distribution and affect the solvability. Thus, one can decompose $\sum_{i=1}^{N} e_i^T(t)Q_i e_i(t)$ into two parts, i.e.,

$$\sum_{i=1}^{N} e_i^T(t)\frac{Q_i}{2}e_i(t)$$

and

$$\sum_{i=1}^{N} e_i^T(t)\frac{Q_i}{2}e_i(t) = \sum_{j=1}^{n} \bar{e}_j^T(t)\frac{\bar{Q}_j}{2}\bar{e}_j(t).$$

Therefore, considering above discussions and Theorems 10.7 and 10.8, we have the following results directly.

Theorem 10.14 *Under Assumption 10.3, if there exist positive definite diagonal matrices* $P_i = diag(p_1^i, p_2^i, \ldots, p_n^i)$, $R_i = diag(r_1^i, r_2^i, \ldots, r_n^i)$, $Q_i = diag(q_1^i, q_2^i, \ldots, q_n^i)$, $\bar{P}_j = diag(p_j^1, p_j^2, \ldots, p_j^N)$, $\bar{Q}_j = diag(q_j^1, q_j^2, \ldots, q_j^N)$, *positive diagonal matrices* Q_{1i} *and* Q_{2i}, *such that the following linear matrix inequalities hold simultaneously,*

$$\Psi_{1i} = \begin{bmatrix} \Psi_0 & P_i A + \Delta Q_{1i} & P_i B & 0 \\ * & -2Q_{1i} & 0 & 0 \\ * & * & -2Q_{2i} & \Delta Q_{2i} \\ * & * & * & -R_i \end{bmatrix} < 0, \quad (10.23)$$

$$\Psi_{2j} = \begin{bmatrix} \bar{Q}_j/2 - d_j \bar{P}_j + 2a_1 c_j \bar{P}_j G & a_2\gamma_j \bar{P}_j G \\ * & -\bar{Q}_j \end{bmatrix} < 0, \quad (10.24)$$

then system (10.4) is globally stable, where $\Psi_0 = Q_i/2 + R_i - 0.5(P_i D + D^T P_i)$, $i = 1, \ldots, N$, $j = 1, \ldots, n$.

Theorem 10.15 *Under Assumption 10.3, if there exist positive definite diagonal matrices* $P_i = diag(p_1^i, p_2^i, \ldots, p_n^i)$, $R_i = diag(r_1^i, r_2^i, \ldots, r_n^i)$, $Q_i = diag(q_1^i, q_2^i, \ldots, q_n^i)$, $\bar{P}_j = diag(p_j^1, p_j^2, \ldots, p_j^N)$, $\bar{Q}_j = diag(q_j^1, q_j^2, \ldots, q_j^N)$, *positive diagonal matrices* Q_{1i} *and* Q_{2i}, *such that the following linear matrix inequalities hold simultaneously,*

$$\Psi_{1i} = \begin{bmatrix} \Psi_1 & P_i A & P_i B & 0 \\ * & -Q_{1i} & 0 & 0 \\ * & * & -Q_{2i} & 0 \\ * & * & * & \Delta Q_{2i}\Delta - R_i \end{bmatrix} < 0, \quad (10.25)$$

$$\Psi_{2j} = \begin{bmatrix} \bar{Q}_j/2 - d_j \bar{P}_j + 2a_1 c_j \bar{P}_j G & a_2 \gamma_j \bar{P}_j G \\ * & -\bar{Q}_j \end{bmatrix} < 0, \tag{10.26}$$

then system (10.4) is globally stable, where $\Psi_1 = Q_i/2 + R_i + \Delta Q_{1i} \Delta - 0.5(P_i D + D^T P_i), i = 1, \ldots, N, j = 1, \ldots, n.$

The following results are some special cases of Theorems 10.7–10.15 if we let $Q_{1i} = Q_1$, $Q_{2i} = Q_2$, $R_i = R$ in Theorems 10.7–10.15. The proof procedures are all omitted.

Corollary 10.16 *Under Assumption 10.3, if there exist positive definite diagonal matrices* $P = diag(p_1, p_2, \ldots, p_n)$, $R = diag(r_1, r_2, \ldots, r_n)$, $Q = diag(q_1, q_2, \ldots, q_n)$, *positive diagonal matrices* Q_1 *and* Q_2, *such that the following linear matrix inequalities hold simultaneously,*

$$\Psi_1 = \begin{bmatrix} \Psi_0 & PA + \Delta Q_1 & PB & 0 \\ * & -2Q_1 & 0 & 0 \\ * & * & -2Q_2 & \Delta Q_2 \\ * & * & * & -R \end{bmatrix} < 0, \tag{10.27}$$

$$\Psi_{2j} = \begin{bmatrix} (q_j - d_j p_j) I_N + 2a_1 c_j p_j G & a_2 \gamma_j p_j G \\ * & -q_j I_N \end{bmatrix} < 0, \tag{10.28}$$

then the coupled system (10.4) is globally stable, where $\Psi_0 = R - 0.5(PD + D^T P), i = 1, \ldots, N, j = 1, \ldots, n.$

Corollary 10.17 *Under Assumption 10.3, if there exist positive definite diagonal matrices* $P = diag(p_1, p_2, \ldots, p_n)$, $R = diag(r_1, r_2, \ldots, r_n)$, $Q = diag(q_1, q_2, \ldots, q_n)$, *positive diagonal matrices* Q_1 *and* Q_2, *such that the following linear matrix inequalities hold simultaneously,*

$$\Psi_1 = \begin{bmatrix} \Psi_1 & PA & PB & 0 \\ * & -Q_1 & 0 & 0 \\ * & * & -Q_{2i} & 0 \\ * & * & * & \Delta Q_2 \Delta - R \end{bmatrix} < 0, \tag{10.29}$$

$$\Psi_{2j} = \begin{bmatrix} (q_j - d_j p_j) I_N + 2a_1 c_j p_j G & a_2 \gamma_j p_j G \\ * & -q_j I_N \end{bmatrix} < 0, \tag{10.30}$$

then the coupled system (10.4) is globally stable, where $\Psi_1 = R + \Delta Q_1 \Delta - 0.5(PD + D^T P), i = 1, \ldots, N, j = 1, \ldots, n.$

Corollary 10.18 *Under Assumption 10.3, if there exist positive definite diagonal matrices* $P = diag(p_1, p_2, \ldots, p_n)$, $R = diag(r_1, r_2, \ldots, r_n)$, $Q = diag(q_1, q_2, \ldots, q_n)$, *positive diagonal matrices* Q_1 *and* Q_2, *such that the following linear matrix inequalities hold simultaneously,*

$$\Psi_1 = \begin{bmatrix} \Psi_0 & PA + \Delta Q_1 & PB & 0 \\ * & -2Q_1 & 0 & 0 \\ * & * & -2Q_2 & \Delta Q_2 \\ * & * & * & -R \end{bmatrix} < 0, \tag{10.31}$$

$$\Psi_{2j} = \begin{bmatrix} (q_j/2 - d_j p_j)I_N + 2a_1 c_j p_j G & a_2 \gamma_j p_j G \\ * & -q_j I_N \end{bmatrix} < 0, \tag{10.32}$$

then the coupled system (10.4) is globally stable, where $\Psi_0 = Q/2 + R - 0.5(PD + D^T P), i = 1, \ldots, N, j = 1, \ldots, n.$

Corollary 10.19 *Under Assumption 10.3, if there exist positive definite diagonal matrices* $P = diag(p_1, p_2, \ldots, p_n^i)$, $R = diag(r_1, r_2, \ldots, r_n)$, $Q = diag(q_1, q_2, \ldots, q_n)$, *positive diagonal matrices* Q_1 *and* Q_2, *such that the following linear matrix inequalities hold simultaneously,*

$$\Psi_1 = \begin{bmatrix} \Psi_1 & PA & PB & 0 \\ * & -Q_1 & 0 & 0 \\ * & * & -Q_2 & 0 \\ * & * & * & \Delta Q_2 \Delta - R \end{bmatrix} < 0, \tag{10.33}$$

$$\Psi_{2j} = \begin{bmatrix} (q_j/2 - d_j p_j)I_N + 2a_1 c_j p_j G & a_2 \gamma_j p_j G \\ * & -q_j I_N \end{bmatrix} < 0, \tag{10.34}$$

then the coupled system (10.4) is globally stable, where $\Psi_1 = Q/2 + R + \Delta Q_1 \Delta - 0.5(PD + D^T P), i = 1, \ldots, N, j = 1, \ldots, n.$

10.4 Illustrative Example

In this section, we will use an illustrative example to show the effectiveness of the obtained result.

Example 10.20 Let us consider the following recurrent neural networks,

$$\frac{\mathrm{d}y(t)}{\mathrm{d}t} = -Dy(t) + Ag(y(t)) + Bg(y(t - \tau_1)) + I, \tag{10.35}$$

where $y(t) = (y_1(t), y_2(t))^T$ is the state vector of neural networks, $g(y_i(t)) = \tanh(y_i(t))$ is the activation function, $I = (-10, -10)^T$ is the external input vector,

$$D = \begin{bmatrix} 12 & 0 \\ 0 & 12 \end{bmatrix}, A = \begin{bmatrix} 2 & -0.1 \\ -5 & 3.0 \end{bmatrix}, B = \begin{bmatrix} -1.5 & -0.1 \\ -0.2 & -2.5 \end{bmatrix},$$

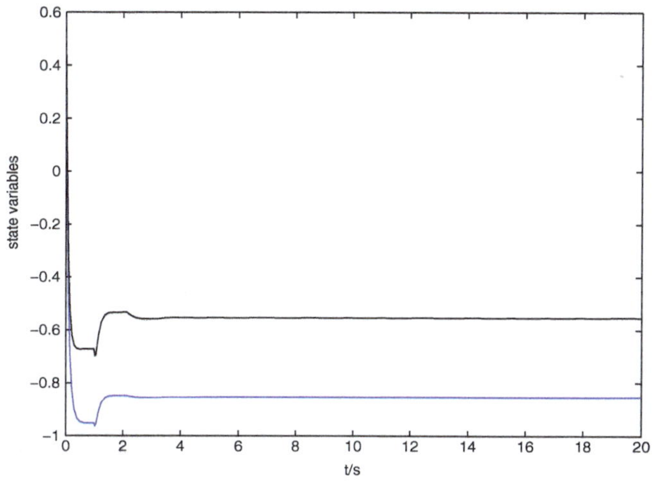

Fig. 10.1 The state trajectories of node system (10.35)

Since $M = D - |A|\Delta - |B|\Delta$ is an M-matrix, according to the stability result of Theorem 2 in [11], the concerned isolated neural networks is globally asymptotically stable, which has a unique equilibrium point, $(-0.8540, -0.5540)$. The state trajectory of system (10.35) is shown in Fig. 10.1.

Now we consider a dynamical system consisting of three linearly coupled identical models (10.35). The state equations of the entire array are the same as system (10.1), where $\tau_1 = \tau_2 = 1$, $C = \begin{bmatrix} 4 & 0 \\ 0 & 4 \end{bmatrix}$, $\Gamma = \begin{bmatrix} 0.3 & 0 \\ 0 & 0.3 \end{bmatrix}$.

In the following, we will discuss four cases of coupling matrix G, which will be used to verify the effectiveness of the proposed results and remarks in this chapter.

Case I: G is nonsymmetric, zero-row-sum and $G_{ij} \geq 0$.

$$G = \begin{bmatrix} -4.7497 & 4.5647 & 0.1850 \\ 4.4470 & -12.6611 & 8.2141 \\ 7.9194 & 6.1543 & -14.0737 \end{bmatrix},$$

By using the MATLAB LMI Control Toolbox, solving the conditions (10.5) and (10.6) in Theorem 10.7, it yields the following feasible solutions,

$$\begin{aligned}
P_1 &= P_2 = P_3 = \mathrm{diag}(3.0546, 0.2753),\ R_1 = R_2 = R_3 = \mathrm{diag}(3.7716, 0.7247),\\
\bar{P}_1 &= \mathrm{diag}(3.0546, 3.0546, 3.0546),\ \bar{P}_2 = \mathrm{diag}(0.2753, 0.2753, 0.2753),\\
\bar{Q}_1 &= \mathrm{diag}(5.4619, 5.0869, 5.9030),\ \bar{Q}_2 = \mathrm{diag}(1.1408, 1.7505, 2.0176),\\
Q_{11} &= Q_{12} = Q_{13} = \mathrm{diag}(7.2276, 0.9042),\\
Q_{21} &= Q_{22} = Q_{23} = \mathrm{diag}(3.0434, 0.7267),
\end{aligned}$$

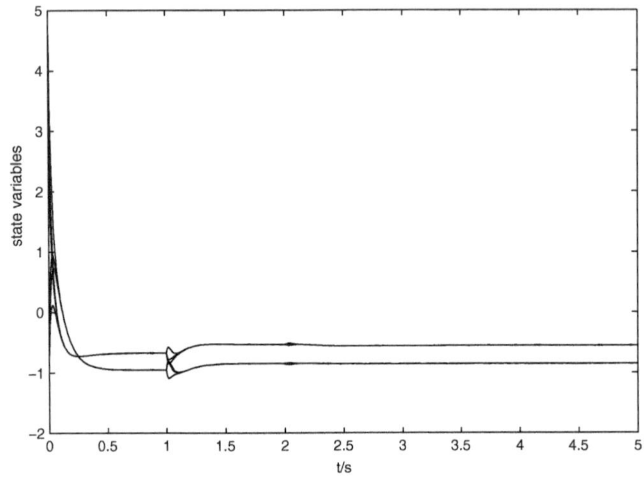

Fig. 10.2 The synchronization state trajectories of coupled system (10.35) with Case I

According to Theorem 10.14, the coupled neural networks can achieve the global synchronization, and the synchronized states converge to (−0.8540, −0.5540), which is the global stable equilibrium point of isolated system (10.35). The synchronization performance is shown in Fig. 10.2.

Case II: G **is nonsymmetric and zero-row-sum.**

$$G = \begin{bmatrix} -2 & -0.25 & 2.25 \\ -0.5 & -0.5 & 1 \\ -0.25 & 3.5 & -3.25 \end{bmatrix},$$

By using the MATLAB LMI Control Toolbox, solving the conditions (10.5) and (10.6) in Theorem 10.7, it yields the following feasible solutions,

$$\begin{aligned} P_1 &= P_2 = P_3 = \text{diag}(3.0546, 0.2753),\ R_1 = R_2 = R_3 = \text{diag}(3.7716, 0.7247), \\ \bar{P}_1 &= \text{diag}(3.0546, 3.0546, 3.0546),\ \bar{P}_2 = \text{diag}(0.2753, 0.2753, 0.2753), \\ \bar{Q}_1 &= \text{diag}(5.0151, 4.9651, 5.0419),\ \bar{Q}_2 = \text{diag}(3.6395, 0.9549, 2.1528), \\ Q_{11} &= Q_{12} = Q_{13} = \text{diag}(7.2276, 0.9042), \\ Q_{21} &= Q_{22} = Q_{23} = \text{diag}(3.0434, 0.7267), \end{aligned}$$

According to Theorem 10.7, the coupled neural networks can achieve the global synchronization, and the synchronized states converge to (−0.8540, −0.5540), which is the global stable equilibrium point of isolated system (10.35). The synchronization performance is shown in Fig. 10.3.

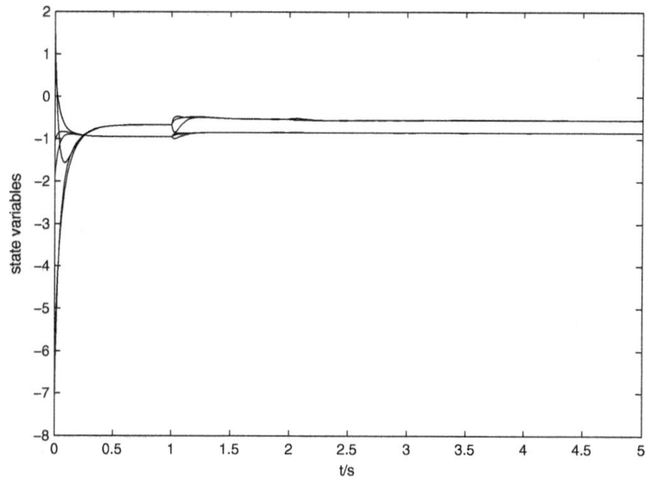

Fig. 10.3 The synchronization state trajectories of coupled system (10.35) with Case II

Case III: G is arbitrary matrix with stable states.

$$G = \begin{bmatrix} -1.4781 & 0.3217 & 0.2334 \\ 0.3619 & -1.8776 & 1.2395 \\ 0.4778 & 0.6805 & 0.1257 \end{bmatrix},$$

By using the MATLAB LMI Control Toolbox, solving the conditions (10.5) and (10.6) in Theorem 10.7, it yields the following feasible solutions,

$$\begin{aligned}
&P_1 = P_2 = P_3 = \mathrm{diag}(3.0546, 0.2753),\ R_1 = R_2 = R_3 = \mathrm{diag}(3.7716, 0.7247),\\
&\bar{P}_1 = \mathrm{diag}(3.0546, 3.0546, 3.0546),\ \bar{P}_2 = \mathrm{diag}(0.2753, 0.2753, 0.2753),\\
&\bar{Q}_1 = \mathrm{diag}(4.8567, 4.8657, 4.7407),\ \bar{Q}_2 = \mathrm{diag}(3.2379, 3.0410, 1.3645),\\
&Q_{11} = Q_{12} = Q_{13} = \mathrm{diag}(7.2276, 0.9042),\\
&Q_{21} = Q_{22} = Q_{23} = \mathrm{diag}(3.0434, 0.7267),
\end{aligned}$$

Since the coupling matrix G does not satisfy the zero-row-sum condition, the coupled neural networks (10.35) can only achieve the global stable state according to Theorem 10.7, and the state trajectories converge to $(-0.6880, -0.4672)$, $(-0.9067, -0.5599)$, and $(-1.2828, -0.7576)$, respectively. This means that the final stable states are different from that of the isolated node system due to the different couple topology. The global stable performance is shown in Fig. 10.4.

Case IV: G is arbitrary matrix with unstable states.

$$G = \begin{bmatrix} 2.2047 & 0.8039 & 0.2302 \\ 1.7129 & -1.1685 & 0.1369 \\ -2.3570 & -1.9590 & -1.0598 \end{bmatrix},$$

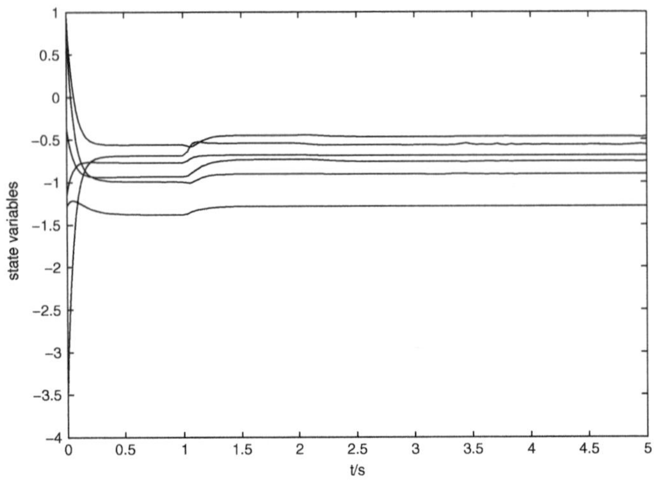

Fig. 10.4 The state trajectories of coupled system (10.35) with Case III

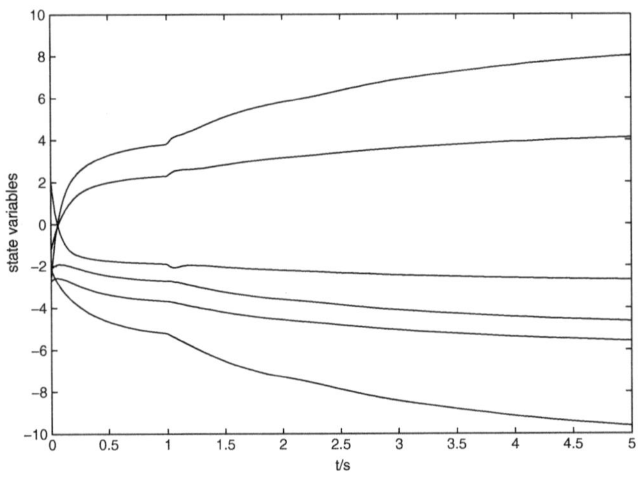

Fig. 10.5 The state trajectories of coupled system (10.35) with Case IV

By using the MATLAB LMI Control Toolbox, solving the conditions (10.5) and (10.6) in Theorem 10.7, it yields infeasible solutions. That is, Theorem 10.7 cannot ensure whether the coupled neural networks stable or not. By simulation, the evolving trajectories are shown in Fig. 10.5.

From these four cases of simulation, we can find that the coupled complex networks can exhibit many complex dynamics by taking different coupling topology matrix G, although the node neural networks are always stable. Meanwhile, the simulation examples also show that the proposed results and remarks are effective.

10.5 Summary

In this chapter, we have established the global stability criteria for arrays of linearly coupled delayed neural networks with nonsymmetric coupling on the basis of LMI method. The derived results are two separate conditions, one is for the dynamics of the node networks, and the other is for the couple configuration. The outstanding feature of the proposed stability results is to bridge the gap of stability theory of recurrent neural networks and the synchronization stability of complex networks with an array of linearly coupled complex networks consisting of N identical delayed neural networks. The relations of coupling matrix G to stability and synchronization are discussed in detail, and some analysis method to the synchronization stability of complex networks are also stated. A numerical example is used to show the effectiveness of the theoretical results and the comments.

References

1. Z. Wang, H. Zhang, Synchronization stability in complex interconnected neural networks with nonsymmetric coupling. Neurocomputing **108**, 84–92 (2013)
2. V. Perez-Munuzuri, V. Perez-Villar, L.O. Chua, Autowaves for image processing on a two-dimensional CNN array of excitable nonlinear circuits: flat and wrinkled labyrinths. IEEE Trans. Circuits Syst. I **40**, 174–181 (1993)
3. F. Hoppensteadt, E. Izhikevich, Pattern recognition via synchronization in phase-locked loop neural networks. IEEE Trans. Neural Netw. **11**(3), 734–738 (2000)
4. G. Chen, J. Zhou, Z. Liu, Global synchronization of coupled delayed neural networks and applications to chaos CNN models. Int. J. Bifurc. Chaos **14**(7), 2229–2240 (2004)
5. X. He, W. Lu, T. Chen, Nonnegative periodic dynamics of delayed Cohen-Grossberg neural networks with discontinuous activations. Neurocomputing **73**(15), 2765–2772 (2010)
6. C.M. Marcus, R.M. Westervelt, Stability of analog neural networks with delays. Phys. Rev. A Gen. Phys. **39**(1), 347–359 (1989)
7. S. Arik, V. Tavsanoglu, On the global asymptotic stability of delayed cellular neural networks. IEEE Trans. Circuits Syst. I Fundam. Theory Appl. **47**(5), 571–574 (2000)
8. Z. Wang, H. Zhang, W. Yu, Robust stability criteria for interval Cohen-Grossberg neural networks with time varying delay. Neurocomputing **72**(6), 1105–1110 (2009)
9. Z. Wang, H. Zhang, B. Jiang, LMI-based approach for global asymptotic stability analysis of recurrent neural networks with various delays and structures. IEEE Trans. Neural Netw. **22**(7), 1032–1045 (2011)
10. C. Zheng, H. Zhang, Z. Wang, Novel exponential stability criteria of high-order neural networks with time-varying delays. IEEE Trans. Syst. Man Cybern. Part B Cybern. **41**(2), 486–496 (2011)
11. T. Chen, L. Rong, Robust global exponential stability of Cohen-Grossberg neural networks with time delays. IEEE Trans. Neural Netw. **15**(1), 203–206 (2004)
12. J. Cao, G. Chen, P. Li, Global synchronization in an array of delayed neural networks with hybrid coupling. IEEE Trans. Syst. Man Cybern Part B Cybern. **38**(2), 488–498 (2008)
13. W. Lu, T. Chen, Synchronization of coupled connected neural networks with delays. IEEE Trans. Circuits Syst. I Regul. Papers **51**(12), 2491–2503 (2004)
14. B. Liu, W. Lu, T. Chen, Global almost sure self-synchronization of Hopfield neural networks with randomly switching connections. Neural Netw. **24**, 305–310 (2011)
15. H. Wang, Q. Song, Synchronization for an array of coupled stochastic discrete-time neural networks with mixed delays. Neurocomputing **74**(10), 1572–1584 (2011)

16. J. Yu, C. Hu, H. Jiang, Z. Teng, Exponential synchronization of Cohen-Grossberg neural networks via periodically intermittent control. Neurocomputing **74**, 1776–1782 (2011)
17. T. Li, A. Song, S. Fei, Synchronization control for arrays of coupled discrete-time delayed Cohen-Grossberg neural networks. Neurocomputing **74**, 197–204 (2010)
18. J. Feng, S. Wang, Z. Wang, Stochastic synchronization in an array of neural networks with hybrid nonlinear coupling. Neurocomputing **74**, 3808–3815 (2011)
19. J. Cao, P. Li, W. Wang, Global synchronization in arrays of delayed neural networks with constant and delayed coupling. Phys. Lett. A **353**(4), 318–325 (2006)
20. S. Wen, Z. Zeng, T. Huang et al., Exponential adaptive lag synchronization of memristive neural networks via fuzzy method and applications in pseudorandom number generators. IEEE Trans. Fuzzy Syst. **22**(6), 1704–1713 (2014)
21. J. Wang, H. Zhang, Z. Wang, H. Liang, Stochastic synchronization for markovian coupled neural networks with partial information on transition probabilities. Neurocomputing **149**, 983–992 (2015)
22. C. Zhang, W. Li, K. Wang, Graph-theoretic method on exponential synchronization of stochastic coupled networks with Markovian switching. Nonlinear Anal. Hybrid Syst. **15**, 37–51 (2015)
23. C. Li, G. Chen, Synchronization in general complex dynamical networks with coupling delays. Phys. Lett. A **343**, 263–278 (2004)
24. J. Cao, Y. Wan, Matrix measure strategies for stability and synchronization of inertial BAM neural network with time delays. Neural Netw. **53**, 165–172 (2014)
25. N. Monshizadeh, H.L. Trentelman, M.K. Camlibel, Stability and synchronization preserving model reduction of multi-agent systems. Syst. Control Lett. **62**, 1–10 (2013)
26. Z. Fei, H. Gao, W.-X. Zheng, New synchronization stability of complex networks with an interval time-varying coupling delay. IEEE Trans. Circuits Syst. II: Express Briefs **56**(6), 499–503 (2009)
27. Z. Duan, G. Chen, Global Robust stability and synchronization of networks with lorenz-type nodes. IEEE Trans. Circuits Syst. II Express Briefs **56**(8), 679–683 (2009)
28. G. Russo, M. di Bernardo, Contraction theory and master stability function: linking two approaches to study synchronization of complex networks. IEEE Trans. Circuits Syst. II Express Briefs **56**(2), 177–181 (2009)
29. C. Wu, L. Chua, Synchronization in an array of linearly coupled dynamical systems. IEEE Trans. Circuits Syst. I Fundam. Theory Appl. **42**(8), 430–447 (1995)
30. C. Wu, L. Chua, On a conjecture regarding the synchronization in an array of linearly coupled dynamical systems. IEEE Trans. Circuits Syst. I Fundam. Theory Appl. **43**(2), 161–165 (1996)
31. Z. Li, G. Chen, Global synchronization and asymptotic stability of complex dynamical networks. IEEE Trans. Circuits Syst. II Express Briefs **53**(1), 28–33 (2006)
32. P. Lu, Y. Yang, Global asymptotic stability of a class of complex networks via decentralised static output feedback control. IEET Control Theory Appl. **4**(11), 2463–2470 (2010)
33. P.P. Menon, C. Edwards, Static output feedback stabilisation and synchronisation of complex networks with H_2 performance. Int. J. Robust Nonlinear Control **20**, 703–718 (2010)
34. J. Lu, D.W.C. Ho, M. Liu, Globally exponential synchronization in an array of asymmetric coupled neural networks. Phys. Lett. A **369**, 444–451 (2007)
35. X. Liu, T. Chen, Synchronization analysis for nonlinearly coupled complex networks with an asymmetrical coupling matrix. Phys. A **387**, 4429–4439 (2008)
36. J. Wu, L. Jiao, Global synchronization and state tuning in asymmetric complex dynamical networks. IEEE Trans. Circuits Syst. II Express Briefs **55**(9), 932–936 (2008)
37. J. Wu, L. Jiao, Synchronization in complex dynamical networks with nonsymmetric coupling. Phys. D **237**, 2487–2498 (2008)
38. J. Wu, L. Jiao, Observer-based synchronization in complex dynamical networks with nonsymmetric coupling. Phys. A **386**, 469–480 (2007)
39. H. Zheng, Y. Jing, X. Zheng, N. Jiang, Synchronization of a class of dynamical complex networks with nonsymmetric coupling based on decentralized control, in *2009 American Control Conference*, Hyatt Regency Riverfront (St. Louis, MO, USA, June 2009), pp. 10–12

40. Q. Song, J. Cao, F. Liu, Pinning-controlled synchronization of hybrid-coupled complex dynamical networks with mixed time-delays. Int. J. Robust. Nonlinear Control **22**, 690–706 (2012)
41. J. Cao, Z. Wang, Y. Sun, Synchronization in an array of linearly stochastically coupled networks with time delays. Phys. A **385**, 718–728 (2007)
42. L. Pecora, T. Carroll, Synchronization in chaotic systems. Phys. Rev. A **64**, 821–824 (1990)
43. M. Boutayeb, M. Darouach, H. Rafaralahy, Generalized state-space observers for chaotic synchronization and secure communication. IEEE Trans. Circ. Syst. I Fundam. Theory Appl. **49**(3), 345–349 (2002)
44. N. Balmforth, C. Tresser, P. Worfolk, W.C. Wu, Master-slave synchronization and the Lorenz equations. Chaos **7**(3), 392–394 (1997)
45. H. Gao, J. Lam, G. Chen, New criteria for synchronization stability of general complex dynamical networks with coupling delays. Phys. Lett. A **360**(2), 263–273 (2006)
46. K. Li, S. Guan, X. Gong, C. Lai, Synchronization stability of general complex dynamical networks with time-varying delays. Phys. Lett. A **372**, 7133–7139 (2008)
47. Y. Liu, Z. Wang, J. Liang, X. Liu, Stability and synchronization of discrete-time markovian jumping neural networks with mixed mode-dependent time delays. IEEE Trans. Neural Netw. **20**(7), 1102–1116 (2009)
48. D. Yue, H. Li, Synchronization stability of continuous/discrete complex dynamical networks with interval time-varying delays. Neurocomputing **73**, 809–819 (2010)
49. G. Chen, X. Wang, X. Li, J. Lu, Some recent advances in complex networks synchronization. *A Chapter in Recent Advance in Nonlinear Dynamics and Synchronization* (Springer, Berlin, 2009)
50. J. Lu, X. Yu, G. Chen, Chaos synchronization of general complex dynamical networks. Phys. A **334**, 281–302 (2004)
51. X. Nie, J. Cao, Multistability of competitive neural networks with time-varying and distributed delays. Nonlinear Anal. Real World Appl. **10**, 928–942 (2009)
52. C. Cheng, K. Lin, C.W. Shih, Multistability in recurrent neural networks. SIAM J. Appl. Math. **66**(4), 1301–1320 (2006)
53. C. Cheng, K. Lin, C. Shih, Multistability and convergence in delayed neural networks. Phys. D **225**, 61–74 (2007)
54. Z. Zeng, J. Wang, Multiperiodicity and exponential attractivity evoked by periodic external inputs in delayed cellular neural networks. Neural Comput. **18**, 848–870 (2006)
55. L. Wang, W. Lu, T. Chen, Multistability and new attraction basins of almost-periodic solutions of delayed neural networks. IEEE Trans. Neural Netw. **20**(10), 1581–1593 (2009)
56. J. Cao, G. Feng, Y. Wang, Multistability and multiperiodicity of delayed Cohen-Grossberg neural networks with a general class of activation functions. Phys. D **237**, 1734–1749 (2008)

Chapter 11
Stabilization of Stochastic RNNs with Stochastic Delays

The research in Chaps. 4–10 is focused on the qualitative analysis of complex neural networks with delays. It is well known that the qualitative analysis of nonlinear dynamical systems is the foundation of controlling the systems. Therefore, in this chapter controller design problem will be studied for a class of stochastic Cohen-Grossberg neural networks with mode-dependent mixed time delays and Markovian switching, in which the neural dynamical networks will be stabilized. The contents in this chapter are from the research result in [1].

11.1 Introduction

In recent decades, neural networks have been successfully applied to various fields such as optimization, image processing, and associative memory design. In such application, it is important to know the stability properties of the designed neural network, these properties include asymptotic stability and exponential stability. However, time delays inevitably exist in neural networks due to various reasons [2]. The existence of *time delay* may lead to some complex dynamic behaviors such as oscillation, divergence, chaos, instability, or other poor performance of the neural networks. Since neural networks usually have a spatial extent, there is a distribution of propagation delays over a period of time. In these circumstances, the signal propagation is not instantaneous and cannot be modeled with discrete-time delays [3]. A more appropriate way is to incorporate discrete and continuously distributed time delays in the neural network model [2, 4]. Stability analysis for neural networks with delays has attracted more and more interests in recent years, for example, see [5–21] and references therein.

On the other hand, the stabilization issue has been an important focus of research in the control fields, and several feedback stabilizing control design approaches have been proposed (see [7, 22–25]). Some interesting results [6, 26–35] on the

Z. Wang et al., *Qualitative Analysis and Control of Complex Neural Networks with Delays*, Studies in Systems, Decision and Control 34,
DOI 10.1007/978-3-662-47484-6_11

stabilization of a wide range and different types of neural networks have been reported in the literature. For a class of discrete-time dynamic neural networks, reference [29] proposes two methods, namely, the gradient projection and the minimum distance projection to investigate the stabilization. For a class of dynamic neural network systems, a global robust stabilizing controller with unknown nonlinearities is developed in [6] via Lyapunov stability and inverse optimality. For a class of linearly coupled stochastic neural networks, some results are derived in [31] on the design of the minimum number of controllers for the pinning stabilization, which are expressed in terms of strict linear matrix inequality (LMI). For a class of neutral neural networks with varying delays, a novel criterion is obtained in [28] for the global stabilization using the Razumikhin's method. For a class of so-called standard neural network models with time delays, a few stabilization criteria are presented [30] which are based on the Lyapunov–Krasovskii stability theory and the LMI approach. For a class of impulsive high-order Hopfield-type neural networks with time-varying delays, some stabilization criteria are reported in [26] by employing the Lyapunov–Razumikhin technique. Very recently, for a class of neural networks with various activation functions and time-varying continuously distributed delays, LMI-based delay-dependent conditions are obtained in [27] for the global exponential stabilization. Despite some good progress on the stability analysis of delayed neural networks with various activation functions [36–38], the stabilization issue has not been fully explored in the existing studies.

Although the stabilization problem for some kinds of neural networks with or without time delays is investigated by some authors, there has been no literature reported on the stabilization of stochastic *Cohen-Grossberg neural networks* with both Markovian jumping parameters and mixed mode-dependent time delays. As well known, mode-dependent time delays are of practical significance since the signal may switch between different modes and also propagate in a distributed way during a certain time period with the presence of an amount of parallel pathways [24]. The purpose of this chapter is to make an attempt to deal with the control problem for a class of stochastic neural networks with mode-dependent delays [1]. By introducing a new Lyapunov–Krasovskii functional that accounts for the mode-dependent mixed delays, stochastic analysis is conducted in order to derive delay-dependent criteria for the exponential stabilization problem. The feedback stabilizing controller is designed to satisfy some exponential stability constraints on the closed-loop poles. The stabilization criteria are obtained in terms of LMI and hence the gain control matrix is easily determined by numerical MATLABs LMI Control Toolbox. Three numerical examples are carried out to demonstrate the feasibility of our delay-dependent stabilization criteria.

Throughout this chapter, the shorthand $\text{col}\{M_1, M_2, \ldots, M_l\}$ denotes a column matrix with the matrices $M_1, M_2, \ldots, M_l$. $\left(\Omega, \mathcal{F}, \{\mathcal{F}_t\}_{t\geq 0}, \mathcal{P}\right)$ denotes a complete probability space with a filtration $\{\mathcal{F}_t\}_{t\geq 0}$ satisfying the usual conditions, i.e., the filtration is right continuous and contains all $\mathcal{P}$-null sets. $\mathcal{L}^p_{\mathcal{F}_0}([-h, 0], \mathbb{R}^n)$ denotes the family of all $\mathcal{F}_0$-measurable $\mathbb{C}\left([-h, 0]; \mathbb{R}^n\right)$-valued random variables

$\xi = \{\xi(\theta) : -h \leq \theta \leq 0\}$ such that $\sup_{-h\leq\theta\leq 0} \mathbb{E}|\xi(\theta)|^p < \infty$, where $\mathbb{E}\{\cdot\}$ stands for the mathematical expectation operator with respect to the given probability measure $\mathcal{P}$.

11.2 Problem Formulation and Preliminaries

We consider the following stochastic neural network with both feedback control law and Markovian jumping parameters described by

$$\begin{aligned} dx(t) = &- \alpha(x(t), \eta_t)\Big[\beta(x(t), \eta_t) - A(\eta_t)f(x(t)) \\ &- B(\eta_t)f(x(t-\tau(t,\eta_t))) \\ &- C(\eta_t)\int_{t-\upsilon(t,\eta_t)}^{t} g(x(s))ds - D(\eta_t)u(t,\eta_t)\Big]dt \\ &+ \Big[E_1(\eta_t)x(t) + E_2(\eta_t)x(t-\tau(t,\eta_t)) \\ &+ E_3(\eta_t)f(x(t)) + E_4(\eta_t)f(x(t-\tau(t,\eta_t))) \\ &+ E_5(\eta_t)\int_{t-\upsilon(t,\eta_t)}^{t} g(x(s))ds\Big]d\omega(t), \end{aligned} \tag{11.1}$$

where $x(t) = [x_1(t), \ldots, x_n(t)]^T$ denotes the neuron state at time t, $u(t) \in L_2([0, s), \mathbb{R}^m)$, $\forall s > 0$, is the control input vector of the neural networks, $\alpha(x(t), \eta_t) = \text{diag}\{\alpha_j(x_j(t), \eta_t), \ldots, \alpha_n(x_n(t), \eta_t)\}$ denotes the amplification function, $\beta(x(t), \eta_t) = \text{diag}\{\beta_j(x_j(t), \eta_t), \ldots, \beta_n(x_n(t), \eta_t)\}$ denotes the appropriately behaved function such that the solution of the model given in (11.1) remains bounded, and $f(x(t)) = [f_1(x_1(t)), \ldots, f_n(x_n(t))]^T$, $g(x(s)) = [g_1(x_1(s)), \ldots, g_n(x_n(s))]^T$ denote the activation functions. $f(x(t-\tau(t,\eta_t))) = [f_1(x_1(t-\tau(t,\eta_t))), \ldots, f_n(x_n(t-\tau(t,\eta_t)))]^T$. $0 \leq \tau(t,\eta_t) \leq \bar{\tau}(\eta_t) \leq \bar{\tau}$, $0 \leq \upsilon(t,\eta_t) \leq \bar{\upsilon}(\eta_t) \leq \bar{\upsilon}$ $(j = 1, \ldots, n)$ are bounded and unknown delays. The matrices $A(\eta_t), B(\eta_t), C(\eta_t) \in \mathbb{R}^{n\times n}$, $D(\eta_t) \in \mathbb{R}^{n\times m}$ are the connection weight matrix, the discretely delayed connection weight matrix, the distributively delayed connection weight matrix and the control input weights, respectively. $E_j(\eta_t)(j = 1, 2, \ldots, 5)$ is known real constant matrix with appropriate dimension, $\omega(t)$ is a one-dimensional Brownian motion defined on complete probability space $(\Omega, \mathcal{F}, \{\mathcal{F}_t\}_{t\geq 0}, \mathcal{P})$ with $\mathbb{E}\{d\omega(t)\} = 0$, $\mathbb{E}\{[d\omega(t)]^2\} = dt$. $\{\eta_t = \eta(t), t \geq 0\}$ is a homogeneous, finite-state Markovian process with right continuous trajectories and taking values in finite set $\wp = \{1, 2, \ldots, N\}$ with given probability space $(\Omega, \mathcal{F}, \{\mathcal{F}_t\}_{t\geq 0}, \mathcal{P})$ and the initial model η_0. It is assumed that the initial condition of neural network (11.1) has the form $x(t) = \varphi(t)$ for $t \in [-\varpi, 0]$, where $\varphi(t) = [\varphi_1(t), \ldots, \varphi_n(t)]^T$, function $\varphi_j(t)(j = 1, 2, \ldots, n)$ is continuous, $\varpi = \max\{\bar{\tau}, \bar{\upsilon}\}$. Let $\aleph = [\pi_{ij}]_{i,j\in\wp}$ denote the transition rate matrix with given probability:

$$P(\eta_{t+\delta} = j|\eta_t = i) = \begin{cases} \pi_{ij}\delta + o(\delta), & i \neq j, \\ \pi_{ii}\delta + o(\delta) + 1, & i = j, \end{cases}$$

where $\delta > 0$, $\lim_{\delta \to 0^+} \frac{o(\delta)}{\delta} = 0$ and π_{ij} is the transition rate from mode i to mode j satisfying $\pi_{ij} \geq 0$ for $i \neq j$ with $\pi_{ii} = -\sum_{j=1, i\neq j}^{N} \pi_{ij}, i, j \in \wp$.

For convenience, each possible value of η_t is denoted by $i (i \in \wp)$ in the sequel. Then we have

$$\begin{aligned} &\alpha_i(x(t)) = \alpha(x(t), \eta_t),\ \beta_i(x(t)) = \beta(x(t), \eta_t), \\ &A_i = A(\eta_t),\ B_i = B(\eta_t),\ C_i = C(\eta_t), \\ &D_i = D(\eta_t),\ \tau_i(t) = \tau(t, \eta_t),\ \upsilon_i(t) = \upsilon(t, \eta_t), \\ &E_{li} = E_l(\eta_t),\ l = 1, \ldots, 5. \end{aligned}$$

In the following, we need the following definitions, assumptions, and lemmas.

Definition 11.1 ([24, 27]) Given $r > 0$, and any initial condition $\varphi \in \mathcal{L}^2_{\mathcal{F}_0}([-\varpi, 0], \mathbb{R}^n)$ with $u(t, \eta_t) = 0$. The zero solution of system (11.1) is said to be *r-exponentially stable* in the mean square, if there exists a positive scalar M such that any solution $x(t, \varphi)$ of the system satisfies the following inequality,

$$\mathbb{E}||x(t, \phi)||^2 \leq M \sup_{-\varpi \leq s \leq 0} \mathbb{E}||\phi(s)||^2 e^{-2rt}, \quad \forall\, t \geq 0.$$

Definition 11.2 ([24, 27]) Given $r > 0$. The system (11.1) is said to be *r-exponentially stabilizable* in the mean square, if there is a feedback control law $u(t, \eta_t) = \overline{U}(\eta_t)x(t)$, such that the following closed-loop system

$$\begin{aligned} \mathrm{d}x(t) =& -\alpha(x(t), \eta_t)\Big[\beta(x(t), \eta_t) - A(\eta_t)f(x(t)) \\ &- B(\eta_t)f(x(t - \tau(t, \eta_t))) \\ &- C(\eta_t)\int_{t-\upsilon(t,\eta_t)}^{t} g(x(s))\mathrm{d}s - D(\eta_t)\overline{U}(\eta_t)x(t)\Big]\mathrm{d}t \\ &+ \Big[E_1(\eta_t)x(t) + E_2(\eta_t)x(t - \tau(t, \eta_t)) \\ &+ E_3(\eta_t)f(x(t)) + E_4(\eta_t)f(x(t - \tau(t, \eta_t))) \\ &+ E_5(\eta_t)\int_{t-\upsilon(t,\eta_t)}^{t} g(x(s))\mathrm{d}s\Big]\mathrm{d}\omega(t), \\ x(t) =& \varphi(t), \quad t \in [-\varpi, 0], \end{aligned}$$

is r-exponentially stable.

Assumption 11.3 ([8]) Each $\alpha_{ji}(\cdot)$ is a continuous function and satisfies $\bar{\alpha}_{ji} \geq \alpha_{ji}(\cdot) \geq \underline{\alpha}_{ji} > 0$, $j = 1, 2, \ldots, n, i = 1, 2, \ldots, N$.

Here, we denote $\underline{\alpha}_i = \min_{1\leq j\leq n}\{\underline{\alpha}_{ji}\}$, $\bar{\alpha}_i = \max_{1\leq j\leq n}\{\bar{\alpha}_{ji}\}$ for simplicity.

Assumption 11.4 Each function $\beta_{ji}(\cdot)$ is locally Lipschitz continuous, $\beta_{ji}(0) = 0$ and there exist constants $\bar{\beta}_{ji} > \underline{\beta}_{ji} \geq 0$ such that

$$\underline{\beta}_{ji}s^2 \leq \beta_{ji}(s)s \leq \bar{\beta}_{ji}s^2,$$

for any $s \in \mathbb{R}$, $j = 1, 2, \ldots, n$, $i = 1, 2, \ldots, N$.

For simplicity, we denote $\Pi_i = \text{diag}\{\bar{\beta}_{1i}, \ldots, \bar{\beta}_{ni}\}$, $\Gamma_i = \text{diag}\{\underline{\beta}_{1i}, \ldots, \underline{\beta}_{ni}\}$.

Assumption 11.5 For $j = 1, 2, \ldots, n$, $f_j(0) = g_j(0) = 0$. Furthermore, there exist constants $\varrho_j^-, \varrho_j^+, \psi_j^-, \psi_j^+$ such that $\varrho_j^- < \varrho_j^+$, $\psi_j^- < \psi_j^+$ and

$$\varrho_j^- \leq \frac{f_j(s)}{s} \leq \varrho_j^+, \quad \psi_j^- \leq \frac{g_j(s)}{s} \leq \psi_j^+,$$

for any $s \in \mathbb{R}$, $j = 1, 2, \ldots, n$.

Remark 11.6 As pointed out in [24], the constants $\varrho_j^-, \varrho_j^+, \psi_j^-, \psi_j^+$ in Assumption 11.5 are allowed to be positive, negative, or zero. Then, those previously used Lipschitz conditions are just the special cases of Assumption 11.5. Hence, the activation functions can be of more general descriptions than those earlier forms.

For notational simplicity, we denote

$$\begin{aligned}
\bar{\Sigma} &= \text{diag}\left\{\varrho_1^+, \varrho_2^+, \ldots, \varrho_n^+\right\}, \\
\Sigma &= \text{diag}\left\{\varrho_1^-, \varrho_2^-, \ldots, \varrho_n^-\right\}, \\
F_1 &= \text{diag}\left\{\varrho_1^-\varrho_1^+, \varrho_2^-\varrho_2^+, \ldots, \varrho_n^-\varrho_n^+\right\}, \\
F_2 &= \text{diag}\left\{\frac{\varrho_1^- + \varrho_1^+}{2}, \frac{\varrho_2^- + \varrho_2^+}{2}, \ldots, \frac{\varrho_n^- + \varrho_n^+}{2}\right\}, \\
F_3 &= \text{diag}\left\{\psi_1^-\psi_1^+, \psi_2^-\psi_2^+, \ldots, \psi_n^-\psi_n^+\right\}, \\
F_4 &= \text{diag}\left\{\frac{\psi_1^- + \psi_1^+}{2}, \frac{\psi_2^- + \psi_2^+}{2}, \ldots, \frac{\psi_n^- + \psi_n^+}{2}\right\}.
\end{aligned}$$

Lemma 11.7 (Jensen integral inequality, see [39]) *For any constant matrix $M > 0$, any scalars a and b with $a < b$, and a vector function $\chi(t) : [a, b] \to \mathbb{R}$ such that the integrals concerned are well defined, then the following inequality holds*

$$\left\langle \int_a^b \chi(s)\mathrm{d}s, M\int_a^b \chi(s)\mathrm{d}s \right\rangle \leq (b-a)\int_a^b \chi(s)^T M\chi(s)\mathrm{d}s,$$

where $\langle A, B\rangle = A^T B$ denotes the inner product.

Lemma 11.8 *Assume that $\nu, \mu, \underline{\vartheta}, \bar{\vartheta}$ are real scalars such that $\nu \leq 1, \nu + \mu \leq 4$, and $\underline{\vartheta} < \bar{\vartheta}$. Let $\vartheta : \mathbb{R} \to (\underline{\vartheta}, \bar{\vartheta})$ be a real function. Then for any nonnegative scalars a, b, the following inequality holds*

$$\begin{aligned}
&-\frac{a}{\vartheta(t)-\underline{\vartheta}}-\frac{b}{\bar{\vartheta}-\vartheta(t)}\\
&\leq \frac{1}{\bar{\vartheta}-\underline{\vartheta}}\max\{-\nu a-\mu b, -\mu a-\nu b\}.
\end{aligned} \tag{11.2}$$

Proof Without loss of generality, we assume that $\nu \leq \mu$. First consider the case that $a \leq b$. It is easy to see that $\max\{-\nu a - \mu b, -\mu a - \nu b\} = -\mu a - \nu b$. Therefore, we have

$$\begin{aligned}
&\left(\vartheta(t)-\underline{\vartheta}\right)\left(\bar{\vartheta}-\vartheta(t)\right)(-\mu a-\nu b)\\
&\quad+\left(\bar{\vartheta}-\underline{\vartheta}\right)\left[\left(\bar{\vartheta}-\vartheta(t)\right)a+\left(\vartheta(t)-\underline{\vartheta}\right)b\right]\\
&=\left(\bar{\vartheta}-\vartheta(t)\right)\left[\bar{\vartheta}+(\mu-1)\underline{\vartheta}-\mu\vartheta(t)\right]a\\
&\quad+\left(\vartheta(t)-\underline{\vartheta}\right)\left[(1-\nu)\left(\bar{\vartheta}-\vartheta(t)\right)+\left(\vartheta(t)-\underline{\vartheta}\right)\right]b\\
&\geq\Big\{\left(\bar{\vartheta}-\vartheta(t)\right)\left[\bar{\vartheta}+(\mu-1)\underline{\vartheta}-\mu\vartheta(t)\right]\\
&\quad+\left(\vartheta(t)-\underline{\vartheta}\right)\left[(1-\nu)\left(\bar{\vartheta}-\vartheta(t)\right)+\left(\vartheta(t)-\underline{\vartheta}\right)\right]\Big\}a\\
&=\frac{a}{4}\left[(\nu+\mu)\left(2\vartheta(t)-\underline{\vartheta}-\bar{\vartheta}\right)^2+(4-\nu-\mu)\left(\bar{\vartheta}-\underline{\vartheta}\right)^2\right]\\
&\geq 0.
\end{aligned}$$

That is

$$\begin{aligned}
&\frac{1}{\bar{\vartheta}-\underline{\vartheta}}\max\{-\nu a-\mu b, -\mu a-\nu b\}\\
&\quad=\frac{1}{\bar{\vartheta}-\underline{\vartheta}}(-\mu a-\nu b)\\
&\quad\geq-\frac{a}{\vartheta(t)-\underline{\vartheta}}-\frac{b}{\bar{\vartheta}-\vartheta(t)}.
\end{aligned}$$

Similarly, we can also conclude that the inequality (11.2) holds for $a > b$. Now, the proof of Lemma 11.8 is completed.

Remark 11.9 If we set $\nu = 1, \mu = 3$, then we get Lemma 3 of [40] from Lemma 11.8. Thus, based on Lemma 11.8, we can get some conditions of exponential stabilization problem with less conservativeness.

11.3 Stabilization Result

As is well known, for stochastic systems, Itô's formula plays an important role in the stability analysis of stochastic systems and we cite some related results here [41]. Consider a general stochastic system

$$\mathrm{d}x(t) = f(x(t), t, \eta_t)\mathrm{d}t + g(x(t), t, \eta_t)\mathrm{d}\omega(t) \tag{11.3}$$

on $t \geq t_0$ with initial value $x(t_0) = x_0 \in \mathbb{R}^n$, where $f : \mathbb{R}^n \times \mathbb{R}^+ \times \wp \to \mathbb{R}^n$ and $g : \mathbb{R}^n \times \mathbb{R}^+ \times \wp \to \mathbb{R}^{n+m}$. Let $\mathbb{C}^{2,1}(\mathbb{R}^n \times \mathbb{R}^+, \mathbb{R}^+)$ denote the family of all nonnegative functions $V(x, t, i)$ on $\mathbb{R}^n \times \mathbb{R}^+$ which are continuously differentiable in t and twice differentiable in x. Let £ be the weak infinitesimal generator of the random process $\{x(t), \eta(t)\}_{t\geq 0}$ along the system (11.3) (see [24, 42, 43]), i.e.,

$$\begin{aligned}\pounds V(x_t, t, i) := \lim_{\delta\to 0^+} \frac{1}{\delta} \sup\Big[\mathbb{E}\big\{V(x_{t+\delta}, t+\delta, \eta(t+\delta))\big|x(t),\\ \eta(t) = i\big\} - V(x_t, t, \eta(t) = i)\Big],\end{aligned}$$

then, by the generalized Itô's formula, one can get

$$\mathbb{E}V(x, t, i) = \mathbb{E}V(x_0, t_0, i) + \mathbb{E}\int_{t_0}^{t} \pounds V(x(s), s, i)\mathrm{d}s.$$

Theorem 11.10 *Given $r > 0$. For any given scalars $\bar{\tau}_i > 0$, $\bar{\upsilon}_i > 0$, $\upsilon_i' < 1$, considering the system (11.1) satisfying Assumptions 11.3–11.5 and $\dot{\tau}_i(t) \leq \tau_i'$, $\dot{\upsilon}_i(t) \leq \upsilon_i'$, the system (11.1) is globally r-exponentially stabilized if there exist symmetric positive definite matrices $P_i \in \mathbb{R}^{n\times n}$, symmetric nonnegative definite matrices $Q_{ji}, R_i, M_i, S_l, Z_i$ ($j = 1, \ldots, 4, l = 1, \ldots, 9$), positive diagonal matrices G_i, U_i, T_i, W_i, H, K, and real matrices X_i satisfying the following inequalities ($i = 1, \ldots, N$)*

$$\sum_{j=1}^{N} \pi_{ij} Q_{lj} < S_l, \quad l = 1, 2, 3, 4, \tag{11.4}$$

$$\sum_{j=1}^{N} \pi_{ij} R_j < S_5, \tag{11.5}$$

$$\sum_{j=1}^{N} \pi_{ij} Z_j < S_6, \tag{11.6}$$

$$\sum_{j=1}^{N} \pi_{ij} \bar{\upsilon}_j R_j < S_7, \tag{11.7}$$

$$\sum_{j=1}^{N} \pi_{ij}\bar{\tau}_j Z_j < S_8, \tag{11.8}$$

$$\sum_{j=1}^{N} \pi_{ij}\bar{\tau}_j Q_{4j} < S_9, \tag{11.9}$$

$$\begin{bmatrix} \Omega_i + \widetilde{\Omega}_i & \mathcal{E}^T \\ \mathcal{E} & \overline{Z}_i \end{bmatrix} < 0, \tag{11.10}$$

$$\begin{bmatrix} \Omega_i + \widehat{\Omega}_i & \mathcal{E}^T \\ \mathcal{E} & \overline{Z}_i \end{bmatrix} < 0, \tag{11.11}$$

where

$$\begin{aligned}
\Omega_i =& \begin{bmatrix} \Omega_{1i} & \Omega_{2i} & \Omega_{4i} & \Omega_{7i} \\ * & \Omega_{3i} & \Omega_{5i} & 0 \\ * & * & \Omega_{6i} & \Omega_{8i} \\ * & * & * & \Omega_{9i} \end{bmatrix}, \\
\widetilde{\Omega}_i =& -\frac{2}{\bar{\tau}_i}\mathbb{I}^T Q_{4i}\mathbb{I}, \quad \widehat{\Omega}_i = -\frac{2}{\bar{\tau}_i}\mathfrak{I}^T Q_{4i}\mathfrak{I}, \\
\mathcal{E} =& [\, E_{1i} \;\; E_{2i} \;\; E_{3i} \;\; E_{4i} \;\; 0 \;\; E_{5i} \;\; 0\;0\;0\;0\;0\;0 \,], \\
\overline{Z}_i =& \frac{\bar{\tau}^2}{2} S_6 + \bar{\tau} S_8 + \bar{\tau}_i Z_i + \widetilde{Z}_i, \\
\widetilde{Z}_i =& \underline{\alpha}_i^{-1}[P_i + H(\Pi_i - \Gamma_i) + K(\bar{\Sigma} - \Sigma)],
\end{aligned}$$

with

$$\begin{aligned}
\Omega_{1i} =& \begin{bmatrix} \Omega_{11i} & \Omega_{12i} & \Omega_{13i} & \Omega_{14i} & \Omega_{15i} & \Omega_{16i} \\ * & \Omega_{22i} & 0 & \Omega_{24i} & 0 & 0 \\ * & * & \Omega_{33i} & \Omega_{34i} & 0 & \Omega_{36i} \\ * & * & * & \Omega_{44i} & 0 & 0 \\ * & * & * & * & \Omega_{55i} & 0 \\ * & * & * & * & * & \Omega_{66i} \end{bmatrix}, \\
\Omega_{2i} =& [\, 0 \;\; 0 \;\; A_i \;\; B_i \;\; 0 \;\; C_i \,]^T G_i^T,
\end{aligned}$$

$$\begin{aligned}
\Omega_{3i} =& -2\bar{\alpha}_i^{-2} G_i + \underline{\alpha}_i^{-2}\Big[\frac{1}{2}\bar{\tau}^2 S_4 + \bar{\tau}_i Q_{4i} + \bar{\tau} S_9\Big], \\
\Omega_{4i} =& \begin{bmatrix} \Omega_{18i} & 0 & \Omega_{1ai} \\ 0 & \Omega_{29i} & 0 \\ \Omega_{38i} & 0 & \Omega_{3ai} \\ 0 & 0 & \Omega_{4ai} \\ 0 & 0 & 0 \\ 0 & 0 & \Omega_{6ai} \end{bmatrix}, \quad \Omega_{7i} = \begin{bmatrix} \Omega_{1bi} & 0 \\ \Omega_{2bi} & \Omega_{2ci} \\ 0 & 0 \\ 0 & 0 \\ 0 & 0 \\ 0 & 0 \end{bmatrix},
\end{aligned}$$

$$\Omega_{5i} = G_i \begin{bmatrix} D_i D_i^T & 0 & -I \end{bmatrix},$$

$$\Omega_{6i} = \begin{bmatrix} \Omega_{88i} & 0 & \Omega_{8ai} \\ * & \Omega_{99i} & 0 \\ * & * & \Omega_{aai} \end{bmatrix}, \quad \Omega_{8i} = \begin{bmatrix} 0 & 0 \\ 0 & \Omega_{9ci} \\ 0 & 0 \end{bmatrix},$$

$$\Omega_{9i} = \mathrm{diag}\{\Omega_{bbi},\ \Omega_{cci}\},$$

$$\mathbb{I} = \begin{bmatrix} 0 & -I & 0 & 0 & 0 & 0 & 0 & 0 & I & 0 & 0 & I \end{bmatrix},$$

$$\mathfrak{I} = \begin{bmatrix} -I & I & 0 & 0 & 0 & 0 & 0 & 0 & 0 & 0 & I & 0 \end{bmatrix},$$

and

$$\begin{aligned}
\Omega_{11i} =& -2P_i\Gamma_i + Q_{1i} + Q_{3i} - \frac{1}{\bar{\tau}_i}Q_{4i} \\
&+ \sum_{j=1}^{N} \pi_{ij}\rho_{ij}^{-1}P_j + \bar{\tau}(S_1 + S_3) - U_iF_1 - W_iF_3 \\
&+ \sum_{j=1}^{N} \bar{\pi}_{ij}\underline{\alpha}_i^{-1}\left[P_i + 2H(\Pi_i - \Gamma_i) + 2K(\bar{\Sigma} - \Sigma)\right],
\end{aligned}$$

$$\Omega_{12i} = \frac{1}{\bar{\tau}_i}Q_{4i},$$

$$\Omega_{22i} = -(1 - \tau_i')Q_{1i} + \sum_{j=1}^{N} \bar{\pi}_{ij}\bar{\tau}_jQ_{1j} - \frac{2}{\bar{\tau}_i}Q_{4i} - T_iF_1,$$

$$\Omega_{13i} = P_iA_i + U_iF_2 - \Gamma_iHA_i - \Sigma KA_i,$$

$$\Omega_{33i} = Q_{2i} - U_i + KA_i + A_i^TK + \bar{\tau}S_2,$$

$$\Omega_{14i} = P_iB_i - \Gamma_iHB_i - \Sigma KB_i,$$

$$\Omega_{24i} = -T_iF_2, \quad \Omega_{34i} = KB_i,$$

$$\Omega_{44i} = -T_i - (1 - \tau_i')Q_{2i} + \sum_{j=1}^{N} \bar{\pi}_{ij}\bar{\tau}_jQ_{2j},$$

$$\Omega_{15i} = W_iF_4, \quad \Omega_{55i} = -W_i + \bar{\upsilon}_iR_i + \frac{\bar{\upsilon}^2}{2}S_5 + \bar{\upsilon}S_7,$$

$$\Omega_{16i} = P_iC_i - \Gamma_iHC_i - \Sigma KC_i,$$

$$\Omega_{36i} = KC_i, \quad \Omega_{66i} = -\frac{1 - \upsilon_i'}{\bar{\upsilon}_i}R_i,$$

$$\Omega_{18i} = (P_i - \Sigma K + \Gamma_iH)D_iD_i^T + \bar{M}_i^T,$$

$$\Omega_{38i} = KD_iD_i^T, \quad \Omega_{88i} = -2M_i, \quad \Omega_{29i} = \frac{1}{\bar{\tau}_i}Q_{4i},$$

$$\begin{aligned}
\Omega_{99i} =& - \frac{1}{\bar{\bar{\tau}}_i} Q_{4i} - Q_{3i} + \sum_{j=1}^{N} \bar{\pi}_{ij} \bar{\tau}_j Q_{3j},\\
\Omega_{1ai} =& \Gamma_i H + \Sigma K, \ \ \Omega_{3ai} = A_i^T H - K,\\
\Omega_{4ai} =& B_i^T H, \ \ \Omega_{6ai} = C_i^T H, \ \ \Omega_{8ai} = D_i D_i^T H,\\
\Omega_{aai} =& - 2H, \ \ \Omega_{1bi} = \frac{1}{\bar{\bar{\tau}}_i} Q_{4i},\\
\Omega_{2bi} =& - \frac{1}{\bar{\bar{\tau}}_i} Q_{4i}, \ \ \Omega_{bbi} = - \frac{1}{\bar{\bar{\tau}}_i} Q_{4i} - Z_i,
\end{aligned}$$

$$\Omega_{2ci} = \frac{1}{\bar{\bar{\tau}}_i} Q_{4i}, \ \ \Omega_{9ci} = - \frac{1}{\bar{\bar{\tau}}_i} Q_{4i}, \ \ \Omega_{cci} = - \frac{1}{\bar{\bar{\tau}}_i} Q_{4i} - Z_i,$$

and $\bar{\pi}_{ij} = \max\{\pi_{ij}, 0\}, \ \bar{M}_i = M_i X_i,$

$$\rho_{ij} = \begin{cases} \bar{\alpha}_i, & j = i \\ \underline{\alpha}_i, & j \neq i \end{cases}.$$

Furthermore, the feedback stabilizing control law is defined by $u_i(t) = D_i^T X_i x(t)$.

Proof From Assumption 11.3, we know that the amplification function $\alpha_i(x(t))$ is nonlinear and satisfies $\alpha_i(x(t))\alpha_i(x(t)) \leq \bar{\alpha}_i^2 I$. Following the way in [15], pre- and postmultiplying the left-hand sides of inequalities (11.10) and (11.11) by $\mathrm{diag}\{I \ \ I \ \ I \ \ I \ \ I \ \ \alpha_i(x(t)) \ \ I \ \ I \ \ I \ \ I \ \ I\}$, respectively, it follows that

$$\begin{bmatrix} \overline{\Omega}_i + \widetilde{\Omega}_i & \mathcal{E}^T \\ \mathcal{E} & \overline{Z}_i \end{bmatrix} < 0, \tag{11.12}$$

$$\begin{bmatrix} \overline{\Omega}_i + \widehat{\Omega}_i & \mathcal{E}^T \\ \mathcal{E} & \overline{Z}_i \end{bmatrix} < 0, \tag{11.13}$$

where

$$\overline{\Omega}_i \doteq \begin{bmatrix} \Omega_{1i} & \overline{\Omega}_{2i} & \Omega_{4i} \\ * & \overline{\Omega}_{3i} & \overline{\Omega}_{5i} \\ * & * & \Omega_{6i} \end{bmatrix},$$

with

$$\begin{aligned}
\overline{\Omega}_{2i} =& [\, 0 \ \ 0 \ \ A_i \ \ B_i \ \ 0 \ \ C_i \,]^T \alpha_i(x(t)) G_i^T,\\
\overline{\Omega}_{3i} =& \bar{\tau}_i Q_{4i} - (G_i + G_i^T) + \frac{\bar{\tau}^2}{2} S_4 + \bar{\tau} S_9,\\
\overline{\Omega}_{5i} =& G_i \alpha_i(x(t)) \Big[\, D_i D_i^T \ \ \ 0 \ \ - I \ \ 0 \ \ 0 \,\Big].
\end{aligned}$$

For any $j = 1, 2, \ldots, n$, from Assumption 11.5 we obtain that

$$\left(f(x_j(t)) - \varrho_j^+ x_j(t)\right)\left(f(x_j(t)) - \varrho_j^- x_j(t)\right) \le 0,$$
$$\left(g(x_j(t)) - \psi_j^+ x_j(t)\right)\left(g(x_j(t)) - \psi_j^- x_j(t)\right) \le 0.$$

Therefore, the following matrix inequalities hold for any positive diagonal matrices U_i, T_i, W_i,

$$\left\langle \begin{bmatrix} x(t) \\ f(x(t)) \end{bmatrix}, \begin{bmatrix} -U_iF_1 & U_iF_2 \\ U_iF_2 & -U_i \end{bmatrix} \begin{bmatrix} x(t) \\ f(x(t)) \end{bmatrix} \right\rangle \ge 0, \tag{11.14}$$

$$\left\langle \begin{bmatrix} x(t-\tau_i(t)) \\ f(x(t-\tau_i(t))) \end{bmatrix}, \begin{bmatrix} -T_iF_1 & T_iF_2 \\ T_iF_2 & -T_i \end{bmatrix} \begin{bmatrix} x(t-\tau_i(t)) \\ f(x(t-\tau_i(t))) \end{bmatrix} \right\rangle \ge 0, \tag{11.15}$$

$$\left\langle \begin{bmatrix} x(t) \\ g(x(t)) \end{bmatrix}, \begin{bmatrix} -W_iF_3 & W_iF_4 \\ W_iF_4 & -W_i \end{bmatrix} \begin{bmatrix} x(t) \\ g(x(t)) \end{bmatrix} \right\rangle \ge 0. \tag{11.16}$$

Denoting

$$\begin{aligned} \iota_i(t) = &- \beta_i(x(t)) + A_i f(x(t)) + B_i f(x(t-\tau_i(t))) \\ &+ C_i \int_{t-\upsilon_i(t)}^{t} g(x(s))\mathrm{d}s + D_i u_i(t), \end{aligned}$$

$$\begin{aligned} \vartheta_i(t) =& \alpha_i(x(t))\iota_i(t), \\ \sigma_i(t) =& E_{1i}x(t) + E_{2i}x(t-\tau_i(t)) + E_{3i}f(x(t)) \\ &+ E_{4i}f(x(t-\tau_i(t))) + E_{5i}\int_{t-\upsilon_i(t)}^{t} g(x(s))\mathrm{d}s, \end{aligned}$$

then system (11.1) can be rewritten as

$$\mathrm{d}x(t) = \vartheta_i(t)\mathrm{d}t + \sigma_i(t)\mathrm{d}\omega(t). \tag{11.17}$$

Define the following Lyapunov–Krasovskii functional:

$$V(x_t, t, i) = \sum_{l=1}^{6} V_{li}(x_t, t), \tag{11.18}$$

where

$$V_{1i}(x_t, t) = \sum_{j=1}^{n} 2p_{ji} \int_0^{x_j(t)} \frac{s}{\alpha_{ji}(s)}\mathrm{d}s$$

$$
\begin{aligned}
&+\sum_{j=1}^{n} 2h_j \int_0^{x_j(t)} \frac{\beta_{ji}(s)-\underline{\beta}_{ji}s}{\alpha_{ji}(s)}\mathrm{d}s \\
&+\sum_{j=1}^{n} 2k_j \int_0^{x_j(t)} \frac{f_j(s)-\varrho_j^{-}s}{\alpha_{ji}(s)}\mathrm{d}s,
\end{aligned}
$$

$$
\begin{aligned}
V_{2i}(x_t,t) &= \int_{t-\tau_i(t)}^{t} \langle x(s), Q_{1i}x(s)\rangle \,\mathrm{d}s \\
&\quad+\int_{t-\tau_i(t)}^{t} \langle f(x(s)), Q_{2i}f(x(s))\rangle \,\mathrm{d}s \\
&\quad+\int_{t-\bar{\tau}_i}^{t} \langle x(s), Q_{3i}x(s)\rangle \,\mathrm{d}s,
\end{aligned}
$$

$$
\begin{aligned}
V_{3i}(x_t,t) &= \int_{-\bar{\tau}_i}^{0}\int_{t+\theta}^{t} \langle \vartheta_i(s), Q_{4i}\vartheta_i(s)\rangle \,\mathrm{d}s\mathrm{d}\theta \\
&\quad+\int_{-\upsilon_i(t)}^{0}\int_{t+\theta}^{t} \langle g(x(s)), R_i g(x(s))\rangle \,\mathrm{d}s\mathrm{d}\theta \\
&\quad+\int_{-\bar{\tau}_i}^{0}\int_{t+\theta}^{t} \langle \sigma_i(s), Z_i\sigma_i(s)\rangle \,\mathrm{d}s\mathrm{d}\theta,
\end{aligned}
$$

$$
\begin{aligned}
V_{4i}(x_t,t) &= \int_{-\bar{\tau}}^{0}\int_{t+\theta}^{t} \big\{ \langle x(s), (S_1+S_3)x(s)\rangle \\
&\qquad + \langle f(x(s)), S_2 f(x(s))\rangle \big\}\mathrm{d}s\mathrm{d}\theta, \\
V_{5i}(x_t,t) &= \int_{-\bar{\tau}}^{0}\int_{\theta}^{0}\int_{t+\lambda}^{t} \langle \vartheta_i(s), S_4\vartheta_i(s)\rangle \,\mathrm{d}s\mathrm{d}\lambda\mathrm{d}\theta \\
&\quad+\int_{-\bar{\upsilon}}^{0}\int_{\theta}^{0}\int_{t+\lambda}^{t} \langle g(x(s)), S_5 g(x(s))\rangle \,\mathrm{d}s\mathrm{d}\lambda\mathrm{d}\theta \\
&\quad+\int_{-\bar{\tau}}^{0}\int_{\theta}^{0}\int_{t+\lambda}^{t} \langle \sigma_i(s), S_6\sigma_i(s)\rangle \,\mathrm{d}s\mathrm{d}\lambda\mathrm{d}\theta, \\
V_{6i}(x_t,t) &= \int_{-\bar{\upsilon}}^{0}\int_{t+\theta}^{t} \langle g(x(s)), S_7 g(x(s))\rangle \,\mathrm{d}s\mathrm{d}\theta \\
&\quad+\int_{-\bar{\tau}}^{0}\int_{t+\theta}^{t} \big\{ \langle \sigma_i(s), S_8\sigma_i(s)\rangle \\
&\qquad + \langle \vartheta_i(s), S_9\vartheta_i(s)\rangle \big\}\mathrm{d}s\mathrm{d}\theta,
\end{aligned}
$$

with $P_i = \mathrm{diag}\{p_{1i}, p_{2i}, \ldots, p_{ni}\}$, $H = \mathrm{diag}\{h_1, h_2, \ldots, h_n\}$, $K = \mathrm{diag}\{k_1, k_2, \ldots, k_n\}$.

For any $\eta(t) = i \in \wp$, it can be shown that

$$\begin{aligned}
&\pounds\left\{\sum_{j=1}^{n} 2p_{ji}\int_0^{x_j(t)} \frac{s}{\alpha_{ji}(s)}\mathrm{d}s\right\}\\
&= \lim_{\Delta\to 0^+}\frac{1}{\Delta}\mathbb{E}\left\{\sum_{j=1}^{n} 2\left(\sum_{l=1}^{N}[\pi_{il}\Delta + o(\Delta)]p_{jl} + p_{ji}\right)\right.\\
&\quad\times\int_0^{x_j(t+\Delta)} \frac{s}{\sum_{l=1}^{N}[\pi_{il}\Delta + o(\Delta)]\alpha_{jl}(s) + \alpha_{ji}(s)}\mathrm{d}s\\
&\quad\left.-\sum_{j=1}^{n} 2p_{ji}\int_0^{x_j(t)} \frac{s}{\alpha_{ji}(s)}\mathrm{d}s\right\}\\
&=\sum_{l=1}^{N}\pi_{il}\sum_{j=1}^{n} 2p_{jl}\int_0^{x_j(t)} \frac{s}{\alpha_{ji}(s)}\mathrm{d}s\\
&\quad+\lim_{\Delta\to 0^+}\frac{1}{\Delta}\mathbb{E}\left\{\sum_{j=1}^{n} 2p_{ji}\left[-\int_0^{x_j(t)} \frac{s}{\alpha_{ji}(s)}\mathrm{d}s\right.\right.\\
&\quad\left.\left.+\int_0^{x_j(t+\Delta)} \frac{s}{\sum_{l=1}^{N}[\pi_{il}\Delta + o(\Delta)]\alpha_{jl}(s) + \alpha_{ji}(s)}\mathrm{d}s\right]\right\}
\end{aligned}$$

$$\begin{aligned}
&=\sum_{l=1}^{N}\pi_{il}\sum_{j=1}^{n} 2\int_0^{x_j(t)} \frac{s\left[p_{jl}\alpha_{ji}(s) - p_{ji}\alpha_{jl}(s)\right]}{\alpha_{ji}^2(s)}\mathrm{d}s\\
&\quad+2\left\langle \iota_i(t), P_i x(t)\right\rangle + \mathrm{trace}\left\langle \sigma_i(t), \alpha_i^{-1}(x(t))P_i\sigma_i(t)\right\rangle,
\end{aligned} \tag{11.19}$$

$$\begin{aligned}
&\pounds\left\{\sum_{j=1}^{n} 2h_j\int_0^{x_j(t)} \frac{\beta_{ji}(s) - \underline{\beta}_{ji}s}{\alpha_{ji}(s)}\mathrm{d}s\right\}\\
&= \lim_{\Delta\to 0^+}\frac{1}{\Delta}\mathbb{E}\left\{\sum_{j=1}^{n} 2h_j\left[-\int_0^{x_j(t)} \frac{\beta_{ji}(s) - \underline{\beta}_{ji}s}{\alpha_{ji}(s)}\mathrm{d}s\right.\right.\\
&\quad+\int_0^{x_j(t+\Delta)} \frac{\sum_{l=1}^{N}[\pi_{il}\Delta + o(\Delta)]\beta_{jl}(s) + \beta_{ji}(s)}{\sum_{l=1}^{N}[\pi_{il}\Delta + o(\Delta)]\alpha_{jl}(s) + \alpha_{ji}(s)}\mathrm{d}s\\
&\quad\left.\left.-\int_0^{x_j(t+\Delta)} \frac{\sum_{l=1}^{N}[\pi_{il}\Delta + o(\Delta)]\underline{\beta}_{jl}s + \underline{\beta}_{ji}s}{\sum_{l=1}^{N}[\pi_{il}\Delta + o(\Delta)]\alpha_{jl}(s) + \alpha_{ji}(s)}\mathrm{d}s\right]\right\}
\end{aligned}$$

$$
\begin{aligned}
&= \lim_{\Delta\to 0^+} \frac{1}{\Delta}\mathbb{E}\left\{\sum_{j=1}^{n} 2h_j\Bigg[-\int_0^{x_j(t)} \frac{\beta_{ji}(s)-\underline{\beta}_{ji}s}{\alpha_{ji}(s)}\mathrm{d}s\right.\\
&\quad + \int_0^{x_j(t+\Delta)} \frac{\sum_{l=1}^{N}[\pi_{il}\Delta+o(\Delta)]\beta_{jl}(s)}{\sum_{l=1}^{N}[\pi_{il}\Delta+o(\Delta)]\alpha_{jl}(s)+\alpha_{ji}(s)}\mathrm{d}s\\
&\quad - \int_0^{x_j(t+\Delta)} \frac{\sum_{l=1}^{N}[\pi_{il}\Delta+o(\Delta)]\underline{\beta}_{jl}s}{\sum_{l=1}^{N}[\pi_{il}\Delta+o(\Delta)]\alpha_{jl}(s)+\alpha_{ji}(s)}\mathrm{d}s\\
&\quad \left. + \int_0^{x_j(t+\Delta)} \frac{\beta_{ji}(s)-\underline{\beta}_{ji}s}{\sum_{l=1}^{N}[\pi_{il}\Delta+o(\Delta)]\alpha_{jl}(s)+\alpha_{ji}(s)}\mathrm{d}s\Bigg]\right\}\\
&= \sum_{l=1}^{N}\pi_{il}\sum_{j=1}^{n} 2h_j\int_0^{x_j(t)} \frac{\left[\beta_{ji}(s)-\underline{\beta}_{ji}s\right]\left[\alpha_{ji}(s)-\alpha_{jl}(s)\right]}{\alpha_{ji}^2(s)}\mathrm{d}s\\
&\quad + 2\left\langle \iota_i(t), H\left(\beta_i(x(t))-\Gamma_i x(t)\right)\right\rangle\\
&\quad + \mathrm{trace}\left\langle \sigma_i(t), \alpha_i^{-1}(x(t))H(\Pi_i-\Gamma_i)\sigma_i(t)\right\rangle,
\end{aligned}
\tag{11.20}
$$

$$
\begin{aligned}
&\pounds\left\{\sum_{j=1}^{n} 2k_j\int_0^{x_j(t)} \frac{f_j(s)-\varrho_j^{-}s}{\alpha_{ji}(s)}\mathrm{d}s\right\}\\
&= \sum_{l=1}^{N}\pi_{il}\sum_{j=1}^{n} 2k_j\int_0^{x_j(t)} \frac{\left[f_j(s)-\varrho_j^{-}s\right]\left[\alpha_{ji}(s)-\alpha_{jl}(s)\right]}{\alpha_{ji}^2(s)}\mathrm{d}s\\
&\quad + 2\left\langle \iota_i(t), K\left(f(x(t))-\Sigma x(t)\right)\right\rangle\\
&\quad + \mathrm{trace}\left\langle \sigma_i(t), \alpha_i^{-1}(x(t))K(\bar{\Sigma}-\Sigma)\sigma_i(t)\right\rangle,
\end{aligned}
\tag{11.21}
$$

where $\alpha_i^{-1}(x(t)) = \mathrm{diag}\left\{\alpha_{1i}^{-1}(x_1(t)),\ \ldots,\ \alpha_{ni}^{-1}(x_n(t))\right\}$.

According to the definition of ρ_{il} and Assumptions 11.3–11.5 we have that

$$
\sum_{l=1}^{N}\pi_{il}\sum_{j=1}^{n} 2p_{jl}\int_0^{x_j(t)} \frac{s}{\alpha_{ji}(s)}\mathrm{d}s \le \left\langle x(t), \sum_{l=1}^{N}\pi_{il}\rho_{il}^{-1}P_l x(t)\right\rangle,
\tag{11.22}
$$

$$
\begin{aligned}
&-\sum_{l=1}^{N}\pi_{il}\sum_{j=1}^{n} 2p_{ji}\int_0^{x_j(t)} \frac{s\alpha_{jl}(s)}{\alpha_{ji}^2(s)}\mathrm{d}s\\
&\le -\pi_{ii}\sum_{j=1}^{n} 2p_{ji}\int_0^{x_j(t)} \frac{s}{\alpha_{ji}(s)}\mathrm{d}s\\
&\le \left\langle x(t), \sum_{l=1}^{N}\bar{\pi}_{il}\underline{\alpha}_i^{-1}P_i x(t)\right\rangle,
\end{aligned}
\tag{11.23}
$$

$$\sum_{l=1}^{N}\pi_{il}\sum_{j=1}^{n}2h_j\int_0^{x_j(t)}\frac{\beta_{ji}(s)-\underline{\beta}_{ji}s}{\alpha_{ji}(s)}\mathrm{d}s$$

$$\leq\sum_{l=1}^{N}\bar{\pi}_{il}\sum_{j=1}^{n}2h_j\int_0^{x_j(t)}\frac{\beta_{ji}(s)-\underline{\beta}_{ji}s}{\alpha_{ji}(s)}\mathrm{d}s$$

$$\leq\left\langle x(t),\,H\sum_{l=1}^{N}\bar{\pi}_{il}\underline{\alpha}_i^{-1}(\Pi_i-\Gamma_i)x(t)\right\rangle, \tag{11.24}$$

$$-\sum_{l=1}^{N}\pi_{il}\sum_{j=1}^{n}2h_j\int_0^{x_j(t)}\frac{\left[\beta_{ji}(s)-\underline{\beta}_{ji}s\right]\alpha_{jl}(s)}{\alpha_{ji}^2(s)}\mathrm{d}s$$

$$\leq-\pi_{ii}\sum_{j=1}^{n}2h_j\int_0^{x_j(t)}\frac{\beta_{ji}(s)-\underline{\beta}_{ji}s}{\alpha_{ji}(s)}\mathrm{d}s$$

$$\leq\left\langle x(t),\,H\sum_{l=1}^{N}\bar{\pi}_{il}\underline{\alpha}_i^{-1}(\Pi_i-\Gamma_i)x(t)\right\rangle, \tag{11.25}$$

$$\sum_{l=1}^{N}\pi_{il}\sum_{j=1}^{n}2k_j\int_0^{x_j(t)}\frac{f_j(s)-\varrho_j^-s}{\alpha_{ji}(s)}\mathrm{d}s$$

$$\leq\sum_{l=1}^{N}\bar{\pi}_{il}\sum_{j=1}^{n}2k_j\int_0^{x_j(t)}\frac{f_j(s)-\varrho_j^-s}{\alpha_{ji}(s)}\mathrm{d}s$$

$$\leq\left\langle x(t),\,K\sum_{l=1}^{N}\bar{\pi}_{il}\underline{\alpha}_i^{-1}(\bar{\Sigma}-\Sigma)x(t)\right\rangle, \tag{11.26}$$

$$-\sum_{l=1}^{N}\pi_{il}\sum_{j=1}^{n}2k_j\int_0^{x_j(t)}\frac{\left[f_j(s)-\varrho_j^-s\right]\alpha_{jl}(s)}{\alpha_{ji}^2(s)}\mathrm{d}s$$

$$\leq-\pi_{ii}\sum_{j=1}^{n}2k_j\int_0^{x_j(t)}\frac{f_j(s)-\varrho_j^-s}{\alpha_{ji}(s)}\mathrm{d}s$$

$$\leq\left\langle x(t),\,K\sum_{l=1}^{N}\bar{\pi}_{il}\underline{\alpha}_i^{-1}(\bar{\Sigma}-\Sigma)x(t)\right\rangle. \tag{11.27}$$

Using the well-known Itô's differential formula [41, 44], we obtain

$$\begin{aligned}\pounds V_{1i}(x_t,t)\leq{}&2\big\langle\iota_i(t),\;P_ix(t)+H\big[\beta_i(x(t))-\Gamma_ix(t)\big]+K\left(f(x(t))-\Sigma x(t)\right)\big\rangle\\&+\mathrm{trace}\big\langle\sigma_i(t),\;\alpha_i^{-1}(x(t))\big[P_i+H(\Pi_i-\Gamma_i)+K(\bar{\Sigma}-\Sigma)\big]\sigma_i(t)\big\rangle\end{aligned}$$

$$
\begin{aligned}
&+\sum_{l=1}^{N}\pi_{il}\rho_{il}^{-1}\big\langle x(t),\, P_l x(t)\big\rangle \\
&+\sum_{l=1}^{N}\bar{\pi}_{il}\Big\langle x(t),\ \underline{\alpha}_i^{-1}\big[P_i+2H\ \times(\Pi_i-\Gamma_i)+2K(\bar{\Sigma}-\Sigma)\big]x(t)\Big\rangle,
\end{aligned}
\tag{11.28}
$$

$$
\begin{aligned}
\pounds V_{2i}(x_t,t) = {}& \langle x(t),\, Q_{1i}x(t)\rangle + \langle f(x(t)),\, Q_{2i}f(x(t))\rangle \\
& -(1-\dot{\tau}_i(t))\big\{\langle x(t-\tau_i(t)),\, Q_{1i}x(t-\tau_i(t))\rangle \\
& +\langle f(x(t-\tau_i(t))),\, Q_{2i}f(x(t-\tau_i(t)))\rangle\big\} \\
& +\langle x(t),\, Q_{3i}x(t)\rangle - \langle x(t-\bar{\tau}_i),\, Q_{3i}x(t-\bar{\tau}_i)\rangle \\
& +\sum_{j=1}^{N}\pi_{ij}\Big[\int_{t-\tau_i(t)}^{t}\big\langle x(s),\, Q_{1j}x(s)\big\rangle\mathrm{d}s \\
& +\int_{t-\tau_i(t)}^{t}\big\langle f(x(s)),\, Q_{2j}f(x(s))\big\rangle\mathrm{d}s + \int_{t-\bar{\tau}_i}^{t}\big\langle x(s),\, Q_{3j}x(s)\big\rangle\mathrm{d}s\Big] \\
& +\sum_{j=1}^{N}\pi_{ij}\tau_j(t)\big[\langle x(t-\tau_i(t)),\, Q_{1i}x(t-\tau_i(t))\rangle \\
& +\langle f(x(t-\tau_i(t))),\, Q_{2i}f(x(t-\tau_i(t)))\rangle\big] \\
& +\sum_{j=1}^{N}\pi_{ij}\bar{\tau}_j\,\langle x(t-\bar{\tau}_i),\, Q_{3i}x(t-\bar{\tau}_i)\rangle,
\end{aligned}
\tag{11.29}
$$

$$
\begin{aligned}
\pounds V_{3i}(x_t,t) = {}& \bar{\tau}_i\,\langle\vartheta_i(t),\, Q_{4i}\vartheta_i(t)\rangle - \int_{t-\bar{\tau}_i}^{t}\langle\vartheta_i(s),\, Q_{4i}\vartheta_i(s)\rangle\,\mathrm{d}s \\
& +\upsilon_i(t)\,\langle g(x(t)),\, R_i g(x(t))\rangle - \int_{t-\upsilon_i(t)}^{t}\langle g(x(t)),\, R_i g(x(t))\rangle\,\mathrm{d}s \\
& +\bar{\tau}_i\,\langle\sigma_i(t),\, Z_i\sigma_i(t)\rangle - \int_{t-\bar{\tau}_i}^{t}\langle\sigma_i(t),\, Z_i\sigma_i(t)\rangle\,\mathrm{d}s \\
& +\sum_{j=1}^{N}\pi_{ij}\Big[\int_{-\bar{\tau}_i}^{0}\int_{t+\theta}^{t}\big\langle\vartheta_i(s),\, Q_{4j}\vartheta_j(s)\big\rangle\mathrm{d}s\mathrm{d}\theta \\
& +\int_{-\upsilon_i(t)}^{0}\int_{t+\theta}^{t}\big\langle g(x(s)),\, R_j g(x(s))\big\rangle\mathrm{d}s\mathrm{d}\theta \\
& +\int_{-\bar{\tau}_i}^{0}\int_{t+\theta}^{t}\big\langle\sigma_i(s),\, Z_j\sigma_j(s)\big\rangle\mathrm{d}s\mathrm{d}\theta\Big] \\
& +\sum_{j=1}^{N}\pi_{ij}\Big[\bar{\tau}_j\int_{t-\bar{\tau}_i}^{t}\langle\vartheta_i(s),\, Q_{4i}\vartheta_i(s)\rangle\,\mathrm{d}s
\end{aligned}
$$

$$+\,\upsilon_j(t)\int_{t-\upsilon_i(t)}^{t}\big\langle g(x(s)),\,R_j g(x(s))\big\rangle\,\mathrm{d}s$$
$$+\,\bar{\tau}_j\int_{t-\bar{\tau}_i}^{t}\big\langle\sigma_i(s),\,Z_j\sigma_i(s)\big\rangle\,\mathrm{d}s\Bigg], \tag{11.30}$$

$$\begin{aligned}\pounds V_{4i}(x_t,t) = {}&\bar{\tau}\,\langle x(t),(S_1+S_3)x(t)\rangle+\bar{\tau}\,\langle f(x(t)),\,S_2 f(x(t))\rangle\\&-\int_{t-\bar{\tau}}^{t}\{\langle x(s),(S_1+S_3)x(s)\rangle+\langle f(x(s)),\,S_2 f(x(s))\rangle\}\,\mathrm{d}s,\end{aligned} \tag{11.31}$$

$$\begin{aligned}\pounds V_{5i}(x_t,t) = {}&\frac{\bar{\tau}^2}{2}\{\,\langle\vartheta_i(t),\,S_4\vartheta_i(t)\rangle+\langle\sigma_i(t),\,S_6\sigma_i(t)\rangle\,\}\\&-\int_{-\bar{\tau}}^{0}\int_{t+\theta}^{t}\langle\vartheta_i(s),\,S_4\vartheta_i(s)\rangle\,\mathrm{d}s\mathrm{d}\theta\\&+\frac{\bar{\upsilon}^2}{2}\langle g(x(t)),\,S_5 g(x(t))\rangle-\int_{-\bar{\upsilon}}^{0}\int_{t+\theta}^{t}\langle g(x(s)),\,S_5 g(x(s))\rangle\,\mathrm{d}s\mathrm{d}\theta\\&-\int_{-\bar{\tau}}^{0}\int_{t+\theta}^{t}\langle\sigma_i(s),\,S_6\sigma_i(s)\rangle\,\mathrm{d}s\mathrm{d}\theta,\end{aligned} \tag{11.32}$$

$$\begin{aligned}\pounds V_{6i}(x_t,t) = {}&\bar{\upsilon}\,\langle g(x(t)),\,S_7 g(x(t))\rangle-\int_{t-\bar{\upsilon}}^{t}\langle g(x(s)),\,S_7 g(x(s))\rangle\,\mathrm{d}s\\&+\bar{\tau}\,\langle\sigma_i(t),\,S_8\sigma_i(t)\rangle+\bar{\tau}\,\langle\vartheta_i(t),\,S_9\vartheta_i(t)\rangle\\&-\int_{t-\bar{\tau}}^{t}\langle\sigma_i(s),\,S_8\sigma_i(s)\rangle\,\mathrm{d}s-\int_{t-\bar{\tau}}^{t}\langle\vartheta_i(s),\,S_9\vartheta_i(s)\rangle\,\mathrm{d}s.\end{aligned} \tag{11.33}$$

Based on Assumption 11.4, we obtain that

$$-x^T(t)P_i\beta_i(x(t))\le -x^T(t)P_i\Gamma_i x(t). \tag{11.34}$$

From Lemma 11.7, it follows that

$$\begin{aligned}&-\int_{t-\upsilon_i(t)}^{t}\langle g(x(s)),\,R_i g(x(s))\rangle\,\mathrm{d}s\\&\le-\frac{1}{\bar{\upsilon}_i}\Big\langle\int_{t-\upsilon_i(t)}^{t}g(x(s))\mathrm{d}s,\,R_i\int_{t-\upsilon_i(t)}^{t}g(x(s))\mathrm{d}s\Big\rangle.\end{aligned} \tag{11.35}$$

For simplicity, we denote

$$\varsigma_{1i}(t)=\int_{t-\bar{\tau}}^{t-\tau_i(t)}\vartheta_i(s)\mathrm{d}s,\quad \varsigma_{2i}(t)=\int_{t-\tau_i(t)}^{t}\vartheta_i(s)\mathrm{d}s.$$

When $0 < \tau_i(t) < \bar{\tau}_i$, from Lemma 11.8 with $\nu = 1$, $\mu = 3$, one can obtain that

$$\begin{aligned}
&-\int_{t-\bar{\tau}_i}^{t} \langle \vartheta_i(s), Q_{4i}\vartheta_i(s)\rangle \,\mathrm{d}s \\
&= -\int_{t-\bar{\tau}_i}^{t-\tau_i(t)} \langle \vartheta_i(s), Q_{4i}\vartheta_i(s)\rangle \,\mathrm{d}s \\
&\quad -\int_{t-\tau_i(t)}^{t} \langle \vartheta_i(s), Q_{4i}\vartheta_i(s)\rangle \,\mathrm{d}s \\
&\le -\frac{1}{\bar{\tau}_i - \tau_i(t)} \langle \varsigma_{1i}(t), Q_{4i}\varsigma_{1i}(t)\rangle - \frac{1}{\tau_i(t)} \langle \varsigma_{2i}(t), Q_{4i}\varsigma_{2i}(t)\rangle \\
&\le \frac{1}{\bar{\tau}_i} \max\big\{ -\langle \varsigma_{1i}(t), Q_{4i}\varsigma_{1i}(t)\rangle - 3\langle \varsigma_{2i}(t), Q_{4i}\varsigma_{2i}(t)\rangle, \\
&\qquad -3\langle \varsigma_{1i}(t), Q_{4i}\varsigma_{1i}(t)\rangle - \langle \varsigma_{2i}(t), Q_{4i}\varsigma_{2i}(t)\rangle \big\}. \qquad (11.36)
\end{aligned}$$

Obviously, from Lemma 11.7, inequality (11.36) holds when $\tau_i(t) = 0$ or $\tau_i(t) = \bar{\tau}_i$. Therefore, inequality (11.36) holds for any t with $0 \le \tau_i(t) \le \bar{\tau}_i$.

On the other hand, by the *Leibniz-Newton formula*, we get

$$x(t) - x(t-\tau_i(t)) - \int_{t-\tau_i(t)}^{t} \vartheta_i(s)\mathrm{d}s - \int_{t-\tau_i(t)}^{t} \sigma_i(s)\mathrm{d}\omega(s) = 0, \qquad (11.37)$$

$$x(t-\tau_i(t)) - x(t-\bar{\tau}_i) - \int_{t-\bar{\tau}_i}^{t-\tau_i(t)} \vartheta_i(s)\mathrm{d}s - \int_{t-\bar{\tau}_i}^{t-\tau_i(t)} \sigma_i(s)\mathrm{d}\omega(s) = 0. \qquad (11.38)$$

It is easy to see that the following equality holds for any positive diagonal matrices G_i with compatible dimensions

$$0 = -2\langle G_i\vartheta_i(t), \vartheta_i(t) - \alpha_i(x(t))\iota_i(t)\rangle. \qquad (11.39)$$

Considering that the feedback stabilizing control law being defined by $u_i(t) = D_i^T X_i x(t)$, if we denote $y_i(t) = X_i x(t)$, then for any symmetric nonnegative definite matrices M_i, we have

$$0 = -2\langle M_i y_i(t), y_i(t) - X_i x(t)\rangle \ (i = 1, 2, \ldots, N). \qquad (11.40)$$

Noticing that the following equality holds

$$\begin{aligned}
-\int_{t-\bar{\tau}_i}^{t} \langle \sigma_i(s), Z_i\sigma_i(s)\rangle \,\mathrm{d}s &= -\int_{t-\bar{\tau}_i}^{t-\tau_i(t)} \langle \sigma_i(s), Z_i\sigma_i(s)\rangle \,\mathrm{d}s \\
&\quad -\int_{t-\tau_i(t)}^{t} \langle \sigma_i(s), Z_i\sigma_i(s)\rangle \,\mathrm{d}s. \qquad (11.41)
\end{aligned}$$

From [14], we have

$$\mathbb{E}\left\{\int_{t-\tau_i(t)}^{t} \langle \sigma_i(s), Z_i\sigma_i(s)\rangle\, \mathrm{d}s\right\}$$
$$=\mathbb{E}\left\langle \int_{t-\tau_i(t)}^{t} \sigma_i(s)\mathrm{d}\omega(s),\ Z_i \int_{t-\tau_i(t)}^{t} \sigma_i(s)\mathrm{d}\omega(s)\right\rangle, \tag{11.42}$$
$$\mathbb{E}\left\{\int_{t-\bar{\tau}_i}^{t-\tau_i(t)} \langle \sigma_i(s), Z_i\sigma_i(s)\rangle\, \mathrm{d}s\right\}$$
$$=\mathbb{E}\left\langle \int_{t-\bar{\tau}_i}^{t-\tau_i(t)} \sigma_i(s)\mathrm{d}\omega(s),\ Z_i \int_{t-\bar{\tau}_i}^{t-\tau_i(t)} \sigma_i(s)\mathrm{d}\omega(s)\right\rangle. \tag{11.43}$$

By (11.4)–(11.9) and (11.14)–(11.43), we obtain

$$\frac{\mathrm{d}\mathbb{E}[V(x(t),t,i)]}{\mathrm{d}t}$$
$$\leq \mathbb{E}\max\left\{\langle\zeta_i(t), (\overline{\Omega}_i+\widetilde{\Omega}_i+\mathcal{E}^T\overline{Z}_i\mathcal{E})\zeta_i(t)\rangle, \langle\zeta_i(t), (\overline{\Omega}_i+\widehat{\Omega}_i+\mathcal{E}^T\overline{Z}_i\mathcal{E})\zeta_i(t)\rangle\right\}, \tag{11.44}$$

where

$$\zeta_i(t)=\mathrm{col}\Big\{x(t)\quad x(t-\tau_i(t))\quad f(x(t))$$
$$f(x(t-\tau_i(t)))\quad g(x(t))\quad \int_{t-\upsilon_i(t)}^{t} g(x(s))\mathrm{d}s$$
$$\vartheta_i(t)\quad y_i(t)\quad x(t-\bar{\tau}_i)\quad \beta_i(x(t))$$
$$\int_{t-\tau_i(t)}^{t}\sigma_i(s)\mathrm{d}\omega(s)\quad \int_{t-\bar{\tau}_i}^{t-\tau_i(t)}\sigma_i(s)\mathrm{d}\omega(s)\Big\}.$$

Next, we prove that the *error system* is exponentially stable in mean square. For convenience, we define

$$\lambda_p=\min_{i\in\wp}\{\lambda_{\min}(P_i)\},$$

$$\lambda_M=\min_{i\in\wp}\{\lambda_{\min}(-\overline{\Omega}_i-\widetilde{\Omega}_i-\mathcal{E}^T\overline{Z}_i\mathcal{E}),\ \lambda_{\min}(-\overline{\Omega}_i-\widehat{\Omega}_i-\mathcal{E}^T\overline{Z}_i\mathcal{E})\}.$$

From (11.12) and (11.13) and the well-known *Schur complements*, it can be easily seen that $\lambda_M > 0$. Furthermore, from (11.44) we have that

$$\frac{\mathrm{d}\mathbb{E}[V(x(t),t,i)]}{\mathrm{d}t}\leq -\lambda_M\mathbb{E}||\zeta_i(t)||^2\leq -\lambda_M\mathbb{E}||x(t)||^2. \tag{11.45}$$

Similar to [45], from (11.18) and the definition of $\vartheta_i(t)$, there exist positive scalars ε_1 and ε_2 such that

$$\mathbb{E}[V(x(t),t,i)] \le \varepsilon_1 \mathbb{E}||\zeta(t)||^2 + \varepsilon_2 \mathbb{E}\int_{t-\bar{\tau}_i}^{t} ||x(s)||^2 \mathrm{d}s.$$

To prove the mean square exponential stability, we modify the *Lyapunov function* candidate (11.18) as $\bar{V}(x(t),t,i) = e^{rt}V(x(t),t,i)$, where r is chosen such that $r(\varepsilon_1 + \bar{\tau}\varepsilon_2 e^{r\bar{\tau}}) \le \lambda_M$.

Then, we have

$$\mathbb{E}[\bar{V}(x(t),t,i)] \ge \lambda_p \mathbb{E}||x(t)||^2.$$

Furthermore, by the Dynkin's formula [14], for any $\eta(t) = i \in \wp, t > 0$, we obtain that

$$\begin{aligned}
\mathbb{E}[\bar{V}(x(t),t,i)] =& \mathbb{E}[\bar{V}(x(0),0,\eta(0))] + \mathbb{E}\int_0^t e^{rs}\left[r\bar{V}(x(s),s,i) + \pounds\bar{V}(x(s),s,i)\right]\mathrm{d}s \\
\le& (\varepsilon_1 + \bar{\tau}\varepsilon_2) \sup_{-\varpi\le s\le 0} \mathbb{E}||x(t)||^2 + r\varepsilon_1 \int_0^t e^{rs}\mathbb{E}||x(s)||^2 \mathrm{d}s \\
&+ r\varepsilon_2 \mathbb{E}\int_0^t e^{rs}\int_{s-\bar{\tau}}^{s} ||x(\theta)||^2 \mathrm{d}\theta \mathrm{d}s - \lambda_M \int_0^t e^{rs}||x(s)||^2 \mathrm{d}s.
\end{aligned}$$

By changing the integration sequence, we get

$$\begin{aligned}
&\int_0^t e^{rs}\int_{s-\bar{\tau}}^{s} ||x(\theta)||^2 \mathrm{d}\theta \mathrm{d}s \\
\le& \int_{-\bar{\tau}}^{0} e^{rs}\int_0^{\theta+\bar{\tau}} ||x(\theta)||^2 \mathrm{d}s\mathrm{d}\theta + \int_0^t e^{rs}\int_0^{\theta+\bar{\tau}} ||x(\theta)||^2 \mathrm{d}s\mathrm{d}\theta \\
\le& \int_{-\bar{\tau}}^{0} (\theta+\bar{\tau})e^{r(\theta+\bar{\tau})}||x(\theta)||^2 \mathrm{d}\theta + \bar{\tau}\int_0^t e^{r(\theta+\bar{\tau})}||x(\theta)||^2 \mathrm{d}\theta \\
\le& \bar{\tau}e^{r\bar{\tau}}\left\{\sup_{-\varpi\le s\le 0} ||x(s)||^2 + \int_0^t e^{r\theta}||x(\theta)||^2 \mathrm{d}\theta\right\}.
\end{aligned}$$

Therefore we have

$$\mathbb{E}||x(t)||^2 \le \epsilon e^{-rt} \sup_{-\varpi\le s\le 0} ||x(s)||^2,$$

or

$$\lim_{t\to\infty} \sup \frac{1}{t} \log(\mathbb{E}||x(t)||^2) \le -r,$$

where $\epsilon = \lambda_p^{-1}(\varepsilon_1 + \bar{\tau}\varepsilon_2 + r\bar{\tau}^2\varepsilon_2 e^{r\bar{\tau}})$.

Consequently, we prove that the error system (11.1) is exponentially stable in mean square. So the system (11.1) is *r-exponentially stabilizable* in the mean square. This ends the proof.

Remark 11.11 The Lyapunov functional (11.18) of this chapter fully uses the information about the amplification function and the mode-dependent time-varying delays, but [15, 20] only use the information about delays when constructing their Lyapunov functionals. Therefore the Lyapunov functional is more general than those in [15, 20], and the stability criteria in this chapter may be less conservativeness.

Remark 11.12 When one of the time-varying delays $\dot{\tau}_i(t)$ is not differentiable or unknown, the result in Theorem 11.10 is no longer applicable. For this case, by setting $Q_{1i} = Q_{2i} = 0$ in Theorem 11.10, one can obtain a result of the mean square exponential stability of system (11.1).

If there are no stochastic disturbances, that is $E_j(\eta_t) = 0$ $(j = 1, \ldots, 5)$, then the neural network (11.1) is simplified to

$$\begin{aligned}\dot{x}(t) = &- \alpha(x(t), \eta_t)\Big[\beta(x(t), \eta_t) - A(\eta_t)f(x(t)) - B(\eta_t)f(x(t - \tau(t, \eta_t)))\\ &- C(\eta_t)\int_{t-\upsilon(t,\eta_t)}^{t} g(x(s))\mathrm{d}s - D(\eta_t)u(t, \eta_t)\Big].\end{aligned} \tag{11.46}$$

For system (11.46), by setting $Z_i = S_6 = S_8 = 0$ in Theorem 11.10 and deleting $\int_{t-\tau_i(t)}^{t} \sigma_i(s)\mathrm{d}\omega(s)$, $\int_{t-\bar{\tau}_i}^{t-\tau_i(t)} \sigma_i(s)\mathrm{d}\omega(s)$ from $\zeta_i(t)$, we can get the following result of the mean square exponential stability.

Corollary 11.13 *Given* $r > 0$. *For any given scalars* $\bar{\tau}_i > 0$, $\bar{\upsilon}_i > 0$, $\upsilon_i' < 1$, *considering the system (11.46) satisfying Assumptions 11.3–11.5 and* $\dot{\tau}_i(t) \leq \tau_i'$, $\dot{\upsilon}_i(t) \leq \upsilon_i'$, *the system (11.46) is globally r-exponentially stabilizable if there exist symmetric positive definite matrices* $P_i \in \mathbb{R}^{n\times n}$, *symmetric nonnegative definite matrices* Q_{li}, R_i, M_i, S_l $(l = 1, \ldots, 4, l = 1, \ldots, 5, 7, 9)$, *positive diagonal matrices* G_i, U_i, T_i, W_i, H, K, *and real matrices* X_i *such that (11.4), (11.5), (11.7), (11.9) and the following inequalities hold,*

$$\begin{bmatrix} \underline{\Omega}_i + \check{\Omega}_i & \mathfrak{E}^T \\ \mathfrak{E} & \widetilde{Z}_i \end{bmatrix} < 0, \tag{11.47}$$

$$\begin{bmatrix} \underline{\Omega}_i + \grave{\Omega}_i & \mathfrak{E}^T \\ \mathfrak{E} & \widetilde{Z}_i \end{bmatrix} < 0, \tag{11.48}$$

where

$$\underline{\Omega}_i = \begin{bmatrix} \Omega_{1i} & \Omega_{2i} & \Omega_{4i} \\ * & \Omega_{3i} & \Omega_{5i} \\ * & * & \Omega_{6i} \end{bmatrix},$$

$$\check{\Omega}_i = -\frac{2}{\bar{\tau}_i}\mathcal{I}^T Q_{4i}\mathcal{I}, \quad \grave{\Omega}_i = -\frac{2}{\bar{\tau}_i}\mathbb{J}^T Q_{4i}\mathbb{J},$$
$$\mathcal{I} = \begin{bmatrix} 0 & -I & 0 & 0 & 0 & 0 & 0 & 0 & I & 0 \end{bmatrix},$$
$$\mathbb{J} = \begin{bmatrix} -I & I & 0 & 0 & 0 & 0 & 0 & 0 & 0 & 0 \end{bmatrix},$$
$$\mathfrak{E} = [\, E_{1i} \;\; E_{2i} \;\; E_{3i} \;\; E_{4i} \;\; 0 \;\; E_{5i} \;\; 0 \;\; 0 \;\; 0 \;\; 0 \,],$$

$i = 1, \ldots, N$, and other parameters are defined in Theorem 11.10. Furthermore, the feedback stabilizing control law is defined by $u_i(t) = D_i^T X_i x(t)$.

11.4 Illustrative Examples

In this section, we provide three numerical examples to demonstrate the feasibility of our delay-dependent stabilization criteria.

Example 11.14 Consider system (11.1) with $N = 2$,

$$\begin{aligned}
\alpha_{ji}(x_j(t)) &= 0.4\sin(x_j(t)) + 0.8,\\
\beta_{ji}(x_j(t)) &= 7.5x_j(t) + 0.5\sin(x_j(t)),\\
f_j(x_j(t)) &= g_j(x_j(t)) = \tanh(x_j(t)), \quad j = 1, 2,\\
\tau_i(t) &= 0.2\sin(t) + 0.2,\\
\upsilon_i(t) &= 0.3\sin(t) + 0.3, \quad i = 1, 2,
\end{aligned}$$

and

$$\begin{aligned}
A_1 &= \begin{bmatrix} 1 & -0.01 \\ 0.1 & 1.2 \end{bmatrix}, \quad A_2 = \begin{bmatrix} 1.1 & -0.01 \\ 0.1 & 1.2 \end{bmatrix},\\
B_1 &= \begin{bmatrix} 5.2 & 1.2 \\ 1.12 & 2.3 \end{bmatrix}, \quad B_2 = \begin{bmatrix} 5.3 & 1.1 \\ 1.11 & 2.3 \end{bmatrix},\\
C_1 &= \begin{bmatrix} 1.2 & 0.11 \\ 0.1 & 1.22 \end{bmatrix}, \quad C_2 = \begin{bmatrix} 1.1 & 0.12 \\ 0.1 & 1.22 \end{bmatrix},\\
D_1 &= D_2 = 0, \quad E_{11} = E_{21} = 0.5I,\\
E_{12} &= E_{22} = 0.4I, \quad E_{l1} = E_{l2} = 0, \; l = 3, 4, 5;\\
\aleph &= \begin{bmatrix} -0.8 & 0.8 \\ 0.3 & -0.3 \end{bmatrix}.
\end{aligned}$$

For this system without external controller, Fig. 11.1a shows the results of time response of $x_1(t)$ and $x_2(t)$.

However, if we set

$$D_1 = \begin{bmatrix} 4 \\ 2.1 \end{bmatrix}, \quad D_2 = \begin{bmatrix} 4 \\ 2 \end{bmatrix},$$

(a)

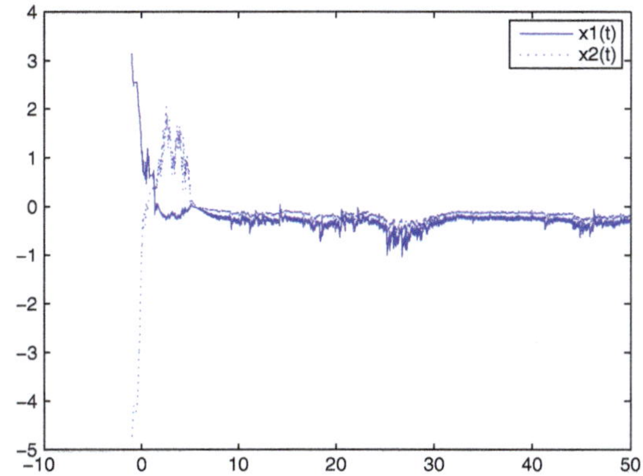

(b)

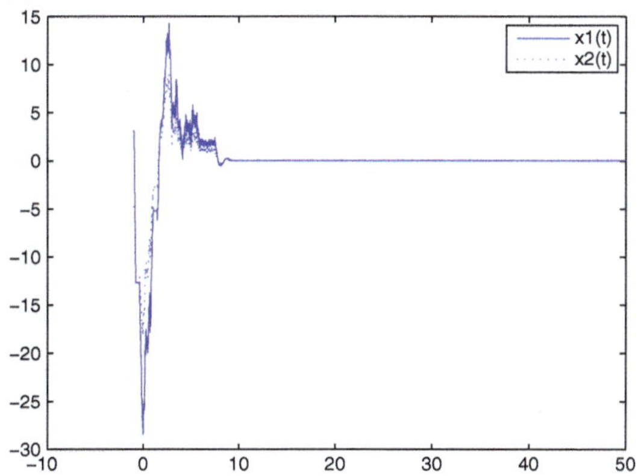

Fig. 11.1 **a** Time response of $x_1(t)$ and $x_2(t)$ without external controller in Example 11.14, **b** Time response of $x_1(t)$ and $x_2(t)$ with external controller $u_1(t)$, $u_2(t)$ in Example 11.14

it is easy to see that Assumptions 11.3–11.5 are satisfied with $\underline{\alpha}_i = 0.4, \bar{\alpha}_i = 1.2, \Pi_i = 8I, \Gamma_i = 7I, \bar{\Sigma} = I, \Sigma = F_1 = F_3 = 0, F_2 = F_4 = 0.5I$, and $\bar{\tau} = \bar{\tau}_i = 0.4, \bar{\upsilon} = \bar{\upsilon}_i = 0.6,\ i = 1, 2$. Using the Matlab LMI Toolbox, the LMIs (11.4)–(11.11) are feasible and the feedback control is

$$u_1(t) = [\ -15.9876 \quad 28.4673\ ]x(t),$$
$$u_2(t) = [\ -9.8136 \quad 17.5622\ ]x(t).$$

The simulation of the solution is given in Fig. 11.1b for $t \in [-0.65, 200]$. It is clear that both $x_1(t)$ and $x_2(t)$ converge exponentially to zeros.

Example 11.15 Consider system (11.46) with $N = 2$,

$$B_1 = \begin{bmatrix} 6.2 & 1.2 \\ 1.12 & 0.3 \end{bmatrix}, \quad B_2 = \begin{bmatrix} 6.3 & 1.1 \\ 1.11 & 0.3 \end{bmatrix},$$

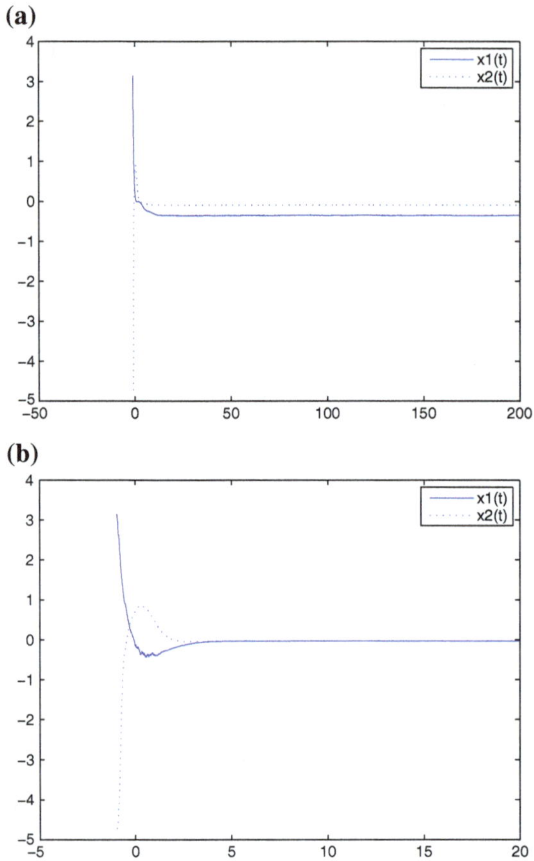

Fig. 11.2 **a** Time response of $x_1(t)$ and $x_2(t)$ without external controller in Example 11.15, **b** Time response of $x_1(t)$ and $x_2(t)$ with external controller $u_1(t)$, $u_2(t)$ in Example 11.15

and other parameters are defined in Example 11.14.

For this system without external controller, Fig. 11.2a shows the results of time response of $x_1(t)$ and $x_2(t)$.

However, if we set $D_1 = D_2 = [\ 4 \ \ 0\ ]^T$, it is easy to see that Assumptions 1-3 are satisfied. Using the Matlab LMI Toolbox, the LMIs (11.4), (11.5), (11.7), (11.9), (11.47) and (11.48) are feasible and the feedback control is

$$u_1(t) = [\ -0.9144 \quad -1.20177\ ]x(t),$$
$$u_2(t) = [\ -1.0149 \quad -0.1481\ ]x(t).$$

The simulation of the solution is given in Fig. 11.2b for $t \in [-0.65, 200]$. It is clear that both $x_1(t)$ and $x_2(t)$ converge exponentially to zeros.

Example 11.16 Consider system (11.46) with $N = 1$,

$$\begin{aligned}\alpha_{j1}(x_j(t)) &= 1,\ \beta_{j1}(x_j(t)) = 8x_j(t),\\ f_j(x_j(t)) &= g_j(x_j(t)) = \tanh(x_j(t)),\quad j = 1, 2,\\ \tau_1(t) &= 8.5,\quad \upsilon_1(t) = 2.5,\end{aligned}$$

and

$$A_1 = \begin{bmatrix} 1 & -0.01 \\ 0.1 & 1.2 \end{bmatrix},\quad B_1 = \begin{bmatrix} 5.2 & 1.2 \\ 1.12 & 2.3 \end{bmatrix},$$
$$C_1 = \begin{bmatrix} 1.2 & 0.11 \\ 0.1 & 1.22 \end{bmatrix},\quad D_1 = \begin{bmatrix} -1.2 \\ 0.2 \end{bmatrix}.$$

For this system, Assumptions 11.3–11.5 are satisfied with $\underline{\alpha}_i = \bar{\alpha}_i = 1$, $\Pi_i = \Gamma_i = 8I$, $\bar{\Sigma} = I$, $\Sigma = F_1 = F_3 = 0$, $F_2 = F_4 = 0.5I$, and $\bar{\tau} = \bar{\tau}_1 = 8.5$, $\bar{\upsilon} = \bar{\upsilon}_1 = 2.5$. It is easy to verify that Theorem 1 of [27] admits no feasible solution. However, using the Matlab LMI Toolbox, the LMIs (11.4), (11.5), (11.7), (11.9), (11.47) and (11.48) are feasible with the following matrices:

$$\begin{aligned}
P_1 &= \begin{bmatrix} 72.4939 & -13.8747 \\ -13.8747 & 103.5930 \end{bmatrix},\\
Q_{11} &= \begin{bmatrix} 16.7250 & -26.0105 \\ -26.0105 & 88.6304 \end{bmatrix},\\
Q_{21} &= \begin{bmatrix} 644.0687 & 178.7808 \\ 178.7808 & 234.9008 \end{bmatrix},\\
Q_{31} &= \begin{bmatrix} 14.2369 & -26.9039 \\ -26.9039 & 81.3660 \end{bmatrix},\\
Q_{41} &= \begin{bmatrix} 0.3839 & -0.0816 \\ -0.0816 & 0.6447 \end{bmatrix},\\
R_1 &= \begin{bmatrix} 175.2573 & -2.4592 \\ -2.4592 & 269.3507 \end{bmatrix},\\
M_1 &= \begin{bmatrix} 95.6377 & -2.9849 \\ -2.9849 & 75.6384 \end{bmatrix},\\
\bar{M}_1 &= \begin{bmatrix} -334.3276 & 58.5190 \\ 58.5190 & -17.4669 \end{bmatrix},\\
T_1 &= \mathrm{diag}\{12.6571,\ 49.6002\},\\
U_1 &= \mathrm{diag}\{125.4140,\ 170.7878\},
\end{aligned}$$

$$\begin{aligned}
W_1 &= \mathrm{diag}\{10.2844,\ 37.0834\},\\
G_1 &= \mathrm{diag}\{9.2136,\ 15.5713\},\\
H &= \mathrm{diag}\{11.6382,\ 15.4633\},\\
K &= \mathrm{diag}\{5.1548,\ 8.9735\},
\end{aligned}$$

and accordingly the feedback control is

$$u_1(t) = [\ 0.3810 \quad 0.2053\ ]x(t).$$

Based on Example 11.16, it is easy to see that the obtained results are better than those in [27]. Hence, the proposed method is an improvement over the existing ones.

11.5 Summary

In this chapter, the problem of designing a feedback control law to exponentially stabilize a class of stochastic Cohen-Grossberg neural networks with both Markovian jumping parameters and mixed mode-dependent time delays has been studied. The mixed time delays consist of both discrete and distributed delays. Using a new Lyapunov–Krasovskii functional that accounts for the mode-dependent mixed delays, a new delay-dependent condition for the global exponential stabilization has been established in terms of linear matrix inequalities. Upon the feasibility of the LMI, all the control parameters can be easily computed and the design of a stabilizing controller can be accomplished.

References

1. C. Zheng, Q. Shan, H. Zhang, Z. Wang, On stabilization of stochastic Cohen-Grossberg neural networks with mode-dependent mixed time-delays and Markovian switching. IEEE Trans. Neural Netw. Learn. Syst. **24**(5), 800–811 (2013)
2. Q. Song, Z. Wang, Neural networks with discrete and distributed time-varying delays: a general stability analysis. Chaos Solitons Fractals **37**, 1538–1547 (2008)
3. X.F. Liao, K. Wong, C.G. Li, Global exponential stability for a class of generalized neural networks with distributed delays. Nonlinear Anal. Real World Appl. **5**, 527–547 (2004)
4. C.H. Lien, L.Y. Chung, Global asymptotic stability for cellular neural networks with discrete and distributed time-varying delays. Chaos Solitons Fractals **34**, 1213–1219 (2007)
5. Z. Liu, H. Zhang, Q. Zhang, Novel stability analysis for recurrent neural networks with multiple delays via line integral-type L-K functional. IEEE Trans. Neural Netw. **21**(11), 1710–1718 (2010)
6. Z. Liu, S.C. Shih, Q. Wang, Global robust stabilizing control for a dynamic neural network system. IEEE Trans. Syst. Man Cybern. A **39**(2), 426–436 (2009)
7. F.O. Souza, L.A. Mozelli, R.M. Palhares, On stability and stabilization of T-S Fuzzy time-delayed systems. IEEE Trans. Fuzzy Syst. **17**(6), 1450–1455 (2009)
8. H. Zhang, Z. Wang, D. Liu, Robust stability analysis for interval Cohen-Grossberg neural networks with unknown time-varying delays. IEEE Trans. Neural Netw. **19**(11), 1942–1955 (2008)
9. H. Zhang, Z. Liu, G.B. Huang, Z. Wang, Novel weighting-delay-based stability criteria for recurrent neural networks with time-varying delay. IEEE Trans. Neural Netw. **21**(1), 91–106 (2010)

10. H. Zhang, T. Ma, G.B. Huang, Z. Wang, Robust global exponential synchronization of uncertain chaotic delayed neural networks via dual-stage impulsive control. IEEE Trans. Syst. Man Cyber. B Cyber. **40**(3), 831–844 (2010)
11. H. Zhang, Z. Wang, Global asymptotic stability of delayed cellular neural networks. IEEE Trans. Neural Netw. **18**(3), 947–950 (2007)
12. H. Zhang, Z. Wang, D. Liu, Robust exponential stability of cellular neural networks with multiple time varying delays. IEEE Trans. Circuits Syst. II **54**(8), 730–734 (2007)
13. H. Zhang, Z. Wang, D. Liu, Global asymptotic stability of recurrent neural networks with multiple time-varying delays. IEEE Trans. Neural Netw. **19**(5), 855–873 (2008)
14. J. Tian, Y. Li, J. Zhao, S. Zhong, Delay-dependent stochastic stability criteria for Markovian jumping neural networks with mode-dependent time-varying delays and partially known transition rates. Appl. Math. Comput. **218**, 5769–5781 (2012)
15. H. Zhang, Y. Wang, Stability analysis of Markovian jumping stochastic Cohen-Grossberg neural networks with mixed time delays. IEEE Trans. Neural Netw. **19**(2), 366–370 (2008)
16. Z.G. Wu, J.H. Park, H. Su, J. Chu, Stochastic stability analysis of piecewise homogeneous Markovian jump neural networks with mixed time-delays. J. Frankl. Inst. **349**, 2136–2150 (2012)
17. S.M. Lee, O.M. Kwon, J.H. Park, A novel delay-dependent criterion for delayed neural networks of neutral type. Phys. Lett. A **374**, 1843–1848 (2010)
18. O.M. Kwon, S.M. Lee, J.H. Park, Improved delay-dependent exponential stability for uncertain stochastic neural networks with time-varying delays. Phys. Lett. A **374**, 1232–1241 (2010)
19. R. Rakkiyappan, P. Balasubramaniam, Dynamic analysis of Markovian jumping impulsive stochastic Cohen-Grossberg neural networks with discrete interval and distributed time-varying delays. Nonlinear Anal.: Hybrid Syst. **3**, 408–417 (2009)
20. P. Balasubramaniam, R. Rakkiyappan, Delay-dependent robust stability analysis for Markovian jumping stochastic Cohen-Grossberg neural networks with discrete interval and distributed time-varying delays. Nonlinear Anal.: Hybrid Syst. **3**, 207–214 (2009)
21. C. Vidhya, P. Balasubramaniam, Robust stability of uncertain Markovian jumping stochastic Cohen-Grossberg type BAM neural networks with time-varying delays and reaction diffusion terms. Neural Parallel Sci. Comput. **19**, 181–196 (2011)
22. C. Hua, X. Guan, Output feedback stabilization for time-delay nonlinear interconnected systems using neural networks. IEEE Trans. Neural Netw. **19**(4), 673–688 (2008)
23. V.N. Phat, New stabilization criteria for linear time-varying systems with state delay and norm-bounded uncertainties. IEEE Trans. Autom. Control **47**(12), 2095–2098 (2002)
24. Z. Wang, Y. Liu, X. Liu, Exponential stabilization of a class of stochastic system with Markovian jump parameters and mode-dependent mixed time-delays. IEEE Trans. Autom. Control **55**(7), 1656–1662 (2010)
25. E. Fridman, Descriptor discretized Lyapunov functional method: analysis and design. IEEE Trans. Autom. Control **51**(5), 890–897 (2006)
26. X. Liu, Q. Wang, Impulsive stabilization of high-order Hopfield-type neural networks with time-varying delays. IEEE Trans. Neural Netw. **19**(1), 71–79 (2008)
27. V.N. Phat, H. Trinh, Exponential stabilization of neural networks with various activation functions and mixed time-varying delays. IEEE Trans. Neural Netw. **21**(7), 1180–1184 (2010)
28. J. Cao, S. Zhong, Y. Hu, Global stability analysis for a class of neural networks with varying delays and control input. Appl. Math. Comput. **189**(2), 1480–1490 (2007)
29. K. Patan, Stability analysis and the stabilization of a class of discrete-time dynamic neural networks. IEEE Trans. Neural Netw. **18**(3), 660–673 (2007)
30. M. Liu, Delayed standard neural network models for control systems. IEEE Trans. Neural Netw. **18**(4), 1376–1391 (2007)
31. J. Lu, D.W.C. Ho, Z. Wang, Pinning stabilization of linearly coupled stochastic neural networks via minimum number of controllers. IEEE Trans. Neural Netw. **20**(10), 1617–1629 (2009)
32. E.N. Sanchez, J.P. Perez, Input-to-state stabilization of dynamic neural networks. IEEE Trans. Syst. Man Cybern. A **33**(4), 532–536 (2003)

33. Z. Chen, D. Zhao, Stabilization effect of diffusion in delayed neural networks systems with Dirichlet boundary conditions. J. Frankl. Inst. **348**(10), 2884–2897 (2011)
34. X. Lou, B. Cui, On robust stabilization of a class of neural networks with time-varying delays, in *Proceedings of IEEE International Conference on Computational Intelligence and Security*, pp. 437–440 (2006)
35. K. Shi, H. Zhu, S. Zhong et al., Less conservative stability criteria for neural networks with discrete and distributed delays using a delay-partitioning approach. Neurocomputing **140**, 273–282 (2014)
36. B. Chen, J. Wang, Global exponential periodicity and global exponential stability of a class of recurrent neural networks with various activation functions and time-varying delays. Neural Netw. **20**(10), 1067–1080 (2007)
37. Z. Guo, L. Huang, LMI conditions for global robust stability of delayed neural networks with discontinuous neuron activations. Appl. Math. Comput. **215**, 889–900 (2009)
38. H. Wu, Global exponential stability of Hopfield neural networks with delays and inverse Lipschitz neuron activations. Nonlinear Anal.: Real World Appl. **10**, 2297–2306 (2009)
39. K. Gu, An integral inequality in the stability problem of time-delay systems, in *Proceedings of the 39th IEEE Conference on Decision and Control, Sydney, Australia*, pp. 2805–2810 (2000)
40. C. Zheng, Q.H. Shan, Z. Wang, Improved stability results for stochastic Cohen-Grossberg neural networks with discrete and distributed delays. Neural Process. Lett. **35**(2), 103–129 (2012)
41. L. Arnold, *Stochastic Differential Equations: Theory and Applications* (Wiley, New York, 1972)
42. X. Mao, Exponential stability of stochastic delay interval systems with Markovian switching. IEEE Trans. Autom. Control **47**(10), 1604–1612 (2002)
43. C. Yuan, J. Lygeros, Stabilization of a class of stochastic differential equations with Markovian switching. Syst. Control Lett. **54**, 819–833 (2005)
44. B. Øksendal, *Stochastic Differential Equations: An Introduction with Applications* (Springer, New York, 1985)
45. Y. Tang, J. Fang, Q. Miao, On the exponential synchronization of stochastic jumping chaotic neural networks with mixed delays and sector-bounded non-linearities. Neurocomputing **72**, 1694–1701 (2009)

Chapter 12
Adaptive Synchronization of Complex Neural Networks

In Chap. 11, stabilization problems were discussed for a class of stochastic Cohen–Grossberg neural networks with mode-dependent mixed time-delays and Markovian switching. In this chapter, synchronization problem will be discussed for a kind of interconnected neural dynamical networks with time-varying coupling connections.

12.1 Introduction

As an important branch of human biology, neuroscience has become a very hot topic nowadays. A lot of efforts have been made by neuroscientists to understand, predict, and control the neural dynamics so as to contribute in this area. At theoretical level, some neural models are constructed to simulate the real neurons and their various electrophysiological characteristics for studying the behavior of neurons and neural networks. The structure of neural networks and its dynamics have significant impacts on the development of artificial intelligence and cognitive science. Whether in biological or artificial neural networks, the constructed neural network model plays an essential role in describing the dynamics and the mechanism of neural networks. In the past decades, many neural network models such as Hopfield neural networks and Cohen–Grossberg neural networks, to name a few, have been widely studied and applications have been found in different areas [1–9]. Most previous studies mainly concentrated on stability analysis, periodic oscillation, and chaotic behavior of neural networks [10].

There is evidence to show that arrays of *complex interconnected neural networks* on the level of linear coupling through synapses or gap junctions can exhibit the small-world properties [11, 12] and many interesting phenomena, such as spatio-temporal chaos [13], auto waves [14, 15], spiral waves [16, 17], etc., which cannot be found in node network or uncoupled network. Moreover, experiment and theoretical results have revealed that a mammalian brain not only displays its storage of associative memory, but also modulates oscillatory neuronal *synchronization* by selective

Z. Wang et al., *Qualitative Analysis and Control of Complex Neural Networks with Delays*, Studies in Systems, Decision and Control 34,
DOI 10.1007/978-3-662-47484-6_12

perceive attention [18, 19]. Circadian rhythms in biological and living systems regulate the functions of cells in the living bodies which are entrained to the day/night cycle via the sophisticated action of various gene regulatory networks [20]. More notably, synchronization has been proposed as a powerful mechanism to explain some of the patterns observed in the brain [21–24]. Thus, investigation of synchronization dynamics of such complex interconnected neural networks is indispensable in theory [25, 26]. Synchronization of complex interconnected neural networks also has many applications. In [14, 27], the authors proposed the so-called "the autowave principles for parallel image processing." In [28], the authors presented an architecture of complex interconnected neural networks to store and retrieve complex oscillatory patterns as synchronization states. In [29], the authors introduced a secure communication system based on interconnected cellular neural networks. Therefore, study on synchronization of complex interconnected or coupled neural networks is an important step for both understanding brain science and designing coupled neural networks for practical applications.

Nowadays, there are many synchronization results for complex interconnected neural networks [30–42]. However, they are mainly suitable for complex neural networks with exact structure and parameters. In the real world, it is impossible to precisely describe the behavior of any physical system through mathematical models. In modeling physical systems, a classical dilemma is the trade-off between model accuracy and tractability. A variety of approximation methods are used, for analysis, simulation, or control design of the "real" systems. It is critical that the approximate model preserves properties and features of interests of the original system, such as stability, Hamiltonian structure, or passivity. One example of approximation is the use of *linearization methods*. Nonlinear behaviors abound in the real world, including saturation, backlash, and dead zone. Linearized models are often used, because methods for analysis and control designs are readily available for linear systems. Another example is model reduction. The need for modeling accuracy may result in large-scale, higher order, and complex mathematical models. Model reduction methods lead to a lower-order, simpler system that can be used to facilitate control designs or speedup simulations. Similarly, in real-world complex networks, it is hard to get the exact estimation of the coupling coefficients and structure.

Therefore, how to design an effective controller to realize the synchronization for complex networks remains a hot topic. In order to handle the *synchronization* problem of complex networks with imprecise models or unknown parameters, many approaches have been proposed, e.g., robust synchronization, switching synchronization, and adaptive synchronization, [43–48]. Among these approaches, as a natural extension of self-synchronization, adaptive synchronization or controlled synchronization has become a major topic in nonlinear sciences due to its suitability of designing adaptive updating laws and achieving better synchronization performance [46–48]. Adaptive synchronization approaches are usually related to the optimal control and system identification [49, 50], and there are many different ways to realize the adaptive synchronization. One of the most simplest ways is to design a linear error feedback controller with fixed gain, for example, the methods in [51]. Another improved method is the linear error feedback with adaptive gain [52–54]. Besides the linear error feedback controller, some complex controllers are also designed, e.g., [55], in which too many adjustable parameters are required to design, and the structure of the controller becomes more complex than the linear feedback controller. This often leads to infeasible performance in applications. When the coupling matrix is time-varying, some local controllers are usually designed by linearized methods [56, 57], in which the Jacobian matrix is required at the *synchronization state*. For global synchronization, the adaptive coupling updating law should also be designed [58]. However, several simplified assumptions are often required in existing adaptive synchronization approaches. For example, it is typically assumed that the network topology is time-invariant, and that a unique coupling gain determines the strength of the coupling between neighboring nodes. Clearly, these assumptions are unrealistic when compared to the natural world. Therefore, how to synchronize a complex neural network with imprecise models or unknown parameters is still a challenging topic.

Based on the above discussions, this chapter aims to present a distributed control strategy to design an adaptive coupling updating law and an adaptive control law simultaneously. The remarkable advantage of the adopted method is to use internal adjusting and external control to adjust the whole networks simultaneously for the purpose of synchronization. The present method can be suitable for large-scale complex neural networks with amounts of nodes due to the distributed updating law.

12.2 Problem Formulation and Preliminaries

In this chapter we will discuss the following complex interconnected neural networks with N identical nodes

$$\begin{aligned}\dot{x}_i(t) = & - Cx_i(t) + Af(x_i(t)) + Bf(x_i(t-\tau(t))) + J(t) \\ & + \sum_{j=1, j\neq i}^{N} d_{ij}G_{ij}(t)\Gamma(x_j(t) - x_i(t)),\end{aligned} \tag{12.1}$$

where $x_i(t) = (x_{i1}(t), \ldots, x_{in}(t))^T$ is the state vector of the ith node, and positive integer n is the dimension of the ith node. Positive integer $N \geq 1$ denotes the number of node network, and $C = \text{diag}(c_1, \ldots, c_n)$ denotes a positive definite diagonal matrix. $A = (a_{rj})_{n\times n}$ and $B = (b_{rj})_{n\times n}$ are the interconnection matrix and delayed interconnection matrix with appropriate dimensions, respectively, where $a_{rj} > 0$ ($a_{rj} < 0$) if the output from the jth unit excites (respectively, inhibits) the rth unit at time t. b_{rj} has the same meaning as a_{rj}, $r, j = 1, \ldots, n$. $f(x_i(t)) = (f_1(x_{i1}(t)), \ldots, f_n(x_{in}(t)))^T$ is the activation function and $J(t)$ is the external input vector of the ith node with appropriate dimensions. $\tau(t) > 0$ is a bounded and time-varying delay, and its change rate satisfies $\dot{\tau}(t) \leq \mu < 1$. $G = (G_{ij}(t))_{N\times N}$ denotes the coupling configuration matrix of the networks with $G_{ij}(t) > 0$ $(i \neq j)$ but not all zeros, and it satisfies the diffusive condition

$G_{ii}(t) = -\sum_{j=1, j\neq i}^{N} G_{ij}(t)$. $D = (d_{ij})_{N\times N}$ is the coupling strength matrix of the networks with $d_{ij} > 0$, where d_{ij} denotes the coupling strength of the networks. Inner-coupling matrix $\Gamma = \text{diag}(\gamma_1, \ldots, \gamma_n)$ is a positive diagonal matrix which describes the individual coupling between two connected nodes of the networks, $i, j = 1, \ldots, N$.

Remark 12.1 In order to study the dynamics of neural networks, mathematical models in the form of differential equations are commonly adopted. Different kinds of electrophysiological characteristics of neurons can be simulated based on a neural network model. For example, if we ignore the coupling effects in complex networks (12.1), i.e., $d_{ij} = 0$ or $G_{ij}(t) = 0$, $i, j = 1, \ldots, N$, complex networks (12.1) are reduced to the following famous *Hopfield neural networks* model [7–9],

$$\dot{y}(t) = -Cy(t) + Af(y(t)) + Bf(y(t-\tau(t))) + J(t), \tag{12.2}$$

which is used to describe how action potentials in neurons are initiated and propagated, where $y(t) = (y_1(t), \ldots, y_n(t))^T$ is the state vector of the network model. Its basic idea is to model the segment of nerve membrane's electrical properties with an equivalent circuit, which can be used in engineering fields such as signal processing and optimization computation [59–64].

Remark 12.2 Neural network (12.2) has potential applications in optimization computation and cognitive computation [63–66]. In different applications, parameters in (12.2) may have different physical meanings. For example, when neural network (12.2) is used in optimization computation, c_r, $r = 1, \ldots, n$, represents the rate with which the rth unit will reset its potential to the resting state in isolation when disconnected from the network and external inputs. $A = (a_{rj})_{n\times n}$ and $B = (b_{rj})_{n\times n}$ denote the synaptic weights and represent the connection strengths of the synaptic connections from the jth neuron to the rth neuron, respectively. $J(t) = (J_1(t), \ldots, J_n(t))^T$ denotes the external input bias term, where $J_r(t)$ represents a firing threshold for the rth neuron. Activation function $f(x_r(t))$ is a nonlinear function and represents the input–output characteristics of the rth neuron. In the circuit implement of neural network (12.2), all these connection matrices C, A, and B are implemented by the capacitors and resistances, and $f(x_r(t))$ is realized by the operational amplifiers, $r = 1, \ldots, n$. More details can be referred to in reference [59].

Remark 12.3 Associative memories are brain-style devices designed to store a set of patterns as stable equilibria such that the stored patterns can be reliably retrieved with the initial probes containing sufficient information about the patterns [60, 61]. When the network (12.2) with space-invariant templates is used to realize associative memory, where the cells are arranged in a two-dimensional array, the meanings of the network parameters are as follows: $A = (a_{rj})_{n\times n}$ denotes the feedback cloning template and $B = (b_{rj})_{n\times n}$ denotes the delay feedback cloning template, respectively. $J(t)$ is the input cloning template. The convergence of the neural network (12.2) resulting from the design procedure can be guaranteed by the obtained stability conditions. More details can be referred to in reference [62].

Remark 12.4 The relations between node network (12.2) and complex networks (12.1) can be briefly stated as follows.

(1) Hopfield neural networks, Cohen–Grossberg neural networks and cellular neural networks can be changed to the neural network model (12.2), and have attracted more and more attention of researchers due to their great perspectives in applications such as associative memory and optimization computation. Previous studies on neural networks mainly aimed at solving the engineering problems, for instance, system modeling, function approximation, and associative memory. However, as a powerful tool of nonlinear modeling and function approximation, neural networks are always regarded as a black-box role. Consequently, neural networks (e.g., model (12.2)) cannot clearly reflect the essence of the complex systems because they are established in a relatively simple connection rule [67]. As pointed out by [68], structural complexity and node diversity are the main features of complex networks. Therefore, existing neural network models (e.g., model (12.2)) have not considered the structural complexity and node diversity.

(2) Recently, arrays of *complex interconnected neural networks* have attracted much attention of researchers in different research fields. They can exhibit many interesting phenomena (for instance, an equilibrium point, a periodic orbit, or a chaotic attractor, among others [69–71]) by choosing different coupling strength $D = (d_{ij})_{N\times N}$ and coupling configuration matrix $G(t)_{N\times N}$ (e.g., refer to model (12.1)). For example, the stability of the synchronous manifold of complex networks (12.1) is determined by the dynamics of the node (12.2), the inner-coupling connection matrix Γ, the outer-coupling configuration matrix $G(t)$, and the coupling strength matrix D. In contrast, the stability of node network (12.2) is only determined by $C, A, B\ f(x(t))$ and $J(t)$, which are all the information of its own networks, and are not related to other nodes. Therefore, the study of synchronization phenomena of complex interconnected neural networks is an important step toward both basic theory of neuroscience and technological practice.

In order to give our main results, we need the following assumption, definitions, and lemma.

Assumption 12.5 The *activation function* $f_i(v)$ is bounded and continuous, which satisfies

$$0 \le \frac{f_i(\eta) - f_i(v)}{\eta - v} \le \delta_i, \tag{12.3}$$

for any $\eta \ne v$, where $\eta, v \in \mathbb{R}$, $\delta_i > 0$, $|f_i(v)| \le G_i^b$, and $G_i^b > 0$ is a positive constant, $i = 1, \ldots, n$. Let $\Delta = \mathrm{diag}(\delta_1, \ldots, \delta_n)$.

Definition 12.6 (*see* [47]) The set $\mathbf{S} = \{(x_1^T(t), x_2^T(t), \ldots, x_N^T(t))^T \in \mathbb{R}^{nN} : x_i(t) = x_j(t), i, j = 1, \ldots, N\}$ is called *synchronization manifold*, where $x_i(t) = (x_{i1}(t), x_{i2}(t), \ldots, x_{in}(t))^T \in \mathbb{R}^n$ is the state of the ith node, $i = 1, \ldots, N$.

Definition 12.7 (*see* [48]) Synchronization manifold **S** is said to be *globally asymptotically stable* for the complex networks (12.1), equivalently, the complex networks (12.1) is globally asymptotically synchronized, if the following conditions hold,

$$\lim_{t\to\infty}\|x_i(t)-x_j(t)\|=0,\ i,\ j=1,2,\ldots,N.$$

Obviously, if complex networks (12.1) realize the *synchronization*, that is,

$$x_1(t)=x_2(t)=\cdots=x_N(t)=s(t),$$

then

$$\sum_{j=1,j\neq i}^{N} d_{ij}G_{ij}(t)\Gamma(x_j(t)-x_i(t))=0.$$

Thus, the *synchronous state* $s(t)$ satisfies the following condition:

$$\dot{s}(t)=-Cs(t)+Af(s(t))+Bf(s(t-\tau(t)))+J(t). \tag{12.4}$$

Synchronous state $s(t)$ can be an equilibrium point, a limit cycle, an aperiodic orbit, or a chaotic orbit [72].

In order to realize the synchronization of complex networks (12.1), an external control $u_i(t)$ should be acted on the networks (12.1), i.e.,

$$\begin{aligned}\dot{x}_i(t)=&-Cx_i(t)+Af(x_i(t))+Bf(x_i(t-\tau(t)))+J(t)\\&+\sum_{j=1,j\neq i}^{N} d_{ij}G_{ij}(t)\Gamma(x_j(t)-x_i(t))+u_i(t),\end{aligned} \tag{12.5}$$

where $i=1,\ldots,N$.

Defining the synchronization error $e_i(t)=x_i(t)-s(t)$, we have the following *error dynamical systems*, $i=1,\ldots,N$:

$$\begin{aligned}\dot{e}_i(t)=&-Cx_i(t)+Af(x_i(t))+Bf(x_i(t-\tau(t)))+J(t)\\&-\Big[-Cs(t)+Af(s(t))+Bf(s(t-\tau(t)))+J(t)\Big]\\&+\sum_{j=1,j\neq i}^{N} d_{ij}G_{ij}(t)\Gamma(x_j(t)-x_i(t))+u_i(t)\\=&-Ce_i+A(f(x_i(t))-f(s(t)))\\&+B(f(x_i(t-\tau(t))-f(s(t-\tau(t)))\\&+\sum_{j=1,j\neq i}^{N} d_{ij}G_{ij}(t)\Gamma(x_j(t)-x_i(t))+u_i(t).\end{aligned} \tag{12.6}$$

In the sequel, we will design the adaptive controller $u_i(t)$ and adaptive coupling updating laws $\dot{G}_{ij}(t)$ to guarantee the synchronization of complex networks (12.5), $i, j = 1, \ldots, N$.

Note that the purpose of designing the adaptive coupling updating laws $\dot{G}_{ij}(t)$ is to estimate the coupling configuration magnitude among the different nodes. That is, we only estimate the magnitudes of $G_{ij}(t)$ for $i \neq j$. For the case of $i = j$, we use the results of $G_{ij}(t)$ with $i \neq j$ to compute $G_{ii}(t)$, which is used to guarantee the diffusive condition of the coupling matrices, i.e., $G_{ii}(t) = -\sum_{j=1, j\neq i}^{N} G_{ij}(t)$, $i, j = 1, \ldots, N$.

In this case, we take the following adaptive control laws:

$$u_i(t) = -\sigma_i \theta_i(t) \Gamma(x_i(t) - s(t)), \tag{12.7}$$

where $\sigma_i = 1$ if node i is chosen to be controlled, otherwise $\sigma_i = 0$. $\theta_i(t)$ is an adaptive control gain which is updated in the following form:

$$\dot{\theta}_i(t) = \sigma_i k_i (x_i(t) - s(t))^T (x_i(t) - s(t)), \tag{12.8}$$

where $\theta_i(0) > 0$ and $k_i > 0$, $i = 1, \ldots, N$.

Combining the methods in [46, 73], the adaptive coupling updating laws $\dot{G}_{ij}(t)$ are selected in the following form:

$$\dot{G}_{ij}(t) = d_{ij} h_{ij} (x_i(t) - x_j(t))^T (x_i(t) - x_j(t)), \tag{12.9}$$

where $G_{ij}(0) \geq 0$ $(i \neq j)$ and $h_{ij} > 0$ are positive constants, $i, j = 1, \ldots, N$.

Lemma 12.8 (see [5]) *Let X, Y and P be real matrices with appropriate dimensions, and P be a positive definite symmetric matrix. Then for any positive scalar $\epsilon > 0$, the following inequality holds:*

$$X^T Y + Y^T X \leq \epsilon^{-1} X^T P^{-1} X + \epsilon Y^T P Y. \tag{12.10}$$

12.3 Adaptive Synchronization Scheme

Now we state our main results in this section.

Theorem 12.9 *Suppose that Assumption 12.5 holds. The complex networks (12.5) are globally asymptotically synchronized if control laws (12.7) and (12.8) and adaptive coupling updating laws (12.9) hold.*

Proof Let us consider the following *Lyapunov function*,

$$V(t) = V_1(t) + V_2(t),$$

where

$$V_1(t) = \sum_{i=1}^{N} e_i^T(t)e_i(t) + \sum_{i=1}^{N}\sum_{j=1,j\neq i}^{N} \frac{1}{2h_{ij}}(G_{ij}(t) - \bar{\alpha}_{ij})^2 + \sum_{i=1}^{N}\sum_{j=1,j\neq i}^{N} \frac{1}{k_i}(\theta_i(t) - \bar{\beta}_i)^2 \tag{12.11}$$

$$V_2(t) = \sum_{i=1}^{N} \int_{t-\tau(t)}^{t} e_i^T(s)Me_i(s)\mathrm{d}s, \tag{12.12}$$

$h_{ij} > 0$ and $k_i > 0$. $\bar{\alpha}_{ij} = \bar{\alpha}_{ji}$ and $\bar{\beta}_i$ are nonnegative constants, and $\bar{\alpha}_{ij} = 0$ if and only if $G_{ij}(t) = 0$. Positive semi-definite matrix M will be defined later, $i, j = 1, \ldots, N$.

The derivative of $V_1(t)$ is as follows:

$$\begin{aligned}
&\dot{V}_1(t) \\
&= 2\sum_{i=1}^{N} e_i^T(t)\dot{e}_i(t) + \sum_{i=1}^{N}\sum_{j=1,j\neq i}^{N} \frac{1}{h_{ij}}(G_{ij}(t) - \bar{\alpha}_{ij})\dot{G}_{ij}(t) \\
&\quad + \sum_{i=1}^{N}\sum_{j=1,j\neq i}^{N} \frac{2}{k_i}(\theta_i(t) - \bar{\beta}_i)\dot{\theta}_i(t) \\
&= 2\sum_{i=1}^{N} e_i^T(t)\Big[- Ce_i(t) + A(f(x_i(t)) - f(s(t))) \\
&\quad + B(f(x_i(t - \tau(t)) - f(s(t - \tau(t))) \\
&\quad + \sum_{j=1,j\neq i}^{N} d_{ij}G_{ij}(t)\Gamma(x_j(t) - x_i(t)) + u_i(t)\Big] \\
&\quad + \sum_{i=1}^{N}\sum_{j=1,j\neq i}^{N} d_{ij}(G_{ij}(t) - \bar{\alpha}_{ij})(x_i(t) - x_j(t))^T\Gamma(x_i(t) - x_j(t)) \\
&\quad + 2\sum_{i=1}^{N}\sum_{j=1,j\neq i}^{N} \sigma_i(\theta_i(t) - \bar{\beta}_i)(x_i(t) - s(t))^T\Gamma(x_i(t) - s(t)).
\end{aligned} \tag{12.13}$$

Applying Assumption 12.5, Lemma 12.8 and the definition of *synchronization error*, we have

$$
\begin{aligned}
&\dot{V}_1(t)\\
=&\sum_{i=1}^{N} e_i^T(t)\Big[-2C+AA^T+\Delta\Delta+BB^T\Big]e_i(t)\\
&+\sum_{i=1}^{N} e_i^T(t-\tau(t))\Delta\Delta e_i^T(t-\tau(t))\\
&+2\sum_{i=1}^{N} e_i^T(t)\Big[\sum_{j=1,j\neq i}^{N} d_{ij}G_{ij}(t)\Gamma(x_j(t)-x_i(t))-\sigma_i\theta_i(t)\Gamma e(t)\Big]\\
&+\sum_{i=1}^{N}\sum_{j=1,j\neq i}^{N} d_{ij}(G_{ij}(t)-\bar{\alpha}_{ij})(x_i(t)-x_j(t))^T\Gamma(x_i(t)-x_j(t))\\
&+2\sum_{i=1}^{N}\sum_{j=1,j\neq i}^{N} \sigma_i(\theta_i(t)-\bar{\beta}_i)e_i^T(t)\Gamma e_i(t)\\
=&\sum_{i=1}^{N} e_i^T(t)\Big[-2C+AA^T+\Delta\Delta+BB^T\Big]e_i(t)\\
&+\sum_{i=1}^{N} e_i^T(t-\tau(t))\Delta\Delta e_i^T(t-\tau(t))\\
&+2\sum_{i=1}^{N} e_i^T(t)\Big[\sum_{j=1,j\neq i}^{N} d_{ij}G_{ij}(t)\Gamma(x_j(t)-x_i(t))-\sigma_i\bar{\beta}_i\Gamma e_i(t)\Big]\\
&+\sum_{i=1}^{N}\sum_{j=1,j\neq i}^{N} d_{ij}(G_{ij}(t)-\bar{\alpha}_{ij})(x_i(t)-x_j(t))^T\Gamma(x_i(t)-x_j(t)).
\end{aligned}
\tag{12.14}
$$

Note that the following condition holds:

$$
\begin{aligned}
&\sum_{i=1}^{N}\sum_{j=1,j\neq i}^{N} d_{ij}(G_{ij}(t)-\bar{\alpha}_{ij})(x_i(t)-x_j(t))^T\Gamma(x_i(t)-x_j(t))\\
=&\sum_{i=1}^{N}\sum_{j=1,j\neq i}^{N} d_{ij}(G_{ij}(t)-\bar{\alpha}_{ij})(e_i(t)-e_j(t))^T\Gamma(x_i(t)-x_j(t))\\
=&2\sum_{i=1}^{N}\sum_{j=1,j\neq i}^{N} d_{ij}(G_{ij}(t)-\bar{\alpha}_{ij})e_i^T(t)\Gamma(x_i(t)-x_j(t))
\end{aligned}
$$

$$= -2\sum_{i=1}^{N}\sum_{j=1,j\neq i}^{N} d_{ij}(G_{ij}(t) - \bar{\alpha}_{ij})e_i^T(t)\Gamma(x_j(t) - x_i(t)). \tag{12.15}$$

Substituting (12.15) into (12.14), it yields

$$\begin{aligned}\dot{V}_1(t) =& \sum_{i=1}^{N} e_i^T(t)\Big[-2C + AA^T + \Delta\Delta + BB^T - 2\sigma_i\bar{\beta}_i\Gamma\Big]e_i(t)\\ &+ \sum_{i=1}^{N} e_i^T(t-\tau(t))\Delta\Delta e_i^T(t-\tau(t))\\ &+ 2\sum_{i=1}^{N}\sum_{j=1,j\neq i}^{N} d_{ij}\bar{\alpha}_{ij}e_i^T\Gamma(x_j(t) - x_i(t)).\end{aligned} \tag{12.16}$$

Meanwhile, the following equality holds:

$$\begin{aligned}&2\sum_{i=1}^{N}\sum_{j=1,j\neq i}^{N} d_{ij}\bar{\alpha}_{ij}e_i^T(t)\Gamma(x_j(t) - x_i(t))\\ &\quad = 2\sum_{i=1}^{N}\sum_{j=1,j\neq i}^{N} d_{ij}\bar{\alpha}_{ij}e_i^T(t)\Gamma(e_j(t) - e_i(t))\\ &\quad = 2\sum_{i=1}^{N}\sum_{j=1}^{N} d_{ij}\bar{\alpha}_{ij}e_i^T(t)\Gamma e_j(t).\end{aligned} \tag{12.17}$$

Combining (12.16) with (12.17), we have

$$\begin{aligned}\dot{V}_1(t) =& \sum_{i=1}^{N} e_i^T(t)\Big[-2C + AA^T + \Delta\Delta + BB^T - 2\sigma_i\bar{\beta}_i\Gamma\Big]e_i(t)\\ &+ \sum_{i=1}^{N} e_i^T(t-\tau(t))\Delta\Delta e_i^T(t-\tau(t))\\ &+ 2\sum_{i=1}^{N}\sum_{j=1}^{N} d_{ij}\bar{\alpha}_{ij}e_i^T(t)\Gamma e_j(t).\end{aligned} \tag{12.18}$$

The derivative of $V_2(t)$ is as follows:

$$\dot{V}_2(t) = \sum_{i=1}^{N}[e_i^T(t)Me_i(t) - e_i^T(t-\tau(t))Me_i(t-\tau(t))(1-\dot{\tau}(t))]. \tag{12.19}$$

If we take $M = \frac{1}{1-\mu}\Delta\Delta$, where $\dot{\tau}(t) \le \mu < 1$, then

$$\begin{aligned}\dot{V}_2(t) =& \sum_{i=1}^{N}\Bigg[e_i^T(t)\frac{1}{1-\mu}\Delta\Delta e_i(t)\\ &- e_i^T(t-\tau(t))\frac{1}{1-\mu}\Delta\Delta e_i(t-\tau(t))(1-\dot{\tau}(t))\Bigg]\\ \le& \sum_{i=1}^{N}\Bigg[e_i^T(t)\frac{1}{1-\mu}\Delta\Delta e_i(t)\\ &- e_i^T(t-\tau(t))\Delta\Delta e_i(t-\tau(t))\Bigg].\end{aligned} \tag{12.20}$$

Therefore,

$$\begin{aligned}\dot{V}(t) =& \dot{V}_1(t) + \dot{V}_2(t)\\ =& \sum_{i=1}^{N} e_i^T(t)\Bigg[-2C + AA^T + \Delta\Delta + BB^T + \frac{1}{1-\mu}\Delta\Delta\Bigg]e_i(t)\\ &+ 2\sum_{i=1}^{N}\sum_{j=1}^{N} d_{ij}\bar{\alpha}_{ij}e_i^T\Gamma e_j - 2\sum_{i=1}^{N} e_i^T(t)\sigma_i\bar{\beta}_i\Gamma e_i(t)\\ =& 2e^T(t)(d_{ij}\bar{\alpha}_{ij})_{N\times N}\otimes\Gamma e^T\\ &- 2e^T(t)(\sigma_i\bar{\beta}_i)_{N\times N}\otimes\Gamma e(t) + e^T(t)I_N\otimes Q_0 e(t)\\ =& 2e^T(t)\Big[(d_{ij}\bar{\alpha}_{ij})_{N\times N} - (\sigma_i\bar{\beta}_i)_{N\times N}\Big]\otimes\Gamma e(t) + e^T(t)I_N\otimes Q_0 e(t),\end{aligned} \tag{12.21}$$

where $Q_0 = -2C + AA^T + \Delta\Delta + BB^T + \frac{1}{1-\mu}\Delta\Delta$.

Since the coupling strength matrix $D = (d_{ij})_{N\times N}$ is a bounded real matrix and $\bar{\alpha}_{ij}$ is positive and symmetric, then the norm of matrix D exits. Also, since σ_i and $\bar{\beta}_i$ are all positive, the eigenvalue distribution of matrix $\Big[(d_{ij}\bar{\alpha}_{ij})_{N\times N} - (\sigma_i\bar{\beta}_i)_{N\times N}\Big]$ is determined by the magnitude of $\bar{\beta}_i$. Therefore, the larger the magnitude of $\bar{\beta}_i$ is, the more negative the eigenvalue of matrix $\Big[(d_{ij}\bar{\alpha}_{ij})_{N\times N} - (\sigma_i\bar{\beta}_i)_{N\times N}\Big]$ is. Since $I_N \otimes Q_0$ is a fixed matrix, then one can choose large enough value $\bar{\beta}_i$ to make the following inequality hold, $i = 1, \ldots, N$:

$$\dot{V}(t) \le -\epsilon\sum_{i}^{N} e_i^T(t)e_i(t) < 0 \ \text{ for } \ e_i(t) \ne 0. \tag{12.22}$$

According to *Lyapunov stability theory*, $e_i(t)$ approaches to zero as time evolves to infinity, $i = 1, \ldots, N$. That is, the synchronization manifold **S** of complex networks (12.5) is globally asymptotically stable under the adaptive control laws (12.7) and (12.8) and adaptive coupling updating laws (12.9). Therefore, complex networks (12.5) are globally asymptotically synchronized. This completes the proof of Theorem 12.9.

Note that if the coupling strength matrix D satisfies the *diffusive condition* $d_{ii} = -\sum_{j=1, j\neq i}^{N} d_{ij}$, or the compound matrix $\bar{G} = (\bar{g}_{ij}) = (d_{ij} g_{ij})$ satisfies the diffusive condition $\bar{g}_{ii} = -\sum_{j=1, j\neq i}^{N} \bar{g}_{ij}$, under the adaptive control laws (12.7) and (12.8) and adaptive coupling updating laws (12.9), the global synchronization is rather easy to be implemented due to the availability of magnitude of $\bar{\beta}_i$, $i = 1, \ldots, N$.

Remark 12.10 Because the proposed method is a distributed adaptive strategy, which only uses the nearest neighbor information of nodes to adjust the coupling strength adaptively, it can work for large-scale networks. One of the remarkable features of the proposed adaptive strategy is that there is no unknown parameters to be designed. Therefore, the proposed method is not confined by the scales of complex networks.

Remark 12.11 From the proof procedure of adaptive synchronization, it is obvious that the parameters $\bar{\alpha}_{ij}$ and $\bar{\beta}_i$ always exist. Especially, if the magnitude of $\bar{\beta}_i$ is large enough, the adaptive adjusting laws (12.7)–(12.9) will always hold to achieve the synchronization. In contrast, if the magnitude of $\bar{\alpha}_{ij}$ is small enough, it will make the implementation of synchronization easier, $i, j = 1, \ldots, N$.

Remark 12.12 For the adaptive control laws (12.7)–(12.9), we can make some remarks as follows.

(1) The external control laws (12.7) and (12.8) can regulate the whole networks by injecting some information flows. There are many ways to design the external control laws such as the linear control law, nonlinear control law, and so on. The internal adjusting laws (12.9) can only regulate the coupling strength or configuration magnitude using the available information of complex networks. If the internal adjusting laws are designed to be too complex, the whole system will become more and more complex and will lose practical meaning. Therefore, the capability of internal adjusting laws is more limited to some degree than the external control laws.

(2) If D is diffusive and the strength $\bar{\alpha}_{ij}$ can be selected to be large enough, then the eigenvalues of matrix $(d_{ij}\bar{\alpha}_{ij})_{N\times N}$ may become much negative, while the maximum eigenvalue is still zero. In this case, if the pinning control (a kind of control methods, which can control a network by pinning part of nodes and significantly reduce the number of controllers for large-scale networks) is not selected on this node, then the synchronization cannot be ensured because the term $I_N \otimes Q_0$ may have positive eigenvalues. If the pinning control is just selected on this node, then the synchronization can be realized. In contrast, the external control laws can be designed elaborately such that the controlled nodes can be selected purposely. Therefore, as far as the adjusting capability is concerned, the external control laws are more powerful than the internal adjusting laws. This is the reason why we use both external

control laws and adaptive coupling updating laws simultaneously to synchronize the networks (12.1).

12.4 Illustrative Example

In this section, we will use two examples to show the effectiveness of the proposed result.

Example 12.13 Let us consider the node neural network (12.2) as follows:

$$\dot{x}(t) = -Cx(t) + Af(x(t)) + Bf(x(t - \tau(t))) + J(t), \tag{12.23}$$

where
$C = \begin{bmatrix} 1 & 0 \\ 0 & 1 \end{bmatrix}, A = \begin{bmatrix} 2 & -0.1 \\ -5 & 3 \end{bmatrix}, B = \begin{bmatrix} -1.5 & -0.1 \\ -0.2 & -2.5 \end{bmatrix},$
$J(t) = (0, 0)^T, \tau(t) = 1.$

Under the initial conditions $x(s) = (0.4, 0.6)^T$, the node neural network (12.23) has a chaotic attractor as shown in Fig. 12.1.

Now we take the state of network (12.23) with initial conditions $x(s) = (0.4, 0.6)^T$ as the desired synchronization state $s(t)$, and then consider the complex interconnected networks (12.5) with five node neural networks (12.23). The adaptive coupling laws of $G_{ij}(t)$ are selected as in the form of (12.9), and the control laws are the same as (12.7) and (12.8). The initial parameters in adaptive coupling laws (12.9) and control laws (12.7) and (12.8) are given as follows: $\sigma_i = 1, i = 1, \ldots, 5$, $k_1 = 1, k_2 = 1, k_3 = 0.5, k_4 = 4, k_5 = 3, \Gamma = \text{diag}(1, 1)$. The initial states of five

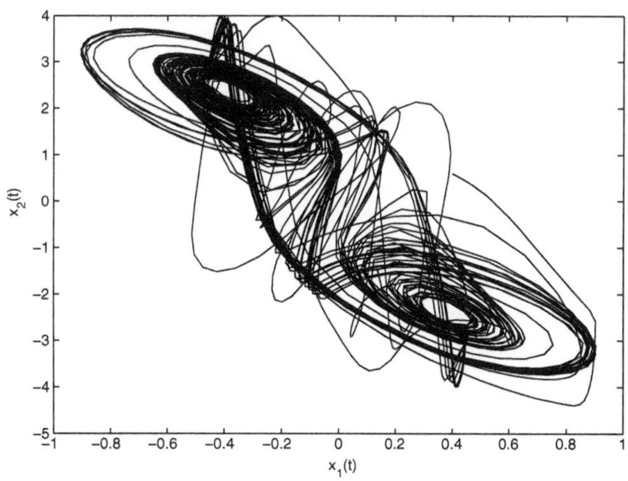

Fig. 12.1 Chaotic node network (12.23)

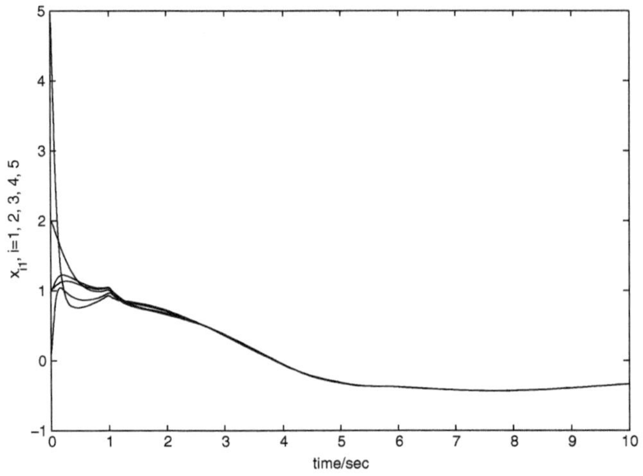

Fig. 12.2 States of complex networks (12.5), $x_{i1}, i = 1, \ldots, 5$

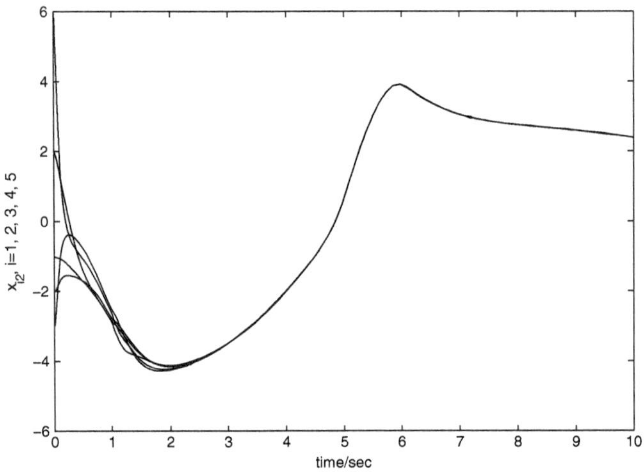

Fig. 12.3 States of complex networks (12.5), $x_{i2}, i = 1, \ldots, 5$

nodes are $x_1(s) = (1, 2)^T$, $x_2(s) = (1, -1)^T$, $x_3(s) = (2, -2)^T$, $x_4(s) = (5, -3)^T$ and $x_5(s) = (0.1, 6)^T$ for $s \in [-1, 0]$, respectively.

The time response curves of state variables x_{i1} and x_{i2} of the complex networks (12.5) are depicted in Figs. 12.2 and 12.3, respectively, $i = 1, \ldots, 5$. The trajectories of synchronization errors $e_i(t) = x_i(t) - s(t)$ are depicted in Figs. 12.4 and 12.5, respectively.

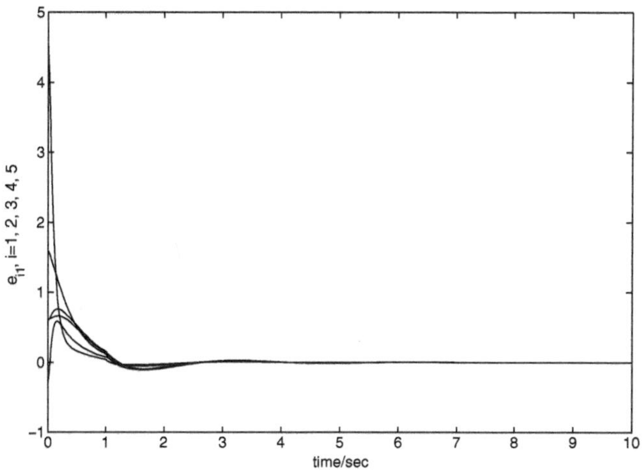

Fig. 12.4 Synchronization errors of complex networks (12.5), $e_{i1}, i = 1, \ldots, 5$

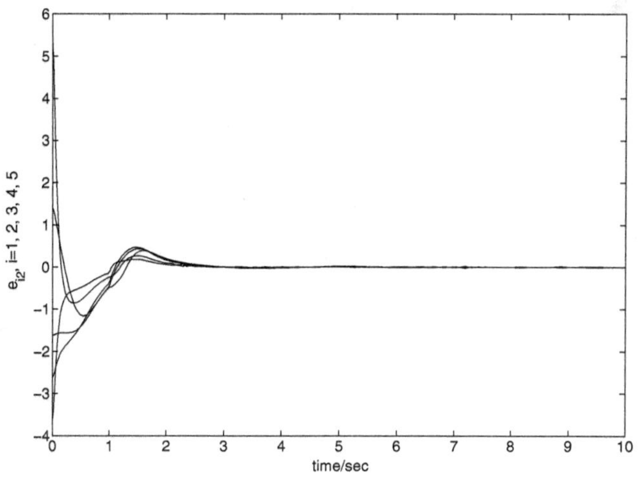

Fig. 12.5 Synchronization errors of complex networks (12.5), $e_{i2}, i = 1, \ldots, 5$

When $h_{ij} = 1$, the initial values of $G = (G_{ij})_{5\times5}$ and D are randomly selected as follows:

$$G_0 = \begin{bmatrix} -1.4258 & 0.6085 & 0.0576 & 0.0841 & 0.6756 \\ 0.6831 & -2.2043 & 0.3676 & 0.4544 & 0.6992 \\ 0.0928 & 0.0164 & -1.2785 & 0.4418 & 0.7275 \\ 0.0353 & 0.1901 & 0.7176 & -1.4214 & 0.4784 \\ 0.6124 & 0.5869 & 0.6927 & 0.1536 & -2.0456 \end{bmatrix},$$

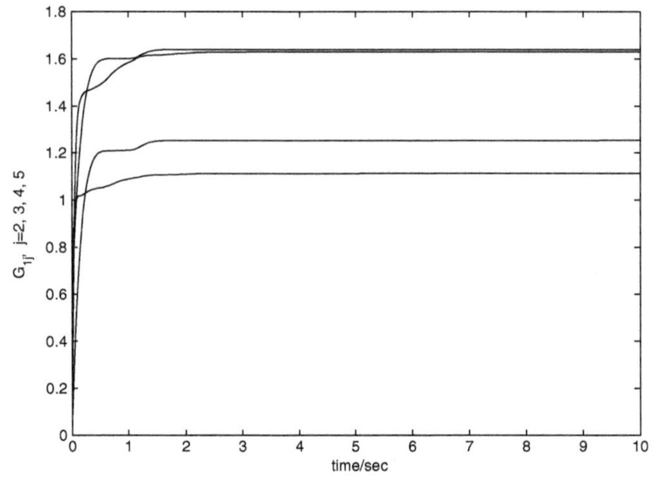

Fig. 12.6 Trajectories of coupling coefficients G_{1j}, $j = 2, 3, 4, 5$

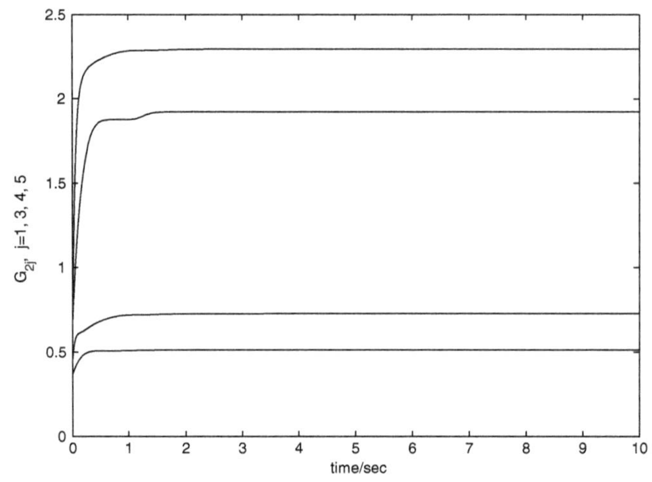

Fig. 12.7 Trajectories of coupling coefficients G_{2j} ,$j = 1, 3, 4, 5$

$$D = (d_{ij})_{5\times 5} = \begin{bmatrix} 0.8699 & 0.6400 & 0.4093 & 0.6084 & 0.5061 \\ 0.7694 & 0.2473 & 0.4635 & 0.1750 & 0.4648 \\ 0.4442 & 0.3527 & 0.6109 & 0.6210 & 0.5414 \\ 0.6206 & 0.1879 & 0.0712 & 0.2460 & 0.9423 \\ 0.9517 & 0.4906 & 0.3143 & 0.5874 & 0.3418 \end{bmatrix},$$

the trajectories of coupling matrix G are shown in Figs. 12.6, 12.7, 12.8, 12.9 and 12.10, respectively. These results show that the coupling coefficients of the

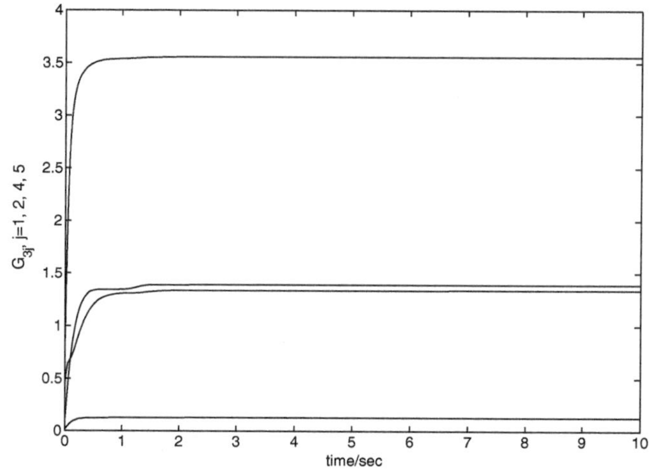

Fig. 12.8 Trajectories of coupling coefficients G_{3j}, $j = 1, 2, 4, 5$

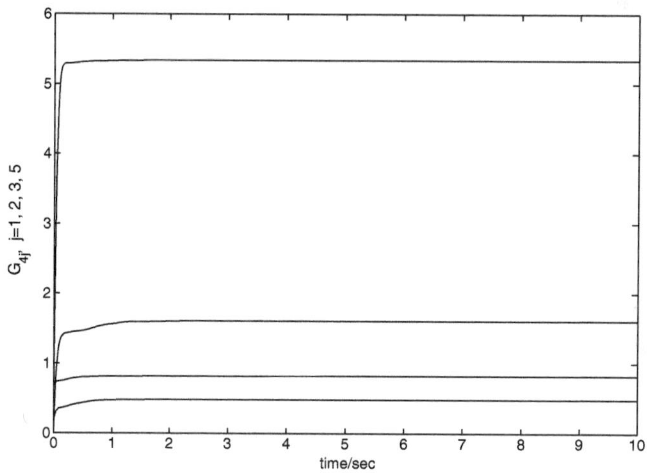

Fig. 12.9 Trajectories of coupling coefficients G_{4j}, $j = 1, 2, 3, 5$

interconnected neural networks are bounded and converge to constants when the synchronization is achieved.

The trajectories of $\theta_i(t)$ are depicted in Fig. 12.11 under the initial conditions $\theta_1(s) = 0.1, \theta_2(s) = 0.2, \theta_3(s) = 0.3, \theta_4(s) = 1$, and $\theta_5(s) = 2$, where s is the initial time instant, $i = 1, \ldots, 5$. Here we take $s = 0$.

Example 12.14 Consider the complex networks (12.5) with 100 nodes. In this case, each node is constituted by network (12.23), and the parameters in (12.23) are the same as those in Example 12.13. Assume that the nodes are coupled through the

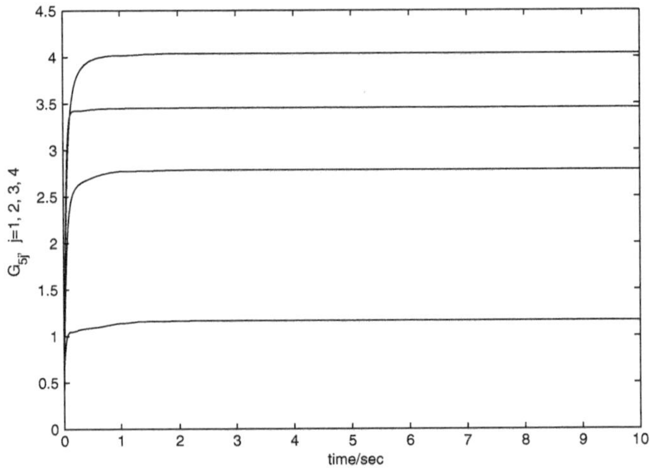

Fig. 12.10 Trajectories of coupling coefficients G_{5j}, $j = 1, \ldots, 4$

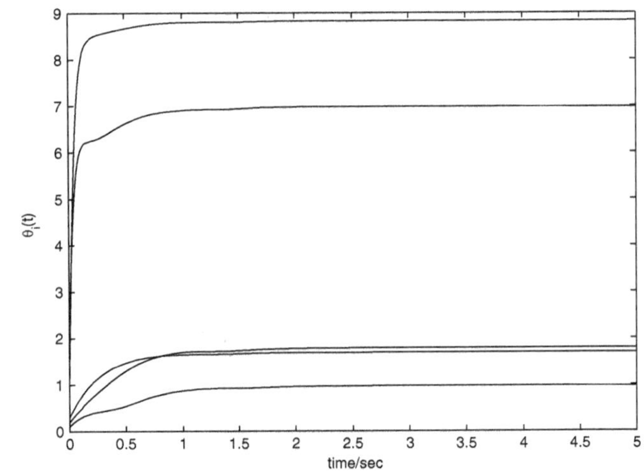

Fig. 12.11 Trajectories of control parameters $\theta_i(t)$, $i = 1, \ldots, 5$

decentralized adaptive strategy (12.9). The simulation starts from null initial coupling gains. The initial states are taken randomly from a normal distribution with standard derivation 40, and the initial parameters are $d_{ij} = 1$, $h_{ij} = h_{ji} = 0.1$, $\Gamma = \text{diag}(1, 1)$ and $N = 100$, respectively. As depicted in Figs. 12.12, 12.13 and 12.14, synchronization is asymptotically achieved and the adaptive control gains asymptotically converge to constant values.

Based on the above simulations, we can see that the proposed control scheme is effective and synchronization is achieved.

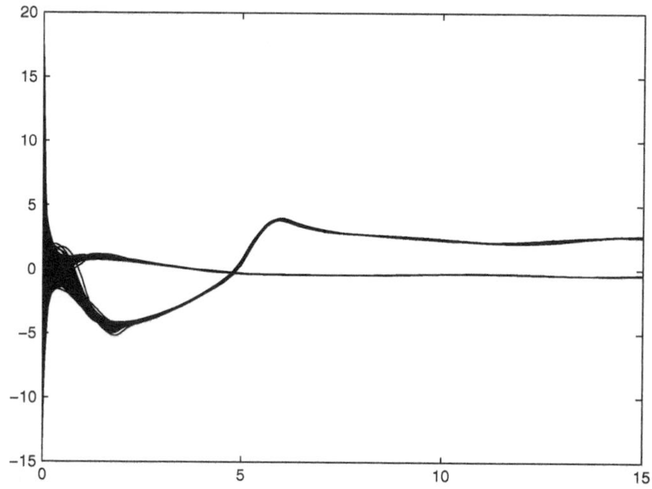

Fig. 12.12 Trajectories of state variables x_{i1} and x_{i2}, $i = 1, \ldots, N$

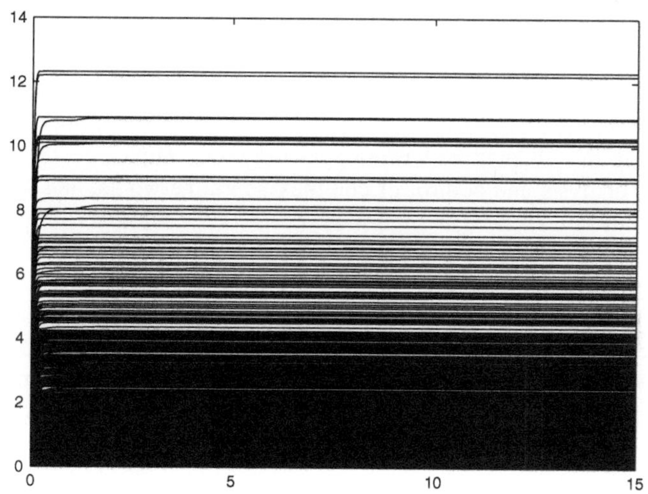

Fig. 12.13 Trajectories of coupling coefficients G_{ij}, $i \neq j$, $i, j = 1, \ldots, N$

12.5 Summary

In this chapter, a distributed adaptive control scheme for synchronization of complex interconnected neural networks has been presented and analyzed, which adaptively tunes the coupling weights of the networks toward reaching synchronization. From the viewpoints of internal adjusting and external control, adaptive coupling updating laws and external control laws are designed, which will greatly improve the

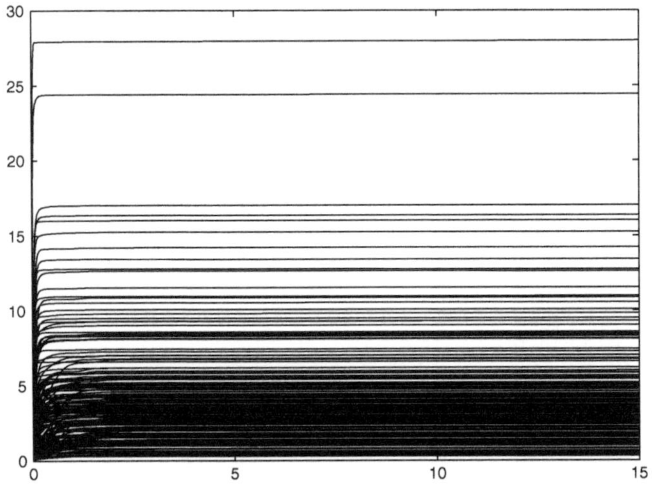

Fig. 12.14 Trajectories of adaptive control gains $\theta_i(t), i = 1, \ldots, N$

synchronization performance both in robustness and fastness. Especially, the distributed control concept is involved in the design of the adaptive coupling updating laws, which can use the neighbor information of the nodes to adaptively adjust the connection strengths. The established result is illustrated by two numerical examples. The contents in this chapter are written purposely for completeness of the research in qualitative analysis and control of complex neural networks, and have not been published in any conferences and journals yet.

Moreover, a fascinating direction could be the exploration of shape, capacity, and evolution of the attractors in the complex interconnected neural networks, which has a strict functional correlation between the attractor features and the network's stored information. Therefore, how to present an architecture of complex interconnected neural networks to store and retrieve complex oscillatory patterns as synchronization states and build internal representations of external input stimuli as attractors of neurons in a complex interconnected neural network are further research directions.

References

1. M. Cohen, S. Grossberg, Absolute stability of global pattern formation and parallel memory storage by competitive neural networks. IEEE Trans. Syst. Man Cybern. **13**(4), 815–826 (1983)
2. D. Tank, J. Hopfield, Simple neural optimization networks: an A/D converter, signal decision circuit and a linear programming circuit. IEEE Trans. Circuits Syst. **33**(4), 533–541 (1986)
3. J. Hopfield, Neurons with graded response have collective computation properties like those of two-stage neurons. Proc. Natl. Acad. Sci. U S A **81**, 3088–3092 (1984)
4. M. Hanna, On the stability of a Tank and Hopfield type neural network in the general case of complex eigenvalues. IEEE Trans. Signal Process. **48**(1), 289–293 (2000)

5. Z. Wang, H. Zhang, B. Jiang, LMI-based approach for global asymptotic stability analysis of recurrent neural networks with various delays and structures. IEEE Trans. Neural Netw. **22**(7), 1032–1045 (2011)
6. H. Zhang, T. Ma, G. Huang, Z. Wang, Robust global exponential synchronization of uncertain chaotic delayed neural networks via dual-stage impulsive control. IEEE Trans. Syst. Man Cybern. Part B-Cybern. **40**(3), 831–844 (2010)
7. H. Zhang, Z. Wang, D. Liu, Global asymptotic stability of recurrent neural networks with multiple time-varying delays. IEEE Trans. Neural Netw. **19**(5), 855–873 (2008)
8. C. Ji, H. Zhang, Y. Wei, LMI approach for global robust stability of Cohen-Grossberg neural networks with multiple delays. Neurocomputing **71**(6), 475–485 (2008)
9. H. Zhang, Z. Liu, G. Huang, Z. Wang, Novel weighting-delay-based stability criteria for recurrent neural networks with time-varying delay. IEEE Trans. Neural Netw. **21**(1), 91–106 (2010)
10. H. Sompolinsky, A. Crisanti, Chaos in random neural networks. Phys. Rev. Lett. **61**(3), 259–262 (1988)
11. D.J. Watts, S.H. Strogatz, Collective dynamics of small-world. Nature **393**, 440–442 (1998)
12. M.E. Newman, D.J. Watts, Renormalization group analysis of the small-world network model. Phys. Lett. A **263**, 341–346 (1999)
13. A. Zheleznyak, L.O. Chua, Coexistence of low- and high-dimensional spatio-temporal chaos in a chain of dissipatively coupled Chua's circuits. Int. J. Bifurc. Chaos **4**(3), 639–672 (1994)
14. V. Perez-Munuzuri, V. Perez-Villar, L.O. Chua, Autowaves for image processing on a two-dimensional CNN array of excitable nonlinear circuits: flat and wrinkled labyrinths. IEEE Trans. Circuits Syst. I **40**(2), 174–181 (1993)
15. J.D. Murray, *Mathematics Biology* (Springer, Berlin, 1989)
16. A. Perez-Munuzuri, V. Perez-Munuzuri, V. Perez-Villar, Spiral waves or a two-dimensional array of nonlinear circuits. IEEE Trans. Circuits Syst. I **40**(8), 872–877 (1993)
17. R.E. Mirollo, S.H. Strogatz, Synchronization of pulse-coupled biological oscillators. SIAM J. Appl. Math. **50**(6), 1645–1662 (1990)
18. P. Fries, J.H. Reynolds, J.H. Rorie, R. Desimore, Modulation of oscillatory neuronal synchronization by selective visual attention. Science **291**(23), 1560–1563 (2001)
19. P.N. Steinmetz, A. Roy, P.J. Fitzgerald, S.S. Hsiao, K.O. Johnson, E. Niebar, Attention modulate synchronized neuronal firing in primate somatosensory cortex. Nature **404**(9), 487–490 (2000)
20. D. Gonze, S. Bernard, C. Walterman, A. Kramer, H. Herzerl, Spontaneous synchronization of coupled circadian oscillators. Biophys. J. **89**, 120–129 (2005)
21. B. Ermentrout, M. Wechselberger, Canars, clusters and synchronization in a weakly coupled interneuron model. SIAM J. Appl. Dyn. Syst. **8**(1), 253–278 (2009)
22. C.W. Lai, C.K. Chen, T.L. Liao, J.J. Yan, Adaptive synchronization for nonlinear FitzHugh-Nagumo neurons in external electrical stimulation. Int. J. Adapt. Control Signal Process. **22**, 833–844 (2007)
23. B. Pesaran, J. Pezaris, M. Sahani, P. Mitra, R. Andersen, Temporal structure in neuronal activity during working memory in macaque parietal cortex. Nature **5**, 805–811 (2002)
24. E.M. Izhikevich, *Dynamical Systems in Neuroscience: The Geometry of Excitability and Bursting* (MIT Press, Cambridge, 2006)
25. S. Wen, Z. Zeng, T. Huang, Y. Zhang, Exponential adaptive lag synchronization of memristive neural networks via fuzzy method and applications in pseudorandom number generators. IEEE Trans. Fuzzy Syst. **22**(6), 1704–1713 (2014)
26. J. Wang, H. Shen, Passivity-based fault-tolerant synchronization control of chaotic neural networks against actuator faults using the semi-Markov jump model approach. Neurocomputing **143**, 51–56 (2014)
27. L.O. Chua, L. Yang, Cellular neural networks: applications. IEEE Trans. Circuits Syst. **35**(10), 1273–1290 (1988)
28. F. Hoppensteadt, E. Izhikevich, Pattern recognition via synchronization in phase-locked loop neural networks. IEEE Trans. Neural Netw. **11**(3), 734–738 (2000)
29. Y. Zhang, Z. He, A secure communication scheme based on cellular neural networks, in *Proceedings of the IEEE International Conference on Intelligent Processing Systems*, pp. 521–524 (1997)

30. W. Lu, T. Chen, Synchronization of coupled connected neural networks with delays. IEEE Trans. Circuits Syst.-I: Regul. Pap. **51**(12), 2491–2503 (2004)
31. W. Wu, T. Chen, Global synchronization criteria of linearly coupled neural network systems with time-varying coupling. IEEE Trans. Neural Netw. **19**(2), 319–332 (2008)
32. T. Chen, W. Wu, W. Zhou, Global μ-synchronization of linearly coupled unbounded time-varying delayed neural networks with unbounded delayed coupling. IEEE Trans. Neural Netw. **19**(10), 1809–1816 (2008)
33. G. Chen, J. Zhou, Z. Liu, Global synchronization of coupled delayed neural networks and applications to chaos CNN models. Int. J. Bifurc. Chaos **14**(7), 2229–2240 (2004)
34. D. Gong, H. Zhang, Z. Wang, D. Ma, Synchronization criteria for an array of neutral-type neural networks with hybrid coupling: a novel analysis approach. Neural Process. Lett. **35**, 29–45 (2012)
35. Q. Ma, S. Xu, Y. Zou, G. Shi, Synchronization of stochastic chaotic neural networks with reaction-diffusion terms. Nonlinear Dyn. **67**, 2183–2196 (2012)
36. G. Zhang, T. Wang, T. Li, S. Fei, Exponential synchronization for delayed chaotic neural networks with nonlinear hybrid coupling. Neurocomputing **85**, 53–61 (2012)
37. Z. Guo, J. Wang, Z. Yan, Global exponential synchronization of two memristor-based recurrent neural networks with time delays via static or dynamic coupling. IEEE Trans. Syst. Man Cybern.: Syst. **45**(2), 235–249 (2015)
38. J. Wang, H. Wu, L. Guo, Novel adaptive strategies for synchronization of linearly coupled neural networks with reaction-diffusion terms. IEEE Trans. Neural Netw. Learn. Syst. **25**(2), 429–440 (2014)
39. J. Zhang, Y. Qiao, J. Miao, L. Duan, Y. Zeng, New global synchronization analysis for complex networks with coupling delay. J. Appl. Math. **2014** Article ID 623249-12 (2014)
40. D. Gong, H. Zhang, Z. Wang, B. Huang, New global synchronization analysis for complex networks with coupling delay based on a useful inequality. Neural Comput. Appl. **22**(2), 205–210 (2013)
41. Z. Wang, C. Cai, J. Wang, H. Zhang, Adaptive fault estimation of coupling connections for synchronization of complex interconnected networks. Lect. Notes Comput. Sci. **7952**, 455–462 (2013)
42. X. Jin, G. Yang, Adaptive pinning synchronization of a class of nonlinearly coupled complex networks. Commun. Nonlinear Sci. Numer. Simul. **18**, 316–326 (2013)
43. J. Liang, Z. Wang, Y. Liu, Robust synchronization of an array of coupled stochastic discrete-time delayed neural networks. IEEE Trans. Neural Netw. **19**(11), 1910–1921 (2008)
44. D. Qi, M. Liu, M. Qiu, Exponential H_∞ synchronization of general discrete-time chaotic neural networks with or without time delays. IEEE Trans. Neural Netw. **21**(8), 1358–1365 (2010)
45. W. Yu, J. Cao, W. Lu, Global synchronization control of switched linearly coupled neural networks with delay. Neurocomputing **73**(4–6), 858–886 (2010)
46. Z. Yuan, J. Cai, M. Lin, Global synchronization in complex networks with adaptive coupling. Math. Probl. Eng. *2010*, Article ID 826721–10 (2010)
47. W. Lu, T. Chen, New approach to synchronization analysis of linearly coupled ordinary differential systems. Phys. D **213**, 214–230 (2006)
48. Z. Li, G. Chen, Global synchronization and asymptotic stability of complex dynamical networks. IEEE Trans. Circuits Syst.-II **53**(1), 28–33 (2006)
49. H. Zhang, Y. Luo, D. Liu, Neural-Network-based near-optimal control for a class of discrete-time affine nonlinear systems with control constraints. IEEE Trans. Neural Netw. **20**(9), 1490–1503 (2009)
50. H. Zhang, Y. Quan, Modeling, identification and control of a class of nonlinear system. IEEE Trans. Fuzzy Syst. **9**(2), 349–354 (2001)
51. Q. Song, J. Cao, On pinning synchronization of directed and undirected complex dynamical networks. IEEE Trans. Circuits Syst.-I: Regul. Pap. **57**(3), 672–680 (2010)
52. S. Liu, X. Li, W. Jiang, Y. Fan, Adaptive synchronization in complex dynamical networks with coupling delays for general graphs. Appl. Math. Comput. **219**, 83–87 (2012)

53. Q. Zhang, J. Lu, J. Lu, C. Tse, Adaptive feedback synchronization of a general complex dynamical network with delayed nodes. IEEE Trans. Circuits Syst.-II: Express Briefs **55**(2), 83–187 (2008)
54. L. Wang, H. Dai, H. Dong, Y. Cao, Y. Sun, Adaptive synchronization of weighted complex dynamical networks through pinning. Eur. Phys. J. B **61**, 35–342 (2008)
55. J. Chen, L. Jiao, J. Wu, X. Wang, Adaptive synchronization between two different complex networks with time-varying delay coupling. Chin. Phys. Lett. **26**, 060505–060508 (2009)
56. B. Wang, Z. Guan, Chaos synchronization in general complex dynamical networks with coupling delays. Nonlinear Anal.: R. World Appl. **11**, 1925–1932 (2010)
57. D. Yue, H. Li, Synchronization stability of continuous/discrete complex dynamical networks with interval time-varying delays. Neurocomputing **73**, 809–819 (2010)
58. X. Wu, Synchronization-based topology identification of weighted general complex dynamical networks with time-varying coupling delay. Phys. A **387**, 997–1008 (2008)
59. H. Tang, K. Tan, Y. Zhang, *Neural Networks: Computational Models and Applications* (Springer, Berlin, 2007)
60. C.L. Giles, C.B. Miller, D. Chen, H.H. Chen, G.Z. Sun, Y.C. Lee, Learning and extracting finite state automata with second-order recurrent neural networks. Neural Comput. **4**, 393–405 (1992)
61. R. Parekh, J. Yang, V. Honavar, Constructive neural-network learning algorithms for pattern classification. IEEE Trans. Neural Netw. **11**(2), 436–451 (2000)
62. A.N. Michel, D. Liu, *Qualitative Analysis and Synthesis of Recurrent Neural Networks* (Marcel Dekker, New York, 2002)
63. H.T. Siegelmann, E.D. Sontag, Analog computation via neural networks. Theor. Comput. Sci. **131**, 331–360 (1994)
64. S.P. Eberhardt, T. Daud, D.A. Kerns, T.X. Brown, A.P. Thakoor, Competitive neural architecture for hardware solution to the assignment problem. Neural Netw. **4**, 431–442 (1989)
65. C. Gros, Cognitive computation with autonomously active neural networks: an emerging field. Cogn. Comput. **1**, 77–90 (2009)
66. G. Indiveri, E. Chicca, R. Douglas, Artificial cognitive systems: from VLSI networks of spiking neurons to neuromorphic cognition. Cogn. Comput. **1**, 119–127 (2009)
67. X. Li, *From Complex to Order: New Advancement of Neural Network Intelligent Control Theory (in Chinese)* (Shanghai Jiaotong University Press, Shanghai, 2006)
68. S.H. Strogatz, Exploring complex networks. Nature **410**, 268–276 (2001)
69. T. Chen, X. Liu, W. Lu, Pinning complex networks by a single controller. IEEE Trans. Circuits Syst. I **54**(6), 1317–1326 (2007)
70. F. Sorrentino, M. Bernardo, F. Garofalo, G. Chen, Controllability of complex networks via pinning. Phys. Rev. E **75**(4), 046103 (2007)
71. W. Yu, G. Chen, J. Lu, On pinning synchronization of complex dynamical networks. Automatica **45**(2), 429–435 (2009)
72. J. Lu, G. Chen, A new chaotic attractor coined. Int. J. Bifurc. Chaos **12**, 659–661 (2002)
73. W. Yu, P. De Lellis, G. Chen, M. di Bernardo, J. Kurths, Distributed adaptive control of synchronization in complex networks. IEEE Trans. Autom. Control **57**(8), 2153–2158 (2012)

Index

Symbols
$\mathcal{K}$-class function, 76
$\mathcal{P}_0$ matrix, 155
$\omega-$limit point, 71
r-exponentially stabilizable, 336, 353
r-exponentially stable, 336

A
Absolute stability, 52, 74, 156, 158
Absolutely stable, 72, 92, 125
Activation function, 17, 96, 101, 126, 136, 176, 207, 240, 241, 247, 252, 280, 305, 306, 365
Amplification function, 113, 120, 123, 126, 304
Artificial neural networks, 5, 11
Asymptotic stability of Sets, 79
Asymptotically stable, 45, 49, 71, 80
Attracting set, 300
Attraction, 18
Attractive, 45, 71
Attractive set, 289–291
Autonomous system, 41

B
Barbalat's lemma, 46, 300, 301
Behaved function, 121, 124
Biological neural networks, 2
Bounded, 75, 81
Boundedness, 55, 84, 121
Brouwer's fixed point theorem, 67, 101, 114

C
Cohen–Grossberg model, 24
Cohen–Grossberg neural networks, 92, 113, 120, 122, 123, 300, 334
Cohen–Grossberg-type neural networks, 124
Competitive connection, 94
Competitive-cooperation dynamical system, 121
Complete stability, 52
Completely stable, 73
Complex interconnected neural networks, 361, 365
Complex networks, 84
Complex-valued neural networks, 111
Computability concept, 4
Computable function, 4, 5
Computational capacity, 28
Computing ability, 26
Computing device, 5
Computing model, 3, 5, 6
Continuous dependence on initial conditions, 70
Contraction mapping principle, 68, 71, 101
Convex combination, 177, 206, 208, 233
Cooperative connection, 94
Coupling matrix, 312

D
Data-driven system, 61
Delay interval contraction method, 107
Delay partitioning approach, 107, 206

Z. Wang et al., *Qualitative Analysis and Control of Complex Neural Networks with Delays*, Studies in Systems, Decision and Control 34,
DOI 10.1007/978-3-662-47484-6

Delay partitioning method, 107, 218
Delay-dependent, 103, 107, 151, 205, 226, 259
Delay-independent, 151, 269
Delay-matrix decomposition method, 106
Descriptor system method, 106
Diffusive condition, 372
Dini derivative, 66, 295, 302, 305
Direct approach, 142
Discontinuous activation functions, 98, 240
Discrete delay, 150
Dissipation, 55
Dissipative, 281
Dissipative system, 277, 290
Dissipativity, 278
Domain of attraction, 56, 60, 92, 109
Dynamic behavior, 75
Dynamic delay interval method, 107
Dynamical system, 28, 41, 277, 288, 312, 313
Dynamical system defined by geometry, 41
Dynamical system defined by maniflow, 42
Dynamical system defined by measure, 42
Dynamical systems theory, 62

E
Energy function, 16–18, 58
Energy supply function, 280
Equilibrium, 114, 130, 208, 230, 314
Equilibrium point, 41, 92
Equilibrium state, 71
Error system, 59, 84, 176, 208, 226, 306, 314, 316, 351, 366
Exponential stability, 226
Exponentially convergent, 71
Exponentially stable, 45, 71, 243

F
Finite-time stability, 56
Fixed point, 48
Forward complete, 76
Free weight matrix method, 104, 107
Free weighting matrix, 278

G
Global dynamics, 74
Global existence and uniqueness, 70
Global stability, 74, 288
Globally asymptotically stable, 71, 72, 77, 91, 192, 366
Globally attractive set, 75
Globally convergent, 71
Globally exponential stability, 226
Globally exponentially stable, 291
Globally stable, 75, 116

H
Hartman–Grobman theorem, 49
Hopfield model, 16, 17, 21, 22, 24
Hopfield neural networks, 16, 20, 112, 123, 155, 174, 278, 364
Hopfield's stability, 58

I
Infinitesimal generator, 66
Initial value problem, 69, 76
Input strictly passive, 280
Input-to-output stable, 77
Input-to-state stability, 53
Input-to-state stable, 77
Integral input-to-state stability, 78
Interval matrix partitioning method, 105
Interval uncertainty, 149, 271
Invariant, 71
Invariant set, 41, 300
Ising model, 17
Isolated systems, 59

J
Jensen integral inequality, 208, 231, 337
Jensen's inequality, 176, 287

K
Kernel function, 117
Krasovskii–LaSalle principle, 65

L
Lagrange stability, 41
LaSalle invariant principle, 73
LaSalle's Principle, 79
Lebesgue-Stieltjes measure, 118
Leibniz–Newton formula, 186, 231, 267, 350
Limit point, 75
Limit set, 75
Linear approximation, 48
Linear growth condition, 128, 132
Linear matrix inequality, 31, 103
Linearization analysis, 94
Linearization method, 362

Lipschitz continuous, 70
Local existence and uniqueness, 69
Lotka-Volterra model, 125
Lyapunov diagonal stability, 108
Lyapunov function, 16, 22, 23, 104, 127, 174, 206, 212, 283, 352, 367
Lyapunov second method, 45, 58
Lyapunov stability, 41, 288, 289
Lyapunov stability theorem, 64
Lyapunov stability theory, 57, 59, 60, 74, 184, 252, 372
Lyapunov stable, 44, 45
Lyapunov's realization, 46
Lyapunov–Krasovskii functional, 65, 180, 186, 206, 317
Lyapunov-Volterra quai-stable, 127

M
M-matrix, 132, 138
Manifold of equilibria, 73
Matrix decomposition method, 105
Memristive neural networks, 111
Multistability, 240, 248

N
Neural optimizer, 91
Neurons, 10
Nonisolated equilibrium point, 73, 74, 125, 126
Nonnegative equilibrium point, 121
Nonsingular M-matrix, 66

O
Orbit stability, 47
Output stable, 54
Output strictly passive, 280
Output-Lagrange input-to-output stable, 77
Output-to-state stable, 77

P
Partial stability, 85
Passive, 261, 280
Passivity, 259, 281
Positive definite, 78
Positive invariant set, 290
Positive semidefinite function, 64
Positive semidefinite matrix, 64
Practical stability, 55
Practically exponentially stable, 80
Practically stable, 80
Proper, 78

Q
Qualitative behavior, 48
Qualitative concepts, 85
Qualitative property, 42
Quasi-Lipschitz condition, 128

R
Razumikhin uniform asymptotical stability theorem, 81
Razumikhin uniform stability theorem, 81
Razumikhin uniformly ultimately bounded theorem, 82
Razumikhin-type stability theorem, 81
Real-valued neural networks, 111
Rectification of dynamical system, 43
Recurrent neural networks, 3, 239
Reference coordinate, 59
Reference frame, 84
Relative motion, 59
Relative stability, 60, 63, 84, 85, 150
Robust stability, 148–150
Robustly absolutely stable, 72
Rough system, 51

S
Schauder fixed-point theorem, 67
Schur complement, 66, 129, 180, 191, 268, 271, 287, 351
Small gain, 54
Solution in the sense of Caratheodory, 69
Solution in the sense of Filippov, 252
Solution of equation, 80
Stability, 48, 260, 288
Stability and ISS, 54
Stability in Hopfield sense, 289
Stability in Lagrange sense, 289
Stability in Lyapunov sense, 289
Stability in the sense of Lyapunov, 252
Stability of fixed point, 48, 84
Stability of Sets, 78
Stability theory, 47, 83
Stability theory problem, 47
Stabilization, 334
Stable, 79, 80, 242, 259
Static neural networks, 226
Stationary state, 73
Steady-state behavior, 76
Storage capacity, 92
Strict inequality, 154

Strict Lyapunov function, 74
Structural stability, 49, 50
Structurally stable systems, 51
Symmetry, 24, 116, 124, 126
Synchronization, 84, 311, 314, 361, 362, 366
Synchronization controller, 314
Synchronization error, 369
Synchronization manifold, 365
Synchronization stability, 59, 82, 83
Synchronization state, 313, 363
Synchronized network, 315
Synchronous state, 366

T
Time delay, 94, 114, 150, 333
Time-varying delay, 100
Total stability, 78
Total variables, 85

U
Ultimately bounded, 81
Unbounded activation function, 96
Uncertainty, 98, 149
Uniformly asymptotically stable, 76, 79, 80
Uniformly bounded, 81
Uniformly exponentially stable, 76
Uniformly stable, 76, 80, 291
Uniformly ultimately bounded, 81
Unique equilibrium, 102
Unstable, 79, 252

V
Very strictly passive, 280

W
Weighting-delay, 174, 178, 195
Weighting-delay method, 175
Wirtinger inequality, 281

Z
Zero solution, 78

Kolibri 360 GmbH
Ferdinand-Jühlke-Straße 7
99095 Erfurt, Deutschland
produktsicherheit@kolibri360.de